A Unified Approach to

The Finite Element Method

and

Error Analysis Procedures

A Unified Approach to

The Finite Element Method

and

Error Analysis Procedures

John O. Dow
University of Colorado

ACADEMIC PRESS

San Diego   London   Boston   New York   Sydney
Tokyo   Toronto

Academic Press
A Division of Harcourt Brace & Company
525 B Street, Suite 1900, San Diego, CA 92101-4495
http: // www.apnet.com

Academic Press
24–28 Oval Road, London NW1 7DX
http: // www.hbuk.co.uk/ap/

Library of Congress Cataloging-in-Publication Data
Dow, John O.
    A unified approach to the finite element method and error analysis
procedures / John O. Dow.
        p.   cm.
    Includes bibliographical references and index.
    ISBN 0-12-221440-4
    1. Finite element method. 2. Error analysis (Mathematics)
    I. Title.
    TA347.F5D69   1998   98-22147
    620′.001′51535—dc21   CIP

Printed in the United States of America

98 99 00 01 02 QW 9 8 7 6 5 4 3 2 1

Contents

Part IV The Strain Gradient Reformulation of the Finite Difference Method

Part V *A Posteriori* Error Analysis Procedures: Pointwise Error Measures and a New Approach for Strain Extraction

Preface

Objective of Book

This book is designed to give analysts and designers confidence in the results of finite element analyses and confidence in their understanding of these results. To accomplish this, the book (1) provides a solid theoretical foundation for the error measures used to evaluate the accuracy of finite element results, (2) expresses the error measures in terms of quantities sought in the solution process so they are directly related to the problem being solved, and (3) develops new error measures that quantify the errors at individual points so critical points are clearly identified.

Importance of Pointwise Error Measures

Pointwise error measures possess two significant advantages over elemental error measures that are based on strain energy: (1) pointwise error measures have a higher resolution than elemental error measures and (2) they estimate errors in quantities of primary interest, namely, stresses and/or strains. Strain-energy-based error measures compute a weighted average of the square error in the strain over an element. As a result, high errors at critical points are submerged in the averaging process. Furthermore, strain energy is a derivative quantity that is not of primary interest in an analysis.

Effect on Finite Element Practice

The structure of a typical finite element analysis can be significantly altered as a result of the theoretical basis provided for the pointwise error measures. Using a component of the error analysis procedure, high-accuracy strain results can be extracted directly from the displacements produced by the finite element analysis. This means that strains no longer need to be extracted from the underlying finite element model as is now done. The focus of finite element modeling then changes from element behavior to the efficient computation of superconvergent strains from the nodal displacements.

A Software Engineer's Dream

The structural change to a finite element analysis just outlined provides the opportunity for creating a universal postprocessor for finite element analysis. Such a postprocessor can be appended to existing finite element packages because the superconvergent strains are independent of the underlying finite element model.

Elemental Errors and Their Elimination

Errors can exist in the strain representations of individual finite elements. These errors are identified by comparing the element's representation of a known condition with the expected result. The errors are corrected with a new element formulation procedure that is an extension of the evaluation procedure presented here.

Discretization Error and Its Elimination

Errors exist in finite element models because discrete representations cannot represent all of the deformations that a continuous problem can exhibit. These errors, called discretization errors, are identified by the pointwise error measures discussed earlier.

They are reduced to an acceptable level by sequentially improving finite element models under the guidance of an error estimator until a specified level of accuracy is achieved.

Theoretical Foundation

The principle of minimum potential energy provides the theoretical foundation for the error measures developed in this book. This principle relates two different approaches for finding the solution to a problem, specifically, (1) the solution can be found by directly minimizing the potential energy contained in the problem (a variational solution) or (2) the solution can be found directly by solving the governing differential equations of equilibrium for the problem (a differential solution). The error measures are formed by comparing an approximate variational solution to an approximate differential solution.

Rationale for Error Measures

The following two characteristics of approximate variational and differential solutions provide the rationale for the success of the error measures just described: (1) both types of solution must converge to the same result and (2) the two solution types represent their stress and strain results differently. Since the two approximations must produce the same result, differences between the two types of stress or strain representations identify the location and magnitude of discretization errors in the underlying finite model.

Practical Application

The pointwise error measures are practical because an approximate finite difference result can be extracted from the displacements of the finite element result. As a validation of this idea, Part V shows that the stress and strain results extracted from the finite element displacements using finite difference templates are identical to the average values of nodal values in finite element models with uniform spacing. Note that this means that practical error measures *do not require the separate solution* of a finite difference approximation to evaluate the errors in a problem. In fact, on the interior of the finite element model, nodal averaging is sufficient to approximate the finite difference result.

Pedagogical Consequences

The level of mathematical knowledge required to fully understand the finite element method and error analysis procedures is reduced as a consequence of the contents of this book. These developments are based on Taylor series expansions and not functional analysis. Thus, the level of mathematical sophistication required to perform meaningful research is reduced to concepts taught in undergraduate calculus.

Research Opportunities

This book offers an important research opportunity, namely, the development of efficient procedures for extracting error measures and superconvergent strains from finite element displacements. The introduction of a significant new research topic and the simplification of the mathematical requirements opens meaningful finite element research to a wider set of researchers.

A Note to Designers

The key to the material presented in this book is contained in a relationship used to form a design tool. This design tool extracts equivalent continuum quantities, such as flexural stiffness (IE), axial stiffness (AE), etc., from the stiffness matrices for frames and trusses. These quantities are sought so candidate designs for space frames and trusses can be compared early in the design cycle. This process relates the nodal displacements to specific states of strain, such as flexure in a beam, using Taylor series expansions. The generalization of this process clarifies the relationship between continuous and discrete representations of structural problems.

Unification I

The following four paragraphs explain why the word ''unified'' appears in the title of this book. This book unifies two levels of error analysis. The procedures for evaluating the strain-modeling capabilities of individual elements and for quantifying the discretization errors in finite element models are nearly identical. Both methods compare the strains produced by the model being evaluated to a more accurate representation formed using the transformation discussed in the previous paragraph. Thus, we can say that these two different levels of error analysis are unified.

Unification II

The finite element and the finite difference methods are reformulated from a common basis. The coordinate transformation that relates the continuum and the discrete representation produces the finite difference operators as a natural consequence. This same transformation is the key to the new element formulation procedure. Thus, we can say that these two approximate solution techniques are unified.

Unification III

The unified formulation procedure for the finite element and finite difference methods result from an inversion of the *a priori* error analysis procedure for evaluating individual finite elements. Therefore, the approximate solution techniques and the error analysis procedures are related or ''unified.''

Unification IV

Two primary approaches for evaluating finite element results exist, namely, (1) smoothing approaches and (2) residual approaches. The accuracy of the stresses or strains is evaluated in the smoothing approach. In the residual approach, the satisfaction of local equilibrium is evaluated. Part V shows that these two error analysis techniques are related on a pointwise basis. Thus, we can say that the two approaches to error analysis are unified.

Lesson Structure

The book is divided into lessons instead of chapters because each lesson is relatively freestanding. That is to say, each lesson has a definitive beginning and end so it can be used independently of a majority of the rest of the book. This is done so parts of the book can be used as supplementary reading for finite element courses or as reference material.

Intended Audience

This book is designed to serve the following readers: (1) beginning students, as a supplementary textbook; (2) advanced students, as a primary textbook; (3) practicing engineers, as a reference source; and (4) researchers in computational mechanics, as a monograph.

Acknowledgments

This book was made possible due to the efforts and contributions of many people, primarily my former and current graduate students and Professor O. C. Zienkiewicz. This book had its beginnings in the ideas contained in Doyle Byrd's master's thesis, where the accuracy of stress concentrations was considered. Procedures for developing finite element stiffness matrices with displacement patterns based on eigenvectors and minimum strain energy content where developed in a thesis by Stein Taugbol. A relationship between the continuous and discrete representations of structures was clearly defined in a series of papers (by J. O. Dow, C. S. Bodley, C. C. Feng, and S. A. Huyer; see references at the end of the General Introduction) that was key to the rest of the developments contained in the book. This relationship was used as the basis for evaluating finite element performance and developing a new approach for formulating finite element stiffness matrices in a thesis by Harold Cabiness and a dissertation by Tom Ho. These ideas were consolidated and extended to correct modeling errors in finite elements in a Ph.D. dissertation by Doyle Byrd. In addition, this dissertation introduced the idea of using finite difference templates to form error measures from the finite element displacements. Shawn Harwood and Mike Jones reformulated the finite difference method using the transformation developed in the series of papers mentioned earlier. Ian Stevenson extended the smoothing approach to error analysis introduced by Doyle Byrd so it was independent of the underlying discrete model. James Hardaway extended the finite difference boundary modeling capabilities introduced by Mike Jones. Mike Schubert developed finite element models of laminated plates using the new approach developed earlier. Joao E. Abdalla applied the new finite element formulation procedure to developing laminated plate elements that could not be correctly derived using the isoparametric approach. Mantu Baishu generalized the finite difference boundary modeling capabilities presented by Mike Jones and James Hardaway. Jubal Hamernick developed and applied the pointwise error measures. Knut Axel Aarnes applied specialized pointwise error measures.

A majority of the figures where drawn by Jason Kintzel, Stephany Spector, and Kenneth Shorter. Jason Kintzel is expanding this work to shells in his thesis.

Doyle Byrd and I spent my sabbatical leave during the 1986–87 academic year at University College of Swansea, University of Wales, at the invitation of Professor Zienkiewicz. Dr. Byrd was supported by a Fulbright Dissertation Research Fellowship. The research we performed in Swansea forms part of the basis of this book. The work also benefited greatly from discussions with Professor Zienkiewicz. Our lively talks produced ideas that added to the contents of the book. I cannot express the depth of my gratitude for the friendship and support of Professor Zienkiewicz over the years.

General Introduction

The finite element method has evolved into one of the most powerful and widely used techniques for finding approximate solutions to the differential equations that occur in engineering and science. In this method, a continuous problem is first broken into a discrete physical representation consisting of a finite number of regions or finite elements. The governing equations for the discrete representation of the continuous problem are formed by combining the stiffness matrices and load vectors for the individual finite elements. These equations are then solved to produce an approximate solution for the continuous system. This process is shown schematically in Fig. 1, where the stiffness matrices of the individual finite elements are combined to form the overall or global model of the problem.

Sources of Modeling Error

The very act of subdividing the continuum provides the opportunity for two types of modeling error to enter the finite element analysis, namely, *discretization errors* and *elemental errors*. Discretization errors can occur because the *finite* element model is, by definition, incapable of duplicating *every* one of the *infinite* deformation patterns that the continuum can assume. Elemental errors, if they exist, most often take the form of anomalies in the strain representations within individual elements, e.g., certain of the strain representations may be distorted or, in some cases, may be absent. These flaws in the strain representations negatively affect the strain energy content of the individual elements. As a result, when the full model is assembled, the elemental errors enter the overall representation. Then, as expected, the elemental errors contribute to the discretization errors.

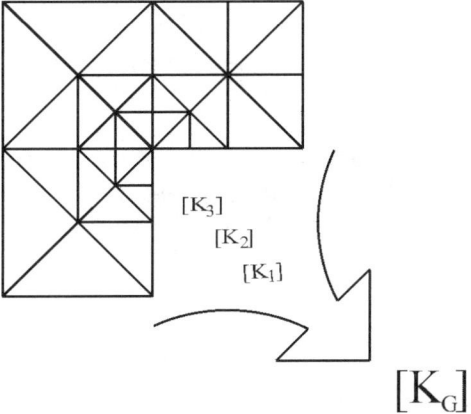

Figure 1. Schematic diagram.

Identification of Discretization Errors

We can see discretization errors explicitly in finite element results as discontinuities or jumps in the stresses or strains between elements and by misrepresentations of the actual boundary stresses. The magnitude of the jumps is indicative of the size of the local discretization error. We can also identify discretization errors by determining whether the finite element solution satisfies equilibrium at every point on the domain of the problem. The amount by which the solution fails to satisfy equilibrium identifies the magnitude of the local discretization error in the finite element model. New methods for quantifying the local discretization error are developed and demonstrated in Part V of this book. These local error measures are based on either stress discontinuities between elements or on the failure to satisfy local equilibrium at selected points on the finite element model.

Identification of Elemental Errors

We can identify elemental errors using a new approach that is developed and demonstrated in Part III of this book. In this approach, an element is given a series of nodal displacements that serve as basis functions for the element. Each set of nodal displacements is associated with a known strain distribution. The modeling capabilities of the element are then evaluated by comparing the actual strains in the element to the expected strain distributions. In this way, the full range of behavior of an individual element can be tested before it is used in an analysis. If we find modeling errors using this *a priori* procedure, we can correct the behavior of the element using a new approach for formulating finite element stiffness matrices. Furthermore, the ability to directly relate the nodal displacements to known strain distributions is used to form both types of discretization error measures discussed in the previous paragraph. The relationship between element formulation, elemental error analysis, and discretization error measures unifies the finite element method and the error analysis procedures.

Need for Error Analysis Procedures

The need to remove errors from finite element results has been recognized since the inception of the method. When the problems were restricted in size due to limited computer capacity and when the method was used mostly by analysts well versed in the theory and limitations of the method, the results could be "validated" by the user with reasonable success. However, in the current environment such a hit-and-miss approach for evaluating the accuracy of finite element results is no longer acceptable.

On one hand, the size and complexity of the problems now being solved has increased to such an extent that even the most knowledgeable user finds it difficult to assess the validity of a typical result. On the other hand, the method is now being employed routinely by nonexperts who cannot be expected to evaluate successfully the accuracy of the finite element results. These conditions clearly identify the need for procedures to evaluate finite element results and to automate the process of improving the finite element model.

Objective and Scope of the Book

The objective of this book is to present and demonstrate methods for identifying and eliminating errors contained in finite element models. Specifically, methods for evaluating three types of errors are elaborated, namely, (1) modeling errors in individual elements, (2) discretization errors in the overall model, and (3) pointwise errors in the final stress or strain results.

These error analysis techniques are not presented as a collection of heuristic techniques gathered from a variety of sources. The three different types of error analysis techniques are developed from a common theoretical foundation in a systematic way. With its strong theoretical foundation, this unified approach makes these procedures useful to the following readers: (1) students studying the finite element method, (2) practicing engineers applying the finite element method, and (3) researchers extending the field of computational mechanics.

If you are a student, the book provides you with a parallel perspective of the finite element method that supplements the text used at any course level. If you are a practicing engineer, the book provides you with a background for better understanding finite element results as well as methods for improving the accuracy of finite element models. If you are a researcher, the common rational basis that links these methods enables you to directly extend them to types of problems not specifically addressed in this book. Thus, this book serves as a text, a reference source, and a research monograph.

Typical Error Analysis Application

The availability of accurate and reliable error estimation techniques enables modification of the finite element method, as shown in Fig. 2. The procedure shown here differs from a simple finite element solution procedure because of its recursive nature. After the problem is solved for the first time, the solution is analyzed for discretization errors on an element-by-element basis. The results of this error analysis are then used to guide the refinement of the mesh to produce a more accurate model. The finite element model is refined in regions of high error and the problem is solved again. The errors in the new solution are estimated and the procedure is repeated until a predefined level of accuracy is achieved. This recursive combination of error analysis and mesh refinement is called *adaptive refinement*.

Significance of Error Analysis Applications

Adaptive refinement provides a level of confidence in the accuracy of finite element results that would not otherwise exist. Both expert and nonexpert users can be assured that the results in critical regions of a problem have been evaluated and that results of a specified accuracy have been achieved. This confidence does not depend on blind faith in the output of a computer program. The progress to the desired level of accuracy can be traced in the steady refinement of the model and the accompanying reduction in the errors.

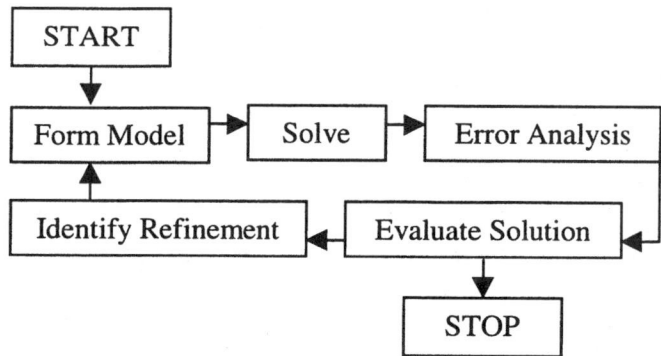

Figure 2. An adaptive refinement scheme.

When adaptive refinement procedures come into routine use, the finite element method can finally be considered a mature technology. This will mean that the accuracy of finite element results can be taken for granted in the same way lights are expected to go on when the wall switch is flipped. Then, the analysis step in the design cycle can drop into the background. This book is designed to assist in the achievement of this goal.

In addition to assisting the individual user, adaptive refinement provides a missing link in the development of a fully automated design procedure. Without adaptive refinement, the smoothly flowing computer-aided design process is interrupted by the need to externally evaluate and validate the analysis results. This bottleneck slows the design process and reduces its efficiency. The importance of removing this gap in computer-aided design is apparent when it is considered that some experts believe that the efficient implementation of computer-aided design and manufacturing processes is crucial to the future economic health of a country.

The Identification of Discretization Errors: *A Posteriori* Error Analysis

Discretization errors in finite element solutions are identified by using two different, but related, approaches, namely, (1) *smoothing techniques* and (2) *residual techniques*. Smoothing techniques form error measures by quantifying error in the finite element stress representations. The name comes from the fact that the discontinuous finite element stresses are compared to some improved or ''smoothed'' stress result to form the error measure. Residual techniques form the error measures by quantifying the failure of the finite element solution to satisfy equilibrium at individual points. The residual is the amount by which the finite element solution fails to satisfy equilibrium on a pointwise basis. Because the errors in the finite element results must be estimated after a problem is solved, discretization error analysis is called an *a posteriori* process.

As an introduction to *a posteriori* error analysis, the original form of the smoothing approach is outlined. This error analysis procedure exploits the fact that errors in finite element results can be explicitly seen as jumps in the finite element stresses, as shown in Fig. 3. We form a smoothed solution from the discontinuous finite element solution by averaging the stresses between the elements at the nodes. We assume that the smoothed

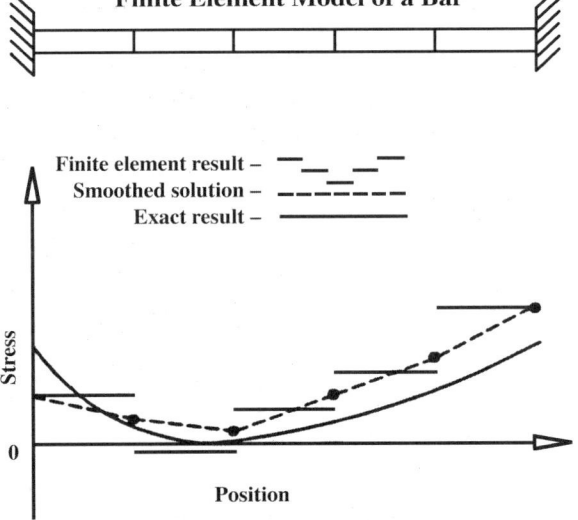

Figure 3. Discretization error measures.

solution more closely approximates the actual solution than does the finite element solution. This assumption is proven in Part V. In this book, the discretization errors associated with each element are estimated as a function of the differences between the finite element solution and the smoothed approximation. These error estimates identify regions of high error and are then used to guide the improvement of the finite element model.

The discretization error analysis procedure just discussed was originated by Zienkiewicz and Zhu (Z/Z). This error estimator provided the first practical approach that could be easily extended to all types of problems. Its importance to the adaptive refinement process cannot be overestimated. Because of its primary importance, the Z/Z approach is presented in detail in the main text.

In Part V, a modification for forming the smoothed result is made to better estimate the errors on the boundary. Instead of forming the smoothed stress distribution on the boundary from the averaged finite element stresses, the actual boundary stresses are incorporated into the smoothed solution. Note that in subsequent versions of the Z/Z procedure, the method for forming the smoothed solution has been modified. Instead of using the averaged result, an extrapolation of the smoothed solution on the interior is used as the smoothed solution on the boundary.

Figure 3 depicts the need for this modification on the boundary. In this problem, the original Z/Z procedure uses the finite element stress as the smoothed approximation on the boundary. As a result, the smoothed stress distribution is closer to the finite element solution than would be the case if the actual boundary conditions were incorporated into the smoothed approximation. This smoothing technique produces a lower estimate of discretization error in the elements on the boundary than does the corrected approach. The improvements produced by the modified smoothing method are greater when the mesh is coarse and the finite element results have not approached the converged value on the boundary. As a result, the finite element model is not improved as rapidly as it could be. However, if no limit is put on the number of applications of the adaptive refinement procedure, the results could be made as accurate as desired.

As mentioned earlier, the discretization errors are increased if modeling errors are present in individual elements. The presence of elemental errors decreases the overall modeling efficiency by requiring a more refined model to achieve a given level of accuracy than is needed. Thus, it is advantageous to eliminate errors in the individual elements, if possible, before they are used in finite element models.

The Identification of Elemental Errors: *A Priori* Error Analysis

Elemental errors are modeling deficiencies contained in individual elements. Typically, these flaws are due to misrepresentations of the strain components. Since these errors are introduced during the element formulation procedure, individual elements need not be evaluated in the context of a specific problem. Because the element is evaluated *before* it is applied to a specific problem, the search for modeling errors in individual elements is called *a priori error analysis*.

Part III develops and applies a procedure for identifying elemental errors. This procedure compares the strain distributions that exist in individual elements with the strain distributions that should exist in the element for sets of nodal displacements that exercise the full range of the element. That is to say, an individual element is analyzed as a single entity for a linearly independent set of displacement patterns. As a result of applying these nodal displacements, all of the strain distributions it is capable of representing are produced. We then compare these strain distributions to an absolute standard.

The errors identified by these tests are used as guides for developing better elements. The changes made to the modeling characteristics of an element are tested with sample problems before they are applied in practice. Naturally, the results of this analysis and testing are generally applicable. From then on, the improved element can be used with confidence since the full range of the elements behavior has been tested. The strain distributions that serve as the absolute standard are discussed in the next section of this introduction.

The results of an *a priori* analysis are shown in Fig. 4. In this figure, two configurations of a six-node "linear" strain element are shown. The upper diagram in Fig. 4 is a linear strain *triangle* since it has three straight edges. The element shown in the lower diagram of Fig. 4 is a six-node "linear" strain element with one edge initially curved. The curved edge has nothing to do with the deformation of the element. Such a configuration would appear around the representation of a hole. The stiffness matrices for these elements were created using the isoparametric procedure, and their characteristics subsequently analyzed.

The two six-node elements are being evaluated for their ability to represent a linear variation of ε_x in the y direction. The errors in the normal strain representation for these two elements are quantified in Fig. 4. As we can see, the element with the straight edges does not contain any strain modeling errors when it is representing this linear strain distribution. However, the element with the curved edge does not accurately represent this strain distribution. A maximum error of -77% occurs at the corner. The error in the strain energy of the element for this condition of strain is 180%.

As is shown in Part III, the stiffness matrices for six-node elements with one or more curved edges contain errors when they are produced by the isoparametric formulation procedure. The word "linear" was placed in quotes earlier in the text because of this flaw

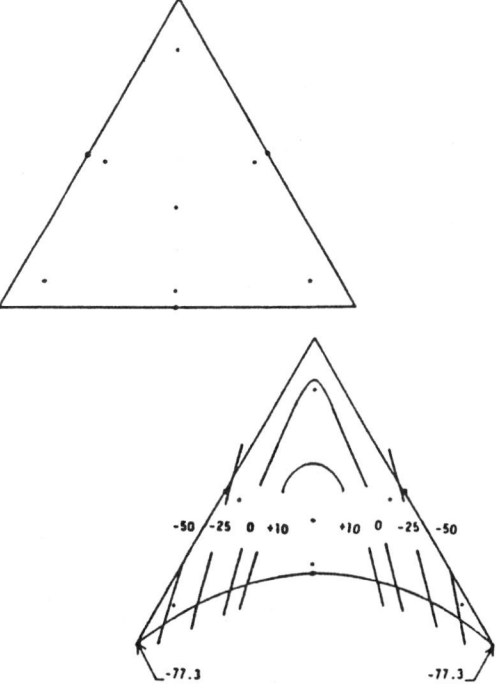

Figure 4. Distorted strain model.

in isoparametric elements with a curved edge. Such elements cannot correctly represent linear strain distributions.

The strain modeling errors in isoparametric elements with curved edges cannot be corrected because the modeling deficiency is inherent in the isoparametric method itself. This modeling error is caused by the kind of mapping that is used in the formulation procedure that transforms the element with the curved edge onto the standard equilateral triangle for easy numerical integration. We see in Part III that strain modeling errors occur in isoparametric elements when the Jacobians associated with the isoparametric transformations are not constant over the domain of the element (see Lesson 8, Appendix A). Note for completeness that the same types of errors occur in higher order ''four-sided'' isoparametric elements if the initial element shapes are not parallelograms.

This elemental error is not ''fatal'' to finite element models that contain curved elements. That is to say, models containing elements with curved edges will converge to an accurate result as they are adaptively refined. The reason that the modeling flaws in the elements with curved edges does not irrevocably damage the finite element model is that the ability of curved isoparametric finite elements to correctly represent the rigid body motions and constant strain states is not affected by the curved edge. As the finite element model is refined, the flawed elements become smaller and a larger percentage of the strain energy is represented by constant strain. That is to say, the accuracy of the finite element model improves when the curved six-node elements act as constant strain elements and not as linear strain elements. As expected, more elements are needed to attain a given level of accuracy with defective elements than if the elements performed as desired. The conclusion is that isoparametric elements with curved edges lead to inefficient models because more elements are needed in the model for a given level of accuracy than would be needed if the linear variations in strain were accurately represented.

Since the errors are inherent to the isoparametric formulation procedure, they can only be eliminated by using a different method for creating finite element stiffness matrices. Part III introduces a new approach for formulating these stiffness matrices. The errors contained in isoparametric elements that are not triangles or parallelograms do not exist in elements created with the new procedure. As we will see, the new approach is computationally competitive with the isoparametric approach. Thus, finite element analysts must decide whether the isoparametric approach with its inherent inaccuracies is worth maintaining as the standard formulation procedure.

Theoretical Background: Strain Gradient Notation

The procedure for evaluating elemental errors just described is made possible by a physically interpretable notation called *strain gradient notation*. In fact, the theoretical and practical bases of the discretization error measures developed and demonstrated in Part V are also provided by this notation. Those who do not like the idea of a nonstandard notation need not use it once they apprehend the insights it provides into modeling behavior. With this understanding of the modeling characteristics of the finite element method, the developments presented in this book can be implemented with standard notation.

In strain gradient notation, the coefficients of the polynomials that approximate the displacements in finite elements are defined as functions of the physical quantities that produce displacements in the continuum, namely, rigid body motions and gradients of the strains. In this notation, the coefficients of the approximation polynomials and the quantities being sought in the analysis are identical. That is to say, the analysis techniques used to estimate the strains (or quantities directly related to strains such as stresses or displacements) are formulated from polynomials written in terms of strain quantities.

Thus, the problem, the modeling techniques, and the error analysis measures are all integrated via strain gradient notation.

The idea that notation can provide a direct link between the analysis technique and the physical problem being solved is now illustrated. We can see this correspondence between the notation and the physical problem by comparing the standard notation with strain gradient notation. For this comparison, complete second-order polynomials representing displacements are used. Polynomials of this order are chosen for this demonstration because they are simple enough for easy exposition and complex enough to show the value of the physically based notation.

In the standard notation, the complete second-order displacement polynomials are written as

---

Standard Second-Order Displacement Polynomials

$$u = a_1 + a_2x + a_3y + a_4x^2 + a_5xy + a_6y^2$$
$$v = b_1 + b_2x + b_3y + b_4x^2 + b_5xy + b_6y^2$$

The arbitrary nature of the coefficients contained in these displacement representations makes it difficult to directly relate these coefficients to the problems being solved, namely, the approximation of the strains in the continuum. In this context, the word *arbitrary* means undefined. For example, without further analysis, it is difficult, if not impossible, to understand the meaning of the coefficient a_5 with respect to the problem being solved. The undefined coefficients take on a specific physical meaning when they are evaluated in terms of strain gradient quantities. The procedure for evaluating these coefficients is developed and demonstrated in Part II.

When the arbitrary coefficients of the displacement polynomials are evaluated in strain gradient notation, we can write the second-order displacement polynomials just given as the following second-order Taylor series expansions:

Second-Order Strain Gradient Displacement Polynomials

$$u = (u_{rb})_0 + (\varepsilon_x)_0 x + (\gamma_{xy}/2 - r_{rb})_0 y + (\varepsilon_{x,x})_0 x^2 + (\varepsilon_{x,y})_0 xy + [(\gamma_{xy,y} - \varepsilon_{y,x})/2]_0 y^2$$
$$v = (v_{rb})_0 + (\gamma_{xy}/2 + r_{rb})_0 x + (\varepsilon_y)_0 y + [(\gamma_{xy,x} - \varepsilon_{x,y})/2]_0 x^2 + (\varepsilon_{y,x})_0 xy + (\varepsilon_{y,y})_0 y^2$$

When we compare these two notations we find that the arbitrary coefficients of the standard notation are given a physical meaning in the strain gradient representation. For example, the arbitrary coefficient a_5 of the xy term becomes $(\varepsilon_{x,y})_0$ in strain gradient notation. The subscript zero indicates that the derivative of ε_x with respect to y is evaluated at the local origin of the Taylor series expansion. This term, which represents the linear variation of ε_x in the y direction, is later seen to be equivalent to flexure in a beam.

As a result of this notation, the role of the term a_5 is now clear. This term quantifies the linear variation of ε_x with respect to y. In Part III, we see that this term is responsible for the modeling error known as "parasitic shear." It would be difficult to make this identification and to correct the strain representation in a rational way without using strain gradient notation.

In mathematical terminology, strain gradient notation is said to be self-referential. This means that the quantities being computed and the notation used to form the computational procedures refer to the same entities, which, in this case, are strain quantities. The self-referential nature of this notation enables the notation to become a direct participant in the analysis process because the symbols are "transparent," i.e., the notation enables us to "see" the relationship between the analysis technique and the problem being solved. The role of strain gradient notation is discussed further in the Introduction to Part II.

The physically based nature of the coefficients of the displacement polynomials relates the problem being solved to the finite element method and to the error analysis procedures. Strain gradient notation clarifies the connection between these components of an integrated analysis process instead of obscuring their common relationship.

Applications of Strain Gradient Notation to Error Analysis

The direct connection between strain gradient notation and the sources of displacement in the continuum provides the insights that make the developments presented in this book possible. The relation between this notation and error analysis procedures are now briefly outlined.

Elemental Error Analysis

The *elemental errors* are identified using two distinct approaches based on strain gradient notation. In the first approach, a direct comparison is made between the polynomial strain representations contained in the element model and the Taylor series representation of the strains in the continuum. This comparison directly identifies the validity of the strain components contained in the finite element strain representation. It also identifies the components that are missing.

Such a comparison has identified the causes of the following types of errors in individual finite elements: (1) parasitic shear, (2) spurious zero-energy modes, (3) shear locking, (4) aspect ratio stiffening, and (5) incorrect coupling in laminated composite plates. Procedures for correcting or controlling these errors are formulated using strain gradient notation.

In the second approach, the element is given a set of nodal displacements that would impose a specific strain distribution on an accurate element or the continuum. The strain components actually generated in the element being analyzed are compared to the strain components that should exist in the continuum. This comparison is done numerically on a point-by-point basis to identify the distribution of the strain modeling errors in the element. The results of two such comparisons are shown in Fig. 4. The strain energies are computed and compared to give a measure of the aggregated errors in the strain representations.

This approach enables the evaluation of any type of element. The evaluation of a single element can be viewed as an extension of the "patch test." In this extension, we evaluate all of the element capabilities, rather than just the ability to represent rigid body motions and constant strain states. This approach has identified the effects of initial element distortion on the strain representations produced in isoparametric elements. The identification of these modeling errors motivated the development of an alternative finite element formulation procedure that produces accurate elements. The elements created with this new procedure do not exhibit any of the modeling deficiencies contained in isoparametric elements.

A *Posteriori* Error Analysis—The Pointwise Approach

The procedure just discussed for evaluating the modeling characteristics of individual elements is extended in Part V to estimate pointwise errors in overall finite element results. We can use this new pointwise approach to evaluate the stress and/or strain representations or the satisfaction of equilibrium in finite element models. The use of pointwise equilibrium evaluation is not discussed further in this general introduction because the procedure is similar in nature to the procedure for evaluating elements using strain comparisons that is discussed next. The topic of residual error estimators is covered extensively in Part V.

Pointwise error measures estimate the errors in the strain representations at individual points in the overall finite element model for a specific analysis of a particular problem. That is to say, the pointwise error estimator evaluates the accuracy of the strain results at a single point produced by a single element acting as part of a total model. Both the magnitude and sign of the errors in the individual stress or strain components are computed. Since the pointwise errors are estimated for the results of an analysis, this is categorized as an *a posteriori* error analysis technique.

The pointwise error estimator is not designed to be applied to every point in a model. It is designed for application at the critical points identified in a finite element analysis. This localized error measure enables us to evaluate the accuracy of the stresses or strains at these points. If the estimated error is not conservative, the finite element model can be refined to produce more acceptable results.

The pointwise errors in the strain components are estimated by comparing the strains produced by the finite element model to a set of strains found from higher order Taylor series representations. The higher order strain representations are derived from the displacements of a group of nodes surrounding the point in question. The relationship of these nodes to the finite element mesh and the point at which the pointwise error measure is applied are shown in Fig. 5. The Taylor series strain expansions used as the control are of an order higher than the strain models in the finite element result because more nodal points are used to form this expansion than are included in the finite element model. The pointwise error measures are formed using strain gradient notation to relate the strains at a point to the nodal displacements of the finite element solution.

Until this notation came into use, it was difficult, if not impossible, to routinely form a set of polynomials that would produce a nonsingular transformation for forming the Taylor series coefficients used in the higher order strain representations. In Part IV, we see that the ability to routinely invert the transformation between the nodal displacements and the Taylor series coefficients enables us to reformulate the finite element and finite difference methods on a common basis. This is the capability that made the development of pointwise error measures possible.

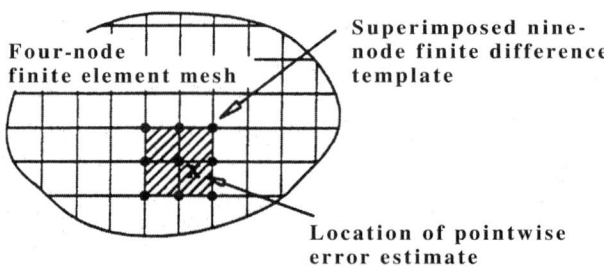

Four-node finite element mesh

Superimposed nine-node finite difference template

Location of pointwise error estimate

Figure 5. Nodal layout for a pointwise error measure.

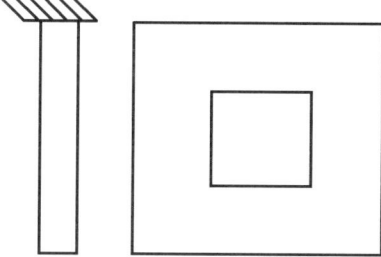

Figure 6. Two sample problems.

A Posteriori Error Analysis—The Smoothing Approach

As mentioned earlier, the Zienkiewicz error estimator underestimates the errors in the elements on the boundary. We can see this underestimation when we compare the global error estimations found with corrected and uncorrected procedures for the problems shown in Fig. 6. These comparisons demonstrate the two types of effect that the underestimation produces in error estimates. The two effects are illustrated in Fig. 7.

In Fig. 7a, the overall error estimate versus the number of degrees of freedom is shown for the beamlike structure shown on the left in Fig. 6. The original analysis procedure produces error estimates that begin low, go up, and then drop. The improved error estimator produces a result that begins much higher than the original result and then drops until it nearly matches the result of the original procedure. The initial low value of the error estimate for the original procedure is due to the underestimation of the errors in the elements on the boundary during the initial representation with the coarse mesh.

In Fig. 7b, the two error estimates produced by the two estimators are compared for the problem with the square hole in the middle shown on the right in Fig. 6. In this case, the corrected error estimator estimates a higher error for every level of mesh refinement.

These results have shown the effect of modifying the discretization error estimator to better approximate the errors in the elements on the boundary. This improvement is

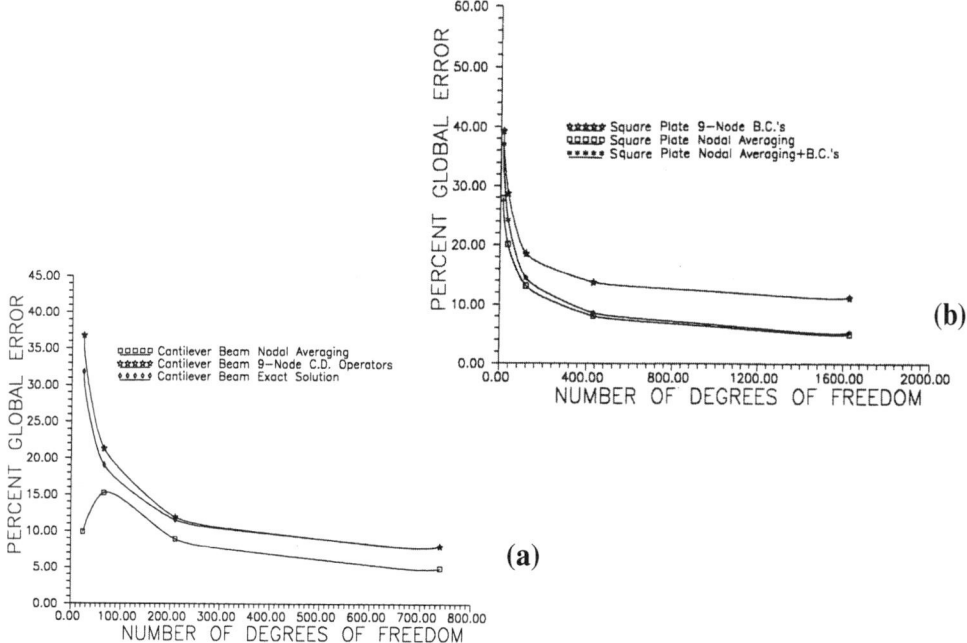

Figure 7. Example of error analysis results.

accomplished by incorporating boundary-modeling concepts from the finite difference method into the finite element error estimator.

To enable this modification to be generally applicable, the ability of the finite difference method to represent complex boundary conditions had to be improved. Remember that the finite element method largely displaced the finite difference method because of the perceived inability of the finite difference method to represent complex boundary conditions. In Part IV, the ability of the finite difference method is extended so that it can represent virtually any boundary condition. This extension is accomplished by reformulating the finite difference method using strain gradient notation. The fact that the reformulation of the finite difference method was accomplished as an ancillary task in improving error analysis procedures demonstrates the power of self-referential notation, i.e., notation that is written in terms of the physical quantities being sought.

New Developments

The material presented in this book contains several developments of recent origin based on strain gradient notation. To be precise, all of the material presented here with a few exceptions is extracted from a series of papers and theses written by the author and his colleagues and his graduate students over the last 15 years. The exceptions are Part I, which provides the background material concerning the principle of minimum potential energy, variational calculus and the Rayleigh-Ritz solution technique, and Lesson 16, which presents the error analysis procedure of Zienkiewicz and Zhu.

The developments presented in book form for the first time are

1. The physically based strain gradient notation

2. Procedures for evaluating individual finite element strain modeling capabilities

3. A new finite element formulation procedure

4. A reformulated version of the finite difference method

5. A generalized approach to finite difference boundary modeling

6. Pointwise strain error measures for evaluating finite element results

7. Pointwise residual or equilibrium error measures

8. A unification of smoothing and residual error measures

9. New strain extraction procedures

10. New error measures for finite difference results

11. A universal error analysis procedure.

Book Structure

The book consists of five parts:

Part I—Problem Definition and Development

Part II—Physically Interpretable Notation

Part III—The Strain Gradient Reformulation of the Finite Element Method

Part IV—The Strain Gradient Reformulation of the Finite Difference Method

Part V—*A Posteriori* Error Analysis Procedures: Pointwise Error Measures and a New Approach for Strain Extraction

Part I develops the problem, relates the finite element and finite difference methods to the problem, and motivates the use of the finite element method. Part II develops the self-referential, or physically based, notation used as the basis for the developments in the rest of the book. Part III reformulates the finite element method, applies it to several elements, and identifies the causes and cures for several elemental modeling errors. Part IV reformulates the finite difference method in terms of strain gradient notation and extends the boundary modeling capabilities of the method. Part V presents the discretization error estimator developed by Zienkiewicz and Zhu and then uses the finite difference developments to improve its ability to estimate errors on the boundary. The finite difference method is then used to develop both pointwise smoothing and residual error estimators. Finally, an alternative procedure for extracting strain results from finite element solutions is proposed. If implemented, this strain extraction procedure would change the face of the finite element method as we know it. The strains would be of an order higher than those extracted directly from the finite element results. Only the displacements of the underlying finite element model would be used to evaluate the strains. As a result, the strain results would be independent of the individual finite elements so elements would be chosen for reasons other than their strain modeling capabilities.

Closure

As the work on the developments presented in this book progressed, a seemingly contradictory situation has emerged. The use of the self-referential strain gradient notation made important research questions in computational mechanics accessible to advanced undergraduates and lower level graduate students while at the same time the notation opened new research areas at all levels.

The reason for this expansion of audience in one direction and expansion of research opportunities in the other is due to the relation of the notation to the problems. The analysis techniques are expressed in terms of the quantities being sought. In this case, strain or strain-dependent quantities are estimated using procedures developed using strain-based notation. The notation does not obscure the problem. The notation is transparent. That is to say, the problem and the notation are one.

Several significant developments in other fields have occurred when self-referential notation was employed. Godel proved his revolutionary theorems in number theory by transforming equations to integers called Godel numbers. Also, the digital computer as we know it today is based on von Neumann's use of the same structure for data and instructions.

The preceding references are not meant to imply that the developments presented here with their basis in a self-referential notation are of the magnitude of those just mentioned. However, this common use of self-referential notation suggests that extensions of the use of strain gradient notation in solid mechanics could be profitably sought.

References and Other Reading

1. Byrd, Doyle. "Applications of Scale-Free Finite Elements." M.S. Thesis. University of Colorado, 1980.

2. Taugbol, Stein. "Generally Applicable Stress Concentration Finite Elements." M.S. Thesis. University of Colorado, 1982.

3. Dow, J. O., Bodley, C. S., and Feng, C. C. "An Equivalent Continuum Representation of Structures Composed of Repeated Elements." *AIAA/ASME/ASCE/AHS 24th Structures, Structural Dynamics and Materials Conference Proceedings,* Lake Tahoe, Nev., May 2–4, 1983, pp. 630–640.

4. Dow, J. O., Su, Z. W., Feng, C. C., and Bodley, C. S. "An Equivalent Continuum Representation of Structures Composed of Repeated Elements." *AIAA Journal,* Vol. 23, No. 10, Oct. 1985, pp. 1564–1569.

5. Dow, J. O., and Huyer, S. A. "An Equivalent Continuum Analysis Procedure for Space Station Lattice Structures." *AIAA/ASME/ASCE/AIIS 28th Structures, Structural Dynamics and Materials Conference Proceedings,* Monterey, Calif., April 6–8, 1987, Part I, pp. 110–122.

6. Dow, J. O., and Huyer, S. A. "Continuum Models for Space Station Structures." *ASCE Journal of Aerospace Engineering,* Vol. 2, No. 4, Oct. 1989, pp. 212–230.

7. Cabiness, Harold. "Strain Gradient-Based Finite Element Evaluation Technique." M.S. Thesis. University of Colorado, 1984.

8. Ho, Tom. "A Generalized Finite Element Evaluation Procedure." Ph.D. Dissertation. University of Colorado, 1984.

9. Byrd, Doyle. "Identification and Elimination of Errors in Finite Element Analyses." Ph.D. Dissertation. University of Colorado, 1988.

10. Harwood, Shawn. "Finite Element Error Analysis Using the Finite Difference Method." M.S. Thesis. University of Colorado, 1989.

11. Zienkiewicz, O. C., and Zhu, J. Z. "A Simple Error Estimator and Adaptive Procedure for Practical Engineering Analysis." *International Journal for Numerical Methods in Engineering,* Vol. 24, 1987, pp. 337–357.

12. Zienkiewicz, O. C., and Zhu, J. Z. "Superconvergent Recovery Techniques and A-Posterori Error Estimation in the Finite Element Method, Parts I and II." *International Journal for Numerical Methods in Engineering,* Vol. 33, 1992, pp. 1331–82.

13. Dow, J. O., Jones, M. S., and Harwood, S. A. "A Generalized Finite Difference Method for Solid Mechanics." *International Journal of Numerical Methods for Partial Differential Equations,* Vol. 6, No. 2, Summer 1990, pp. 137–152.

14. Dow, J. O., Jones, M. S., and Harwood, S. A. "A New Approach to Boundary Modeling for Finite Difference Applications in Solid Mechanics." *International Journal for Numerical Methods in Engineering,* Vol. 30, No. 1, July 1990, pp. 99–113.

15. Stevenson, Ian. "A Generalized Adaptive Refinement Procedure for Finite Element and Finite Difference Analyses." M.S. Thesis. University of Colorado, 1990.

16. Hardaway, James. "Finite Difference Boundary Modeling." M.S. Thesis. University of Colorado, 1991.

17. Schubert, Mike. "Formulation of Laminated Plate Finite Elements Using Strain Gradient Notation." M.S. Thesis. University of Colorado, 1991.

18. Abdalla, Joao E. "Qualitative and Discretization Error Analysis of Laminated Composite Plate Models." Ph.D. Dissertation. University of Colorado, 1992.

19. Baishya, Mantu. "A Generalized Finite Difference Method for Plane Stress Problems." Ph.D. Dissertation. University of Colorado, 1993.

20. Hamernick, Jubal. "A Unified Approach to Error Analysis in the Finite Element Method." Ph.D. Dissertation. University of Colorado, 1993.

21. Sandor, Matt. "Submodel Boundary Identification in a Global/Local Adaptive Refinement Technique." M.S. Thesis. University of Colorado, 1993.

22. Aarnes, Knut Axel. "Extension and Application for a Pointwise Error Estimator Based on the Finite Difference Method." M.S. Thesis. University of Colorado, 1993.

Part I

Problem Definition and Development

Introduction

The objective of Part I is twofold: (1) to show the relationship between the finite element method and the problems the method is designed to solve and (2) to show the relationship between the finite element method and other solution techniques, particularly the finite difference method. The presentation of these two relationships has four purposes: (1) to present motivations for the finite element method; (2) to provide both the theoretical basis for the finite element method and a new approach for forming finite element stiffness matrices; (3) to identify two distinct categories of solution techniques, namely, the variational and the differential approaches; and (4) to provide the theoretical background for the error analysis procedures presented in Part V.

The relationship between the variational and the differential solution techniques provides the theoretical and practical basis for the error analysis techniques developed in Part V. The theoretical basis can be stated briefly as follows: if the variational and the differential solution techniques are developed using the same approximations, i.e., the same underlying polynomial representations and the same discrete model, any differences between the results produced by the two solution techniques for the same problem must be due to errors in the model. These errors occur because of the reduction of the continuous problem to a discrete problem with a finite number of degrees of freedom. The differences between the two results can be used to identify the locations of the errors and to quantify the magnitudes local errors in the discrete model.

These local error measures can be used to evaluate the accuracy of the solutions. The analysis can be terminated if it is sufficiently accurate. If the error is excessive, the model can be improved and the analysis can continue until a result of sufficient accuracy is found. This process, which continually improves the model under the direction of error measures until an acceptable result is produced, is called *adaptive refinement*.

In Part V, this approach to forming error measures is used to compute strain energy errors in individual finite elements and to compute pointwise errors in stresses and/or strains. The error measures are formed by comparing the finite element result, which is a variational solution, to a finite difference result, which is a differential solution. The finite difference result is computed from the nodal displacements produced by the finite element analysis. This detail concerning the source of the approximate finite difference result is included at this point to make it clear that the problem need be solved *only once* to compute the error measures. A separate finite difference analysis need not be performed.

The developments in Part II provide the polynomial expansions that enable both the variational and differential solution techniques used in developing error measures to be formulated from the same polynomial basis. Since both the finite element and the finite difference methods are based on the same polynomial expansions, they can represent the same domains. The use of the same polynomial representations for both methods expands the use of the finite difference method, without limiting the use of the finite element method.

Statics and Dynamics Problems

The variational and differential forms of the equations of motion that describe the continuum are directly related through *Hamilton's principle*. In the case of static equilibrium problems, Hamilton's principle reduces to the *principle of minimum potential energy.* This book uses the principle of minimum potential energy to develop and evaluate models that represent the deformation characteristics of the continuum; therefore, problems of static equilibrium are the primary focus.

The fact that dynamics problems are not specifically discussed here is not a significant limitation. The finite element representation of dynamic problems consists of a stiffness model plus a representation of the mass properties of the continuum. The mass properties can be formed using a subset of the procedures that are used to form the stiffness properties.

This correctly implies that the representation of stiffness properties in the finite element method is more difficult to construct than the mass representation. The fact that the mass properties can be computed more easily than the stiffness properties does not imply that the equations of motion are easier *to solve* than the equations of static equilibrium. In general, the additional difficulties with dynamics problems begin with the solution algorithms and not with the finite element models. As a result, the developments presented here can be applied to dynamics problems. However, as mentioned earlier, dynamics problems are not explicitly discussed.

Variational and Differential Forms

As stated, the principle of minimum potential energy directly relates the variational and differential forms of equilibrium problems in solid mechanics. Specifically, the principle of minimum potential energy states that a function that minimizes the potential energy expression is a solution to the differential equations of equilibrium. The solution found by directly minimizing the potential energy function is called a variational solution. A solution found by directly solving the differential equations of equilibrium is called a differential solution.

The finite element and the finite difference methods find approximate solutions to continuum problems using variational and differential approaches, respectively. The basic differences between these two numerical solution techniques are now briefly outlined.

The Finite Element Method

The finite element method represents the continuum with an assemblage of discrete elements that represent the potential energy of the continuous problem. Once the potential energy expression for the discrete system is formed, it is minimized to find the approximate solution. Thus, we can say that the finite element method approximates the actual physical system being analyzed. Actual pieces of the structure are represented by individual finite elements. The finite element method is a variational solution technique because it directly minimizes the potential energy of the problem.

The structure of the finite element method is illustrated in Fig. 1. The continuous problem is broken into a discrete number of finite regions, as indicated. This collection of connected subregions is a direct approximation of the physical problem. The potential energy associated with this representation is formed from the potential energy expressions for the individual finite elements that make up the overall model. The development of the potential energy expressions for the individual finite elements constitutes the majority of the "theoretical" work associated with the finite element method. The assembly of the potential energy function for the individual elements to form the overall potential energy model is a simple process. The nodal displacements of different elements at common nodes are given the same displacements. This enforcement of nodal compatibility is a simple step to implement. Therefore, it is not explicitly discussed.

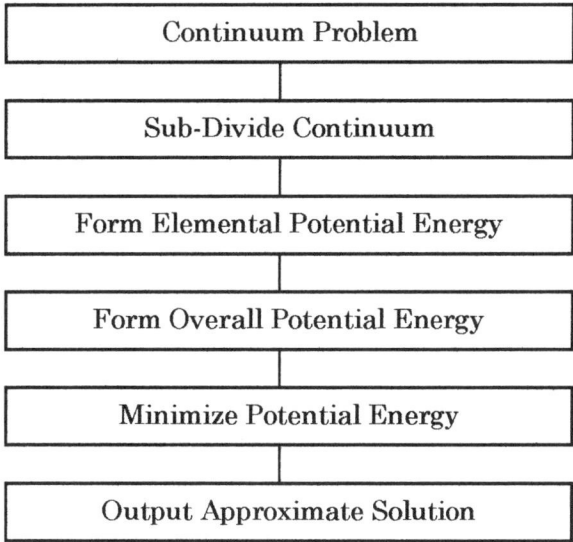

Figure 1. The finite element method.

The resulting potential energy expression for the assembled discrete system is a set of algebraic equations. These equations are minimized to form an approximate solution to the differential equation of the continuous system. The error analysis procedures presented in Part V are designed to evaluate how well the variational solution produced by the finite element method satisfies the differential equations of equilibrium and the boundary conditions for the continuous problem.

The Finite Difference Method

In contrast to the finite element method, which models the physical system, the finite difference method forms a representation of the differential equations of equilibrium that describe the problem being solved. This approximation is accomplished by replacing the derivatives contained in the differential equations with approximations of these derivatives in the form of difference equations. The resulting representation is a set of simultaneous algebraic equations. Since the finite difference method is directly applied to the differential equation, it is a differential form of solution.

An example of a finite difference approximation of a derivative is shown for a one-dimensional case in Fig. 2. In this figure, the first derivative of the displacement with respect to a change in position on the x axis is approximated at two different points. The differential lengths contained in the definition of the derivative are replaced by *difference* quantities in the approximation of the derivatives. When such derivative approximations

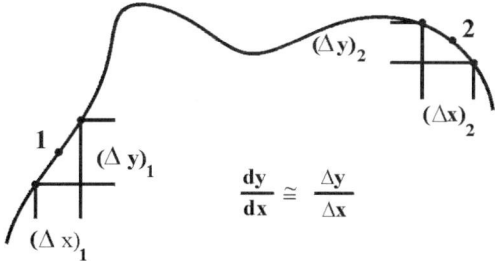

Figure 2. Approximations of derivatives.

are used to replace the derivatives in an equilibrium equation, the differential equation is reduced to an algebraic equation.

This process is repeated at a finite number of points over the continuum to form the overall finite difference representation of the differential equations. The resulting set of algebraic equations is solved to approximate the exact solution of the differential equations of equilibrium. This process is shown schematically in Fig. 3.

When we compare Fig. 3 to Fig. 1, the essential differences between the finite element and the finite difference methods are apparent. The finite element method approximates the physical system via potential energy models of the individual elements, while the finite difference method approximates the differential equations of equilibrium, i.e., the mathematical representation, that describe the physical system via representations of the derivatives.

This concrete comparison of the two methods highlights a primary difference between the finite element and finite difference methods. The finite element method, a variational approach, produces a solution that minimizes the potential energy of an approximation of the physical system. The finite difference method, a differential approach, finds a solution that satisfies an approximation to the differential equations of equilibrium for the system. The details of these differences are discussed in Part IV.

Error Analysis Developments

As mentioned earlier, the availability of two different solution techniques is exploited to develop procedures for evaluating the errors in finite element results. This is accomplished by using concepts from the finite difference method to improve existing error analysis methods and to develop new procedures for identifying errors in finite element results.

If the error analysis procedures are to find wide usage, the modeling capability of the finite difference method must be extended so it can represent the same range of boundary conditions as the finite element method. The extension of the capability of the finite difference method is not as simple as it might seem. The difficulty of this task becomes obvious when we consider that the finite difference method was largely replaced by the finite element method because of the perception that the finite difference method could

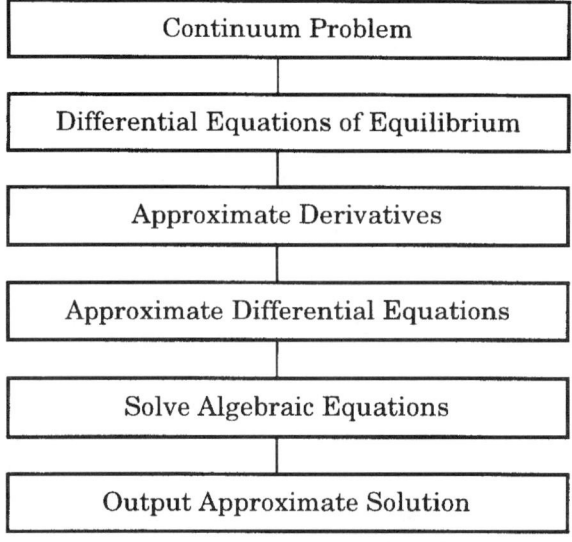

Figure 3. The finite difference method.

not represent as wide a range of boundary conditions on complex regions as the finite element method.

In Parts III and IV, we reformulate both the finite element and the finite difference methods using the same physically based, self-referential notation. This means that the two methods can represent the same geometries and boundary conditions. Thus, the error analysis procedures developed in Part V can be applied to any problem that the finite element can solve.

Contents of Part I

The previous discussion has alluded to the close relationship between the variational and differential forms of the problems being solved. The four lessons of Part I elaborate the details of this relationship. Of particular interest is the relationship of the differential equations and the boundary conditions to the minimization of the potential energy. The differential equations of equilibrium and the associated boundary conditions are the conditions that a solution must satisfy to minimize the potential energy expression. This knowledge provides the background necessary to formulate the finite-difference-based procedures for evaluating the errors in finite element results.

Lesson Outlines

Lesson 1 introduces the *principle of minimum potential energy*. This principle ultimately provides both the variational and differential forms of the governing differential equations that describe the physical problem. The two components of the potential energy function for plane elasticity problems, namely, the strain energy expression and the work function, are formulated. The strain energy expressions for continuous systems are found to be *functionals*, i.e., functions of functions. A functional cannot be minimized using procedures from differential calculus. Functionals must be minimized using techniques from the *calculus of variations*, which is the subject of Lesson 2.

Lesson 2 introduces and develops elements of the *calculus of variations*. The calculus of variations provides methods for finding the extreme values of a functional. The derivation depends on the idea of a variation, which is a generalization of a virtual displacement. The primary result is the derivation of the necessary and sufficient conditions for finding a minimum of the potential energy function. The conditions are found to be the Euler-Lagrange equations and the associated boundary equations. The application of these conditions to the functional produces the equilibrium equations and the boundary conditions that make up the differential form of the problem.

Lesson 3 uses the concepts developed in the first two lessons to derive the variational and differential forms of the governing equations for *plane stress problems*. Plane stress problems are used in this book to illustrate the process of forming the differential form of the equilibrium problem for multidimensional problems because it solves an important class of problems and it is simple enough for convenient exposition.

Lesson 4 finds approximate variational solutions to several problems by using the *Rayleigh-Ritz solution technique*. In this early variational approach to solving continuous problems, a solution is postulated in the form of a linear combination of trial functions that extend over the *full region*. The difficulty in finding functions that satisfying the boundary conditions on complex regions provides a major motivation for the finite element method. The Rayleigh-Ritz criteria for evaluating solutions are introduced. The Rayleigh-Ritz criteria state that if the *geometric boundary conditions*, i.e., the displacements and the slopes, are satisfied by an approximate variational solution, the exact solution to the problem provides an upper bound to the strain energy for the approximate solution. That is to say, if two or more approximate solutions are compared, the one with the greatest strain energy content is the best solution. We then see that if the natural boundary conditions are also satisfied, the approximate solution is more accurate. The examples presented clearly identify the role of boundary conditions in approximate solutions and provide the motivation for introducing the finite difference boundary models into the finite element error estimator.

Lesson 1

Principle of Minimum Potential Energy

Purpose of Lesson   To introduce the principle of minimum potential energy and the concept of a functional (a function of a function).

The equations of statics that govern the behavior of discrete systems result when the potential energy function is minimized using the procedures of differential calculus. When the principle is extended to continuous systems, the potential energy function that must be minimized is found to be a functional. The emergence of the need to minimize a functional provides the motivation for Lesson 2, which introduces the calculus of variations, the branch of mathematics used to identify the conditions required to minimize the potential energy functional.

The principle of minimum potential energy directly relates the potential energy function and the governing equations of the physical systems being studied. An understanding of this relationship identifies two independent approaches for solving the equilibrium equations. The finite element method finds approximate solutions that directly minimize the potential energy. The finite difference method finds approximate solutions that satisfy the governing equations and indirectly minimize the potential energy. Later lessons show that the finite element and finite difference methods can be formulated from a common basis that enables finite element error analysis procedures to be improved using concepts from the finite difference method.

■   ■   ■

There are two fundamental approaches to classical mechanics. Vector mechanics defines physical systems in terms of force and momentum vectors. These vectors are related to the physical system by Newton's laws. Analytic mechanics defines the same physical systems in terms of two scalar quantities: kinetic energy and potential energy (see Notes 1 and 2 at the end of the lesson). These scalars are related to the physical system by *Hamilton's principle*. The expected relationship between the two approaches is succinctly stated as follows.

Hamilton's Principle: The displacements that satisfy Newton's laws can be found by minimizing the *action integral*, which is given as

$$A = \int_{t_1}^{t_2} L \, dt = \int_{t_1}^{t_2} (T - V) \, dt \qquad (1.1)$$

where L is the Lagrangian.

T is the kinetic energy.

V is the potential energy.

When Hamilton's principle is applied and the action integral is explicitly minimized, the Newtonian equations of motion result. These equations are essentially the same as those produced by the direct application of Newton's second law. The analytic approach provides us with two advantages. The process of minimizing the action integral produces the possible boundary conditions that a specific problem must satisfy and it provides an alternative approach for solving the resulting governing differential equations. These two features are discussed in detail in subsequent lessons.

Since we are focusing on solution techniques that concern the spatial domain of elasticity problems, only statics problems are considered further in this presentation. This means that the kinetic energy term in the Lagrangian is taken as zero, which leaves the potential energy function as the function to be minimized. This focus reduces Hamilton's principle to the *principle of minimum potential energy*, which can be stated as follows.

The Principle of Minimum Potential Energy: The displacements that satisfy the equilibrium equations can be found by minimizing the potential energy function, which is given as

$$V = \int_{\Omega} (U - W)\, d\Omega \qquad (1.2)$$

where U is the strain energy.

   W is the work done by the applied loads.

The principle of minimum potential energy reduces the problem of formulating the governing equations to four discrete steps, namely: (1) the formulation of the strain energy function, (2) the formulation of the work function, (3) the transformation of the two functions to a common set of independent or global coordinates, and (4) the minimization of the potential energy function. The results of this process are the governing equations and a set of boundary terms. The governing equations are the same equilibrium equations that are produced by applying Newton's second law to a free-body diagram of the system. The boundary terms identify admissible boundary conditions and are discussed in subsequent lessons. This procedure is shown schematically in Fig. 1.

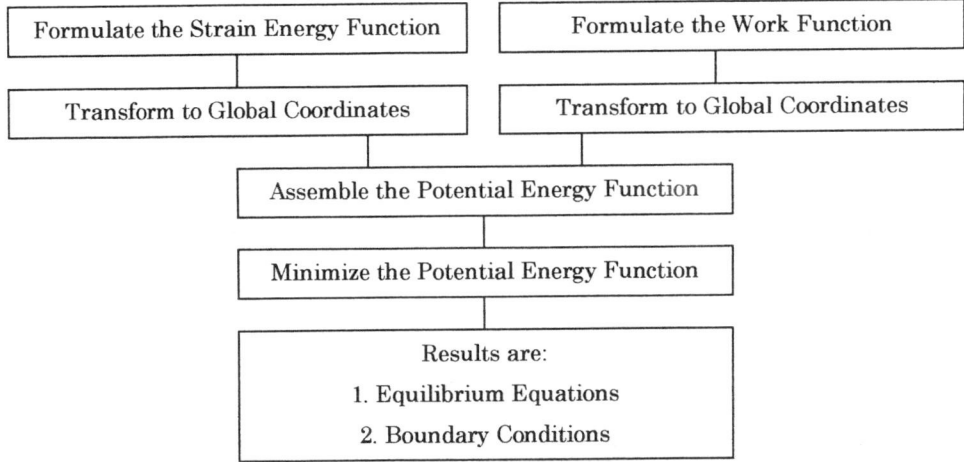

Figure 1. Steps in applying the principle of minimum potential energy.

Discrete Problems

We now apply the principle of minimum potential energy to a problem with a finite number of degrees of freedom. This example demonstrates the use of the transformation to global coordinates from local or relative coordinates indicated in the second level of Fig. 1 and verifies that the process does, indeed, produce the equations of equilibrium. For future reference, we should note that this example problem is analogous to the problem that must be solved in order to determine the equivalent nodal loads when forces are applied between nodal points in the finite element method.

Example 1   Equilibrium Position of a Two-Degree-of-Freedom System

Consider the weightless and undeformable bar shown in Fig. 2 supported by two linear springs and loaded with the point load shown. The two independent global degrees of freedom for this system are the vertical displacements of the ends of the bar, x_1 and x_2. We find the equilibrium position of this bar by applying the principle of minimum potential energy, as outlined in Fig. 1.

The first step is to determine the strain energy expression for the system. The strain energy is contained in the two springs K_1 and K_2. The strain energy in each spring has the following form:

Strain Energy in Relative Coordinates

$$U_i = \frac{1}{2}(K_i \Delta_i^2) \tag{1.3}$$

where K_i is the spring constant.

Δ_i is a relative or local displacement of the spring.

The use of relative coordinates enables the strain energy of an individual stiffness element to be written without reference to the global coordinate system. In this way, the strain energy expressions for these elements can be written so that they are independent of a specific problem. However, the procedure for minimizing the potential energy for a specific problem requires that the function be expressed in terms of the linearly independent global coordinates. We put the strain energy expression for this example in global coordinates as shown in Fig. 3 by utilizing the following coordinate transformation.

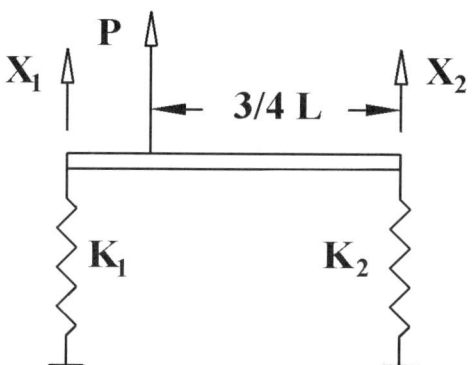

Figure 2. Two-degree-of-freedom system.

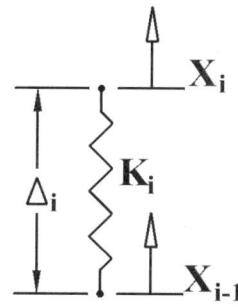

Figure 3. Transformation between relative and global coordinates.

Local to Global Coordinate Transformation

$$\Delta_i = x_i - x_{\text{ground}} \tag{1.4}$$

where x_i is the global displacement of the ith coordinate.

x_{ground} is the global displacement of the ground.

In this example, we include the boundary conditions of the problem at this point instead of at the end of the formulation, as shown in Fig. 1, to reduce the size of the expressions. From Fig. 2, we can see that the displacement of the ground point is zero. Thus, we can write the strain energy for this system in terms of global coordinates as

Strain Energy Transformed to Global Coordinates

$$\begin{aligned} U &= \frac{1}{2}\left(K_1\Delta_1^2 + K_2\Delta_2^2\right) \\ &= \frac{1}{2}\left(K_1 x_1^2 + K_2 x_2^2\right) \end{aligned} \tag{1.5}$$

The next step in applying the principle of minimum potential energy requires the formulation of the expression for the work function. The only external force present in the system is the force P applied at the quarter point. The local displacement of this force is defined as Δ_{force}. The work due to this force is $P\Delta_{\text{force}}$. We use the sign convention that the work is positive if a force and its corresponding displacement are both in the same direction. In this case, the local displacement of the force is a function of the two global displacements, x_1 and x_2, as shown in Fig. 4. The transformation from the local coordinate to global coordinates can be written as

Local to Global Coordinate Transformation

$$\Delta_{\text{force}} = \frac{3}{4}x_1 + \frac{1}{4}x_2 \tag{1.6}$$

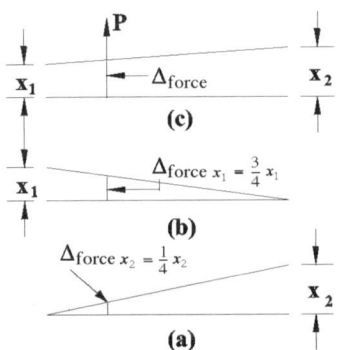

Figure 4. Displacement of the applied force.

The meaning of the two terms on the right-hand side of Eq. 1.6 is illustrated in Fig. 4. In this figure, it is seen that the force is located between the two global coordinates. Equation 1.6 states that the total motion of the force is a linear combination of the two motions produced by the coordinate displacements. Thus, Eq. 1.6 can be viewed as an interpolation function.

We can now write the potential energy function for this problem in terms of the global coordinates as

---

Potential Energy Function

$$V = \frac{1}{2}(K_1 x_1^2 + K_2 x_2^2) - P\left(\frac{3}{4}x_1 + \frac{1}{4}x_2\right) \tag{1.7}$$

This expression is a function of the two independent global coordinates, x_1 and x_2. We minimize this function using the procedures from differential calculus for finding extreme values (minimum, maximum, and saddle points) of multivariable functions. We take the partial derivatives of the potential energy function given by Eq. 1.7 and force these derivatives to zero. For the problems considered in this book, the potential energy functions always consist of a strain energy expression that has a positive definite quadratic form unless otherwise specified. As a result, the unique equilibrium point is found by the minimization process. Thus, the potential energy function is minimized when the following conditions hold:

Minimum Potential Energy Criteria

$$\frac{\partial V}{\partial x_i} = 0, \qquad i = 1, \ldots n \tag{1.8}$$

When the partial derivatives of Eq. 1.7, as defined by Eq. 1.8, are taken with respect to the two independent coordinates and set to zero, the result is

---

Equilibrium Conditions

$$\frac{\partial V}{\partial x_1} = K_1 x_1 - \frac{3}{4}P = 0$$
$$\frac{\partial V}{\partial x_2} = K_2 x_2 - \frac{1}{4}P = 0 \tag{1.9}$$

The desired goal of finding the equilibrium position of the bar is accomplished by solving these two equations for x_1 and x_2. The two components of Eq. 1.9 can be seen to be the equilibrium equations found by summing the moments about the left and right ends of the bar, respectively.

This application of the principle of minimum potential energy has demonstrated several concepts. The use of relative coordinates has been operationally demonstrated for both the strain energy and the work function. Expressing the strain energy in terms of the relative coordinates, Δ_i, and the work function in terms of the displacement under the load, Δ_{force}, was a straightforward process. After determining the transformation from

relative to global coordinates, we transformed both of these expressions to the independent global coordinates so the required partial derivatives could be taken. This produced the desired equilibrium equations.

This example has demonstrated the important idea that the vector equations of Newtonian mechanics are produced by minimizing the scalar expression for potential energy. In the case of the finite element method, the partial derivatives of the potential energy function produces the equilibrium equations at the nodal points.

Continuous Problems

We now extend the use of the principle of minimum potential energy to continuous problems. The strain energy for the general three-dimensional elasticity problem can be written as

---

Strain Energy Expression

$$U = \frac{1}{2} \int_\Omega \left(\sigma_x \varepsilon_x + \sigma_y \varepsilon_y + \sigma_z \varepsilon_z + \tau_{xy} \gamma_{xy} + \tau_{xz} \gamma_{xz} + \tau_{yz} \gamma_{yz} \right) d\Omega \qquad (1.10)$$

The potential energy functions for a longitudinal bar and an Euler-Bernoulli beam is now formulated. Both the area, A, and Young's modulus, E, are constant for this problem. Only the stresses and strains in the x direction are represented in these mathematical models, so the strain energy function reduces to

---

Strain Energy for a Bar and a Beam

$$U = \frac{1}{2} \int_0^L E\varepsilon_x^2 A \, dx$$

where $\qquad (1.11)$

$$\sigma_x = E\varepsilon_x$$

Example 2 Potential Energy Function for a Longitudinal Bar

We now formulate the potential energy function for a bar deformed under its own weight, as shown in Fig. 5. The unknown in this problem is the *function* describing the displacement along the length of the bar, denoted by $u(x)$. Our first task is to formulate the strain energy expression in terms of the function $u(x)$. This is accomplished by introducing the definition of the normal strain in the x direction, ε_x, from linear elasticity, which is

---

Strain Definition

$$\varepsilon_x = \frac{\partial u}{\partial x} = u_{,x} \qquad (1.12)$$

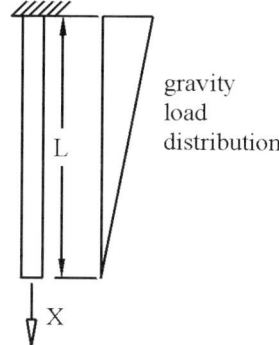

Figure 5. A bar hanging under its own weight.

When this expression is substituted into the strain energy expression, the result is

Strain Energy in a Bar

$$U = \frac{1}{2} \int_0^L AE(u_{,x})^2 \, dx \tag{1.13}$$

In this notation, the subscript on the u indicates that a derivative with respect to x has been taken.

The first step in formulating the work function is to develop an expression for the applied load due to the weight of the bar. This load increases linearly from zero at the free end to a maximum of mgL at the support. This distribution can be written in equation form as

Applied Load Due to Weight

$$P(x) = mgL\left(1 - \frac{x}{L}\right) \tag{1.14}$$

where m is the mass per unit length.

   g is the gravitational constant.

The load at each point on the bar is displaced by an amount $u(x)$. This displacement is due to the deformation of the bar, which is caused by the applied load. Thus, the work function can be written as

Work Function Due to Self-Weight

$$W = \int_0^L P(x) u \, dx \tag{1.15}$$

This function is expressed in terms of the independent function $u(x)$ so no coordinate transformation is required to put the strain energy and the work functions in terms of the same unknown.

The potential energy function can now be written by combining the strain energy expression and the work function as

---

Potential Energy Function

$$V = \int_0^L \left[\frac{1}{2} AE(u_{,x})^2 - P(x)u \right] dx \qquad (1.16)$$

Equation 1.16 contains the function that we must minimize to get the equilibrium equation and the boundary conditions that apply to the problem. Note that the boundary conditions shown in Fig. 5 do not enter into the formation of the potential energy function. The role of the boundary conditions in continuous systems is discussed in the next lesson.

When we inspect the potential energy expression given by Eq. 1.16, we see that this function is itself dependent on an unknown function. That is to say, we must find the specific function $u(x)$ that minimizes the potential energy. A function that is a function of a function is called a *functional*. The procedures from differential calculus for minimizing a function do not apply to a functional. The field of study that deals with the minimization of functionals is the calculus of variations. The topics from the calculus of variations required to minimize potential energy functionals of the type encountered in this book are discussed in the next lesson.

Potential Energy Functional

Let us study the form of the potential energy functional just developed in Eq. 1.16. The integrand contains three quantities that are dependent on x, namely, P, u, and $u_{,x}$. The notation of the calculus of variations generalizes this integrand by writing it as $F(x,u,u_{,x})$. Thus, Eq. 1.16 can be generalized as

---

Generalized Functional Form for a Bar

$$V = \int_0^L F(x, u, u_{,x}) \, dx \qquad (1.17)$$

The x is included in the notation to identify the independent spatial coordinate. The quantities u and $u_{,x}$ are included to identify the unknown function and the order of the derivatives that are contained in the functional. The order of the highest derivative present is important because the boundary conditions that are available for a particular problem depend on the order of this derivative. The function P is not explicitly included because it is a known quantity.

We see in the next lesson that the general results developed using the calculus of variations depend only on the number of independent coordinates and the order of the highest derivatives of the unknown functions in the integrand. Thus, the results apply to a wide range of problems. In Lesson 3, we see that the functional just developed for bar problems is the one-dimensional analog of the functional for the plane stress problem.

To demonstrate the formulation of a functional with a higher order derivative, we now develop the potential energy function for an Euler-Bernoulli beam.

Example 3   Potential Energy Functional for an Euler-Bernoulli Beam

The functional for the Euler-Bernoulli beam loaded with an arbitrary lateral load is now developed. The independent function to be determined is the lateral displacement in the y direction denoted by $v(x)$. An example of such a problem is shown in Fig. 6. As was the case in the bar problem, the boundary conditions do not enter the potential energy function. The boundary conditions are considered in the next lesson.

The only strain represented in an Euler-Bernoulli beam is the normal strain, ε_x. The displacement in the x direction is given as a function of the slope of the neutral axis of the beam by the following expression

---

Kinematic Relation in a Beam

$$u = -yv_{,x} \tag{1.18}$$

where y is the distance from the neutral axis.

$v_{,x}$ is the slope of the neutral axis.

When the definition of displacement given by Eq. 1.18 is substituted into the definition of ε_x given by Eq. 1.12, the result is

---

Strain-Displacement Relation in a Beam

$$\varepsilon_x = -yv_{,xx} \tag{1.19}$$

Substituting this expression for the normal strain into Eq. 1.10 for the strain energy function gives

---

Strain Energy Function

$$U = \frac{1}{2} \int_0^L \int_A Ey^2 (v_{,xx})^2 \, dA \, dx \tag{1.20}$$

When this expression is integrated over the area, the result is

---

Strain Energy Expression

$$U = \frac{1}{2} \int_0^L EI (v_{,xx})^2 \, dx \tag{1.21}$$

where I is the moment of inertia of the cross section.

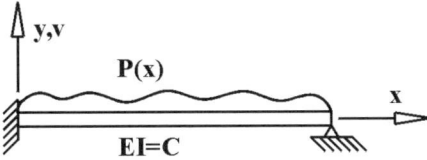

Figure 6. An Euler beam loaded with an arbitrary load.

The only applied force is the arbitrary lateral load. This applied force is displaced by the lateral displacement $v(x)$, so the work function can be written as

---

Work Function

$$W = \int_0^L P(x)v\,dx \tag{1.22}$$

Since both the strain energy and the work function are expressed in terms of the independent variable $v(x)$, the potential energy functional can be written directly as

---

Potential Energy Functional

$$V = \int_0^L \left[\frac{1}{2}EI(v_{,xx})^2 - P(x)v \right] dx \tag{1.23}$$

This is the functional for the beam problem. In this case, the general form of the integrand is $F(x, v, v_{xx})$. When we compare this general functional form to the functional for the bar problem given in Eq. 1.17, we see that the integrand for the beam problem has a second-order derivative of the function to be determined, $v(x)$. It, however, does not contain a first-order derivative, as was the case for the bar problem.

Generalized Functional Form

We can generalize the notation of the integrand of the functional a bit further. Let us simply denote the unknown function that we seek by y. This notation, along with the lack of a comma to indicate a derivative, corresponds to the standard notation used in much of the variational calculus literature. With this notation, we can represent the functionals for both the beam problem and the bar problem with one expression, namely,

---

Generalized Functional Form

$$V = \int_0^L F(x, y, y_x, y_{xx})\,dx \tag{1.24}$$

The y_{xx} term is not present in the functional for the bar problem and the y_x term is not present in the functional for the beam problem. In the next lesson, we develop the criteria for minimizing the generalized functional given by Eq. 1.24. As a result, we will have the criteria needed to minimize the functionals for both the bar and the beam problems. This concept is extended to two dimensions in Lesson 3, where the functional and the associated Euler-Lagrange equations and boundary terms are developed for the plane stress problem.

Closure

This lesson has introduced the principle of minimum potential energy and illustrated the relationship between the potential energy function and the equations of statics. We saw that the equilibrium equations for discrete systems are produced when the potential energy function is minimized using the procedures of differential calculus. That is to say, the equilibrium equations result when the partial derivatives of the potential energy function are taken and set to zero. This capability is used in later lessons to form finite element stiffness matrices.

The principle of minimum potential energy was extended to continuous systems where we found that the potential energy expression has the form of a functional, i.e., an expression that is a function of a function. In other words, the unknowns being sought to minimize the potential energy functional are functions. This means that the procedures of differential calculus that seek a maximum or a minimum at a point do not apply to the minimization of a functional.

The need to minimize the potential energy functional for continuous systems provides the motivation for the subject of the next lesson, the calculus of variations. The calculus of variations extends the derivative process so that functionals can be minimized. The process can be viewed as a formalization of the idea of a virtual displacement, i.e., a variation is an extension of the idea of a virtual displacement.

The presentation of the calculus of variations in the next lesson provides two valuable contributions that are used later in the analysis of the errors contained in finite element results. The relationship between the potential energy functional and the governing partial differential equations is clearly shown and the difference between the two classes of boundary conditions that are involved in specifying a continuous problem are identified.

Notes

1. The introduction of Lanczos's classic book, *The Variational Principles of Mechanics* (Ref. 1), originally written in 1949, discusses vectorial and analytic mechanics. Chapter 10 contains a historical survey of analytic mechanics.

2. One of the first treatments of analytic mechanics written for an engineering audience is Langhaar's *Energy Methods in Applied Mechanics* (Ref. 2). It contains functionals for several classes of engineering problems. A more recent treatment of energy methods directed toward engineers is *Energy and Finite Element Methods in Structural Mechanics* by Shames and Dym (Ref. 3).

References and Other Reading

1. Lanczos, C. *The Variational Principles of Mechanics*, 3rd ed. University of Toronto Press, Toronto, 1966.

2. Langhaar, H. L. *Energy Methods in Applied Mechanics*, John Wiley and Sons, Inc., New York, 1962.

3. Shames, I. H., and Dym, C. L. *Energy and Finite Element Methods in Structural Mechanics*, McGraw-Hill Book Co., New York, 1985.

Lesson 1   Problems

1. Form the potential energy function for the problem of Example 1 for the case where the point load on the bar is replaced with a distributed load of constant value q. *Hint*: Are the boundary conditions explicitly contained in the potential energy functional?

2. Redo Example 1 with the global coordinates consisting of the displacement and rotation of the center of the bar.

3. Form the strain energy function in relative coordinates Δ_i for the system shown in the figure. Transform this function to the global coordinates, u_i. Then form the potential energy function and find the equilibrium equations.

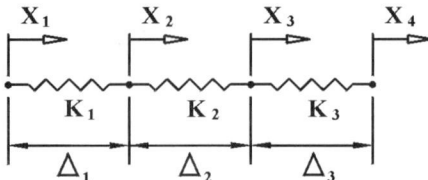

4. Form the potential energy function for the continuous bar of Example 2 if the distributed load due to gravity is replaced by a point load applied to the free end. *Hint*: This problem is designed to introduce a nonzero natural boundary condition.

5. Form the potential energy function for the problem of Example 2 if the free end of the bar is restrained in the same manner as the other end.

6. Form the potential energy function for the problem of Example 3 if both ends of the beam are clamped, i.e., no rotation and no displacement.

Lesson 2

Elements of the Calculus of Variations

Purpose of Lesson   To introduce the concepts or elements of the calculus of variations and use them to develop the necessary conditions for minimizing functionals.

The principles of variational calculus are used here to identify the conditions that must hold if a functional is to be minimized. We show that minimization is accomplished when the Euler-Lagrange equations and the boundary terms are forced to zero. The process of satisfying these conditions produces the governing differential equations and boundary conditions for a problem. In the case of the elasticity problems studied, the resulting governing differential equations are equilibrium equations.

The developments of this lesson define the relationship among the functional, the governing differential equations, and the associated boundary conditions. The solution of the governing differential equations that satisfies the boundary conditions minimizes the functional. Thus we have two distinct methods for solving the governing differential equations, namely, solving the equations directly or solving them indirectly by minimizing the functional.

A solution found by minimizing the functional from which the governing differential equations are derived is called a variational solution of that equation. A solution found by directly solving the differential equation is called a classical or analytic solution. The finite element method minimizes an approximation of the functional, therefore, the finite element method is a variational technique. In contrast, the finite difference method approximates the differential equation and then solves the approximation, so it is a classical approach. The similarities and differences between these two approximate solution techniques are discussed in later lessons.

■   ■   ■

In the previous lesson, we saw that the potential energy functions for discrete systems are polynomials that can be minimized using the procedures of differential calculus. When we extended the principle of minimum potential energy to continuous systems, we discovered that a functional (a function of a function) must be minimized. For the problems studied in the last lesson, this requires that we find a function that minimizes the following functional:

General Functional Form

$$V = \int_0^L F(x, y, y_x, y_{xx}) \, dx \tag{2.1}$$

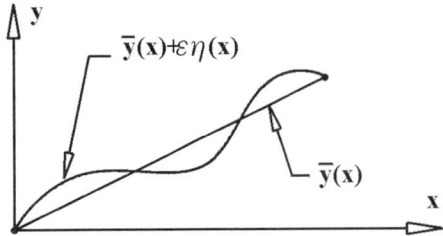

Figure 1. The variation of a minimizing function.

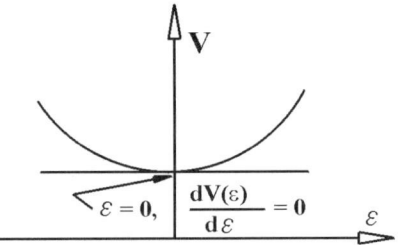

Figure 2. Variation of V in the neighborhood of $\eta = 0$.

The branch of mathematics that concerns itself with finding the extreme values of a functional (the minima, the maxima, or the saddle points) is the calculus of variations. The approach for finding the minimizing functions is somewhat analogous to that for determining the minimizing point of a function.

In the case of a function, the extreme values are found by taking the first partial derivatives of the function and setting them to zero. Then the points that yield minima are identified. In some cases, however, it is known *a priori* that a minimum is obtained. This is the situation that holds in the problems studied in this book.

In the case of a functional, "functional derivatives" are taken and set to zero. The final result of this operation is the Euler-Lagrange equations and the boundary terms. When the functional for a particular problem is substituted into these equations, the differential equations and the boundary conditions that the minimizing function must satisfy for this problem are produced. For the case of a potential energy functional, the equilibrium equations result from this process.

The key element in the procedures of variational calculus revolves around the introduction of a variation or imaginary perturbation into the functional. A variation is typically called a virtual displacement in mechanics. The idea of a variation is shown in Fig. 1. We assume that the minimizing function $\bar{y}$ is known. We then give this function an imaginary perturbation or variation that must satisfy the essential boundary conditions of the problem. A problem may contain two types of boundary conditions: *essential* and *natural* boundary conditions. The difference between the two types of boundary condition is discussed later (see Note 1). The variation consists of the difference between the two curves in Fig. 1 (see Note 2). Appendix A presents a detailed discussion of virtual displacements in the context of discrete rigid body systems.

The perturbation in the minimization function causes a change or variation in the value of the functional away from the minimum value, as shown in Fig. 2. We then take a functional derivative of the potential energy expression and set it to zero. A functional derivative is analogous to the standard derivative. Instead of finding the change in a function when an independent variable is given a small change or perturbation, a functional derivative finds the change in the scalar value of the functional when the defining function is given a variation. The use of a functional derivative is demonstrated in the next section. The Euler-Lagrange equations and the boundary integrals result when the functional derivative is simplified. These are the conditions that the minimization function must satisfy.

Development of the Euler-Lagrange Equations and the Boundary Terms

We now derive the Euler-Lagrange equations and the boundary terms for the general functional given in Eq. 2.1. These results are then applied to bar and beam problems to demonstrate their use.

The first step in the development requires that we clearly define the meaning of a variation. Theoretically defined, a variation is a small imaginary change away from the minimizing function being sought. A variation can be defined operationally by identifying a new function y as

---

A Variation Away from Equilibrium

$$y = \bar{y} + \varepsilon \eta(x) \tag{2.2}$$

In this expression, y is an arbitrary function that does not minimize the functional. This new function consists of the sum of the unknown minimizing function $\bar{y}$ and the variation or perturbation. The variation consists of two components. The parameter ε controls the magnitude of the perturbation. The function $\eta(x)$ controls the shape and boundary conditions of the variation. Each of these components provides an important contribution to the analysis. The parameter that controls the magnitude enables us to take the functional derivative. The totally arbitrary function $\eta(x)$ enables us to identify admissible boundary conditions for a problem, as is shown in the subsequent development. As mentioned earlier, virtual displacements in the context of discrete rigid body systems are discussed in Appendix A of this lesson. Note that the restrictions on virtual displacements are the same for both rigid body and continuous systems.

The next step is to incorporate the virtual displacement into the general expression for the functional. This is accomplished by recognizing that the derivatives of y with respect to x can be written in terms of the virtual displacement as

---

Variations in Derivatives

$$y_{,x} = \bar{y}_{,x} + \varepsilon \eta_{,x}$$
$$y_{,xx} = \bar{y}_{,xx} + \varepsilon \eta_{,xx} \tag{2.3}$$

Note that a slight change in notation has occurred. In the standard literature and as shown in Eq. 2.1, the derivatives are often denoted as $y_{,x}$ and $y_{,xx}$. From this point forward, these derivatives are denoted as $y_{,x}$ and $y_{,xx}$. In general, the comma denotes a derivative. This notation is convenient when we wish to express the derivatives of strains. For example, the term $\varepsilon_{x,x}$ represents the derivative of ε_x with respect to x. This notation is used extensively in the lessons that follow, starting in the next part of the book. When the derivative expressions just formed in Eq. 2.3 are substituted into the functional given by Eq. 2.1, the result is

---

A Functional with a Variation

$$V = \int_0^L F\left(x, (\bar{y} + \varepsilon \eta), (\bar{y}_{,x} + \varepsilon \eta_{,x}), (\bar{y}_{,xx} + \varepsilon \eta_{,xx})\right) dx \tag{2.4}$$

The next step in the procedure is to formalize the process of minimizing the functional illustrated in Fig. 2. This requires that we find the rate of change of the general functional with respect to ε, the magnitude of the perturbation, and set the derivative to zero. This is accomplished by recognizing that each of the dependent variables, y, $y_{,x}$, and $y_{,xx}$, in Eqs. 2.1 or 2.4 are functions of ε and by using this fact to form the total derivative of the general functional with respect to ε using the chain rule. This operation produces the functional derivative, which is given as (see Note 3)[1]

Functional Derivative

$$\frac{dV}{d\varepsilon} = \int_0^L \left(\frac{\partial F}{\partial y}\frac{\partial y}{\partial \varepsilon} + \frac{\partial F}{\partial y_{,x}}\frac{\partial y_{,x}}{\partial \varepsilon} + \frac{\partial F}{\partial y_{,xx}}\frac{\partial y_{,xx}}{\partial \varepsilon} \right) dx$$

$$= \int_0^L \left(\frac{\partial F}{\partial y}\eta + \frac{\partial F}{\partial y_{,x}}\eta_{,x} + \frac{\partial F}{\partial y_{,xx}}\eta_{,xx} \right) dx = 0$$

(2.5)

We can simplify the functional derivative by recognizing that the function η and its derivatives, $\eta_{,x}$ and $\eta_{,xx}$, are not independent of each other. This simplification is accomplished by integrating the last two terms of Eq. 2.5 by parts to eliminate the $\eta_{,x}$ and $\eta_{,xx}$ terms from the integrals. The one-dimensional expression for integration by parts is given in Fig. 3.

We now integrate the second and third terms of the functional derivative given in Eq. 2.5 in turn. The result of integrating the functional derivative by parts identifies the source of the individual components in the Euler-Lagrange equations and the boundary terms. The two components of the integral term containing $\eta_{,x}$ that correspond to the components of Fig. 3 are

$$v = \frac{\partial F}{\partial y_{,x}}$$

$$du = \eta_{,x}\, dx$$

(2.6)

The application of integration by parts produces the following result:

Integration by Parts

$$\int_0^L \frac{\partial F}{\partial y_{,x}}\eta_{,x}\, dx = \left(\frac{\partial F}{\partial y_{,x}}\eta \right)_0^L - \int_0^L \frac{d}{dx}\left(\frac{\partial F}{\partial y_{,x}} \right)\eta\, dx$$

(2.7)

$$\int v\, du = uv - \int u\, dv$$

Figure 3. Integration by parts.

---

[1] An alternative approach for developing this expression is presented in Appendix B.

The first expression on the right-hand side of Eq. 2.7 is part of the boundary term. This is as would be expected because this expression is evaluated at the two boundaries. The integrand of the second expression on the right-hand side of Eq. 2.7 is a component of the Euler-Lagrange equation. The meaning of these terms is discussed when the functional derivative is simplified.

The two components of the third integral term of Eq. 2.5 that correspond to the components of Fig. 3 are

$$v = \frac{\partial F}{\partial y_{,xx}}$$
$$du = \eta_{,xx} \, dx$$

(2.8)

The integration of this term produces

Integration by Parts

$$\int_0^L \frac{\partial F}{\partial y_{,xx}} \eta_{,xx} \, dx = \left(\frac{\partial F}{\partial y_{,xx}} \eta_{,x} \right)_0^L - \int_0^L \frac{d}{dx} \left(\frac{\partial F}{\partial y_{,xx}} \right) \eta_{,x} \, dx$$

(2.9)

This integration has produced a boundary term and an integral. The integral contains $\eta_{,x}$ as a multiplier. This is eliminated by a second application of integration by parts, to give

---

Second Integration by Parts

$$\int_0^L \frac{\partial F}{\partial y_{,xx}} \eta_{,xx} \, dx = \left(\frac{\partial F}{\partial y_{,xx}} \eta_{,x} \right)_0^L - \left[\frac{d}{dx} \left(\frac{\partial F}{\partial y_{,xx}} \right) \eta \right]_0^L + \int_0^L \frac{d^2}{dx^2} \left(\frac{\partial F}{\partial y_{,xx}} \right) \eta \, dx$$

(2.10)

The final form of the integration of the third term of the functional derivative given by Eq. 2.10 contains two boundary terms and an integral. All of the integrals contained in Eqs. 2.5, 2.7, and 2.10 now contain the term η. As a result, the functional derivative can be written in its simplified form as

---

Functional Derivative

$$\int_0^L \left[\frac{\partial F}{\partial y} - \frac{d}{dx} \left(\frac{\partial F}{\partial y_{,x}} \right) + \frac{d^2}{dx^2} \left(\frac{\partial F}{\partial y_{,xx}} \right) \right] \eta \, dx$$

$$+ \left[\frac{\partial F}{\partial y_{,x}} \eta - \frac{d}{dx} \left(\frac{\partial F}{\partial y_{,xx}} \eta \right) + \frac{\partial F}{\partial y_{,xx}} \eta_{,x} \right]_0^L = 0$$

(2.11)

Equation 2.11 consists of two distinct components, an integral and a set of boundary terms. Because η is arbitrary, it means that both the integral term and the boundary term must individually be zero. The integrand contains the functional portion of the perturbation, η, as a multiplier. Since η can take on any value at any point on the domain, the coefficient of the perturbation function must vanish everywhere on the domain. When we set this coefficient to zero, we obtain the Euler-Lagrange equation for this functional:

Euler-Lagrange Equation

$$\frac{\partial F}{\partial y} - \frac{d}{dx}\left(\frac{\partial F}{\partial y_{,x}}\right) + \frac{d^2}{dx^2}\left(\frac{\partial F}{\partial y_{,xx}}\right) = 0 \tag{2.12}$$

When a functional is substituted into Eq. 2.12, a differential equation results. The solution of this differential equation is the function that minimizes the functional for the problems encountered here.

The second term of the functional derivative given by Eq. 2.11 is the boundary term. It must also vanish. This term can be rearranged and written as follows:

Boundary Term

$$\left\{\left[\frac{\partial F}{\partial y_{,x}} - \frac{d}{dx}\left(\frac{\partial F}{\partial y_{,xx}}\right)\right]\eta + \frac{\partial F}{\partial y_{,xx}}\eta_{,x}\right\}_0^L = 0 \tag{2.13}$$

The boundary term given by Eq. 2.13 cannot be simplified further. As a result, the coefficient of η and the coefficient of $\eta_{,x}$ in Eq. 2.13 must individually be zero for the total expression to vanish. The components of the boundary term make it possible to define the admissible boundary conditions for the problem being considered. This is demonstrated in the examples that follow.

One-Dimensional Example Problems

We now demonstrate the use of the Euler-Lagrange equation and explicate the meaning of the boundary terms. This is done by finding the conditions that minimize the functions associated with the bar and beam problems in the previous lesson.

Example 1   Longitudinal Bar Problem

We now apply the results just developed to the problem of the longitudinal deformation of a bar. The functional for that problem was found in Lesson 1 to be

Potential Energy Functional

$$V = \int_0^L \left[\frac{1}{2}AE(u_{,x})^2 - P(x)u\right]dx \tag{2.14}$$

When we compare this function to the general form of the functional given as Eq. 2.1, we see that u, the longitudinal displacement of the bar, is equivalent to y, the general dependent variable. Also, the functional for the bar problem does not contain a second-order derivative term. This reduces the applicable Euler-Lagrange equation given by Eq. 2.12 and the boundary term given by Eq. 2.13 to the following:

Euler-Lagrange Equation

$$\frac{\partial F}{\partial u} - \frac{d}{dx}\left(\frac{\partial F}{\partial u_{,x}}\right) = 0 \tag{2.15}$$

Boundary Terms

$$\left[\left(\frac{\partial F}{\partial u_{,x}}\right)\eta\right]_0^L = 0 \tag{2.16}$$

Let us evaluate the individual terms of the Euler-Lagrange equation by taking the appropriate derivatives of the integrand of Eq. 2.14, to give

$$\frac{\partial F}{\partial u} = -P(x)$$
$$\frac{d}{dx}\left(\frac{\partial F}{\partial u_{,x}}\right) = \frac{d}{dx}(AEu_{,x}) = AEu_{,xx} \tag{2.17}$$

When the terms of Eq. 2.17 are combined, the governing differential equation for the problem results:

Governing Differential Equation

$$AEu_{,xx} + P(x) = 0 \tag{2.18}$$

This is the equation of equilibrium for this system. Note that if the cross-sectional area were a function of x, the second term of the Euler-Lagrange equation given by Eq. 2.17 would become

$$\frac{d}{dx}\left(\frac{\partial F}{\partial u_{,x}}\right) = \frac{d}{dx}(AEu_{,x}) = A_{,x}Eu_{,x} + AEu_{,xx} \tag{2.19}$$

The resulting governing differential equation is the following:

Governing Differential Equation

$$A_{,x}Eu_{,x} + AEu_{,xx} + P(x) = 0 \tag{2.20}$$

This simple example gives a hint of the power of the variational approach in formulating the governing differential equations for elastic systems. These governing equations hold regardless of the boundary conditions of the problem. This is shown by the fact that no reference to the boundary conditions was made during the formulation of Eqs. 2.18 and 2.20. Note that the highest order of the derivative present in the governing differential equation is twice that of the highest order of derivative in the functional for this problem, given by Eq. 2.14. This increase in the order of the derivative is discussed later.

Let us now interpret the meaning of the boundary terms as they apply to the problem shown in Fig. 4. When the functional is substituted into the boundary expression given by Eq. 2.16, the result is

$$\left[\left(\frac{\partial F}{\partial u_{,x}}\right)\eta\right]_0^L = [(AEu_{,x})\eta]_0^L = 0 \tag{2.21}$$

When Eq. 2.21 is evaluated at the two limits, the two boundary terms are given as

---

Boundary Terms

$$AEu_{,x}(0) \cdot \eta(0) = 0$$
$$AEu_{,x}(L) \cdot \eta(L) = 0 \tag{2.22}$$

We can observe that each of the boundary terms contains two components: one that is a function of η and one that is a function of the first derivative of u.

We will see that the η terms are associated with boundary conditions that are specified by the geometry of the problem. For example, if the end of the bar is fixed or given an initial displacement, the boundary condition is required to equal zero or the initial displacement. Such a boundary condition is called a forced, a geometric, or an *essential* boundary condition.

Furthermore, we will see that the u terms are associated with boundary conditions that are required to satisfy equilibrium. If a boundary is free, we will find the stresses on the boundary to equal zero, thus satisfying equilibrium. In the problems given at the end of the lesson, a boundary with an applied load is posed. The stresses are found to be such that equilibrium is satisfied. These boundary conditions are called *natural* boundary conditions.

Thus, each boundary term requires that either an essential or a natural boundary condition be satisfied. We now clarify these concepts with the example of a bar with a fixed end and a free end. The fixed end is an essential boundary condition and the free end is a natural boundary condition. A variation or virtual displacement is applied to this problem, as shown in Fig. 5.

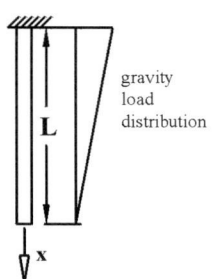

Figure 4. Longitudinal bar example.

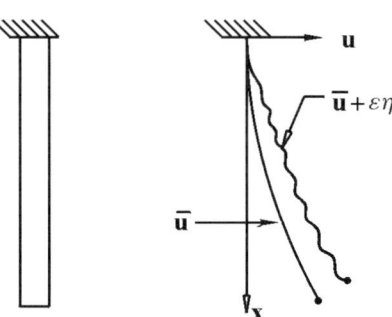

Figure 5. Variation for the bar problem.

The top of the bar is fixed so the displacement $u(0)$ is specified as zero. Let us show that this boundary condition satisfies the boundary term at $x=0$. Since the displacement is specified and the variation must satisfy the boundary conditions, η must equal zero at $x=0$. Thus, the boundary term at $x=0$ given in Eq. 2.22 is satisfied because one of its components, the η term, is zero. As a result of this argument, geometric or essential boundary conditions satisfy the requirement that the boundary term be zero if the functional is to be minimized. Thus the boundary condition for the fixed end of the bar is $u(0)=0$.

The bottom of the bar is free, so the displacement is not specified at $x = L$. Thus, the variation $\eta(L)$ need not be zero at the free boundary. Thus, for the boundary term to be zero, the other component of the boundary term given by the second equation of Eq. 2.22 must be zero, i.e., $AEu_{,x}(L)=0$.

This expression can be interpreted physically by recalling that $u_{,x}$ is the definition of the normal strain in the x direction. Substituting this expression for strain into the preceding boundary term and recalling that $\sigma_x = E\,\varepsilon_x$ gives the following expression:

Boundary Force on Bar

$$AE\varepsilon_x = A\sigma_x = 0 \tag{2.23}$$

This expression can be interpreted to mean that the stress and, hence, the applied load must be zero at the free end of the bar shown in Fig. 5. Since this is the condition that prevails at the free end of the bar, this boundary term is satisfied for this problem.

The results of this analysis can be summarized for a bar with a variable cross section and general boundary conditions as follows:

Governing Differential Equation

$$A_{,x}Eu_{,x} + AEu_{,xx} + P(x) = 0$$

Boundary Conditions

at $x = 0$

$$u(0) = 0 \quad \text{or} \quad u_{,x}(0) = 0$$

at $x = L$

$$u(L) = 0 \quad \text{or} \quad u_{,x}(L) = 0$$

$$\tag{2.24}$$

The governing differential equation expresses the equilibrium requirements on the domain of the problem. The general boundary conditions can be interpreted as follows. At the end $x=0$, the displacement can be specified to be zero or the applied force can be specified to be zero. Thus, four different problems are possible, namely, both ends can be free, both ends can be fixed, one end can be fixed and the other end can be free, and vice versa. If a function satisfies the governing differential equation and any of the four sets of boundary conditions, the function minimizes the functional for the problem.

This example has explicated several important concepts concerning variations and boundary conditions. The bar problem has shown that essential boundary conditions are those imposed by constraints on the displacements. This means that the variation at a forced boundary condition must be zero if the boundary term is to be satisfied. In other words, a variation cannot violate the constraints on a problem.

If no geometric constraint is imposed on the bar, the variation need not be zero at the end. Thus, the other component of the boundary term must be zero to force the boundary term to zero. This specifies the natural boundary conditions that must hold if the solution is to minimize the functional. These boundary conditions can be interpreted physically as requiring no force or stress at the free end. Thus, the idea of a natural boundary condition can be extended to mean that equilibrium is satisfied at boundaries whose displacements are not constrained. This concept opens the door for including other boundary conditions such as applied loads or boundaries with springs attached to them.

Example 2   Euler-Bernoulli Beam Problem

We now find the governing differential equation and boundary conditions for an Euler-Bernoulli beam. In Lesson 1 we found the potential energy functional for this problem to be

Potential Energy Functional

$$V = \int_0^L \left[\frac{1}{2} EI(v_{,xx})^2 - P(x)v \right] dx \tag{2.25}$$

This expression does not contain a $v_{,x}$ term, so the form of the Euler-Lagrange equation given by Eq. 2.12 and the form of the boundary term given by Eq. 2.13 reduce to

Euler-Lagrange Equation

$$\frac{\partial F}{\partial v} + \frac{d^2}{dx^2} \left(\frac{\partial F}{\partial v_{,xx}} \right) = 0 \tag{2.26}$$

Boundary Term

$$\left[-\frac{d}{dx} \left(\frac{\partial F}{\partial v_{,xx}} \right) \eta + \frac{\partial F}{\partial v_{,xx}} \eta_{,x} \right]_0^L = 0 \tag{2.27}$$

When the individual terms contained in the Euler-Lagrange equation given by Eq. 2.26 are evaluated for the functional given by Eq. 2.25, the result is

$$\frac{\partial F}{\partial v} = -P(x)$$
$$\frac{d^2}{dx^2} \left(\frac{\partial F}{\partial v_{,xx}} \right) = \frac{d^2}{dx^2} (EIv_{,xx}) = EIv_{,xxxx} \tag{2.28}$$

When the terms of Eq. 2.28 are combined according to Eq. 2.26, the following governing differential equation results:

Governing Differential Equation

$$EIv_{,xxxx} - P(x) = 0 \tag{2.29}$$

This fourth-order equation is identical to the equilibrium equation produced by static analysis. Note that the derivatives present in the equation are two orders higher than the order of the derivatives that appear in the functional for this problem given in Eq. 2.25. The significance of this doubling of the order of the derivatives in the governing differential equation is discussed later.

When we form the boundary expressions, we find them to be

$$\left[\frac{d}{dx}\left(\frac{\partial F}{\partial v_{,xx}}\right)\eta\right]_0^L = \left[\frac{d}{dx}(EIv_{,xx})\eta\right]_0^L = [(EIv_{,xxx}\eta]_0^L = 0 \tag{2.30}$$

$$\left[\frac{\partial F}{\partial v_{,xx}}\eta_{,x}\right]_0^L = [(EIv_{,xx})\eta_{,x}]_0^L = 0 \tag{2.31}$$

The boundary expression given by Eq. 2.30 can be evaluated to give the following two boundary terms:

Boundary Terms

$$EIv_{,xxx}(0) \cdot \eta(0) = 0$$
$$EIv_{,xxx}(L) \cdot \eta(L) = 0 \tag{2.32}$$

The first boundary term states that either the shear (a function of $v_{,xxx}$) or the displacement at $x = 0$ must be zero. The second boundary term says the same thing concerning the boundary at $x = L$.

The second boundary expression in Eq. 2.31 can be evaluated to give the following two boundary terms:

Boundary Terms

$$EIv_{,xx}(0) \cdot \eta_{,x}(0) = 0$$
$$EIv_{,xx}(L) \cdot \eta_{,x}(L) = 0 \tag{2.33}$$

These boundary terms can be interpreted to mean that the moment (a function of $v_{,xx}$) or the slope, $v_{,x}$, at the two boundaries must be zero.

The contents of Eqs. 2.32 and 2.33 can be summarized to produce the following two sets of possible boundary conditions for an Euler-Bernoulli beam:

Boundary Conditions

$$v_{,xxx}(0) = 0 \quad \text{or} \quad v(0) = 0$$
$$v_{,xx}(0) = 0 \quad \text{or} \quad v_{,x}(0) = 0$$

and $\tag{2.34}$

$$v_{,xxx}(L) = 0 \quad \text{or} \quad v(L) = 0$$
$$v_{,xx}(L) = 0 \quad \text{or} \quad v_{,x}(L) = 0$$

The two forced or essential boundary conditions, v and $v_{,x}$, are deduced from the virtual displacement terms η and $\eta_{,x}$. The argument is as follows. If the end displacement of the beam is constrained to zero, the variation cannot violate this constraint. So, if η is equal to zero, then the end displacement v must be zero. A similar argument holds for the slope of the beam that is associated with the virtual term $\eta_{,x}$.

The two natural boundary conditions, $v_{,xx}$ and $v_{,xxx}$, come directly from the terms containing η and $\eta_{,x}$, respectively. The argument is as follows. If the variations are not constrained, then η and $\eta_{,x}$ need not be zero. Thus, the other components of the boundary term must be zero. The four types of simple boundary conditions that are physically possible or admissible are shown in Fig. 6. Others are mathematically and physically possible but do not arise in practice or are not considered here.

We can see the physical significance of the natural boundary conditions by referring to the derivation of the beam equations from static considerations. The moment is given by $EIv_{,xx}$. So, if the boundary is free to rotate and $\eta_{,x} \neq 0$, then $v_{,xx}$ must be zero. This defines the moment as zero. Similarly, the shear is given by $EIv_{,xxx}$. So, if the boundary is free to displace and $\eta \neq 0$, then $v_{,xxx}$ must be zero, as must be the shear.

Closure

We have accomplished our purpose of developing the conditions that must be satisfied to minimize a functional. The Euler-Lagrange equations provide the differential equations and the boundary term provides the boundary conditions that the minimizing function must satisfy. We can use these conditions to solve for the minimizing function.

Conversely, if we find the function that minimizes the functional, we have the solution to the differential equation. This approach is exploited in finding what are called variational solutions. This is the approach used in the finite element method.

The variational solutions are called weak solutions because the order of the derivatives in the functional and, hence, the number of boundary conditions that *must* be represented by the solution is less than that required for the solution to the differential equation itself. Another way to look at this is to say that the constraints on the variational solutions are weaker than those for classical solutions. We see in Lesson 4 that the variational solutions are more accurate if they also satisfy the natural boundary conditions. This idea is exploited in the development of procedures for evaluating the errors in finite element solutions in the last part of this book. As expected, the classical solutions are called strong solutions.

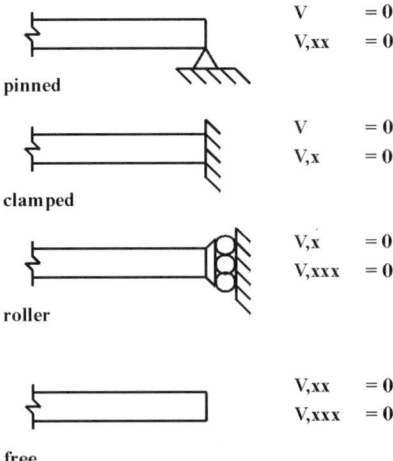

Figure 6. Admissible boundary conditions.

We extend the contents of Lessons 1 and 2 to two dimensions in Lesson 3 with an application to the plane stress problem. The effect of these considerations on the solution techniques are discussed in Lesson 4.

Notes

1. The formal mathematical names for essential and natural boundary conditions are Dirichlet and von Neumann boundary conditions, respectively. The formal names do not directly refer to their nature, so they must be memorized. Estimates suggest that close to 100 different kinds of boundary conditions can be applied to the ends of a beam. This means that the treatment of boundary conditions must be understood if the many situations that occur in practice are to be successfully represented.

2. The perturbation or variation that served as the heart of the development here is also known as a virtual displacement in mechanics. The idea and application of virtual displacements is one of the most subtle and powerful concepts in analysis (see Ref. 1, particularly the preface).

3. The functional derivative is discussed in Section 2.3 and applied in Section 3.3 in the Weinstock book (Ref. 4). The order of the derivative of the perturbation can be reduced by a procedure based on the expansion of the potential energy functional as a Taylor series. This is presented in Chapter 2 of the Lanczos book (Ref. 1). The idea of the functional derivative is developed by the Taylor series approach in Appendix B of this lesson.

Appendix A
Virtual Displacements for Discrete Systems

The objective of this appendix is to present the concept of an admissible virtual displacement in the context of discrete rigid body systems. An admissible virtual displacement is an imaginary displacement imposed on a system that conforms to the constraints on the problem.[2] The constraints on a virtual displacement are the same regardless of the problem type, i.e., these constraints apply to continuous as well as to discrete systems. In the remainder of this discussion, the term *virtual displacement* is synonymous with an admissible virtual displacement unless otherwise specified.

Admissible virtual displacements for a fixed (cantilevered) connection and for a pinned connection are shown in Fig. A.1. A flexible beam with a fixed boundary is shown in Fig. A.1a. This continuous system can deform but no displacement or slope is allowed at the boundary. A rigid beam with a pinned connection is shown in Fig. A.1b. This rigid body can rotate but it cannot deform or displace for an admissible virtual displacement. These virtual displacements conform to the constraints on their respective systems.

Inadmissible virtual displacements are shown in Fig. A.2. A rigid or a flexible beam with a fixed boundary is shown in Fig. A.2a. The element is given a virtual displacement. It is shown in a displaced and rotated position, but the motions shown violate the constraints of the connection. Thus, this is not an admissible virtual displacement. Figure A.2b illustrates an inadmissible virtual displacement applied to a pinned connection. The constraint on the displacement has been violated. The virtual displacements shown in Fig. A.2 are inadmissible because they violate the constraints of their respective boundaries. A virtual displacement is denoted here as $\overline{\delta r}$.

The characteristics of a virtual displacement are explicated by deriving the principle of virtual work for discrete systems and by using a virtual displacement in a problem. The derivation identifies the theoretical reason for the constraints on a virtual displacement. The example problem clearly demonstrates the meaning of the constraints in practice.

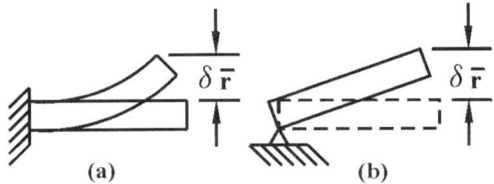

Figure A.1. Admissible virtual displacements.

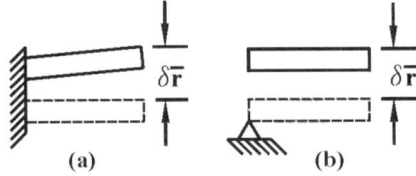

Figure A.2. Inadmissible virtual displacements.

---

[2] The idea that a virtual displacement is imaginary does not have any significance in statics applications. However, when the idea of a virtual displacement is extended to dynamics problems, it is necessary that the virtual displacement be independent of time. This means that the virtual displacement can be only an experiment of the mind, i.e., it is a product of the imagination.

Derivation of the Principle of Virtual Work

The principle of virtual work states that the work done by the applied forces must be zero when a virtual displacement is imposed on a system in equilibrium. As shown previously, a virtual displacement must not violate the geometric constraints imposed on a system. In the process of deriving this alternative method for solving statics problems, the theoretical basis for this restriction on virtual displacements will be seen. The derivation proceeds in a step-by-step manner as follows:

Step 1: Assume that the system being analyzed is in equilibrium, i.e.,

Statement of Equilibrium

$$\sum \overline{F} = 0 \tag{A.1}$$

Step 2: Divide the forces into applied forces $\overline{F}_a$ and constraint forces $\overline{F}_c$, i.e.,

Separation of Force Types

$$\sum \overline{F}_a + \sum \overline{F}_c = 0 \tag{A.2}$$

Step 3: Impose a virtual displacement $\overline{\delta r}$ on the system and form the expression for virtual work, i.e.,

Apply a Virtual Displacement

$$\sum \overline{F}_a \cdot \overline{\delta r}_a + \sum \overline{F}_c \cdot \overline{\delta r}_c = 0 \tag{A.3}$$

Step 4: Apply the definition of a virtual displacement. An admissible virtual displacement is an imaginary displacement that does not allow motion at a constraint. Thus, the virtual displacement at the constraint $\overline{\delta r}_c$ must be zero. This means that the work of the constraint forces are zero, i.e.,

Definition of Virtual Displacement

$$\sum \overline{F}_c \cdot \overline{\delta r}_c \equiv 0 \tag{A.4}$$

Step 5: When this definition is inserted into Eq. A.3, the statement of the principle of virtual work results, i.e.,

Principle of Virtual Work

$$\sum \overline{F}_a \cdot \overline{\delta r}_a = 0 \tag{A.5}$$

This expression states that the work done by the applied forces must equal zero when a virtual displacement is imposed on a system in equilibrium. Equation A.5 is the mathematical statement of the principle of virtual work. As just seen, the derivation hinges on the need for the work of the constraint forces to be zero. This requires that the virtual displacements do not violate the geometric constraints of the problem.

Example Problem

The principle of virtual work is now applied to a simple problem to demonstrate the use of an admissible virtual displacement. The problem chosen for this demonstration is that of the teeter-totter shown in Fig. A.3. A teeter-totter consists of a rigid beam and a pinned connection. A known force is applied to the right end of the rigid beam. The objective of the problem is to find the unknown force on the left end required to produce equilibrium.

The system is given an admissible virtual displacement, as shown in Fig. A.4. In this figure, the pivot has been replaced by the constraint forces at the pin. The pivot restricts the displacement of the beam to zero so the virtual displacement at this point must be zero. Since the beam is a rigid element, it cannot deform. It is equally important to note that a pivot does not constrain the rotation. The virtual displacement causes the force on the right end to move upward and the force on the left to move downward. The constraint forces do no work. This validates the fact that the virtual displacement shown in Fig. A.4 is admissible as defined by Eq. A.3.

Considering an example of a virtual displacement for this problem that is not admissible is instructive. An inadmissible virtual displacement is shown in Fig. A.5. As we can see, the virtual displacement violates the constraints by exhibiting a movement at the pivot. The constraint forces would do work in this case. This would violate the condition of a virtual displacement given by Eq. A.3.

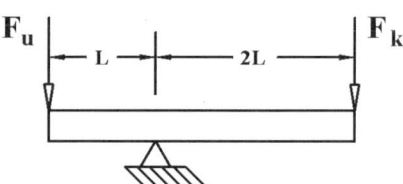

Figure A.3. Teeter-totter.

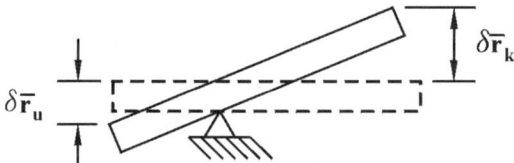

Figure A.4. Admissible virtual displacement.

Figure A.5. Inadmissible virtual displacement.

We now apply the principle of virtual work as expressed by Eq. A.5 to this problem. The virtual work due to the known force $\overline{F}_k$ is

$$(VW)_{\text{applied}} = -\overline{F}_k \cdot \overline{\delta r}_k \qquad (A.6)$$

This quantity is negative because the force and the displacement are in opposite directions. The virtual work due to the unknown force $\overline{F}_u$ is

$$(VW)_{\text{unknown}} = +\overline{F}_u \cdot \overline{\delta r}_u \qquad (A.7)$$

This quantity is positive because the force and the displacement are in the same direction. Since the bar is rigid and the lever arms have a proportion of 2:1, the virtual displacements are related by the following coordinate transformation:

Coordinate Transformation

$$\overline{\delta r}_k = 2\overline{\delta r}_u \qquad (A.8)$$

When Eqs. A.6, A.7, and A.8 are used to form an expression for the virtual work that corresponds to Eq. A.5, the result is

---

Principle of Virtual Work

$$(\overline{F}_u - 2\overline{F}_k) \cdot \overline{\delta r}_u = 0 \qquad (A.9)$$

Since the virtual displacement $\overline{\delta r}_u$ can have any value, the coefficient multiplying this term must be zero. When this expression is rearranged, the result is

---

Solution

$$\overline{F}_u = 2\overline{F}_k \qquad (A.10)$$

Thus, we have found the obvious result that the unknown force is twice the known force. The use of the concept of a virtual displacement has produced a powerful alternative approach to solving statics problems. Note that the principle of virtual work does not explicitly introduce the constraint forces into the equations to be solved. This is a general characteristic of most energy methods unless a specific effort is made to include these forces.

Closure

The derivation and problem solution have defined and demonstrated the characteristics of an admissible virtual displacement. A virtual displacement must not violate the geometric constraints of the problem. Specifically, this means that the constraint forces must do no work when a virtual displacement is applied to a system. This was stated in equation form by Eq. A.4 in this appendix. To see other examples of statics problems for discrete systems solved using the principle of virtual work see Chapter 10 of *Vector Mechanics for Engineers* by Beer and Johnson (Ref. 8).

Appendix B
An Alternative Development of the Functional Derivative

The more common approach for deriving the Euler-Lagrange equation(s) and the boundary terms utilizes the *definition of the derivative* and the *Taylor series expansion* of the general functional rather than the functional derivative. The results of the two approaches are identical. The functional derivative approach was chosen for the primary presentation because of its compact nature. The results of the functional derivative are now derived using the more common approach. The expression for the general functional needed to solve bar and beam problems is reproduced here for convenience as

---

General Functional Form

$$V = \int_0^L F(x, y, y_{,x}, y_{,xx})\, dx \tag{B.1}$$

In Lanczos's book (Ref. 1) on variational principles, he shows that the integration of the functional and the variation (virtual displacement) are commutative. We simply accept that the integration operation and the derivative with respect to ε are interchangeable. The effect of this interchangeability means that we can directly take the derivative of the integrand of the general functional. The result of this operation is

---

Definition of a Derivative

$$\frac{dF}{d\varepsilon} = \lim_{\varepsilon \to 0} \frac{\Delta F}{\varepsilon} \tag{B.2}$$

The expression for ΔF, the change in the integrand of the general functional when the system is given a virtual displacement, can be written as

---

Definition of ΔF

$$\begin{aligned} \Delta F = F(x,\ (\bar{y} + \varepsilon\eta),\ (\bar{y}_{,x} + \varepsilon\eta_{,x}),\ (\bar{y}_{,xx} + \varepsilon\eta_{,xx})) \\ - F(x, \bar{y}, \bar{y}_{,x}, \bar{y}_{,xx}) \end{aligned} \tag{B.3}$$

This notation can be simplified by utilizing the definition of the variational approximation of the solution as

---

Definition of a Variation

$$y = \bar{y} + \varepsilon\eta \tag{B.4}$$

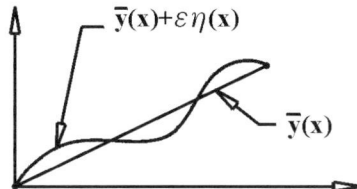

Figure B.1. The variation of a minimizing function.

The meaning of Eq. B.4 is shown in Fig. B.1. A variation is added to the actual equilibrium position $\bar{y}$ to give an approximation of the solution. That is to say, we assume we know the solution at equilibrium and then give it a perturbation. The perturbation consists of a "shape" with an arbitrary magnitude. At this point, we do not know the actual solution, but this strategy enables us to find the actual solution.

When Eq. B.3 is rewritten with the notation given in Eq. B.4, the result is

---

Perturbated Integrand

$$\Delta F = F(x, y, y_{,x}, y_{,xx}) - F(x, \bar{y}, \bar{y}_{,x}, \bar{y}_{,xx}) \tag{B.5}$$

The first term on the right-hand side of Eq. B.5 can be expanded in a Taylor series about the actual solution $\bar{y}$ as

---

Taylor Series Expansion

$$
\begin{aligned}
F(x, y, y_{,x}, y_{,xx}) = {} & F(x, \bar{y}, \bar{y}_{,x}, \bar{y}_{,xx}) \\
& + \varepsilon \Bigg\{ \left[\frac{\partial F(x, \bar{y}, \bar{y}_{,x}, \bar{y}_{,xx})}{\partial y} \right] \eta + \left[\frac{\partial F(x, \bar{y}, \bar{y}_{,x}, \bar{y}_{,xx})}{\partial y_{,x}} \right] \eta_{,x} \\
& + \left[\frac{\partial F(x, \bar{y}, \bar{y}_{,x}, \bar{y}_{,xx})}{\partial y_{,xx}} \right] \eta_{,xx} \Bigg\} + \varepsilon^2(\cdots)
\end{aligned}
\tag{B.6}
$$

The higher order terms in ε are neglected after this point since these terms approach zero in the definition of the derivative as the limit is imposed. The result of this operation, after reintroducing the integral, becomes

---

Functional Derivative

$$
\begin{aligned}
\frac{dV}{d\varepsilon} &= \int_0^L \left(\frac{dF}{d\varepsilon} \right) dx \\
&= \int_0^L \left(\frac{\partial F}{\partial y} \eta + \frac{\partial F}{\partial y_{,x}} \eta_{,x} + \frac{\partial F}{\partial y_{,xx}} \eta_{,xx} \right) dx = 0
\end{aligned}
\tag{B.7}
$$

This expression is identical to the result of the functional derivative presented in the main text. The application of integration by parts produces the Euler-Lagrange equations and the boundary terms.

References and Other Reading

1. Lanczos, C. *The Variational Principles of Mechanics*, 3rd. ed. University of Toronto Press, Toronto, 1966. A classic book on the subject originally published in 1949. The introduction discusses vectorial and analytic mechanics. Chapter 10 consists of a historical survey of analytic mechanics.

2. Langhaar, H. L. *Energy Methods in Applied Mechanics*, John Wiley and Sons, Inc., New York, 1962.

3. Shames, I. H., and Dym, C. L. *Energy and Finite Element Methods in Structural Mechanics*, McGraw-Hill Book Co., New York, 1985.

4. Weinstock, R. *Calculus of Variations: With Applications to Physics and Engineering*, Dover Publications, Inc., New York, 1974. This book, originally published in 1952, contains the use of the functional derivative (he calls it the integration of an integral by a parameter).

5. Forray, M. J. *Variational Calculus in Science and Engineering*, McGraw-Hill Book Co., New York, 1968.

6. Feynman, R. P., Leighton, R. B., and Sands, M. *The Feynman Lectures on Physics*, Vol. II Chapter 19, Principle of Least Action, Addison-Wesley Publishing Co., Reading, Mass., 1965. A note included with this lecture indicates that it is included as part of the lecture series for the sake of entertainment.

7. Gelfand, I. M., and Fomin, S.V. *Calculus of Variations*, Prentice-Hall, Inc., Englewood Cliffs, N.J., 1963.

8. Beer, F. P., and Johnson, E. R, Jr. *Vector Mechanics for Engineers—Statics*, 5th ed. McGraw-Hill Book Co., New York, 1988. See Chapter 10, Method of Virtual Work.

Lesson 2   Problems

1. Find the governing differential equation and boundary conditions for a bar with a uniform cross section that is fixed at both ends. *Hint:* See Fig. 6 in the lesson.

2. Find the governing differential equation and boundary conditions for a beam with a uniform cross section that is fixed at one end and simply supported at the other end.

3. Find the differential equation and boundary conditions for a bar of uniform cross section that is fixed at one end and has an applied load at the "free" end. *Hint:* This problem is designed to introduce the idea of a nonzero natural boundary condition.

4. Find the governing differential equation and boundary conditions for a cantilever beam with a constant cross section and axial load applied at the free end. *Hint:* Neglect axial deformations and add a term to the work function due to the work done by the axial load. The length changes because of the lateral deformation. Look in the index of almost any finite element book under the listing "Geometric stiffness matrix."

5. Outline the procedure for finding the Euler-Lagrange equations and the boundary terms for a rectangular two-dimensional region. Objective: Identify the changes needed to apply the ideas of this lesson to two dimensions. See Lesson 3.

Lesson **3**

Derivation of the Plane Stress Problem

Purpose of Lesson   To derive the governing differential equations and associated boundary conditions for the two-dimensional plane stress problem by applying the principle of minimum potential energy and the techniques of variational calculus to the potential energy functional.

We first form an expression for the potential energy as a function of the two independent displacements, u and v. This defines the functional form that must be minimized. We then use the calculus of variations to identify the conditions necessary to minimize a functional in two dimensions. As was the case in one dimension, the conditions that must be satisfied are the Euler-Lagrange equations and the boundary terms. When the functional is substituted into the Euler-Lagrange equations and the boundary terms, the governing differential equation and the boundary conditions for the plane stress problem result. This development clearly identifies the relationship between the variational and differential forms of two-dimensional problems as two independent ways to minimize the potential energy function and to solve the governing differential equations.

The plane stress problem is chosen for detailed study because it is both complex enough to require subtle analysis and small enough so that the concepts discussed are not obscured by the size and complexity of the equations that must be manipulated. This two-dimensional problem provides a demonstration that can readily be extended to other problems.

■   ■   ■

The primary focus of this lesson is to establish the relationship between the variational and differential forms of the two-dimensional plane stress problem. This problem applies to important classes of practical problems in engineering such as the in-plane deformations of flat structural elements. These elements represent building walls and bulkheads in aerospace vehicles and ships.

Such a problem is shown in Fig. 1, where one portion of the boundary of the region Ω is fixed, two portions are loaded, and the remainder of the boundary is free of loads and constraints. For small displacements, the in-plane deformations are uncoupled from the out-of-plane deformations. The out-of-plane deformations belong to the allied study of flat plate bending. The procedures presented here apply equally well to the plate-bending problem, as discussed in Lesson 12. The development of the plane stress problem follows the steps outlined in Fig. 1 of Lesson 1.

Formulation of the Strain Energy Functional

The development of the variational form of the plane stress problem begins with the formulation of the strain energy expression. The strain energy is first formulated in terms of stresses and strains. Then, this expression is transformed to a function of strains using

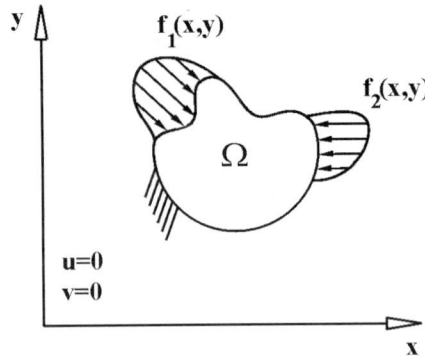

Figure 1. Example plane stress problem.

the constitutive relations. Finally, the strain-displacement relations are used to transform the strain energy expression to a function of u and v, with the displacements in the x and y directions, respectively. The resulting strain energy expression is the functional that must be minimized to solve the equilibrium equations.

A plane stress problem is a specialization of the three-dimensional elasticity problem that contains no through thickness stresses. Specifically, this means that the following conditions hold:

Definition of Plane Stress

$$\sigma_z = 0$$
$$\tau_{xz} = 0$$
$$\tau_{yz} = 0$$

(3.1)

This reduces the general three-dimensional strain energy expression of Lesson 1 to

---

Strain Energy for Plane Stress

$$U = \frac{1}{2}\int_\Omega \left(\sigma_x \varepsilon_x + \sigma_y \varepsilon_y + \tau_{xy}\gamma_{xy}\right)d\Omega$$

(3.2)

Our first goal is to obtain an expression for the strain energy as a function of the strains in preparation for a transformation to displacement quantities. To achieve this aim, we must form the stress-strain relation for plane stress. This is accomplished by inverting the reduced constitutive relation that results when the definition of plane stress is substituted into the three-dimensional constitutive relation. The reduced constitutive relation is given as

---

Reduced Constitutive Relation

$$\begin{Bmatrix} \varepsilon_x \\ \varepsilon_y \\ \gamma_{xy} \end{Bmatrix} = \frac{1}{E}\begin{bmatrix} 1 & -\nu & 0 \\ -\nu & 1 & 0 \\ 0 & 0 & 2(1+\nu) \end{bmatrix}\begin{Bmatrix} \sigma_x \\ \sigma_y \\ \tau_{xy} \end{Bmatrix}$$

(3.3)

When this expression is inverted, the stresses are given as the following functions of the strains:

Hooke's Law for Plane Stress

$$\left\{ \begin{array}{c} \sigma_x \\ \sigma_y \\ \tau_{xy} \end{array} \right\} = \frac{E}{(1 - \nu^2)} \begin{bmatrix} 1 & \nu & 0 \\ \nu & 1 & 0 \\ 0 & 0 & \frac{1-\nu}{2} \end{bmatrix} \left\{ \begin{array}{c} \varepsilon_x \\ \varepsilon_y \\ \gamma_{xy} \end{array} \right\} \qquad (3.4)$$

The stress-strain relation given by Eq. 3.4 is used to eliminate the stresses from Eq. 3.2, the strain energy expression for plane elasticity, to give

---

Strain Energy for Plane Stress

$$U = U_{\varepsilon_x} + U_{\varepsilon_y} + U_{\gamma_{xy}}$$

where

$$U_{\varepsilon_x} = \frac{1}{2} \int_\Omega \varepsilon_x \left[\frac{E}{1 - \nu^2} (\varepsilon_x + \nu \varepsilon_y) \right] d\Omega$$

$$U_{\varepsilon_y} = \frac{1}{2} \int_\Omega \varepsilon_y \left[\frac{E}{1 - \nu^2} (\varepsilon_y + \nu \varepsilon_x) \right] d\Omega \qquad (3.5)$$

$$U_{\gamma_{xy}} = \frac{1}{2} \int_\Omega \gamma_{xy} \left[\frac{E}{1 - \nu^2} \frac{(1 - \nu)}{2} \gamma_{xy} \right] d\Omega$$

When Eq. 3.5 is simplified, the result is

Strain Energy Expression

$$U = \frac{1}{2} \int_\Omega \left(\frac{E}{1 - \nu^2} \right) \left[\varepsilon_x^2 + 2\nu \varepsilon_x \varepsilon_y + \varepsilon_y^2 + \left(\frac{1 - \nu}{2} \right) \gamma_{xy}^2 \right] d\Omega \qquad (3.6)$$

The next step is to transform Eq. 3.6 to a function of the two independent displacements, u and v. This transformation is accomplished using the strain displacement relations from linear elasticity. These relationships are written in two forms as

---

Strain-Displacement Relation

$$\varepsilon_x = \frac{\partial u}{\partial x} \qquad \varepsilon_y = \frac{\partial v}{\partial y} \qquad \gamma_{xy} = \frac{\partial u}{\partial y} + \frac{\partial v}{\partial x}$$

or in alternative notation $\qquad\qquad\qquad\qquad\qquad (3.7)$

$$\varepsilon_x = u_{,x} \qquad \varepsilon_y = v_{,y} \qquad \gamma_{xy} = u_{,y} + v_{,x}$$

The second form of the strain-displacement relations is expressed in the notation that is used throughout the developments presented here. The comma is used to denote differentiation, as can be seen when the two forms are compared.

When Eq. 3.6 is transformed using Eq. 3.7, the final form of the strain energy expression as a function of the independent displacements is

---

Strain Energy as a Function of Displacements

$$U = \frac{1}{2}\left(\frac{E}{1-\nu^2}\right)\int_{\Omega}\left[\left(u_{,x}^2 + 2\nu u_{,x}v_{,y} + v_{,y}^2\right)\right.$$
$$\left. + \left(\frac{1-\nu}{2}\right)\left(u_{,y}^2 + 2u_{,y}v_{,x} + v_{,x}^2\right)\right]d\Omega$$

(3.8)

We have accomplished our goal of expressing the strain energy as a function of the displacements in the x-y plane. This expression for the strain energy of the plane stress problem makes up one component of the functional for the plane stress problem.

Formulation of the Work Function

The next step in the process of forming the potential energy functional is to develop an expression for the work done by the applied loads when the region being studied deforms. The loading in the two coordinate directions is assumed to be arbitrary and to be given as $P_x(x,y)$ and $P_y(x,y)$. The loading is expressed in terms of force per unit area. This produces a work function of the following form:

Work Function

$$W = \int_{\Omega}[P_x(x,y)u + P_y(x,y)v]\,d\Omega$$

(3.9)

Equation 3.9 identifies the work done by the body forces applied to the region being studied. These loads are shown to be part of the differential equations of equilibrium. The gravity load fits this category. Loads on the boundary can be introduced at this point. The boundary loads are seen in the problem as part the boundary conditions and are considered later.

The Potential Energy Functional

The potential energy for the plane stress problem is formed by combining Eqs. 3.8 and 3.9. The unknowns in both of these expressions are the displacement functions, u and v, so both components of the potential energy are functions of functions. This defines the strain energy expression as a functional, which is given as

Potential Energy Functional

$$
\begin{aligned}
V = \int_\Omega \Bigg\{ &\frac{1}{2}\left(\frac{E}{1-\nu^2}\right)\left[\left(u_{,x}^2 + 2\nu u_{,x} v_{,y} + v_{,y}^2\right)\right. \\
&+ \left.\left(\frac{1-\nu}{2}\right)\left(u_{,y}^2 + 2u_{,y} v_{,x} + v_{,x}^2\right)\right] \\
&- (P_x u + P_y v) \Bigg\} \, d\Omega
\end{aligned}
$$
(3.10)

The functional for the plane stress problem is represented by the following general form:

General Form of the Functional

$$
V = \int_\Omega F(x, y, u, v, (u_{,x}), (u_{,y}), (v_{,x}), (v_{,y})) \, d\Omega
$$
(3.11)

This functional has two independent variables, x and y, and two dependent functions, u and v. The first derivatives of the unknown displacements are also present in the functional. This expression is the two-dimensional analog of the functional for the bar problem given in Lesson 2 as Eq. 2.14. This general form also applies to the plane strain problem.

Development of the Euler-Lagrange Equations and Boundary Terms

Now that the functional to be minimized has been defined, we can proceed with the derivation of the Euler-Lagrange equations and the boundary terms associated with this general form. In this case, two Euler-Lagrange equations result when the functional derivative is taken and set to zero because there are two independent functions to be determined. The boundary terms are integrals for this case because the boundaries are lines (areas because of the thickness) instead of points, as they were for one-dimensional problems. The derivation for this two-dimensional case is similar to that for the one-dimensional case. The only significant difference is the use of the two-dimensional analog of integration by parts, namely, Green's theorem, to reduce the functional derivative to a useable form.

There are two independent functions to be given a virtual displacement in this case. The variations can be written as

Virtual Displacement in Two Dimensions

$$
\begin{aligned}
u &= \bar{u} + \varepsilon \eta_x(x, y) \\
v &= \bar{v} + \varepsilon \eta_y(x, y)
\end{aligned}
$$
(3.12)

The terms $\bar{u}$ and $\bar{v}$ represent the solutions that minimize the potential energy functional and, consequently, satisfy the governing differential equations. Each is given a virtual displacement consisting of two components. The η_x and η_y terms satisfy the geometric boundary conditions in the two coordinate directions as is required by an admissible virtual displacement. The ε term controls the magnitude of the two independent virtual displacements.

Since the first step in the derivation is to take the functional derivative of Eq. 3.11 with respect to ε, the functions in this equation must be expressed in terms of ε. The displacements are expressed in this form in Eq. 3.12. The four derivatives contained as independent variables in the functional can be formed from Eq. 3.12. When the appropriate derivatives are taken, the result is

---

First Derivatives with Variation

$$u_{,x} = \bar{u}_{,x} + \varepsilon\eta_{x,x}$$

$$u_{,y} = \bar{u}_{,y} + \varepsilon\eta_{x,y}$$

$$v_{,x} = \bar{v}_{,x} + \varepsilon\eta_{y,x}$$

$$v_{,y} = \bar{v}_{,y} + \varepsilon\eta_{y,y}$$

(3.13)

When these variations of the independent functions are included in the general form of the functional, the result is

---

Functional with Variations

$$V = \int_{\Omega} F(x, y, (\bar{u} + \varepsilon\eta_x), (\bar{v} + \varepsilon\eta_y), (\bar{u}_{,x} + \varepsilon\eta_{x,x}),$$
$$(\bar{u}_{,y} + \varepsilon\eta_{x,y}), (\bar{v}_{,x} + \varepsilon\eta_{y,x}), (\bar{v}_{,y} + \varepsilon\eta_{y,y}))\, d\Omega$$

(3.14)

The functional derivative of this expression is given as

Functional Derivative

$$\frac{\partial V}{\partial \varepsilon} = \int_{\Omega} \left(\frac{\partial F}{\partial u}\frac{\partial u}{\partial \varepsilon} + \frac{\partial F}{\partial v}\frac{\partial v}{\partial \varepsilon} + \frac{\partial F}{\partial u_{,x}}\frac{\partial u_{,x}}{\partial \varepsilon} + \frac{\partial F}{\partial u_{,y}}\frac{\partial u_{,y}}{\partial \varepsilon} + \frac{\partial F}{\partial v_{,x}}\frac{\partial v_{,x}}{\partial \varepsilon} + \frac{\partial F}{\partial v_{,y}}\frac{\partial v_{,y}}{\partial \varepsilon} \right) d\Omega \quad (3.15)$$

The partial derivatives of the variational terms can be determined explicitly from Eqs. 3.12 and 3.13. When the derivatives of these expressions are taken with respect to ε, the result is

Derivatives of Variational Terms

$$\frac{\partial u}{\partial \varepsilon} = \eta_x \qquad \frac{\partial v}{\partial \varepsilon} = \eta_y$$

$$\frac{\partial u_{,x}}{\partial \varepsilon} = \eta_{x,x} \qquad \frac{\partial v_{,x}}{\partial \varepsilon} = \eta_{y,x} \qquad (3.16)$$

$$\frac{\partial u_{,y}}{\partial \varepsilon} = \eta_{x,y} \qquad \frac{\partial v_{,y}}{\partial \varepsilon} = \eta_{y,y}$$

When Eq. 3.15 is simplified using Eq. 3.16, the result is

Functional Derivative

$$\frac{\partial V}{\partial \varepsilon} = \int_\Omega \left(\frac{\partial F}{\partial u} \eta_x + \frac{\partial F}{\partial v} \eta_y + \frac{\partial F}{\partial u_{,x}} \eta_{x,x} + \frac{\partial F}{\partial u_{,y}} \eta_{x,y} + \frac{\partial F}{\partial v_{,x}} \eta_{y,x} + \frac{\partial F}{\partial v_{,y}} \eta_{y,y} \right) d\Omega \qquad (3.17)$$

As was the case for the one-dimensional problem, this expression can be simplified further by eliminating the derivatives of the η's. In the one-dimensional case, this reduction is accomplished by integrating the functional derivative by parts. However, integration by parts applies only to the one-dimensional case. The two-dimensional analog of integration by parts is Green's theorem. Thus, these expressions can be put in the desired form by integrating the functional derivative with Green's theorem. See Appendix A for a derivation of the form of Green's theorem used to simplify the functional derivative (see Note 1).

Green's theorem is typically developed in advanced calculus. The material presented here does not anticipate that the readers are familiar with advanced calculus. However, the contents of the previous lesson enable us to interpret the two-dimensional results developed here using Green's theorem. Thus, the plane stress problem being solved can be clearly understood. In Part III, Green's theorem is used to speed the computation of finite element stiffness matrices. For completeness, note that this reduction is accomplished in the case of the three-dimensional problem using Stokes's theorem.

Only a single form of Green's theorem is used. However, for the sake of clarity and the reader's convenience, the four applications of the single form needed for the simplification of Eq. 3.17 are presented explicitly as

A Specialized Form of Green's Theorem

$$\int_\Omega \eta_{x,x} \frac{\partial F}{\partial u_{,x}} d\Omega = -\int_\Omega \eta_x \frac{\partial}{\partial x}\left(\frac{\partial F}{\partial u_{,x}}\right) d\Omega + \oint_\Omega \eta_x \frac{\partial F}{\partial u_{,x}} dy$$

$$\int_\Omega \eta_{x,y} \frac{\partial F}{\partial u_{,y}} d\Omega = -\int_\Omega \eta_x \frac{\partial}{\partial y}\left(\frac{\partial F}{\partial u_{,y}}\right) d\Omega + \oint_\Omega \eta_x \frac{\partial F}{\partial u_{,y}} dx$$

$$\int_\Omega \eta_{y,x} \frac{\partial F}{\partial v_{,x}} d\Omega = -\int_\Omega \eta_y \frac{\partial}{\partial x}\left(\frac{\partial F}{\partial v_{,x}}\right) d\Omega + \oint_\Omega \eta_y \frac{\partial F}{\partial v_{,x}} dy$$

$$\int_\Omega \eta_{y,y} \frac{\partial F}{\partial v_{,y}} d\Omega = -\int_\Omega \eta_y \frac{\partial}{\partial y}\left(\frac{\partial F}{\partial v_{,y}}\right) d\Omega - \oint_\Omega \eta_y \frac{\partial F}{\partial v_{,y}} dx$$

$$(3.18)$$

As mentioned, the application of Green's theorem to two-dimensional problems is analogous to the use of integration by parts in the one-dimensional case. The order of differentiation of the perturbation function is reduced by one and a boundary term is formed. The boundary term is denoted by $\oint_\Omega$, a line integral, around the boundary of the region in a counterclockwise direction.

When these integrations are substituted into the functional derivative given by Eq. 3.17, the derivatives of the functional components of the variation are eliminated. When this expression is set to zero in accordance with the principle of minimum potential energy, the result is

---

Functional Derivative with Reduced-Order Variations

$$\int_\Omega \left[\frac{\partial F}{\partial u} - \frac{\partial}{\partial x}\left(\frac{\partial F}{\partial u_{,x}}\right) - \frac{\partial}{\partial y}\left(\frac{\partial F}{\partial u_{,y}}\right) \right] \eta_x \, d\Omega$$

$$+ \int_\Omega \left[\frac{\partial F}{\partial v} - \frac{\partial}{\partial x}\left(\frac{\partial F}{\partial v_{,x}}\right) - \frac{\partial}{\partial y}\left(\frac{\partial F}{\partial v_{,y}}\right) \right] \eta_y \, d\Omega$$

$$+ \eta_x \left[\oint_\Omega \frac{\partial F}{\partial u_{,x}} \, dy - \oint_\Omega \frac{\partial F}{\partial u_{,y}} \, dx \right]$$

$$+ \eta_y \left[\oint_\Omega \frac{\partial F}{\partial v_{,x}} \, dy - \oint_\Omega \frac{\partial F}{\partial v_{,y}} \, dx \right] = 0$$

(3.19)

The Euler-Lagrange equations and boundary terms are extracted from Eq. 3.19. This is possible because each of the four terms contained in Eq. 3.19 is independent of the other three terms. The first two terms are integrals over the region Ω, so they are independent of the last two terms, which are integrals on the boundary. The two integrals over Ω are independent of each other because η_x and η_y are independent functions. The two boundary integrals are independent of each other for the same reason. Thus, each of the four terms of Eq. 3.19 must independently equal zero.

The integrands of the first two terms of Eq. 3.19 are the source of the Euler-Lagrange equations for this functional. When these expressions are extracted and set to zero, the result is

---

Euler-Lagrange Equations

$$\frac{\partial F}{\partial u} - \frac{\partial}{\partial x}\left(\frac{\partial F}{\partial u_{,x}}\right) - \frac{\partial}{\partial y}\left(\frac{\partial F}{\partial u_{,y}}\right) = 0$$

$$\frac{\partial F}{\partial v} - \frac{\partial}{\partial x}\left(\frac{\partial F}{\partial v_{,x}}\right) - \frac{\partial}{\partial y}\left(\frac{\partial F}{\partial v_{,y}}\right) = 0$$

(3.20)

The boundary terms result when the last two components of Eq. 3.19 are set to zero to give

---

Boundary Terms

$$\eta_x \left[\oint_\Omega \frac{\partial F}{\partial u_{,x}} \, dy - \oint_\Omega \frac{\partial F}{\partial u_{,y}} \, dx \right] = 0$$

$$\eta_y \left[\oint_\Omega \frac{\partial F}{\partial v_{,x}} \, dy - \oint_\Omega \frac{\partial F}{\partial v_{,y}} \, dx \right] = 0$$

(3.21)

When the functional given by Eq. 3.10 is substituted into Eqs. 3.20 and 3.21, the governing differential equations and the boundary conditions for the plane stress problem result. The governing differential equations of equilibrium and the boundary conditions are formulated in turn in the next two sections.

Governing Equations for the Plane Stress Problem

The two differential equations of equilibrium for the plane stress problem are now formed. This is accomplished by computing the individual terms using the two Euler-Lagrange equations given by Eq. 3.20. Using the first equation contained in Eq. 3.20 the components of the equilibrium equations in the x direction are found to be

---

Component Terms from the First Euler-Lagrange Equation

$$\frac{\partial F}{\partial u} = -P_x(x, y)$$

$$\frac{\partial}{\partial x} \left(\frac{\partial F}{\partial u_{,x}} \right) = \frac{\partial}{\partial x} \left(\frac{E}{1 - \nu^2} \right) (u_{,x} + \nu v_{,y}) = \frac{E}{1 - \nu^2} (u_{,xx} + \nu v_{,xy})$$

$$\frac{\partial}{\partial y} \left(\frac{\partial F}{\partial u_{,y}} \right) = \frac{\partial}{\partial y} \left(\frac{E}{1 - \nu^2} \right) \frac{(1 - \nu)}{2} (u_{,y} + v_{,x}) = \left(\frac{E}{1 - \nu^2} \right) \frac{(1 - \nu)}{2} (u_{,yy} + v_{,xy})$$

(3.22)

When these individual terms are combined according to Eq. 3.20, the equilibrium equation in the x direction is given as

---

Equilibrium Equation in the x Direction

$$\left(\frac{E}{1 - \nu^2} \right) \left[u_{,xx} + \nu v_{,xy} + \left(\frac{1 - \nu}{2} \right) (u_{,yy} + v_{,xy}) \right] = p_x$$

(3.23)

Using the second equation contained in Eq. 3.20 the components of the equilibrium equations in the y direction are found to be

---

Component Terms from the Second Euler-Lagrange Equation

$$\frac{\partial F}{\partial v} = -P_y(x, y)$$

$$\frac{\partial}{\partial x}\left(\frac{\partial F}{\partial v_{,x}}\right) = \frac{\partial}{\partial x}\left(\frac{E}{1-v^2}\right)\frac{(1-v)}{2}(u_{,y} + v_{,x}) = \left(\frac{E}{1-v^2}\right)\frac{(1-v)}{2}(u_{,xy} + v_{,xx})$$

$$\frac{\partial}{\partial y}\left(\frac{\partial F}{\partial v_{,y}}\right) = \frac{\partial}{\partial y}\left(\frac{E}{1-v^2}\right)(vu_{,x} + v_{,y}) = \frac{E}{1-v^2}(vu_{,xy} + v_{,xy})$$

$$(3.24)$$

When these terms are combined according to Eq. 3.20, the equilibrium equation in the y direction equation is given as

---

Equilibrium Equation in the y Direction

$$\left(\frac{E}{1-v^2}\right)\left[v_{,yy} + vu_{,xy} + \left(\frac{1-v}{2}\right)(v_{,xx} + u_{,xy})\right] = p_y \qquad (3.25)$$

The two equilibrium equations for the plane stress problem given by Eqs. 3.23 and 3.25 are combined here for convenient later reference as

---

Governing Differential Equations

$$\left(\frac{E}{1-v^2}\right)\left[u_{,xx} + vv_{,xy} + \left(\frac{1-v}{2}\right)(u_{,yy} + v_{,xy})\right] = p_x$$

$$\left(\frac{E}{1-v^2}\right)\left[v_{,yy} + vu_{,xy} + \left(\frac{1-v}{2}\right)(v_{,xx} + u_{,xy})\right] = p_y$$

$$(3.26)$$

These second-order partial differential equations are the equilibrium equations for the plane stress problem in the x and y directions, respectively.

Boundary Conditions for the Plane Stress Problem

The boundary terms are now evaluated for the plane stress problem and interpreted in terms of their physical meaning. When the two boundary terms given by Eq. 3.21 are evaluated using the functional given by Eq. 3.10, the result is

Boundary Terms for Plane Stress Problems

$$\eta_x \left[\oint_\Omega \left(\frac{E}{1 - v^2} \right) (u_{,x} + vv_{,y}) \, dy - \oint_\Omega \left(\frac{E}{1 - v^2} \right) \frac{(1 - v)}{2} (u_{,y} + v_{,x}) \, dx \right] = 0$$

$$\eta_y \left[\oint_\Omega \left(\frac{E}{1 - v^2} \right) \frac{(1 - v)}{2} (u_{,y} + v_{,x}) \, dy - \oint_\Omega \left(\frac{E}{1 - v^2} \right) (vu_{,x} + v_{,y}) \, dx \right] = 0$$

(3.27)

Equation 3.27 is transformed to expressions in terms of stress quantities so a physical interpretation of the equations can be made. The first step in this process is to transform Eq. 3.27 to a function of strains. When the strain displacement relations given by Eq. 3.7 are applied, the result is

Boundary Terms as Functions of Strain Quantities

$$\eta_x \left[\oint_\Omega \left(\frac{E}{1 - v^2} \right) (\varepsilon_x + v\varepsilon_y) \, dy - \oint_\Omega \left(\frac{E}{1 - v^2} \right) \frac{(1 - v)}{2} (\gamma_{xy}) \, dx \right] = 0$$

$$\eta_y \left[\oint_\Omega \left(\frac{E}{1 - v^2} \right) \frac{(1 - v)}{2} (\gamma_{xy}) \, dy - \oint_\Omega \left(\frac{E}{1 - v^2} \right) (v\varepsilon_x + \varepsilon_y) \, dx \right] = 0$$

(3.28)

The terms in this equation containing the strain quantities correspond to the definitions of the stresses in the stress-strain relations given by Eq. 3.4. When these relationships are applied, the result is

Boundary Terms

$$\eta_x \left[\oint_\Omega (\sigma_x) \, dy - \oint_\Omega (\tau_{xy}) \, dx \right] = 0$$

$$\eta_y \left[\oint_\Omega (\tau_{xy}) \, dy - \oint_\Omega (\sigma_y) \, dx \right] = 0$$

(3.29)

Each of these boundary terms contains a component due to a virtual displacement, η_x and η_y, and a component due to a surface traction or boundary load. Thus, the boundary term can be satisfied by either a geometric boundary condition or by a natural boundary condition. That is to say, the boundary terms are zero because either the perturbation terms η_x and η_y, are zero on the boundary or the boundary integrals are zero.

When the perturbation terms η_x and η_y are zero, the definition of an admissible virtual displacement means that the displacement on the boundary is specified. Most often the displacement is specified as zero but this need not be the case.

If the displacement is not specified, η_x and η_y need not be zero, so the line integral component of the boundary term must be zero. The integral components of the boundary terms are statements of equilibrium. The integrands represent either normal or tangential stresses existing in the continuum so the line integral represents the boundary load. The case represented by Eq. 3.29 requires that the normal and tangential loads on the boundary of the region must equal zero. That is to say, when the integral component of the boundary terms in Eq. 3.29 are satisfied, the boundary represented is that of a free surface. Other boundary conditions are specified when boundary loads are contained in the work function component of the potential energy expression, as discussed earlier.

The boundary terms are now evaluated and interpreted with the example of a rectangular region.

Boundary Conditions for a Rectangular Plane Stress Region

The meaning of the boundary terms is now given specific interpretation by examining the possible boundary conditions for the rectangular region shown in Fig. 2. By having the boundaries aligned with the coordinate axes, the boundary terms are simplified. This simplification occurs because one of the differential terms is zero on each edge of the region, as noted on Fig. 2. For example, on the two vertical edges, the x coordinate is constant so $dx = 0$ (see Note 2). This reduces the two boundary terms on the vertical edges to the following:

Boundary Terms of a Vertical Surface

$$\eta_x \left[\oint_\Omega (\sigma_x) \, dy \right] = 0$$

$$\eta_y \left[\oint_\Omega (\tau_{xy}) \, dy \right] = 0$$

(3.30)

When we examine the first equation of Eq. 3.30, we see that it will have a value of zero under two conditions. If the displacement u is specified, the term η_x must be zero to satisfy the requirements of an admissible virtual displacement. If the boundary displacement is not specified, η_x need not be zero. This means that σ_x, the normal stress on the boundary, must be zero to satisfy the boundary term. This condition is satisfied by a free boundary that corresponds to the problem being studied in this lesson (see Eq. 3.29). This analysis of the first boundary term contained in Eq. 3.30 identifies the following two possible boundary conditions on the vertical edges: $u = 0$ or $\sigma_x = 0$.

A similar analysis of the second boundary term of Eq. 3.30 produces the following two possible boundary conditions: $v = 0$ or $\tau_{xy} = 0$. Thus, we can conclude that the

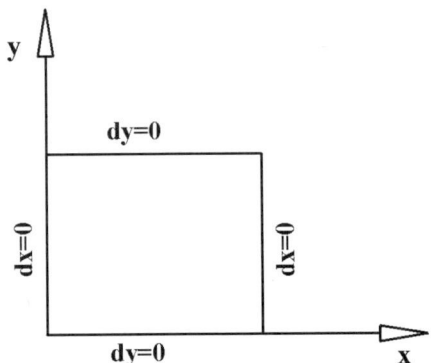

Figure 2. A rectangular plane stress region.

following are possible boundary conditions on a vertical surface in a plane stress problem:

Boundary Conditions on a Vertical Surface

$$u = 0 \quad \text{or} \quad \sigma_x = 0$$
$$v = 0 \quad \text{or} \quad \tau_{xy} = 0$$

(3.31)

Similarly, the boundary terms for the case where $dy = 0$ produces the following boundary conditions:

Boundary Conditions on a Horizontal Surface

$$u = 0 \quad \text{or} \quad \tau_{xy} = 0$$
$$v = 0 \quad \text{or} \quad \sigma_y = 0$$

(3.32)

If a function is to minimize the potential energy functional, one of the boundary conditions from each of these four sets of constraints must be satisfied.

Closure

We have just derived the governing differential equations and boundary conditions for the plane stress problem using the principle of minimum potential energy and techniques from the calculus of variations. As a result of this derivation, we have two independent approaches available for finding the solution to the equilibrium problem for the plane stress problem.

We can find the solution to the equilibrium problem either directly by solving the differential equations of equilibrium or indirectly by minimizing the potential energy functional. The differences between the two forms of the plane stress problem and the associated approximate solution techniques provide the basis for the development of the error analysis procedures developed in Part V.

For example, an opportunity exists for estimating the errors on the boundary in finite element solutions because the natural boundary conditions are treated differently in the variational and differential approaches. The natural boundary conditions are not explicitly contained in the finite element method. This means that the stresses on the boundary in a finite element solution need not satisfy the boundary conditions until the solution has fully converged. However, the natural boundary conditions are explicitly contained in the finite difference method.

This difference in treatment of the boundary conditions enables us to compare the boundary stresses produced by the finite element solution to the finite difference representation of the actual boundary stresses. As a result, we can better estimate the errors on the boundary of the finite element representations on a pointwise basis. In addition, we will see that this boundary stress error estimation procedure can be extended to estimate pointwise errors in the stresses or strains at any point in the region represented by a finite element model.

The reduction in accuracy of variational solutions that occurs when the natural boundary conditions are not satisfied is demonstrated in Lesson 4. The demonstrations contained in Lesson 4 highlight the need for the ability to estimate boundary errors in variational solutions. The identification of this need motivates and guides the development of the pointwise error measures presented in Part V.

Notes

1. This note contains a derivation of integration by parts. The starting point is the derivative of two functions multiplied by each other, i.e., the derivative of uv. This derivative can be written as $d(u*v) = du*v + u*dv$. It can be rearranged as $du*v = d(u*v) - u*dv$. When this expression is integrated, the result is the following: $\int v*du = u*v - \int u*dv$. This is the statement of integration by parts.

2. The example for which the boundary conditions are formed in this lesson is for a rectangular region. This was done to clearly demonstrate the use of the boundary terms to form the boundary conditions. The treatment of curved boundaries is discussed in Parts III and IV.

Appendix A

Derivation of Green's Theorem

The objective of this presentation is to provide a constructive derivation of the form of Green's theorem that is used to simplify the integrals that emerge when the calculus of variations is applied to two-dimensional problems. In this form, we can view Green's theorem as a two-dimensional analog of integration by parts.

Since a constructive derivation identifies the characteristics and limitations of the equation or concept formed, this presentation is designed to assist the reader in applying Green's theorem during the formulation of Euler-Lagrange equations and boundary integrals.

At a basic level, we can view Green's theorem as accomplishing two things:

1. The theorem relates the ideas of a ''single'' integral and a ''double'' integral to each other, i.e., an integral of the form $\int\int F \, dA$ is directly related to integrals of the form $\int F' \, dx$ and $\int F' \, dy$. The relationship between F and F' is discussed later.

2. The theorem modifies the form of a ''single'' integral so it can be interpreted as a ''line'' integral, i.e., integrals of the form $\int F' \, dy$ and $\int F' \, dx$ are directly related to integrals of the form $\oint F \, ds$.

The ideas contained in Green's theorem are illustrated by comparing two different ways of computing the area contained in a region between the two curves shown in Fig. A.1. This presentation focuses on the integration of areas because the applications are related to solid mechanics. Often, Green's theorem is illustrated with the case of fluid mechanics problems. Furthermore, because we are interested only in individual finite elements, only simple closed regions are considered.

Double-Integration Approach

The area of the simple closed region Ω shown in Fig. A.1 can be found using a ''double'' integral, which can be written as

---

Area as a Double Integral

$$A_\Omega = \int \int_\Omega dA$$

$$= \int \int_\Omega dy \, dx \tag{A.1}$$

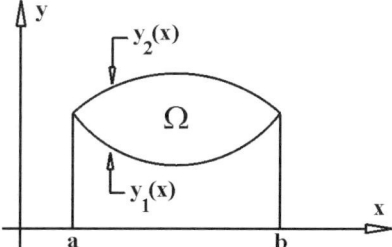

Figure A.1. A region in two dimensions.

The processes underlying Eq. A.1 can be viewed as subdividing the region Ω into "little" or differential squares as shown in Fig. A.2. The area is found from the number of squares and partial squares contained in the region.

Single-Integration Approach

The area in the region Ω can be found using a "single" integral, which can be written as

---

Area as a Single Integral

$$A_\Omega = \int_a^b y \, dx$$

where $y(x) = y_2(x) - y_1(x)$.

(A.2)

The process underlying Eq. A.2 can be viewed as subdividing the region into "little strips" or differential rectangles as shown in Fig. A.3. The area can be found by summing the areas of the individual rectangles.

It is instructive to note that the integrand of Eq. A.1, i.e., unity, is equal to the derivative of the integrand of Eq. A.2, i.e., y, with respect to y. That is to say, unity is the derivative of y with respect to y. Although not important in itself, this observation becomes important when the idea of replacing a double integral by a single integral is expanded for integrands that are functions. Green's theorem is a generalization of this concept.

Line Integral Interpretation of the Single-Integration Approach

The "double" integral given by Eq. A.1 can be interpreted differently than it was in Fig. A.2. We can interpret the integral given by Eq. A.1 in two different ways. These two additional interpretations lead to the idea of Green's theorem.

In the first interpretation, the double integral can be seen as subtracting two areas from each other to form the area of the region of interest. This idea is seen when Eq. A.1 is evaluated by first integrating it with respect to y and then integrating it with respect

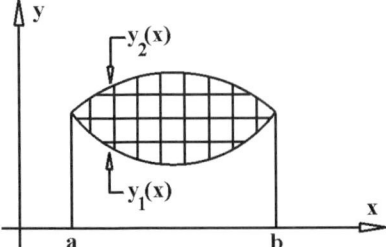

Figure A.2. A region in two dimensions subdivided into differential elements.

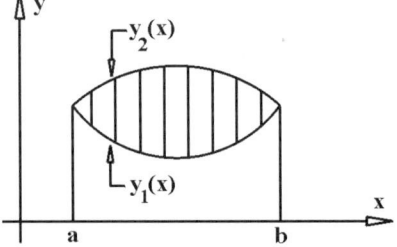

Figure A.3. A region in two dimensions subdivided into differential rectangles.

to x. The result of these two integrations can be written as the following:

$$
\begin{aligned}
A_\Omega &= \int_a^b \int_{y_1(x)}^{y_2(x)} dy\, dx \\
&= \int_a^b [y_2(x) - y_1(x)]\, dx \\
&= \int_a^b y_2(x)\, dx - \int_a^b y_1(x)\, dx
\end{aligned}
\tag{A.3}
$$

The meaning of Eq. A.3 is shown in Fig. A.4. The first integral finds the area under the y_2 curve and the x axis. The second integral evaluates the area under the y_1 curve and the x axis. The areas between the curves and the coordinate axes are sometimes called projected areas. The difference between the two integrals or the two projected areas is the area of the region of interest. Note that the single integral given by Eq. A.2 can be interpreted in the same way.

We can rearrange Eq. A.3 into the following form:

$$
A_\Omega = -\int_b^a y_2(x)\, dx - \int_a^b y_1(x)\, dx
\tag{A.4}
$$

Equation A.4 differs from Eq. A.3 in that the order of integration limits in the first integral has been switched. This change is compensated for by the introduction of a negative sign. This modification enables Eq. A.4 to be interpreted as a line integral, as is discussed next.

Equation A.4 can be interpreted as a line integral in the following way. We can start at point b and integrate the function y_2 on the boundary from b to a. Then we can continue traveling in the counterclockwise direction around the boundary and integrate the second function y_1 on the boundary from a to b. Thus, we have traversed the closed curve around the boundary and returned to our starting point at b. The idea of integrating the function ''on the boundary'' actually means that we are finding the areas under the curves as shown in Fig. A.4.

In the notation in common use, Eq. A.4 can be written as a line integral as

Area as a Line Integral

$$
A_\Omega = -\oint_\Omega y\, dx
\tag{A.5}
$$

The use of the line integral indicates that we can start at any point on the boundary and integrate in the counterclockwise direction and get the identical result. This derivation

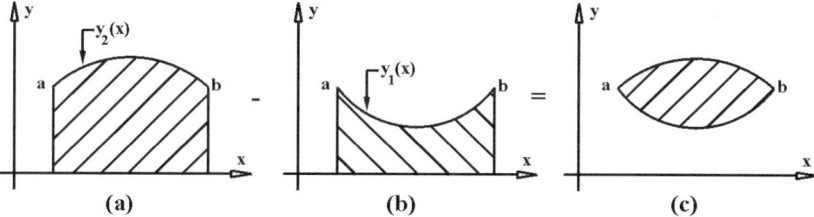

Figure A.4. A region in two dimensions seen as the difference between two projected areas.

shows us how to perform this calculation. That is to say, we can use either Eq. A.3 or Eq. A.4 to evaluate the line integral. We demonstrate the process later.

Equating the Double Integral and a Line Integral

Since both Eq. A.1 and Eq. A.5 compute the area of the region of interest, we can equate these two integrals. This can be written as follows:

$$A_\Omega = \int_\Omega dA = -\oint_\Omega y\, dx \tag{A.6}$$

We see later that Eq. A.6 is a specialized form of Green's theorem. This specialization finds the area of a region contained in a closed curve. As noted earlier, the integrand of the area integral is equal to the derivative of the integrand of the line integral. That is to say,

$$\frac{d(\text{Integrand of } \oint)}{dy} = \frac{d(y)}{dy} = 1$$

As a result of this observation, we can deduce the general form for equating a double integral over a region by a line integral around that region. This general form can be written as

---

A Component of Green's Theorem

$$\int_\Omega \frac{\partial L}{\partial y}\, dA = -\oint_\Omega L\, dx \tag{A.7}$$

This equation says that a surface integral (a double integral) can be evaluated using a line integral evaluated on the boundary of the region if the integrands of the two integrals are related as shown. We will now derive Eq. A.7.

Derivation of Eq. A.7 — A Component of Green's Theorem

Let us begin the derivation by evaluating the left-hand side of Eq. A.7. This is done as follows:

$$\begin{aligned}
\int_\Omega \frac{\partial L}{\partial y}\, dA &= \int_a^b \int_{y_1(x)}^{y_2(x)} \frac{\partial L}{\partial y}\, dy\, dx \\
&= \int_a^b \{L[x, y_2(x)] - L[x, y_1(x)]\}\, dx \\
&= \int_a^b L[x, y_2(x)]\, dx - \int_a^b L[x, y_1(x)]\, dx
\end{aligned} \tag{A.8}$$

Equation A.8 is analogous to Eq. A.3 and it can be interpreted in a similar manner. The first integral evaluates a quantity between the y_2 curve and the x axis. This can be viewed as computing a "weighted" area integral. The meaning of the integral is illustrated in Fig. A.5 by the three-dimensional plot. The third dimension represents the quantity multiplying the differential area at each point, which is equal to the function L. In the case where an area is being computed, the multiplier is a constant value of unity.

The second integral evaluates a quantity under the y_1 curve and the x axis. This weighted area quantity is illustrated in Fig. A.6.

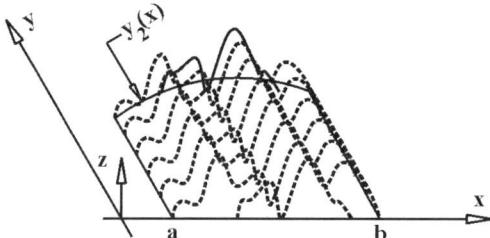

Figure A.5. A region in two dimensions with a weighting function.

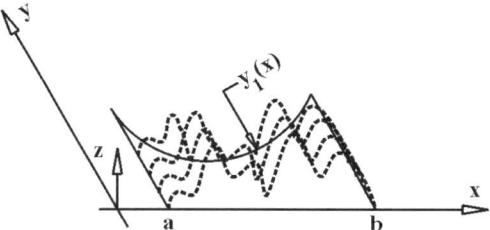

Figure A.6. A region in two dimensions with a weighting function.

The right-hand side of Eq. A.8 is represented schematically in Fig. A.7. The weighted area under curve y_2 is subtracted from the weighted area under curve y_1 to produce the weighted area of the region of interest.

Now that we have in interpretation for Eq. A.8, let us write this equation in a form that can be interpreted as a line integral. This modified view of Eq. A.8 is the following:

$$\int_{\Omega} \frac{\partial L}{\partial y} \, dA = -\int_{b}^{a} L[x, y_2(x)] \, dx - \int_{a}^{b} L[x, y_1(x)] \, dx \tag{A.9}$$

Equation A.9 differs from Eq. A.8 in that the order of the integration limits has been changed on the first term and, correspondingly, a negative sign has been introduced to this term to account for the change in the limits. This modification enables Eq. A.9 to be interpreted as a line integral, as was done with Eq. A.4.

When Eq. A.9 is written as a line integral, it becomes

A Component of Green's Theorem

$$\int_{\Omega} \frac{\partial L}{\partial y} \, dA = -\oint_{\Omega} L \, dx \tag{A.10}$$

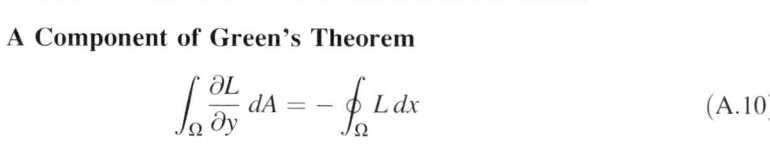

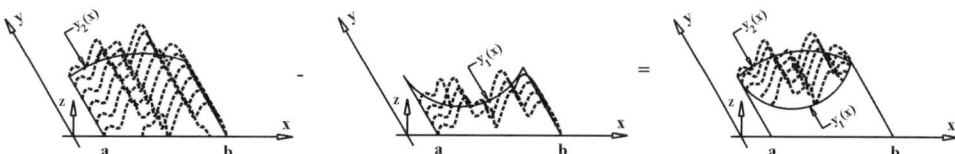

Figure A.7. A schematic representation of Eq. A.8.

We have proven the relationship given by Eq. A.7. Furthermore, we have a procedure available to us for computing the line integral contained in Eq. A.10. That is to say, Eq. A.9 provides a "recipe" for computing the line integral.

By a similar analysis, we can get a similar expression containing a partial derivative of a function with respect to x. This analogous expression can be written as

---

A Second Component of Green's Theorem

$$\int_\Omega \frac{\partial M}{\partial x}\, dA = \oint_\Omega M\, dy \qquad (A.11)$$

This equation does not contain a minus sign. The reason can be seen when the expression is derived. In this case, the counterclockwise direction of integration produces a dy that is positive on the right-hand part of the boundary curve. Similarly, the counterclockwise direction of integration yields a dy that is negative on the left-hand side of the curve. Thus, the order of integration is changed on the second integral to produce a line integral. Since this is the opposite of the case for Eq. A.10, the sign of Eq. A.11 differs from that of Eq. A.10.

When Eqs. A.9 and A.10 are combined, the result is the standard form of Green's theorem, namely,[1]

One Standard Form of Green's Theorem

$$\int_\Omega \left(\frac{\partial M}{\partial x} - \frac{\partial L}{\partial y} \right) dA = \oint L\, dx + M\, dy \qquad (A.12)$$

This derivation of the standard form of Green's theorem has demonstrated several things. First, we see that Green's theorem consists of two separate entities. We can retrieve the two components given by Eqs. A.10 and A.11 by setting L or M to zero in Eq. A.12. Furthermore, as a result of the derivation, we see how to compute a line integral. Since the derivation provides a "recipe" for computing the line integral, the derivation can be seen to be a constructive derivation.

Special Forms of Green's Theorem

We now specialize Green's theorem so that it is the two-dimensional analog of integration by parts. This is the form used to simplify the integrals involved in forming the Euler-Lagrange equations and the boundary integrals. The first step in this process is to define the functions contained in Green's theorem in the following form:

$$M = \eta G$$
$$L = \eta F \qquad (A.13)$$

---

[1] The signs in Eq. A.12 vary from reference to reference. The signs depend on whether Eq. A.10 is subtracted from Eq. A.11 or vice versa. The choice depends on the application.

Taking the derivatives of M and L in the left-hand side of Eq. A.12 gives the following:

$$\int_\Omega \left(\frac{\partial M}{\partial x} - \frac{\partial L}{\partial y} \right) d\Omega = \oint L \, dx + M \, dy$$

$$\int_\Omega \left[\left(\frac{\partial \eta}{\partial x} G + \eta \frac{\partial G}{\partial x} \right) - \left(\frac{\partial \eta}{\partial y} F + \eta \frac{\partial F}{\partial y} \right) \right] d\Omega = \oint \eta F \, dx + \eta G \, dy \tag{A.14}$$

Equation A.14 can be rearranged into the form of integration by parts to give

$$\int_\Omega \left(\frac{\partial \eta}{\partial x} G - \frac{\partial \eta}{\partial y} F \right) d\Omega = - \int_\Omega \left(\eta \frac{\partial G}{\partial x} - \eta \frac{\partial F}{\partial y} \right) d\Omega + \oint \eta F \, dx + \eta G \, dy$$

$$= - \int_\Omega \eta \left(\frac{\partial G}{\partial x} - \frac{\partial F}{\partial y} \right) d\Omega + \oint \eta (F \, dx + G \, dy) \tag{A.15}$$

Each of the separate components can be formed by repeating the procedure just presented for Eqs. A.10 and A.11. Or the same result can be formed by setting the functions F and G to zero, one at a time. When this is done, the result is the following:

Green's Theorem as an Analogy to Integration by Parts

$$\int_\Omega \frac{\partial \eta}{\partial x} G \, d\Omega = - \int_\Omega \eta \frac{\partial G}{\partial x} \, d\Omega + \oint \eta G \, dy$$

$$\int_\Omega \frac{\partial \eta}{\partial y} F \, d\Omega = - \int_\Omega \eta \frac{\partial F}{\partial y} \, d\Omega - \oint \eta F \, dx \tag{A.16}$$

As we can see, these expressions are analogous to the process of integrating by parts. The order of differentiation of the function η is reduced by one and a boundary term is formed. The function η can be taken as a perturbation function when Green's theorem is applied to the calculus of variations.

We have accomplished the objective of this appendix. A constructive derivation of Green's theorem has been presented and the theorem has been cast in a form that is analogous to integration by parts. In addition to its use in deriving the governing differential equations and boundary conditions for planar problems, Green's theorem is used in Part III in the derivation of finite element stiffness matrices.

References and Other Reading

1. Langhaar, H. L. *Energy Methods in Applied Mechanics*, John Wiley and Sons, Inc., New York, 1962.

2. Shames, I. H., and Dym, C. L. *Energy and Finite Element Methods in Structural Mechanics*, McGraw-Hill Book Co., New York, 1985.

3. Weinstock, R. *Calculus of Variations: With Applications to Physics and Engineering*, Dover Publications, Inc., New York, 1974.

4. Forray, M. J. *Variational Calculus in Science and Engineering*, McGraw-Hill Book Co., New York, 1968.

5. Thomas, G. B, Jr., and Finney, R. L. *Calculus and Analytic Geometry*, 8th ed. Addison-Wesley Publishing Co., New York, 1992. This undergraduate calculus book presents two forms of Green's theorem in terms of fluid mechanics applications.

Lesson 3   Problems

1. Derive the functional for the plane strain problem. *Hint*: Obtain or form the plane strain constitutive relationship and form an analogous strain energy functional.

2. Derive the Euler-Lagrange equation for the following functional:

$$V = \int_\Omega F(x, y, w, w_{,x}, w_{,y}, w_{,xx}, w_{,xy}, w_{,yy})\, d\Omega$$

Hint: Since the functional contains second derivatives, two applications of Green's theorem are needed to eliminate the derivative terms. Note that this functional contains only one unknown function, namely, w. We find later that this is the form of the functional for a Kirchhoff plate, an out-of-plane bending model. This problem is analogous to the process presented in Lesson 2 for extending the idea of the Euler-Lagrange equations to the beam problem.

3. Add a boundary stress to the right-hand edge of the rectangular region shown in Fig. 2. Form the boundary conditions for this condition. *Hint*: Add a term to the work function due to the motion of the loaded boundary and follow the result through the existing derivation. Only the boundary term on the loaded edge should be affected. This is a nonhomogeneous boundary condition. It is analogous to the application of boundary forces in Lesson 2.

Lesson **4**
Rayleigh-Ritz Variational Solution Technique

Purpose of Lesson To develop and apply the Rayleigh-Ritz solution technique to demonstrate the need for accurate boundary models in variational methods and to identify the modeling limitations that motivate the use of the finite element method.

The need to satisfy the geometric boundary conditions in variational solutions is formalized in the Rayleigh-Ritz criterion, which states that the strain energy in the approximate solution is less than or equal to the strain energy in the exact solution if the geometric boundary conditions are satisfied. If the geometric boundary conditions are not satisfied, the upper bound given by the Rayleigh-Ritz criterion does not hold.

In the previous lesson, we found that the natural boundary conditions are not explicitly included in variational solutions. In this lesson, we see the deleterious effect on the solutions if the natural boundary conditions are not satisfied. This apparent deficiency is turned to advantage in Part V, where error estimation techniques are developed that compare the actual natural boundary conditions to those contained in the finite element solution.

■   ■   ■

Variational and differential representations of one- and two-dimensional elasticity problems were presented in the previous lessons. The variational formulation is explicitly defined by the statement of the principle of minimum potential energy. That is to say, a variational method finds solutions to the equations of equilibrium by directly minimizing the potential energy functional (see Note 1). The finite element method is a variational approach that approximates the solution of the differential equation by minimizing an approximation of the potential energy functional.

The differential formulation requires that we directly solve the differential equation of equilibrium. As a study of differential equations shows, these solutions are often difficult, if not impossible, to find on complex regions. Approximate solution techniques are often utilized to overcome this limitation. The finite difference method is a differential approach that approximates the exact solution by solving a discrete approximation of the differential equations.

In this lesson, we demonstrate some of the characteristics and limitations of variational approaches that later guide us in the development of error analysis procedures. This is accomplished by introducing and applying the Rayleigh-Ritz method. This variational approach seeks a bounded approximate solution in the form of a linear combination of a finite number of functions, each of which must satisfy the geometric boundary conditions (see Note 2). The Rayleigh-Ritz approximations have the following form:

Rayleigh-Ritz Trial Solutions

$$y = \alpha_1 \phi_1 + \alpha_2 \phi_2 + \cdots + \alpha_n \phi_n$$

where α_i are the unknown coefficients or generalized
coordinates. (4.1)

ϕ_i are the admissible trial functions.

Trial functions that satisfy the geometric boundary conditions are called admissible functions. Although the approximate solution need not satisfy the natural boundary conditions to satisfy the upper bound, more accurate solutions are found if the natural boundary conditions are satisfied. The effect of not satisfying the natural boundary conditions is demonstrated in the examples that follow.

If the approximate Rayleigh-Ritz solution is to converge to the exact solution as more terms are added, the set of approximating trial functions must, in general, satisfy the following two requirements: (1) the set of functions from which the trial functions are chosen must be capable of representing every possible deformation that the body can take under a given loading and (2) no lower order terms can be missing from the set of functions contained in the approximate solution. These conditions are known as the completeness requirements (see Note 3). An analog of this requirement is discussed later in the context of the finite element method.

However, if we have additional knowledge about the problem, this restriction can be relaxed. For example, if the solution to a problem is known to be symmetric, the approximate solution need not contain any nonsymmetric trial functions because they would make no contribution to the deformations. Simply put, the completeness requirement means that no lower order terms needed to represent the solution can be missing from the approximate solution if the solution is to converge. This idea is demonstrated in Example 4 of this lesson.

The coefficients to be determined in the Rayleigh-Ritz process are called "generalized coordinates." This designation distinguishes them from the pointwise displacement or rotation coordinates that describe the motion of an individual point. The Rayleigh-Ritz coefficients do not describe the displacement of a particular point. They indicate the level of "participation" of an individual component function in the approximate solution. In other words, the generalized coordinates scale the magnitude of the individual trial functions.

The function of the generalized coordinates is shown graphically in Fig. 1. The total approximate solution is shown as the sum of three trial functions. The contribution of each trial function is controlled by the generalized coordinate, α_i. For example, let us say the three α_i's shown in Fig. 1 have the values of 6.2, 0.0, and 1.8, respectively. In this case the final approximation is dominated by ϕ_1. The trial function ϕ_2 makes no contribution and ϕ_3 makes a minor contribution.

Application of the Rayleigh-Ritz Solution Technique

The procedure for determining the unknown coefficients contained in the Rayleigh-Ritz approximation is outlined as follows. After the functional is formulated, the most difficult task is often the identification of a set of trial functions that satisfy the geometric boundary conditions and that will, ideally, satisfy the natural boundary conditions. Once

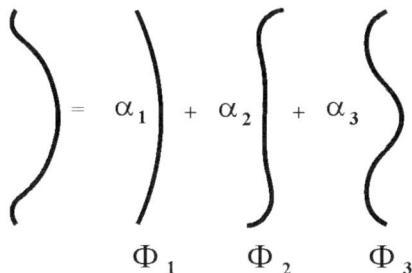

Figure 1. Visualization of a Rayleigh-Ritz solution.

formed, the admissible trial solution is substituted into the potential energy functional and the result is integrated.

This operation reduces the functional to a quadratic polynomial in terms of the finite number of unknown generalized coordinates, which are represented by the α_i's contained in Eq. 4.1. The resultant polynomial is then minimized using the methods of elementary calculus by setting the partial derivatives taken with respect to the unknown coefficients to zero, i.e.;

Function Minimization

$$\frac{\partial V}{\partial \alpha_i} = 0 \qquad (i = 1, \ldots, n) \tag{4.2}$$

The minimization procedure produces a set of linear algebraic equations in the unknown generalized coordinates. There are as many equations as there are unknowns. The unknown coefficients or generalized coordinates are found by solving these equations. The final result is an approximation that is bounded by the actual strain energy in accordance with the Rayleigh-Ritz criterion. A step-by-step description of the Rayleigh-Ritz procedure is shown in Fig. 2.

The Rayleigh-Ritz Criterion

The strain energy contained in the solution found by the Rayleigh-Ritz technique is bounded if the trial functions are admissible. That is to say, the strain energy contained in the approximate solution will be less than or equal to the strain energy contained in the exact solution when the individual components of an approximate solution satisfy the geometric boundary conditions. This upper bound is called the *Rayleigh-Ritz criterion.*

For the special case of a system loaded with a single point load, the Rayleigh-Ritz criterion can also be expressed in terms of the displacement of the point load. In such a system, the strain energy is equal to the work done by the point load. Thus, the displacement under the point load is directly proportional to the strain energy contained in the deformed structure. Hence, the displacement under the point load in a problem with only one load will be less than or equal to the exact result. Examples presented later show that no such bounds exist for displacements elsewhere in a structure. Furthermore, an example shows that no bound on the displacements exists for systems loaded with other than a single point load.

The availability of the Rayleigh-Ritz criterion has important implications when we attempt to assess the accuracy of an approximate solution. If we compute a sequence of approximate solutions for any type of loading, the approximation with the largest strain

1. Form the appropriate functional

2. Identify the boundary conditions

3. Form the trial functions

4. Substitute the trial function into the functional

5. Integrate the functional

6. Take the partial derivatives with respect to the unknown coefficients and set them to zero

7. Solve the resulting equations for the unknown coefficients

8. Form the approximate solution

Figure 2. Rayleigh-Ritz method flow chart.

energy content is the best solution. If the system is loaded with a single point load, the approximation with the largest displacement under the point load provides the most accurate solution.

It is important to note that these upper bounds do not imply that the stresses or strains at individual points are similarly bounded. They can be larger or smaller than the exact results. These characteristics are demonstrated in examples that follow. The fact that no bound exists on the stresses and strains motivates the development of the pointwise error measures presented in Part V.

Euler-Bernoulli Beam Examples

The Rayleigh-Ritz method is first demonstrated with the case of a simply supported beam loaded in the center with a concentrated load, as shown in Fig. 3. The geometric boundary conditions for this problem are $v(0) = 0$ and $v(L) = 0$. The natural boundary conditions are $v_{,xx}(0) = 0$ and $v_{,xx}(L) = 0$. That is to say, the displacements and the moments are zero at the constraints.

This example is used to demonstrate the deterioration in the accuracy of the approximate solution that occurs when the natural boundary conditions are not satisfied. Toward this aim, the problem is first solved using a trial function that satisfies both the geometric and natural boundary conditions. Then the problem is solved using an admissible function that satisfies only the geometric boundary conditions. We see that by

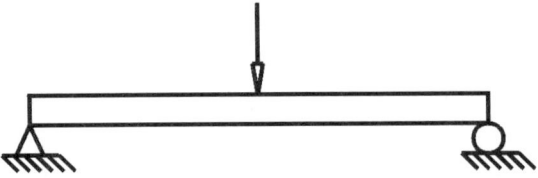

Figure 3. A simply supported beam with a concentrated load at midspan.

not satisfying the natural boundary conditions, this second approximation severely underestimates the actual displacements. The Rayleigh-Ritz criterion still holds because the strain energy content in the second case is even lower when both the geometric and natural boundary conditions are satisfied.

The two approximate solutions are also compared to the exact solution. The exact solution for a simply supported beam with a concentrated load in the center is easily found by directly solving the differential equation as

Exact Solution

$$v_{\text{exact}} = \frac{PL^3}{48EI}\left[3\left(\frac{x}{L}\right) - 4\left(\frac{x}{L}\right)^3\right] \qquad \left(0 \le x \le \frac{L}{2}\right) \tag{4.3}$$

Example 1   Essential and Natural Boundary Conditions Satisfied

The potential energy expression for the problem shown in Fig. 3 is

Potential Energy Functional

$$V = \frac{EI}{2}\int_0^L (v_{,xx})^2\, dx - Pv(x)\big|_{x=L/2} \tag{4.4}$$

The applied load in the problem shown in Fig. 3 is a point load. It is incorporated into the potential energy functional given by Eq. 4.4 by adding a term to the functional consisting of the load and the displacement of the point at which the load is applied. This is to say, the applied load is not part of an integral term.

The assumed solution for this example contains only a single trial function consisting of a sine function. The assumed solution takes on the following form:

Assumed Solution

$$v_{\text{approx.}} = \alpha_1 \sin\frac{\pi x}{L} \tag{4.5}$$

When we evaluate this assumed solution at the boundaries, we can see that it satisfies the geometric or essential boundary conditions:

Geometric Boundary Conditions

$$v(0) = \alpha_1 \sin 0 = 0$$
$$v(L) = \alpha_1 \sin \pi = 0 \tag{4.6}$$

Since this assumed trial function satisfies the geometric boundary conditions, the result is bounded by the Rayleigh-Ritz criterion. The approximate displacement under the point load does not exceed the exact result.

When we form the natural boundary conditions by taking two derivatives of the assumed solution and the derivatives are evaluated on the boundaries, we see that the assumed solution satisfies the natural boundary conditions:

Natural Boundary Conditions

$$v_{,xx}(0) = -\alpha_1 \left(\frac{\pi}{L}\right)^2 \sin 0 = 0$$

$$v_{,xx}(L) = -\alpha_1 \left(\frac{\pi}{L}\right)^2 \sin \pi = 0$$

(4.7)

The satisfaction of the natural boundary conditions in this case means that no moments are imposed on the free ends of the beam, so they are free to rotate. This freedom of rotation can be seen by examining the expression for the slope on the boundary, which is given by the first derivative of the assumed solution. The slopes or rotation at the supports are found to be

---

Boundary Slopes

$$v_{,x}(0) = \alpha_1 \frac{\pi}{L} \cos 0 = \alpha_1 \frac{\pi}{L}$$

$$v_{,x}(L) = \alpha_1 \frac{\pi}{L} \cos \pi = -\alpha_1 \frac{\pi}{L}$$

(4.8)

As we can see, the rotations at the supports are not fixed, i.e., they are not constants. This is shown by the presence of the unknown coefficients or generalized coordinate in the expression for the rotations in Eq. 4.8. Since the generalized coordinate is a function of the applied load, the rotation is a function of the load, as it should be if the beam is free to rotate at the constraints.

We can make another interpretation of these results by remembering that the moment and the slope form a complementary pair of boundary conditions for a beam. Either the moment or the slope must be specified. In this case, the moment is specified to be zero, so the slope boundary condition does not constrain the system.

The approximate solution is now computed. The strain energy functional is reduced to a polynomial by substituting the trial solution into Eq. 4.4 to give

---

Reduced Strain Energy Function

$$V = \alpha_1^2 \frac{EI}{2} \left(\frac{\pi}{L}\right)^4 \int_0^L \sin^2 \frac{\pi x}{L} \, dx - P\alpha_1$$

$$= \alpha_1^2 \frac{EI}{2} \left(\frac{\pi}{L}\right)^4 \left(\frac{L}{2}\right) - P\alpha_1$$

(4.9)

When we take the derivative of the approximate strain energy function given by Eq. 4.9 with respect to α_1 and set to zero, as specified by Eq. 4.2, we find that the generalized coordinate is

Minimization Result

$$\alpha_1 = \frac{2PL^3}{EI\pi^4} = \frac{PL^3}{48.7EI} \tag{4.10}$$

When we substitute the generalized coordinate given by Eq. 4.10 into Eq. 4.5, the Rayleigh-Ritz approximation of the deformation of the beam with a concentrated load in the center is

Rayleigh-Ritz Approximate Solution

$$v_{\text{approx.}} = \frac{PL^3}{48.7EI} \sin \pi \frac{x}{L} \tag{4.11}$$

The deflection under the load given by the approximate solution is $PL^3/48.7EI$. The exact result is $PL^3/48EI$. The approximate deflection under the load is 98.6% of the actual result. Since the approximate deflection and, hence, the approximate strain energy is smaller than the exact values, this result satisfies the Rayleigh-Ritz criterion. The displacement along the length of the beam given by the approximate solution as a percentage of the actual result is presented in Table 1. This table shows that the approximate displacements are larger than the exact displacements over a significant portion of the beam even though the approximate solution satisfies the Rayleigh-Ritz criterion, i.e., the displacement under the applied load is less than the actual displacement. This result emphasizes the fact that the Rayleigh-Ritz criterion is based on the energy content of the problem and not on the displacements at each point in the region being solved.

The normal strains along the length of the beam are given as a percentage of the actual values in Table 2. These quantities exceed the actual values along portions of the beam. Thus, we can see that there is no bound on these quantities. The analyst must keep this fact in mind when assessing the value of the stresses and strains in all approximate solutions.

Table 1. Approximate displacement vs position (exact percentage).

X/L	Exact percentage
0.0	100
0.1	103
0.2	102
0.3	101
0.4	99
0.5	99
0.6	99
0.7	101
0.8	102
0.9	103
1.0	100

Table 2. Approximate strains vs position (exact percentage).

X/L	Exact percentage
0.0	100
0.1	125
0.2	119
0.3	109
0.4	96
0.5	81
0.6	96
0.7	109
0.8	119
0.9	125
1.0	100

Example 2 Only Essential Boundary Conditions Satisfied

The problem solved in Example 1 is now solved again using a trial solution that satisfies only the geometric boundary conditions. This trial function does not satisfy the natural boundary conditions. The purpose of this example is to show the deterioration in the results that occurs when the natural boundary conditions are not satisfied. The assumed solution chosen for this example is the following:

Assumed Solution

$$v_{\text{approx.}} = \alpha_1 \left(1 - \cos \frac{2\pi x}{L} \right) \tag{4.12}$$

When we evaluate the assumed solution at the boundaries, we can see that the geometric or essential boundary conditions are satisfied as follows:

Geometric Boundary Conditions

$$v(0) = \alpha_1 (1 - \cos 0) = 0$$
$$v(L) = \alpha_1 (1 - \cos 2\pi) = 0 \tag{4.13}$$

Since this assumed trial function satisfies the essential boundary conditions, the result is governed by the Rayleigh-Ritz criterion. The strain energy of the approximation will be less than the exact strain energy. For the case of a single point load, this means that the approximate displacement under the point load will not exceed the exact result.

As noted in the problem statement, the assumed solution used in this example does not satisfy the natural boundary conditions. That is to say, the bending moments at the boundaries produced by this assumed solution are not equal to zero. We can see this by evaluating second derivatives of the approximation function at the supports to get

---

Natural Boundary Conditions

$$v_{,xx}(0) = \alpha_1 \left(\frac{2\pi}{L} \right)^2 \cos 0 = \alpha_1 \left(\frac{2\pi}{L} \right)^2$$

$$v_{,xx}(L) = \alpha_1 \left(\frac{2\pi}{L} \right)^2 \cos 2\pi = \alpha_1 \left(\frac{2\pi}{L} \right)^2 \tag{4.14}$$

The moments are not zero because the generalized coordinate α_1 is not equal to zero. We can identify the consequences of this modeling deficiency by studying its effect on the constraints of the problem and on the approximate deformation.

When the essential and natural boundary conditions were derived in Lesson 2, we saw that the two types of boundary conditions formed complementary pairs. That is to say, if a geometric boundary condition did not apply, a natural boundary condition had to be satisfied. This raises the parallel question of what happens to the complimentary geometric boundary condition if a specified natural boundary condition is not satisfied by a trial function.

As we just saw, the natural boundary conditions for the simply supported beam being studied are not satisfied by this trial solution. The complimentary essential boundary condition for this case is the slope of the beam at the boundaries. For this case, we find the slopes to be

Geometric Boundary Conditions

$$v_{,x}(0) = -\alpha_1 \frac{2\pi}{L} \sin 0 = 0$$

$$v_{,x}(L) = -\alpha_1 \frac{2\pi}{L} \sin 2\pi = 0$$

(4.15)

As we can see, this approximate solution forces the slopes to zero. Thus, by not satisfying the natural boundary conditions, this approximation enforces two essential boundary conditions that are not consistent with the definition of a simply supported beam. The enforcement of this boundary condition overconstrains the problem because the ends of a simply supported beam should be free to rotate and allow for further deformation.

The effect of this modeling error on the displacements is now determined by finding the approximate solution. When we substitute the assumed solution into the strain energy functional given by Eq. 4.4, according to the Rayleigh-Ritz procedure, the result is

Reduced Strain Energy Function

$$V = \alpha_1^2 \frac{EI}{2} \left(\frac{2\pi}{L}\right)^4 \int_0^L \cos^2 \frac{2\pi x}{L} \, dx - 2P\alpha_1$$

$$= \alpha_1^2 \frac{EI}{2} \left(\frac{2\pi}{L}\right)^4 \left(\frac{L}{2}\right) - 2P\alpha_1$$

(4.16)

When we take the derivative of V with respect to α_1 and set it to zero, we find the generalized coordinate α_1 to be

Result of Minimization

$$\alpha_1 = \frac{2PL^3}{8EI\pi^4}$$

$$= \frac{1}{8}\left(\frac{PL^3}{48.7EI}\right)$$

(4.17)

The Rayleigh-Ritz approximation for this case is

Rayleigh-Ritz Approximate Solution

$$v_{\text{approx.}} = \frac{1}{8}\left(\frac{PL^3}{48.7EI}\right)\left(1 - \cos \frac{2\pi x}{L}\right)$$

(4.18)

We find the deformation under the load at $x = L/2$ to be

Maximum Deformation

$$v_{max.} = \frac{1}{4}\left(\frac{PL^3}{48.7EI}\right) \tag{4.19}$$

This approximation is equal to one-fourth of the approximate solution given in Eq. 4.11 for the previous example. The deformation under the single point load satisfies the Rayleigh-Ritz criterion because this solution is less than the actual result. This example forcefully demonstrates the effect of not satisfying the natural boundary conditions. The problem is implicitly overconstrained and the result suffers.

This example highlights an implicit problem with the Rayleigh-Ritz procedure. Although it is guaranteed that the strain energy will be bounded when admissible trial functions are used, there is no guaranteed method for determining how close the approximate strain energy content is to the exact value. We know that the solution with the largest amount of strain energy is the best solution due to the Rayleigh-Ritz criterion. However, we do not know how close we are to the exact solution unless the exact solution is available.

We obviously do not know the exact solutions when the approach is used in practice or we would not be attempting to find an accurate approximate solution. The question of the accuracy of approximate solutions is addressed in Part V. We find that satisfactory error estimation procedures are available for the finite element method.

Example 3   A Fixed-Fixed Beam

The problem of an Euler-Bernoulli beam with both ends fixed is now solved. As shown in Fig. 4, the constraints of this problem require that the displacements and slopes at both boundaries be zero. Thus, two geometric boundary conditions are specified at each end of the beam. Both of these constrains must be satisfied by the trial function if the Rayleigh-Ritz criterion is to hold.

These are exactly the boundary conditions satisfied by the trial function used in the previous example. The satisfaction of these geometric boundary conditions is demonstrated in Eqs. 4.13 and 4.15. This means that the result just found to give an unsatisfactory result for a simply supported beam provides satisfactory results for a beam with fixed ends. In fact, this result is equal to 98.6% of the exact solution. This result satisfies the Rayleigh-Ritz criterion and illustrates that the satisfaction of all of the boundary conditions is required for accurate solutions.

The results of these example problems have shown that the failure to satisfy the natural boundary conditions imposes additional constraints on the problem. The effect of

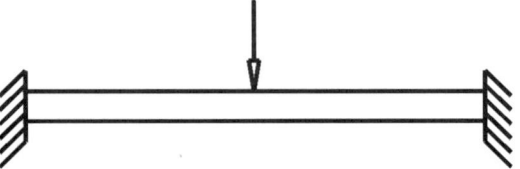

Figure 4. A fixed-end beam with a concentrated load at midspan.

the additional constraints then reduces the deformations produced by the approximate solution. As expected, these constraints also lower the strain energy. These ideas are extended in the following discussion.

Rayleigh-Ritz Criterion Revisited

The examples just solved provide the basis for an intuitive physical argument to support the Rayleigh-Ritz criterion. The Rayleigh-Ritz criterion can be interpreted as follows. Let us assume that we have an exact solution that is represented by the following infinite series of *orthogonal* functions:

Exact Series Solution

$$y_{exact} = \alpha_1 \phi_1 + \alpha_2 \phi_2 + \alpha_3 \phi_3 + \cdots \tag{4.20}$$

An approximate Rayleigh-Ritz solution can be written as a truncated form of the exact solution as

---

Approximate Series Representation

$$y_{approx.} = \alpha_1 \phi_1 + \alpha_2 \phi_2 + \cdots + \alpha_n \phi_n + \mathrm{O}\left[\phi_{n+1} + \phi_{n+2} + \cdots\right] \tag{4.21}$$

When we compare these two equations, we see that the approximate form of the equation consists of the first n terms of the exact solution. This is equivalent to saying that each of the higher order trailing terms of the series has been totally constrained and does not add to the overall displacement. Since these terms do not contribute to the strain energy, the approximation contains less strain energy than the exact result, which is a statement of the Rayleigh-Ritz criterion (see Note 3).

Note that this simple explanation of the Rayleigh-Ritz criterion holds only if the trial functions are orthogonal. If the trial functions are not orthogonal, the coefficients of the leading terms would not be the same for both the exact and the approximate representations. However, the basic idea behind the Rayleigh-Ritz criterion of constrained higher modes of deformation holds regardless of whether or not the admissible approximation functions are orthogonal.

The use of orthogonal functions to represent the exact and approximate solutions also provides an opportunity to gain an intuitive feel for the idea of completeness. We saw earlier that if an approximate solution was to converge to the exact solution as more terms were added, the set of approximate functions must be complete, namely, no lower order terms must be missing from the trial function. This idea can be made concrete by relating it to the forms of the exact and the approximate solutions presented earlier.

Let us assume that we form an approximate solution from the exact solution by eliminating the first term from the exact solution. This, in essence, constrains the deformation given by the first mode of deformation to zero. If we remove the contribution of this function from the deformation, the series obviously cannot converge to the correct solution because the higher order orthogonal terms cannot represent the missing function. Again, this argument holds only in this simple form for orthogonal functions, but the concept is generally applicable. Thus, if an approximation is to represent the deformations accurately, it must contain all of the lower order terms.

Example 4 A Mostly Qualitative Example

In the examples just solved, we saw that the Rayleigh-Ritz criterion does not, in general, place a bound on the displacements. This is now demonstrated for the case of a simply supported beam loaded with a uniformly distributed load, as shown in Fig. 5. The maximum displacement for this problem is $PL^4/(76.8EI)$. The assumed solution for this problem consists of the first n terms of the following series:

Assumed Solution

$$v = \alpha_1 \sin \frac{\pi x}{L} + \alpha_3 \sin \frac{3\pi x}{L} + \alpha_5 \sin \frac{5\pi x}{L} + \cdots \tag{4.22}$$

Each of the components of the solution satisfies both the essential and natural boundary conditions for a simply supported beam. Note that this series does not contain any sine terms that are an even multiple of π. They need not be included because the solution for this problem will obviously be symmetric and the sine terms that are even multiples of π are unsymmetric. When we minimize the functional, we find the generalized coordinates to have the following general form:

Generalized Coordinates

$$\alpha_n = \frac{4PL^4}{\pi^5 EI} \left(\frac{1}{n^5} \right) \quad (n = 1, 3, 5, \ldots) \tag{4.23}$$

Several observations concerning the characteristics of Rayleigh-Ritz solutions can be made from Table 3. This table contains the maximum displacement and the strain energy content as a percentage of the exact result versus the number of terms contained in the approximate solution. We can see that the column of maximum displacements oscillates as more terms are added to the approximate solution. The column of strain energy is converging asymptotically toward the exact strain energy. The key point to note is that the Rayleigh-Ritz criterion is satisfied in every case and that the maximum deformations are larger than those contained in the exact result in some approximate results.

The reason for both of these characteristics is obvious if we study the approximate solution. First, the sign of the terms contained in Eq. 4.22 oscillates from plus to minus. As additional terms are added, the first new term reduces the maximum deflection, the second term adds to the maximum deflection, etc. Second, the strain energy is independent of the sign of the additional terms since it is a function of the square of the sine terms. That is, since strain energy is always positive, the total strain energy quantity continues to grow, albeit, slowly. Such a quantity is said to be positive definite.

Figure 5. A simply supported beam with a uniformly distributed load.

Table 3. Approximate maximum displacement and strain energy as percentage of the exact result.

N	Maximum displacement	Strain energy
1	100.386	99.8555
2	99.973	99.9925
3	100.005	99.9989
4	99.999	99.9997

Extension to Two Dimensions

As mentioned earlier, the most difficult task in applying the Rayleigh-Ritz procedure is the selection of approximation functions that satisfy the essential and natural boundary conditions. This statement can be given substance by considering Fig. 6. Figure 6a shows a rectangular two-dimensional region. Regardless of the boundary conditions, a set of functions adequate for this simple case can be envisioned. However, we need only add a cutout to the region, as shown in Fig. 6b, to complicate the task. If we succeed in developing a function that satisfies the boundary conditions for this case, we need only add other cutouts or change the outline of the region to make it impossible to define a set of admissible functions.

The finite element method is a variant of the Rayleigh-Ritz method that enables us to treat very general boundary conditions in a simple manner. Complicated regions can be routinely treated. Since the finite element method is a version of the Rayleigh-Ritz method, the discussion of the characteristics of Rayleigh-Ritz solutions holds for finite element results. The extension of these characteristics to the finite element method is discussed in detail in later lessons.

Closure

This lesson has presented the Rayleigh-Ritz procedure as a vehicle for illustrating the characteristics and some of the limitations of approximate variational procedures. A major deficiency of variational solutions is the lack of knowledge concerning how closely the approximate solution approaches the exact result. Although the strain energy contained in the approximate solution is bounded if the geometric boundary conditions are satisfied, the accuracy of the strain energy estimate is not known. More importantly, the accuracy of the stress and strain results at individual points is not known. In fact, these quantities are not even bounded. Finally, the Rayleigh-Ritz approach is rarely applicable to problems of practical interest because of the difficulty in satisfying the boundary conditions for complex problems.

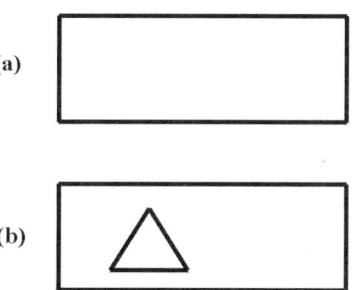

Figure 6. Two-dimensional regions.

The remainder of this book focuses on methods for removing these limitations. The finite element method is a variant of the Rayleigh-Ritz method that eliminates most of the difficulties with complex geometries and boundary conditions. The finite element method reduces the formulation of admissible trial functions to a routine process. This powerful numerical method is developed in Part III using the notation developed in Part II.

However, when the finite element method stands alone, the user does not know the accuracy of the approximate results. Again, the strain energy is bounded in many cases, but the accuracy of the stresses and strains is unknown.

Part V develops the procedures for estimating the amount of error contained in the approximate solutions. The results of error analysis procedures are used to identify regions of high error in finite element models and then these results are used to guide improvements in the model to produce more accurate results. The sequence of error identification and subsequent model improvement is called adaptive refinement. The use of adaptive refinement can produce results with a predefined accuracy in the strain energy and/or in the stresses and strains.

Procedures for estimating the errors in stresses and strains at individual point are also developed in Part V. In addition to estimating the magnitude of the error, these error measures identify the sign of the error. Decisions concerning whether to continue the model refinement can be made depending on the results of the pointwise error estimates. The stress and strain error measures are designed for application after the adaptive refinement has identified and improved the results in critical regions.

When these error estimation procedures are widely implemented, the finite element method becomes ''transparent'' to the designer. That is to say, accurate results are produced as a matter of course. At this stage, the method can be considered a mature technology.

Notes

1. The relationship between the functional and the governing differential equations can be viewed as a chicken and egg problem. We can start with the functional and apply the Euler-Lagrange equations to obtain the differential equations. Or we can start with the differential equation and find a functional from which it is derived. This second possibility has made the variational approaches and the finite element method generally applicable. Because of this generality, the developments presented in this book are not limited solely to planar problems of elasticity.

2. Note that each of the components of the approximate solution must satisfy the geometric boundary conditions. If the overall solution satisfies the essential boundary conditions, but the individual functions that make up the solution do not satisfy the boundary conditions, incorrect results are possible (see Ref. 7).

3. A detailed discussion of the question of completeness is beyond the scope and the needs of this book. Let it suffice to say that if the set of approximation functions is complete, they can represent every possible deformation that the body can take on under a given loading. For example, if a problem is symmetric, the complete set of approximating functions need not contain nonsymmetric functions. One of the features of the finite element method is that it enables us to systematically address the convergence problem.

References and Other Reading

1. Langhaar, H. L. *Energy Methods in Applied Mechanics*, John Wiley and Sons, Inc., New York, 1962.

2. Shames, I. H., and Dym, C. L. *Energy and Finite Element Methods in Structural Mechanics*, McGraw-Hill Book Co., New York, 1985.

3. Weinstock, R. *Calculus of Variations: With Applications to Physics and Engineering*, Dover Publications, Inc., New York, 1974. This book was originally published in 1952. It contains the use of the functional derivative (he calls it the integration of an integral by a parameter).

4. Forray, M. J. *Variational Calculus in Science and Engineering*, McGraw-Hill Book Co., New York, 1968.

5. Crandall, S. H. *Engineering Analysis: A Survey of Numerical Procedures*, McGraw-Hill Book Co., New York, 1956.

6. Temple, G., and Bickley, W. G. *Rayleigh's Principle and Its Application to Engineering*, Dover Publications, Inc., 1956. This book, originally published in 1933, contains proofs of Rayleigh's principle for both discrete and continuous systems.

7. Storch, J., and Strang, G. "Paradox Lost: Natural Boundary Conditions in the Ritz-Galerkin Method," *International Journal for Numerical Methods in Engineering*, Vol. 26, 1988, pp. 2255–2266.

Lesson 4 Problems

These problems consider the application of the Rayleigh-Ritz method to the problem of a simply supported beam with a uniformly distributed load, p, as shown in the figure.

The exact solution for this problem is readily available by directly solving the differential equation using the Fourier series approach. The exact solution is

$$v_{\text{exact}} = \frac{4L^4 p}{\pi^5 EI} \sum_n \frac{1}{n^5} \sin \frac{n\pi x}{L} \qquad (n = 1, 3, 5, \ldots)$$

The potential energy expression for this problem is

$$V = \int_0^L \left[\frac{1}{2} EI(v_{,xx})^2 - pv \right] dx$$

The assumed solution for this problem will have the following form:

$$v_{\text{approx.}} = \sum_i \alpha_i \phi_i \qquad (i = 1, 3, \ldots)$$

$$\phi_i = \sin \frac{i\pi x}{L}$$

1. What boundary conditions does this approximation satisfy? Is the solution likely to be symmetric? Why or why not? Is the approximate solution symmetric? Why or why not?

2. Find the approximate solution using the Rayleigh-Ritz method for one, two, and three terms of the approximate solution. *Hint*: Expand the squared term first. The integrals are then simple to evaluate because of the orthogonality condition.

3. Provide a table and plot both the actual displacements (6 to 10 terms of the exact solution) and the approximate displacements vs the length of the beam. Is each point of the approximation smaller than the exact result? Does each point of the approximation

improve as more terms are added to the approximation? Does this result correspond to the assumption on which the Rayleigh-Ritz method is based? Would the first term of the exact solution satisfy the Rayleigh-Ritz criterion?

4. Provide a table and plot the strain energy density, the integrand of the potential energy, for both the exact and the approximate solutions vs length. Does this information provide guidance concerning the location and magnitude of the error in the approximate solution? Estimate the total energy contained in the beam for each case.

5. Plot the strains on the top of the beam for the exact and the approximate solutions vs the length. Also provide a table. Does this information provide guidance concerning the location and magnitude of the error in the approximate solution? Does the maximum strain in the approximations ever exceed the maximum strain for the exact solution? Does the Rayleigh-Ritz criterion provide any guidance for this question?

Part II

Physically Interpretable Notation

Introduction

The interpolation polynomials that model the displacements in the continuum are central to the finite element method. These displacement approximations are differentiated according to the strain-displacement relations to produce the strain representations. The resulting strain models are then used in the formulation of the individual stiffness matrices.

In the standard approach for forming finite element stiffness matrices, the arbitrary coefficients of the displacement polynomials have no specific meaning.[1] Consequently, these arbitrary quantities provide no direct assistance in evaluating the accuracy of the equations being derived. The coefficients are passively carried along in the formulation process.

In Lesson 5, the arbitrary coefficients contained in the displacement representation are evaluated in terms of the physical phenomena that produce displacements in the continuum. That is to say, the coefficients of the interpolation polynomials are expressed in terms of rigid body displacements, constant strain representations, and gradients of strains in the coordinate directions. We see that the physically defined coefficients provide insights into the modeling capabilities of finite elements that did not previously exist. These insights enable the notation to play an active role in the formulation and evaluation of finite element stiffness matrices.

For example, the physically interpretable notation clearly presents the strain modeling characteristics of individual finite elements. The strain-based notation enables us to see modeling deficiencies just as an x-ray picture identifies a broken bone. Furthermore, this knowledge enables us to correct the modeling deficiencies that are identified during the formulation process. As a result, heuristic numerical corrections such as reduced-order Gaussian quadrature integration need not be used in an attempt to improve element performance. In fact, the physically based notation is used in Part III to identify the reasons that reduced-order integration can introduce additional errors in higher order finite elements.

The use of physically based notation enables us to formulate analysis procedures and to evaluate errors in finite element models in ways that were never before possible with the standard notation. Specifically, the direct connection between the physically interpretable notation and the problems being solved enables us to

1. Identify and remove errors contained in individual finite elements (Part III).

2. Formulate finite element stiffness matrices in a new way that provides insights into the element's modeling characteristics (Part III).

3. Formulate finite difference operators and finite element stiffness matrices from the same polynomial basis (Part IV).

4. Develop pointwise error measures for stresses and strains in finite element models (Part V).

Self-Referential Notation in Computational Mechanics

The developments presented in this book largely result from reversing the generalization process common to mathematics. Instead of stripping the content from equations, content

---

[1] See Appendix A of Lesson 8.

is poured into abstract algebraic equations (see Ref. 1). Several advantages accrue from this process because a close relationship develops between the physical systems and the equations used to model them.

As a result of this close relationship, the understanding of the physical processes being modeled can be used by an analyst to interpret the equations that emerge in the development of finite element stiffness matrices. For this reason, we can say that the notation is transparent because the "mathematics" do not obscure the problem being solved. In mathematical terms, the notation is *self-referential*. The notation and the problem are close to being one and the same thing.

Standard Notation

In their standard forms, the finite element and finite difference methods are based on arbitrary polynomials whose coefficients have no designated physical interpretation. For example, the four-node finite element, which is widely used to analyze plane elasticity problems, is based on the following polynomial displacement representation:

Arbitrary Displacement Polynomials

$$u = a_1 + a_2x + a_3y + a_5xy$$
$$v = b_1 + b_2x + b_3y + b_5xy$$

where u is the displacement in the x direction.   (II.1)

v is the displacement in the y direction.

a_i and b_i are arbitrary coefficients.

The arbitrary nature of the coefficients contained in these displacement approximations makes it difficult to relate them to the physical system being modeled. For example, without further manipulation, we cannot identify the physical meaning of the coefficient of the xy term[2] a_5. The use of self-referential notation provides the physical interpretation of the various coefficients in a seamless manner.

The physically interpretable notation developed here expresses the displacements u and v in terms of the physical causes of the displacements. Specifically, the displacements are expressed as functions of the rigid body motions, the strains, and the gradients of the strains. For example, the evaluation of the arbitrary coefficients performed in Lesson 5 replaces the coefficient a_5 by the physically interpretable expression $\varepsilon_{x,y}$. This term represents the rate of change of the normal strain ε_x in the y direction. When beams are being discussed, this type of strain pattern is called a flexural deformation.

Physically Interpretable Notation

The displacement polynomials for a four-node finite element expressed in physically interpretable notation are the following:

---

[2] The subscript on the fourth coefficient is 5 instead of 4 so that the notation in this introduction matches the notation in the lessons that follow. In the main text, the fourth coefficient is associated with the x^2 term. The fifth term is then associated with the xy term.

Physically Interpretable Displacement Polynomials

$$u = (u_{rb})_0 + (\varepsilon_x)_0 x + (\gamma_{xy} + r_{rb})_0 y + (\varepsilon_{x,y})_0 xy$$
$$v = (v_{rb})_0 + (\gamma_{xy} - r_{rb})_0 x + (\varepsilon_y)_0 y + (\varepsilon_{y,x})_0 xy$$

(II.2)

The displacement expressions given by Eq. II.2 are identical in capability to those given by Eq. II.1. However, the expressions contained in Eq. II.2 provide an understanding of the behavior of individual finite elements that is not available when Eq. II.1 is used. As a result of the physically interpretable nature of these representations, we can evaluate the modeling capabilities of the polynomials and identify some of the inherent limitations contained in these models.

By inspecting these displacement representations, we can see that the coefficient of the xy term of the displacement in the x direction is indeed $\varepsilon_{x,y}$, as mentioned earlier. Furthermore, Eq. II.2 reveals that the polynomial can represent the three rigid body motions, the three constant strain terms and two of the six linear variations in the strains. These terms have the following subscripts: rb; x, y, and xy; and x,y and y,x, respectively.

The fact that the linear portions of the representations are not complete produces negative effects on the modeling capabilities of the four-node finite element. The subscript 0 indicates that the terms on the right-hand side of Eq. II.2 are Taylor series coefficients.

When the strain expressions are formed from Eq. II.2, the result is the following:

Strain Representations

$$\varepsilon_x = (\varepsilon_x)_0 + (\varepsilon_{x,y})_0 y$$
$$\varepsilon_y = (\varepsilon_y)_0 + (\varepsilon_{y,x})_0 x$$
$$\gamma_{xy} = (\gamma_{xy})_0 + (\varepsilon_{x,y})_0 x + (\varepsilon_{y,x})_0 y$$

(II.3)

We see immediately that the resulting strain models contain errors. The normal strain in the x direction is missing the x term and the normal strain in the y direction is missing the y term. The effect of these omissions is shown in Lesson 9 to make four-node finite elements overly stiff. The missing terms are needed so that the Poisson effect associated with linear terms present in the strain expressions can be activated.

The expression for the shear strain contains two errors of commission and two errors of omission. The two flexure terms contained in the shear strain expressions should not be present because shear is uncoupled from the normal strains in the actual strain-displacement relation. These erroneous normal strain terms are called *parasitic shear* terms. Instead of these two erroneous linear terms, the shear model should contain the two linear terms expected from the Taylor series expansion representing gradients of the shear strain in the x and y directions, namely, $\gamma_{xy,x}$ and $\gamma_{xy,y}$. In Lesson 8 we see that the errors just identified cannot be clearly seen with the standard notation.

The two erroneous flexure terms can be removed from the shear strain models for the four-node element using reduced-order Gaussian quadrature integration. However, the parasitic shear terms contained in eight- and nine-node isoparametric elements cannot be successfully removed using the same approach. The attempt to remove these modeling errors from the higher order elements using reduced-order Gaussian quadrature integration introduces a more severe error into the elements. As we will see, this process can introduce ''spurious'' or unnecessary zero-energy modes of deformation. The parasitic shear terms can be eliminated from these elements without introducing this other error using the new element formulation procedure based on strain gradient notation that is developed in Part III.

The modeling errors contained in the strain models of Eq. II.3 are obvious because of the transparent nature of the notation. The notation is expressed in terms of one of the quantities being sought in the analysis, namely, the strains. As a result of this transparency, the modeling characteristics of elements created using this notation are available for inspection. Thus, when the modeling characteristics do not match our understanding of the physical problem, the errors in the representation are clearly seen.

The foregoing short discussion of strain gradient notation in the context of the four-node finite element is not complete. It is designed to reinforce the idea that physically interpretable notation provides valuable insights into the modeling behavior of finite element representations. Further results concerning the four-node element, higher order plane stress elements, and plate elements are discussed in detail in Part III.

Contents of Part II

The three lessons contained in Part II develop and apply the physically based, strain gradient notation. This notation enables the equations formed in the development of finite element stiffness matrices to be directly related to the physical behavior of the problems being analyzed. Finite element stiffness matrices are developed using a new approach in Part III. This notation enables errors to be identified and corrected in individual elements in ways that were not possible before the notation and the physical processes were connected.

Furthermore, the reformulation of the finite element method can be extended to encompass the finite difference method. This puts the two methods on the same basis so both approaches can represent the same geometries and boundary conditions. The reformulation of the finite difference method is presented in Part IV. The availability of a variational and a differential solution technique that can be applied to the same discrete models makes it possible to develop the practical error measures presented in Part V.

Lesson Outlines

Lesson 5 derives the physically interpretable *strain gradient notation* and introduces some of the more obvious advantages and new capabilities resulting from this notation.

Lesson 6 applies the notation to *represent the displacements of two-dimensional discrete structures* in terms of strain gradient quantities. The techniques developed in Lesson 6 extend directly to the identification of the basis sets of deformation patterns for finite elements and finite difference operators. This lesson also reinforces the idea of the generalized coordinate nature of the individual strain gradient terms.

Lesson 7 uses the physically interpretable characteristics of the notation to develop the well-known *strain transformations*. These concepts are used later to identify the modeling characteristics and deficiencies of individual finite elements.

References and Other Reading

1. Langer, S. K. *Symbolic Logic*, 2nd ed. Dover Publications, Inc., New York, 1953. This book is written by a philosopher who was a student of A. N. Whitehead. Langer illustrates the abstraction process in mathematics with the example of algebra. Algebra is defined as arithmetic stripped of content. Among other goals, this book is designed to prepare the reader to approach the classic book on the foundations of mathematics *Mathematica Principia* by Russell and Whitehead. Langer also

wrote a book entitled *Philosophy in a New Key* that is directly related to the use of symbols that has some bearing on this work. Of particular interest with respect to the research process itself is Chapter 1 of *Philosophy in a New Key* and the primary reference for this chapter by C. D. Burns, *The Sense of the Horizon.*

Lesson **5**

Strain Gradient Notation

Purpose of Lesson   To derive the displacement interpolation polynomials that serve as the basis for the development of finite elements and finite difference operators in terms of a physically interpretable notation.

The displacements in the continuum are due to rigid body motions and deformations. The deformations can be expressed in terms of strain quantities, i.e., the strains and derivatives of the strains. In two dimensions, the strain gradient coefficients are the following: the two rigid body displacements, $(u_{rb})_0$ and $(v_{rb})_0$; the rigid body rotation, $(r_{rb})_0$; the three constant strain components, $(\varepsilon_x)_0$, $(\varepsilon_y)_0$, and $(\gamma_{xy})_0$; and derivatives of the strains, such as $(\varepsilon_{x,x})_0$.

The direct connection between this *strain gradient notation* and the physical processes it represents provides insights that enable the modeling characteristics of the resulting polynomials to be identified and new computational capabilities to be developed. Specifically, the physically based notation enables the following new capabilities: (1) identification of modeling errors in individual finite elements (Part III), (2) formulation of elements without these errors (Part III), (3) reformulation of the finite difference method so it applies to irregular meshes and geometries (Part IV), (4) the improvement of existing error analysis procedures (Part V), and (5) the development of pointwise error measures (Part V). The use of strain gradient notation is demonstrated with five applications.

■   ■   ■

The potential energy functionals and the governing differential equations that represent the variational and differential forms of many elastic systems are expressed in terms of the unknown displacements u and v in the x and y coordinate directions. The finite element and the finite difference methods for finding approximate solutions to these representations are based on displacement polynomials with the following general form:

Arbitrary Displacement Polynomials

$$u = a_1 + a_2 x + a_3 y + a_4 x^2 + a_5 xy + a_6 y^2 + a_7 x^3$$
$$+ a_8 x^2 y + a_9 xy^2 + a_{10} y^3$$
$$v = b_1 + b_2 x + b_3 y + b_4 x^2 + b_5 xy + b_6 y^2 + b_7 x^3$$
$$+ b_8 x^2 y + b_9 xy^2 + b_{10} y^3$$

(5.1)

The arbitrary nature of the coefficients contained in the polynomials given in Eq. 5.1 makes it difficult to relate these representations directly to the physical system being modeled. In other words, the physical meaning of the individual coefficients of the displacement approximation is not obvious by simply reading or inspecting the equation. The objective of this lesson is to replace the arbitrary coefficients with physically interpretable expressions that relate directly to the causes of displacements and deformations in the continuum. The complete third-order polynomials given in Eq. 5.1 are chosen for this study because they are the lowest order expressions that contain all of the salient features of strain gradient notation.

For example, the coefficient a_s will be replaced by the expression $(\varepsilon_{x,y})_0$. This quantity can be interpreted physically. We immediately recognize the term as the y derivative of the normal strain ε_x. When this expression is related to the displacements in a simple beam, we see that it represents a flexural deformation. Thus, any approximation polynomial representing the displacement in the y direction that contains this xy coefficient is capable of representing a flexural deformation. The subscript 0 indicates that the strain gradient coefficients are Taylor series terms.

The arbitrary coefficients of Eq. 5.1 are put in strain gradient notation by evaluating them in terms of the physical phenomena that produce motion in the continuum. That is to say, the arbitrary coefficients are expressed in terms of rigid body motions and deformations. The deformations are represented as a function of strain quantities. Since the displacement approximations are expressed in terms of strains, which are the quantities being sought in the analysis, the equations developed from these approximations can be directly related to the problem being solved. This direct connection between the notation and the problem being solved provides the insights that made the developments presented here possible (see Note 1).

Interpretation of Strain Gradient Coefficients

Before we evaluate the arbitrary coefficients of Eq. 5.1 in physical terms, the flavor of strain gradient notation is introduced with two examples. The idea of the rigid body terms is introduced with the case of the motion of an undeformed block in the plane. The rigid body terms for a planar system are

---

Rigid Body Motions

$$u_{rb}, \quad v_{rb}, \quad \text{and} \quad r_{rb}$$

The first term denotes the rigid body displacement in the x direction. If a body is subjected to this motion there is no strain and every point moves the same amount in the x direction. A rigid body motion in the x direction is shown in Fig. 1 by the movement of the block to the right. We consider this a zeroth-order strain. The second rigid body term denotes a similar motion in the y direction. This is shown in Fig. 1 by the upward movement of the block. The third term is a rigid body rotation around the z axis. The rigid body rotation is shown in Fig. 1 by the rotation of the block about its original position. Finally, a general rigid body motion is shown in Fig. 1 by the movement of the block along the diagonal. This motion consists of a linear combination of the three independent rigid body motions.

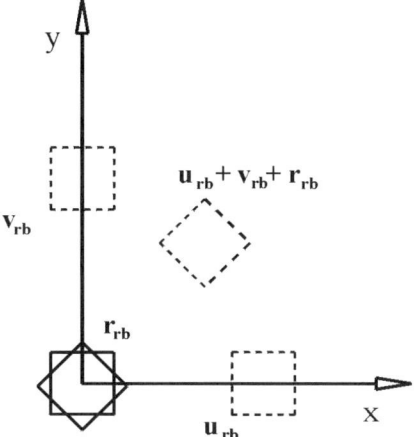

Figure 1. Rigid body motion.

The meaning of the strain gradient coefficients that represent the deformations of the continuum is now introduced. We begin by referring to a Taylor series expansion of the normal strain term ε_x, which is written as

---

Taylor Series Expansion of ε_x

$$\varepsilon_x(x,y) = (\varepsilon_x)_0 + (\varepsilon_{x,x})_0 x + (\varepsilon_{x,y})_0 y$$
$$+ \frac{1}{2}(\varepsilon_{x,xx})_0 x^2 + (\varepsilon_{x,xy})_0 xy + \cdots \qquad (5.2)$$

This equation approximates the normal strain in the x direction at a point in the neighborhood of a local origin as a constant plus higher order terms that depend on the location of the point. The left-hand side of this equation contains x and y in parentheses to emphasize that the strain can be evaluated at any point in the neighborhood of the local origin.

The constant term $(\varepsilon_x)_0$ on the right-hand side of the equation represents the strain at the origin. The coefficients of the higher order terms represent the gradients of the strains at the origin. The fact that these terms represent quantities at the origin is emphasized by the use of the subscript zero on the right-hand side of Eq. 5.2. For example, the higher order term $(\varepsilon_{x,x})_0$ represents the rate of change of (ε_x) in the x direction at the local origin. In these developments, the local origin is usually a point such as the centroid of a finite element or the central node of a finite difference operator.

The meaning of this notation can be demonstrated by expressing the strain at the two points A and B in the plane, as shown in Fig. 2. Point A is located away from the origin along the x axis. The value of the strain at this point is approximated as the constant term of the Taylor series plus all of the x terms. Since $y = 0$, the terms containing y are eliminated.

Point B is a general point in the plane. It is located away from the origin in both the x and y directions. As a result, none of the Taylor series terms are automatically excluded from the strain representation. Thus, the strain at point B is represented by the full series given as Eq. 5.2.

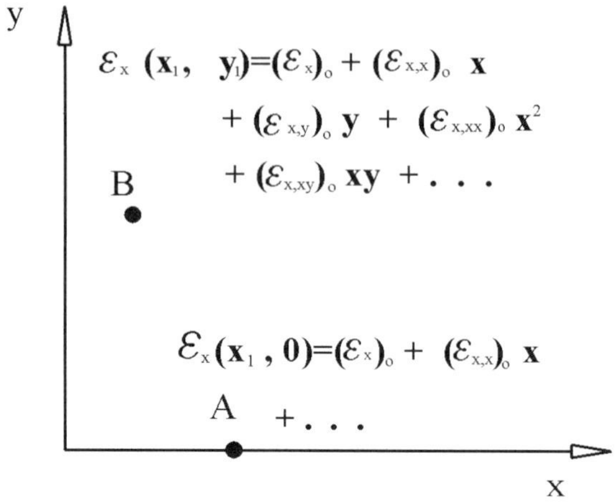

$$\mathcal{E}_x(x_1, y_1) = (\mathcal{E}_x)_0 + (\mathcal{E}_{x,x})_0\, x$$
$$+ (\mathcal{E}_{x,y})_0\, y + (\mathcal{E}_{x,xx})_0\, x^2$$
$$+ (\mathcal{E}_{x,xy})_0\, xy + \ldots$$

B

$$\mathcal{E}_x(x_1, 0) = (\mathcal{E}_x)_0 + (\mathcal{E}_{x,x})_0\, x$$
$$+ \ldots$$

A

Figure 2. Taylor series representation of strain at a point.

The fact that the rates of change are evaluated at the local origin enables the coefficients of the interpolation polynomials to be easily determined. This is seen as we evaluate the coefficients of Eq. 5.1 as functions of the rigid body quantities and the coefficients of the strain expansions, e.g., Eq. 5.2.

Coefficient Evaluation

We now evaluate the 20 coefficients of Eq. 5.1 in terms of the strain gradient quantities. This task is not as formidable as it might appear because the coefficients of the different order terms can be evaluated one set at a time. That is to say, the constant coefficients are evaluated as a separate set, the first-order coefficients are evaluated as a separate set, etc. (see Note 2). As we see in the following development, the coefficients of the different order terms are independent of each other because they are evaluated at the local origin.

Zeroth-Order Strain Gradient Coefficient

The two zeroth-order terms, a_1 and b_1, can be evaluated immediately in terms of the rigid body displacements, $(u_{rb})_0$ and $(v_{rb})_0$. This is accomplished by inserting $x = 0$ and $y = 0$ into Eq. 5.1. All of the terms except for the leading constant terms are eliminated because they are functions of x and y. When this substitution is made and the displacements at the origin are recognized as the rigid body displacements, the leading constants of the displacement polynomials are found to be

---

Zeroth-Order Strain Gradient Coefficients

$$a_1 = (u_{rb})_0$$
$$b_1 = (v_{rb})_0$$

(5.3)

Thus, we have evaluated the two constant or zeroth-order terms, a_1 and b_1, of the displacement approximations as the rigid body displacements. We now evaluate the coefficients of the first-order or linear terms.

First-Order Strain Gradient Coefficient

These four coefficients are evaluated as a function of the rotation and the three strain components at the origin. The rotation around the z axis is the rotation-displacement relation from the small displacement theory of elasticity, namely,

---

Rigid Body Rotation

$$r_{rb} = \frac{1}{2}\left(\frac{\partial v}{\partial x} - \frac{\partial u}{\partial y}\right)$$

(5.4)

Similarly, the strain-displacement relations are

Strain Expressions

$$\varepsilon_x = \frac{\partial u}{\partial x} \qquad \varepsilon_y = \frac{\partial v}{\partial y} \qquad \gamma_{xy} = \frac{\partial u}{\partial y} + \frac{\partial v}{\partial x}$$

(5.5)

When the expressions for the displacements, as given by Eq. 5.1, are differentiated, substituted into Eqs. 5.4 and 5.5, and evaluated at the origin (the remaining terms are zero because they are functions of x and y), the equations containing the first-order coefficients are

$$(r_{rb})_0 = \frac{1}{2}(b_2 - a_3)$$
$$(\varepsilon_x)_0 = a_2$$
$$(\varepsilon_y)_0 = b_3$$
$$(\gamma_{xy})_0 = a_3 + b_2$$

(5.6)

As was noted earlier, these equations are independent of the previously determined zeroth-order terms and they do not contain any second-order or higher order terms. Thus, the first-order terms are evaluated on their own. The independent nature of the coefficients of the various order terms holds for each set of terms. This makes the evaluation of the arbitrary coefficients in terms of strain gradient quantities a relatively simple task.

These expressions are solved for the arbitrary coefficients to give

---

First-Order Strain Gradient Coefficients

$$a_2 = (\varepsilon_x)_0$$
$$a_3 = (\gamma_{xy}/2 - r_{rb})_0$$
$$b_2 = (\gamma_{xy}/2 + r_{rb})_0$$
$$b_3 = (\varepsilon_y)_0$$

(5.7)

These are the coefficients of the four first-order or linear terms contained in Eq. 5.1. These four strain quantities correspond to the three states of constant strain and the rigid body rotation. We now evaluate the coefficients of the second-order or quadratic terms.

Second-Order Strain Gradient Coefficient

The six coefficients of the second-order or quadratic terms of Eq. 5.1 are determined in terms of the first derivatives of the strain components with respect to x and y. These strain gradient terms are defined by taking derivatives of Eq. 5.5 to give

Derivatives of Strains

$$\varepsilon_{x,x} = \frac{\partial \varepsilon_x}{\partial x} = \frac{\partial^2 u}{\partial x^2} \qquad \varepsilon_{x,y} = \frac{\partial \varepsilon_x}{\partial y} = \frac{\partial^2 u}{\partial x \partial y}$$

$$\varepsilon_{y,x} = \frac{\partial \varepsilon_y}{\partial x} = \frac{\partial^2 v}{\partial y \partial x} \qquad \varepsilon_{y,y} = \frac{\partial \varepsilon_y}{\partial y} = \frac{\partial^2 v}{\partial y^2} \qquad (5.8)$$

$$\gamma_{xy,x} = \frac{\partial \gamma_{xy}}{\partial x} = \frac{\partial^2 u}{\partial y \partial x} + \frac{\partial^2 v}{\partial x^2} \qquad \gamma_{xy,y} = \frac{\partial \gamma_{xy}}{\partial y} = \frac{\partial^2 u}{\partial y^2} + \frac{\partial^2 v}{\partial x \partial y}$$

The six second-order coefficients are found by substituting Eq. 5.1 into Eq. 5.8 and evaluating the resulting expressions at the origin. The result of this operation is

$$\begin{aligned}
(\varepsilon_{x,x})_0 &= 2a_4 & (\varepsilon_{x,y})_0 &= a_5 \\
(\varepsilon_{y,x})_0 &= b_5 & (\varepsilon_{y,y})_0 &= 2b_6 \\
(\gamma_{xy,x})_0 &= a_5 + 2b_4 & (\gamma_{xy,y})_0 &= 2a_6 + b_5
\end{aligned} \qquad (5.9)$$

When the arbitrary coefficients are evaluated in terms of the strain gradient quantities, the coefficients of the second-order terms are found to be

Second-Order Coefficients

$$\begin{aligned}
a_4 &= (\varepsilon_{x,x})_0/2 & b_4 &= (\gamma_{xy,x} - \varepsilon_{x,y})_0/2 \\
a_5 &= (\varepsilon_{x,y})_0 & b_5 &= (\varepsilon_{y,x})_0 \\
a_6 &= (\gamma_{xy,y} - \varepsilon_{y,x})_0/2 & b_6 &= (\varepsilon_{y,y})_0/2
\end{aligned} \qquad (5.10)$$

Equation 5.10 shows that the six second-order terms of Eq. 5.1 are functions of the six available first derivatives of the three strain components. These terms represent the linear variations of the strains with respect to the distance from the local origin.

Third-Order Strain Gradient Coefficient

The 8 third-order coefficients of Eq. 5.1 remain to be determined. They are found as a function of the second derivatives of the strains. When the derivatives of the 6 second-order strain gradient terms given in Eq. 5.8 are taken with respect to x and y, 12 terms are produced. However, the symmetric terms such as $\varepsilon_{x,xy}$ and $\varepsilon_{x,yx}$ are equal to each other because of the commutativity of the derivative for a continuous function. When the duplicate terms are eliminated, the following 9 terms remain:

Second Derivative of Strain Components

$$
\begin{array}{ccc}
\varepsilon_{x,xx} & \varepsilon_{x,xy} & \varepsilon_{x,yy} \\
\varepsilon_{y,xx} & \varepsilon_{y,xy} & \varepsilon_{y,yy} \\
\gamma_{xy,xx} & \gamma_{xy,xy} & \gamma_{xy,yy}
\end{array}
\tag{5.11}
$$

As can be seen in Eq. 5.11, there are 9 second-order strain gradient terms available to evaluate the eight remaining coefficients in Eq. 5.1. The presence of this extra equation indicates that a dependency relation exists among these terms. In the next section, this relationship is shown to contain the compatibility equation for plane elasticity.

The coefficients associated with the second-order strain gradients of ε_x are evaluated first. This is accomplished by taking the appropriate derivatives of the displacement polynomial for u, as given by Eq. 5.1. When this is done, three of the eight arbitrary coefficients to be determined are found as

Third-Order Coefficients

$$
\begin{aligned}
a_7 &= (\varepsilon_{x,xx}/6)_0 \\
a_8 &= (\varepsilon_{x,xy}/2)_0 \\
a_9 &= (\varepsilon_{x,yy}/2)_0
\end{aligned}
\tag{5.12}
$$

Similarly, the coefficients associated with the second derivatives of ε_y are found from the displacement polynomial for v in Eq. 5.1 to give

Third-Order Coefficients

$$
\begin{aligned}
b_8 &= (\varepsilon_{y,xx}/2)_0 \\
b_9 &= (\varepsilon_{y,xy}/2)_0 \\
b_{10} &= (\varepsilon_{y,yy}/6)_0
\end{aligned}
\tag{5.13}
$$

The only coefficients that remain to be evaluated are a_{10} and b_7. They are evaluated using the second derivatives of γ_{xy}. The resulting equations are

$$
\begin{aligned}
(\gamma_{xy,xx})_0 &= 2a_8 + 6b_7 \\
(\gamma_{xy,xy})_0 &= 2a_9 + 2b_9 \\
(\gamma_{xy,yy})_0 &= 6a_{10} + 2b_8
\end{aligned}
\tag{5.14}
$$

The final two coefficients being sought, a_{10} and b_7, are contained in the first and third equations of Eq. 5.14. These remaining arbitrary coefficients are evaluated by substituting the strain gradient expressions for a_8 and b_8 given in Eqs. 5.12 and 5.13 to give

Third-Order Coefficients

$$
\begin{aligned}
a_{10} &= (\gamma_{xy,yy} - \varepsilon_{y,xy})_0/6 \\
b_7 &= (\gamma_{xy,xx} - \varepsilon_{x,xy})_0/6
\end{aligned}
\tag{5.15}
$$

Table 1. The 20 coefficients for the two-dimensional displacement functions.

i	Term	a_i for $u(x, y)$	b_i for $v(x, y)$
1	1	$(u_{rb})_0$	$(v_{rb})_0$
2	x	$(\varepsilon_x)_0$	$(\Upsilon_{xy}/2 + r_{rb})_0$
3	y	$(\gamma_{xy}/2 - r_{rb})_0$	$(\varepsilon_y)_0$
4	x^2	$(\varepsilon_{x,x}/2)_0$	$[(\gamma_{xy,x} - \varepsilon_{x,y})/2]_0$
5	xy	$(\varepsilon_{x,y})_0$	$(\varepsilon_{y,x})_0$
6	y^2	$[(\gamma_{xy,y} - \varepsilon_{y,x})/2]_0$	$(\varepsilon_{y,y}/2)_0$
7	x^3	$(\varepsilon_{x,xx}/6)_0$	$[(\gamma_{xy,xx} - \varepsilon_{x,xy})/6]_0$
8	$x^2 y$	$(\varepsilon_{x,xy}/2)_0$	$(\varepsilon_{y,xx}/2)_0$
9	xy^2	$(\varepsilon_{x,yy}/2)_0$	$(\varepsilon_{y,xy}/2)_0$
10	y^3	$[(\gamma_{xy,yy} - \varepsilon_{y,xy})/6]_0$	$(\varepsilon_{y,yy}/6)_0$

We have now evaluated the 20 unknown coefficients of Eq. 5.1 in terms of strain gradient quantities. These coefficients are collected and presented in Table 1.

Table 1 represents one form of strain gradient displacement polynomials. The coefficients of u and v are given in the third and fourth columns, respectively. For example, the first four terms of each displacement polynomial, written in terms of strain gradient coefficients, are

Example Strain Gradient Displacement Polynomials

$$u(x, y) = (u_{rb})_0 + (\varepsilon_x)_0 x + (\gamma_{xy}/2 - r_{rb})_0 y + (\varepsilon_{x,x}/2)_0 x^2$$
$$v(x, y) = (v_{tb})_0 + (\gamma_{xy}/2 + r_{rb})_0 x + (\varepsilon_y)_0 y + [(\gamma_{xy,x} - \varepsilon_{x,y})/2]_0 x^2$$
(5.16)

Equation 5.16 demonstrates that we have accomplished the purpose of this lesson. The coefficients of the displacement polynomial are expressed in terms of physically interpretable quantities that are the source of the displacements in the continuum. Appendix A contains the strain gradient coefficients of higher order two-dimensional and three-dimensional polynomials. The coefficients for the three-dimensional case are developed as noted in the references.

Applications of Self-Referential Strain Gradient Notation

We now apply the displacement polynomials expressed in strain gradient notation to five examples to demonstrate a hint of the power of this physically interpretable notation. Each of the five capabilities developed or demonstrated here is used directly or extended in later lessons. The strain gradient version of the interpolation polynomials is now used to

1. Find the compatibility relationship for two-dimensional elasticity.
2. Identify a basis set of independent strain gradient quantities to describe the displacements in a rectangular region.
3. Identify a strain modeling error inherently contained in incomplete displacement polynomials.

4. Identify the displacement field associated with an individual strain state.

5. Compute the strain energy in a region due to a given strain state.

Application 1 — Formulation of the Compatibility Equation

When the final two coefficients of the third-order terms were evaluated using Eq. 5.14, only the first and third of the three equations were utilized. The second equation was not used in the evaluation of the coefficients. However, this equation is no less significant to the representation than the eight equations directly used to evaluate the eight third-order coefficients of the displacement polynomials. This equation is reproduced here for convenience as

$$(\gamma_{xy,xy})_0 + 2a_9 + 2b_9 \tag{5.17}$$

When the arbitrary coefficients a_9 and b_9 from Table 1 are substituted into Eq. 5.17, the result is

---
Compatibility Equation

$$(\gamma_{xy,xy})_0 = (\varepsilon_{x,yy} + \varepsilon_{y,xx})_0 \tag{5.18}$$

This equation is the compatibility equation for plane elasticity. The compatibility equation expresses a dependency relation that must be satisfied by the strains in the continuum if continuous displacements are to exist. If this condition is not satisfied, "holes" or "overlaps" can appear in the fabric of the continuum. This compatibility equation is used later in the process of correcting errors that are inherent in finite elements based on incomplete displacement polynomials.

In this derivation, the compatibility equation emerged as a natural consequence of the process of evaluating the arbitrary coefficients of the displacement polynomials in terms of strain gradient quantities. This relation is simply a dependency relation that results from the existence of one extra equation when the "symmetric" derivatives are formed (see Note 3).

This derivation differs from the approach seen in most presentations. In most other derivations, the following path is followed. First, the fact that the number of strains exceeds the number of displacements is recognized and it is concluded that a dependency relation must exist between the strains. Then, several strain quantities are differentiated and artificially combined to produce the compatibility equations (see Refs. 5 and 6, for example). The six compatibility equations of three-dimensional elasticity are produced when the three-dimensional strain gradient coefficients are evaluated for third-order polynomials. As expected, higher order compatibility equations are produced when higher order strain gradient coefficients are evaluated. The existence of the higher order compatibility equations is noted for completeness, however, these equations are not explicitly used here.

Application 2 — Identification of Polynomial Modeling Capabilities

The use of strain gradient notation enables us to identify the independent strain states that a displacement polynomial is capable of representing. This is contrasted to the standard form of the displacement polynomials given in Eq. 5.1, where no direct information concerning the modeling capabilities is provided by the arbitrary polynomials. The information concerning the modeling characteristics of displacement polynomials developed here is used in the final two applications of this lesson to compute quantities that later enable us to develop an alternative way of forming finite elements.

This difference in capabilities between strain gradient notation and standard notation is demonstrated for the case of a displacement polynomial representation with eight independent variables. The polynomials are complete linear functions that are augmented with the symmetric quadratic term. As we see later, these polynomials represent the displacements in a four-node plane stress finite element. These displacement representations are written in terms of arbitrary coefficients as

---

Displacement Polynomials with Arbitrary Coefficients

$$u = a_1 + a_2 x + a_3 y + a_5 xy$$
$$v = b_1 + b_2 x + b_3 y + b_5 xy$$

(5.19)

The eight independent variables contained in these polynomials are the four a's and the four b's. However, these independent variables shed little, if any, light on the strain states that this polynomial is capable of representing. The reason for the choice of terms contained in Eq. 5.19 is discussed later.

When the same polynomial representations are written in strain gradient notation, we can directly identify the strain states they are capable of representing. This can be demonstrated by rewriting Eq. 5.19 in strain gradient notation. When this is done, the result is

---

Displacement Polynomial with Strain Gradient Coefficients

$$u(x,y) = (u_{rb})_0 + (\varepsilon_x)_0 x + (\gamma_{xy}/2 - r_{rb})_0 y + (\varepsilon_{x,y})_0 xy$$
$$v(x,y) = (v_{rb})_0 + (\gamma_{xy}/2 + r_{rb})_0 x + (\varepsilon_y)_0 y + (\varepsilon_{y,x})_0 xy$$

(5.20)

These displacement polynomials expressed in terms of strain gradient quantities contain the following eight independent variables:

Strain State Representations

$$(u_{rb})_0 \qquad (v_{rb})_0 \qquad (r_{rb})_0$$
$$(\varepsilon_x)_0 \qquad (\varepsilon_y)_0 \qquad (\gamma_{xy})_0$$
$$(\varepsilon_{x,y})_0 \qquad (\varepsilon_{y,x})_0$$

(5.21)

The presence of these coefficients means that the displacement polynomials given in Eq. 5.20 can represent the eight independent strain states given in Eq. 5.21. That is to say, the interpolation polynomial being discussed can represent the three rigid body motions, the three constant strain states, and two second-order strain gradient terms. The presence of the first six terms means that the four-node finite element will converge to the correct answer as each element becomes infinitely small. Briefly, these six terms represent the constant terms in the Taylor series representations of the displacements and strains. As the dimension of the finite element approaches zero, the higher order terms go to zero. Hence, they are not needed. The two higher order terms that correspond to flexure deformations are discussed in detail later.

This comparison of the two types of notation has shown that the use of strain gradient notation brings a clarity to the modeling process that does not exist in the standard approach. Later, we see that the ability to identify the modeling capabilities leads to a procedure for formulating finite elements that is different from the standard approach. Furthermore, as demonstrated in the next application, the use of strain gradient notation enables errors in the modeling capabilities to be identified.

Application 3 — Identification of a Strain Modeling Error

In the previous application, we identified the strain states that the polynomials given in Eqs. 5.19 and 5.20 are capable of representing. Let us now investigate the strain models provided by these representations. When the displacement polynomials given by Eq. 5.19 are substituted into the definitions of strain given by Eq. 5.5, the results are

---

Strain Models with Arbitrary Coefficients

$$(\varepsilon_x)_0 = a_2 + a_5 y$$
$$(\varepsilon_y)_0 = b_3 + b_5 x$$
$$(\gamma_{xy})_0 = (a_3 + a_5 x) + (b_2 + b_5 y)$$

$$(5.22)$$

The two normal strains both contain a constant term and a linear term. The shear strain contains a constant term and linear terms in the two coordinate directions. We cannot, with confidence, say much more about these models because of the lack of information provided by the arbitrary coefficients. However, when we formulate the strain expressions in terms of strain gradient quantities, the modeling characteristics of the strain representations are clearly shown. This can be seen when the strain expressions are formed using Eq. 5.20 as

---

Strain Models with Strain Gradient Coefficients

$$\varepsilon_x = (\varepsilon_x)_0 + (\varepsilon_{x,y})_0 y$$
$$\varepsilon_y = (\varepsilon_y)_0 + (\varepsilon_{y,x})_0 x$$
$$\gamma_{xy} = (\gamma_{xy})_0 + (\varepsilon_{x,y})_0 x + (\varepsilon_{y,x})_0 y$$

$$(5.23)$$

The two normal strain models correctly represent the normal strains as far as they go. That is to say, the constant terms and the linear terms contained in the expressions are the correct Taylor series terms, but one of the linear terms is missing. The shear strain expression contains two obvious errors. The coefficient representing the constant term is the correct Taylor series term, but the two linear coefficients are incorrect. The coefficients of the two linear terms should not be normal strain quantities. The exact nature of these modeling deficiencies can best be seen by comparing the strain expressions of Eq. 5.23 to the complete linear Taylor series strain representations, which are

---

Complete Linear Taylor Series Strain Models

$$\varepsilon_x = (\varepsilon_x)_0 + (\varepsilon_{x,x})_0 x + (\varepsilon_{x,y})_0 y$$
$$\varepsilon_y = (\varepsilon_y)_0 + (\varepsilon_{y,x})_0 x + (\varepsilon_{y,y})_0 y \qquad (5.24)$$
$$\gamma_{xy} = (\gamma_{xy})_0 + (\gamma_{xy,x})_0 x + (\gamma_{xy,y})_0 y$$

When the expressions for ε_x given in Eqs. 5.23 and 5.24 are compared, the coefficient that is missing is identified. A similar comparison for ε_y identifies the term missing from this representation. The effect of the missing terms is discussed in detail in Lesson 7.

When the expressions for the shear strain given in Eqs. 5.23 and 5.24 are compared, the correct coefficients are seen. The expressions actually present in Eq. 5.23 are normal strain coefficients. Since the required shear strain terms are not present, the shear strain model given in Eq. 5.23 cannot represent linear shear strain variations. Furthermore, the normal strain terms erroneously contribute to the shear strain. Thus, the shear strain model given in Eq. 5.22 is doubly wrong.

The use of strain gradient notation has clearly identified the modeling characteristics of these strain models in a way that is *not possible* with the standard notation. This has significant consequences when we study modeling errors in finite elements.

The erroneous normal strain terms in the shear strain expression are known as parasitic shear terms. This is an accurate description of this error because an erroneous shear strain is produced when a linear variation exists in the normal strain. Later, we see that this modeling deficiency is caused by the use of incomplete displacement polynomial approximations.

Note that the existence of parasitic shear was recognized long before strain gradient notation was developed. In fact, procedures that attempt to correct this error are available. They consist of modifying the numerical integration procedures used in the computation of finite elements. However, these selectively reduced integration techniques are successful for only low-order elements. We see in later lessons that these approaches fail for eight- and nine-node elements. However, any element containing parasitic shear can be accurately corrected during the formulation stage with guidance provided by strain gradient notation.

This application has demonstrated the effectiveness of this physically interpretable notation when analyzing the strain modeling characteristics of the polynomials used in the finite element and finite difference methods.

Application 4 — Identification of Strain Gradient Defined Displacement Fields

In addition to providing insights into the modeling processes, this notation enables us to compute useful quantities that are unavailable without strain gradient notation. For example, the strain gradient form of the polynomials enables us to determine the displacements associated with a particular strain state or strain gradient coefficient. This is demonstrated for the square region shown in Fig. 3. This region can be viewed as representing a differential volume or a finite element with the origin at the center.

For instance, we can represent the rigid body motions shown in Fig. 1 as follows. The rigid body motion in the x direction occurs when the other rigid body motions are zero and there are no deformations. These conditions are represented in strain gradient notation by requiring that all of the strain gradient coefficients in Table 1 be zero except for $(u_{rb})_0$. Thus, the rigid body motion in the x direction is represented by the following displacement relations:

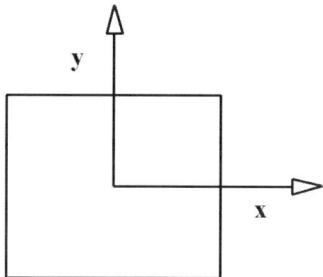

Figure 3. Representative region.

Rigid Body Displacement

$$u = (u_{rb})_0$$
$$v = 0$$

(5.25)

Similarly, the rigid body rotation shown in Fig. 1 is represented in strain gradient notation by requiring all of the coefficients in Table 1 to be zero except for $(r_{rb})_0$. This gives the following displacement expressions for a rigid body rotation about the z axis:

Rigid Body Rotation

$$u = -(r_{rb})_0 y$$
$$v = (r_{rb})_0 x$$

(5.26)

Finally, the general rigid body motion shown in Fig. 1 is a linear combination of the three rigid body motions and can be expressed as

---

General Rigid Body Motion

$$u = (u_{rb})_0 - (r_{rb})_0 y$$
$$v = (v_{rb})_0 + (r_{rb})_0 x$$

(5.27)

These rigid body motions could have been defined without much difficulty using arbitrary coefficients. However, the deformation modes would be difficult, if not impossible, to identify correctly. As a simple example, let us determine the displacement pattern associated with the constant shear strain state for the region shown in Fig. 3.

The expressions describing these displacements are found by setting all of the strain gradient coefficients of Eq. 5.20 (or in Table 1) to zero except for $(\gamma_{xy})_0$. This reduces the displacement polynomials to

---

Displacements Due to Shear Strain

$$u(x, y) = \frac{1}{2}(\gamma_{xy})_0 y$$
$$v(x, y) = \frac{1}{2}(\gamma_{xy})_0 x$$

(5.28)

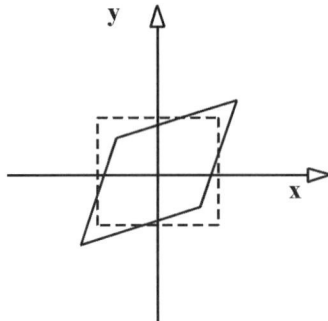

Figure 4. Displacement due to shear strain.

This deformation pattern is shown in Fig. 4 and corresponds to the shape that would be expected.

Similar displacement patterns can be formed for any of the strain gradient coefficients. The capability of directly relating the displacements to the strain states is used later to form finite elements, to evaluate the accuracy of finite elements, and to generate finite difference operators.

Application 5 — Computation of Strain Energy Content

Another quantity that is used in subsequent developments to evaluate and formulate finite elements is the strain energy content associated with a specific strain state. This quantity is immediately available in terms of strain gradient notation by forming the strain representation in terms of strain gradient quantities, substituting this representation into the strain energy expression, and integrating the result.

To demonstrate this capability, we compute the strain energy content of a region that is experiencing a constant shear strain. The strain energy contained in a region due to any shear strain is

---

Shear Strain Energy Expression

$$U_{\text{shear}} = \frac{1}{2} \int_{\Omega} G(\gamma_{xy})^2 \, d\Omega \tag{5.29}$$

---

where G is the shear modulus and Ω represents the area being considered.

When the polynomial expression for constant shear strain is substituted and the result is integrated, the result is

---

Strain Energy Content

$$U_{\text{shear}} = \frac{1}{2} G \int_{\Omega} (\gamma_{xy})_0^2 \, d\Omega$$
$$= \frac{1}{2} G (\gamma_{xy})_0^2 (\text{Area})_{\Omega} \tag{5.30}$$

Thus the strain energy for this case requires only that the area integral for the region be determined. In later developments, we see that the integrals that must be evaluated to compute the strain energy for most practical finite elements (up to nine-node elements) are quite simple. We also see that numerical integration is often not required. The integrals, regardless of shape, can be evaluated directly with Green's theorem. Because fewer integrations are required and they can be evaluated in closed form, finite elements formulated using the strain gradient approach are computationally competitive with other formulations.

As was the case in the previous applications, the ability to compute the strain energy for individual strain states assists in analyzing the performance of individual finite elements and produces techniques for developing finite elements.

Closure

We have developed a notation that directly relates the displacement interpolation polynomials to the linear elastic problems that the mathematical formulation is designed to represent. The coefficients of these polynomials are written in terms of the physical causes of displacements in the continuum, namely, rigid body motions and strain quantities. This physically interpretable notation enables us to clearly understand the modeling characteristics of our approximations and to compute useful quantities that were not previously available using the standard approximation polynomials with arbitrary coefficients. We now summarize these capabilities.

Identification of Modeling Characteristics

The ability of strain gradient notation to identify the modeling characteristics of our approximations was demonstrated in Applications 2 and 3 in this lesson. In Application 2, we identified the strain states that a four-node finite element is capable of representing. We found that the displacement polynomials could represent rigid body motions, three constant strains and two of the six linear strain states.

When the strain representations produced by these displacement polynomials where formed in terms of strain gradient notation, we found that the limitations of the strain models could easily be identified. The constant strains were modeled correctly. However, each of the two normal strain models contained only one of the two linear strain gradient terms. More importantly, we could see that the shear strain contained a significant error. Instead of representing the linear variation in the shear strains, the shear strain model contained linear variations of the normal strains. This error is caused by the use of incomplete polynomials and is called parasitic shear.

In later developments, we use strain gradient notation to show that parasitic shear cannot be removed from higher order finite elements using the techniques now in standard use. However, we show that the error can be removed by an alternative finite element formulation procedure developed using strain gradient notation.

The remaining two lessons of Part II extend Applications 1 and 2. Lesson 6 presents methods for identifying the linearly independent sets of strain gradient variables required to represent specific finite elements and finite difference operators. In Lesson 7, we further study the strain modeling characteristics of the strain gradient polynomials. In particular, the effect of coordinate rotations on the incomplete strain models that are present in all rectangular elements is studied.

New Computational Capabilities

The second major advantage of strain gradient notation is that it enables us to compute quantities that are not available when the displacement polynomials are expressed in terms of arbitrary coefficients. Using this notation, we are able (1) to identify the displacement fields produced by a specific strain gradient coefficient and (2) to compute the strain energy contained in a region when it is representing a specific strain state. These two capabilities were demonstrated in Applications 4 and 5.

In Part III, the capabilities demonstrated in Applications 4 and 5 are exploited to develop a new approach for formulating finite elements. The new approach lets us identify errors in finite elements, eliminate these errors, and create elements with specific characteristics. These capabilities have also enabled the finite difference method to be reformulated and the boundary condition models to be validated. These developments are used in finite element error analysis procedures discussed in later lessons.

Notes

1. This notation has a very special character. The coefficients of the displacement interpolation polynomials are expressed in terms of the physical causes of motion in linear elastic systems. This enables us to directly relate the various equations that occur in the derivation of analysis techniques to the physical system being modeled. The notation is ''transparent.'' That is to say, we can look at an equation written in this notation and ''see'' the physical processes it represents. If it does not relate to the physical world as we believe it to be, the equation is either wrong or it is telling us our perception is incorrect. Either way, the notation provides desirable information.

This type of notation is called self-referential. Self-referential notation was originally used to develop the famous proofs by Godel. The ''layman's'' presentation of Godel's proofs by Nagel and Newman noted in Ref. 1 gives a good view of the use of Godel's use of self-referential notation in number theory. Godel's work provided the inspiration for strain gradient notation.

2. This note is motivated by a misunderstanding that has shown up in several reviews of papers developing finite elements with this notation. The misunderstandings revolve around the use of a specific order of polynomial. The questions that arise essentially ask whether the model would not be better if higher order polynomials were used in the derivation. The answer is no for two reasons that become apparent as the notation becomes more familiar. The first reason is that the leading coefficients of the displacement approximations do not change as additional terms are appended to the series. The second reason is that the order of the polynomial required to develop a finite element or a finite difference operator is totally dependent on the geometry and number of degrees of freedom contained in the discrete model being developed. Thus the order of the polynomial required is defined by the problem.

3. When the arbitrary coefficients for the three-dimensional interpolation polynomial are evaluated, there are six equations that are not needed to evaluate the 60 coefficients up to the third order. These six redundant equations produce the six compatibility equations of three-dimensional elasticity.

As an aside, note that if the fourth-order terms are added to the two-dimensional interpolation polynomials 2 redundant equations appear. Similarly, if the quartic terms are added to the three-dimensional interpolation polynomials, 12 redundant equations arise. These equations result in higher order compatibility equations and they are simply the derivatives of the standard compatibility equations.

Appendix A
Strain Gradient Coefficients for Two and Three Dimensions

This appendix contains the strain gradient coefficients for displacement polynomials of the fourth order. The polynomials for both two and three dimensions are given. The two-dimensional case is given separately for convenience.

Table A.1. The coefficients for the fourth-order three-dimensional displacement functions.

i	Term	a_i for $u(x,y,z)$	b_i for $v(x,y,z)$	c_i for $w(x,y,z)$
1	1	$(u)_0$	$(v)_0$	$(w)_0$
2	x	$(\varepsilon_x)_0$	$(\gamma_{xy}/2 + r)_0$	$(\gamma_{xz}/2 - q)_0$
3	y	$(\gamma_{xy}/2 - r)_0$	$(\varepsilon_y)_0$	$(\gamma_{yz}/2 + p)_0$
4	z	$(q + \gamma_{xz}/2)_0$	$(\gamma_{yz}/2 - p)_0$	$(\varepsilon_z)_0$
5	x^2	$(\varepsilon_{x,x}/2)_0$	$((\gamma_{xy,x} - \varepsilon_{x,y})/2)_0$	$((\gamma_{xz,x} - \varepsilon_{x,z})/2)_0$
6	xy	$(\varepsilon_{x,y})_0$	$(\varepsilon_{y,x})_0$	$((-\gamma_{xy,z} + \gamma_{yz,x} + \gamma_{xz,y})/2)_0$
7	xz	$(\varepsilon_{x,z})_0$	$((\gamma_{xy,z} + \gamma_{yz,x} - \gamma_{xz,y})/2)_0$	$(\varepsilon_{z,x})_0$
8	y^2	$((\gamma_{xy,y} - \varepsilon_{y,x})/2)_0$	$(\varepsilon_{y,y}/2)_0$	$((\gamma_{yz,y} - \varepsilon_{y,z})/2)_0$
9	yz	$((\gamma_{xy,z} - \gamma_{yz,x} + \gamma_{xz,y})/2)_0$	$(\varepsilon_{y,z})_0$	$(\varepsilon_{z,y})_0$
10	z^2	$((\gamma_{xz,z} - \varepsilon_{z,x})/2)_0$	$((\gamma_{yz,z} - \varepsilon_{z,y})/2)_0$	$(\varepsilon_{z,z}/2)_0$
11	x^3	$(\varepsilon_{x,xx}/6)_0$	$((\gamma_{xy,xx} - \varepsilon_{x,xy})/6)_0$	$((\gamma_{xz,xx} - \varepsilon_{x,xz})/6)_0$
12	$x^2 y$	$(\varepsilon_{x,xy}/2)_0$	$(\varepsilon_{y,xx}/2)_0$	$((\gamma_{xz,xy} - \varepsilon_{x,yz})/2)_0$
13	$x^2 z$	$(\varepsilon_{x,xz}/2)_0$	$((\gamma_{xy,xz} - \varepsilon_{x,yz})/2)_0$	$(\varepsilon_{z,xx}/2)_0$
14	y^3	$((\gamma_{xy,yy} - \varepsilon_{y,xy})/6)_0$	$(\varepsilon_{y,yy}/6)_0$	$((\gamma_{yz,yy} - \varepsilon_{y,yz})/6)_0$
15	$y^2 x$	$(\varepsilon_{x,yy}/2)_0$	$(\varepsilon_{y,xy}/2)_0$	$((\gamma_{yz,xy} - \varepsilon_{y,xz})/2)_0$
16	$y^2 z$	$((\gamma_{xy,yz} - \varepsilon_{y,xz})/2)_0$	$(\varepsilon_{y,yz}/2)_0$	$(\varepsilon_{z,yy}/2)_0$
17	z^3	$((\gamma_{xz,zz} - \varepsilon_{z,xz})/6)_0$	$((\gamma_{yz,zz} - \varepsilon_{z,yz})/6)_0$	$(\varepsilon_{z,zz}/6)_0$
18	$z^2 x$	$(\varepsilon_{x,zz}/2)_0$	$((\gamma_{yz,xz} - \varepsilon_{z,xy})/2)_0$	$(\varepsilon_{z,xz}/2)_0$
19	$z^2 y$	$((\gamma_{xz,yz} - \varepsilon_{z,xy})/2)_0$	$(\varepsilon_{y,zz}/2)_0$	$(\varepsilon_{z,yz}/2)_0$
20	xyz	$(\varepsilon_{xyz})_0$	$(\varepsilon_{y,xz})_0$	$(\varepsilon_{z,xy})_0$

(Continued)

Table A.1. (Continued).

i	Term	a_i for $u(x,y,z)$	b_i for $v(x,y,z)$	c_i for $w(x,y,z)$
21	x^4	$(\varepsilon_{x,xxx}/24)_0$	$((\gamma_{xy,xxx} - \varepsilon_{x,xxy})/24)_0$	$((\gamma_{zx,xxx} - \varepsilon_{x,xxz})/24)_0$
22	$x^3 y$	$(\varepsilon_{x,xxy}/6)_0$	$(\varepsilon_{y,xxy}/4)_0$	$((\gamma_{xz,xxy} - \varepsilon_{x,xyz})/6)_0$
23	$x^3 z$	$(\varepsilon_{x,xxz}/6)_0$	$((\gamma_{xy,xxz} - \varepsilon_{x,xyz})/6)_0$	$(\varepsilon_{z,xxx}/6)_0$
24	$x^2 y^2$	$(\varepsilon_{x,xyy}/4)_0$	$(\varepsilon_{y,xyy}/4)_0$	$((\gamma_{xz,xyy} - \varepsilon_{x,yyz})/4)_0$
25	$x^2 z^2$	$(\varepsilon_{x,xzz}/4)_0$	$((\gamma_{yz,xxz} - \varepsilon_{x,xxz})/4)_0$	$(\varepsilon_{z,xxz}/4)_0$
26	$x^2 yz$	$(\varepsilon_{x,xyz}/2)_0$	$(\varepsilon_{y,xxz}/2)_0$	$(\varepsilon_{z,xxy}/2)_0$
27	xy^3	$(\varepsilon_{x,yyy}/6)_0$	$(\varepsilon_{y,xyy}/6)_0$	$((\gamma_{yz,xyy} - \varepsilon_{y,xyz})/6)_0$
28	xz^3	$(\varepsilon_{x,zzz}/6)_0$	$((\gamma_{yz,xzz} - \varepsilon_{z,xyz})/6)_0$	$(\varepsilon_{z,xzz}/6)_0$
29	$xy^2 z$	$(\varepsilon_{x,yyz}/2)_0$	$(\varepsilon_{y,xyz}/2)_0$	$(\varepsilon_{z,xyy}/2)_0$
30	xyz^2	$(\varepsilon_{x,yzz}/2)_0$	$(\varepsilon_{y,xzz}/2)_0$	$(\varepsilon_{z,xyz}/2)_0$
31	y^4	$((\gamma_{xy,yyy} - \varepsilon_{y,xyy})/24)_0$	$(\varepsilon_{y,yyy}/24)_0$	$((\gamma_{yz,yyy} - \varepsilon_{y,yyz})/24)_0$
32	$y^3 z$	$((\gamma_{xy,yyz} - \varepsilon_{y,xyz})/6)_0$	$(\varepsilon_{y,yyz}/6)_0$	$(\varepsilon_{z,yyy}/6)_0$
33	$y^2 z^2$	$((\gamma_{xy,yzz} - \varepsilon_{y,xzz})/4)_0$	$(\varepsilon_{y,yzz}/4)_0$	$(\varepsilon_{z,yyz}/4)_0$
34	yz^3	$((\gamma_{xz,yzz} - \varepsilon_{z,xyz})/6)_0$	$(\varepsilon_{y,zzz}/6)_0$	$(\varepsilon_{z,yzz}/6)_0$
35	z^4	$((\gamma_{xz,zzz} - \varepsilon_{z,xzz})/24)_0$	$((\gamma_{yz,zzz} - \varepsilon_{z,yzz})/24)_0$	$(\varepsilon_{z,zzz}/24)_0$

Table A.2. The coefficients for the fourth-order two-dimensional displacement functions.

i	Term	a_i for $u(x,y)$	b_i for $v(x,y)$
1	1	$(u)_0$	$(v)_0$
2	x	$(\varepsilon_x)_0$	$(\gamma_{xy}/2 + r)_0$
3	y	$(\gamma_{xy}/2 - r)_0$	$(\varepsilon_y)_0$
4	x^2	$(\varepsilon_{x,x}/2)_0$	$((\gamma_{xy,x} - \varepsilon_{x,y})/2)_0$
5	xy	$(\varepsilon_{x,y})_0$	$(\varepsilon_{y,x})_0$
6	y^2	$((\gamma_{xy,y} - \varepsilon_{y,x})/2)_0$	$(\varepsilon_{y,y}/2)_0$
7	x^3	$(\varepsilon_{x,xx}/6)_0$	$((\gamma_{xy,xx} - \varepsilon_{x,xy})/6)_0$
8	x^2y	$(\varepsilon_{x,xy}/2)_0$	$(\varepsilon_{y,xx}/2)_0$
9	xy^2	$(\varepsilon_{x,yy}/2)_0$	$(\varepsilon_{y,xy}/2)_0$
10	y^3	$((\gamma_{xy,yy} - \varepsilon_{y,xy})/6)_0$	$(\varepsilon_{y,yy}/6)_0$
11	x^4	$(\varepsilon_{x,xxx}/24)_0$	$((\gamma_{xy,xxx} - \varepsilon_{x,xxy})/24)_0$
12	x^3y	$(\varepsilon_{x,xxy}/6)_0$	$(\varepsilon_{y,xxx}/6)_0$
13	x^2y^2	$(\varepsilon_{x,xyy}/4)_0$	$(\varepsilon_{y,xxy}/4)_0$
14	xy^3	$(\varepsilon_{x,yyy}/6)_0$	$(\varepsilon_{y,xyy}/6)_0$
15	y^4	$((\gamma_{xy,yyy} - \varepsilon_{y,xyy})/24)_0$	$(\varepsilon_{y,yyy}/24)_0$

References and Other Reading

1. Nagel, E., and Newman, J. R. *Godel's Proof*, New York University Press, New York, 1958. This book is an expansion of an article that appeared in the June 1956 issue of *Scientific American*. It contains figures that show sentences from logic concerning number theory being transformed to integers. This use of self-referential notation is what inspired the development and use of strain gradient notation.

2. Dow, J. O., Bodley, C. S., and Feng, C. C. "An Equivalent Continuum Representation of Structures Composed of Repeated Elements," *Proceedings of the 24th AIAA/ASME/ASCE/AHS Structures, Structural Dynamics, and Materials Conference*, Lake Tahoe, Nev., May 2–4, 1983, pp. 630–640. This is the first presentation of strain gradient notation.

3. Dow, J. O., Su, Z. W., Feng, C. C., and Bodley, C. S. "Equivalent Representation of Structures Composed of Repeated Elements," *AIAA Journal*, Vol. 23, No. 10, Oct. 1985, pp. 1564–1569. This is an expansion of Ref. 2. It is cited because it is more accessible than Ref. 2.

4. Dow, J. O., and Huyer, S. A. "Continuum Models of Space Station Structures," *ASCE Aerospace Journal*, Vol. 2, No. 4, Oct. 1989, pp. 212–230. This paper contains more information concerning the identification of linearly independent strain states for a given structure. This provides valuable assistance in the development of finite element stiffness matrices and finite difference operators.

5. Boresi, A. P. *Elasticity in Engineering Mechanics*, Prentice-Hall, Englewood Cliffs, N.J., 1965. The development of the compatibility equations is presented on page 90.

6. Boresi, A. P., Schmidt, R. J. and Sidebottom, O. M. *Advanced Mechanics of Materials*, 5th ed. John Wiley and Sons, Inc., New York, 1993. The development of the compatibility equations is presented on page 65.

Lesson 5   Problems

1. Develop the shear strain expression using the complete second-order polynomial. Compare your result to Eqs. 5.23 and 5.24.

2. What is the deformation pattern of a square region if it is subjected to the strain state $(\varepsilon_{xy})_0$? That is to say, plot u and v for the case where $(\varepsilon_{xy})_0$ has a value and all of the other coefficients are equal to zero. Use two reference systems. Put the origin for one reference system at the center of the square and the other at the lower left-hand corner.

3. What is the strain distribution for the strain state considered in Problem 2? *Hint*: Substitute the polynomial for $u(x, y)$ into the definition of ε_x given in Eq. 5.5 (also see Eq. 5.24). Note that this is the strain distribution expected in an Euler-Bernoulli beam.

4. Derive an expression for the strain energy for the strain state considered in Problems 2 and 3. *Hint*: The integral that defines the second moment of inertia around the x axis appears.

5. What strain states can be represented by the complete third-order polynomial for u? *Hint*: Use only the x terms. The same question can be asked as follows. Find the strain gradient coefficients for a third-order polynomial in x.

6. Show that the strain gradient representation is indeed a Taylor series representation. *Hint*: Substitute the definition of the strain components and rigid body rotations as functions of displacements into the coefficients of the strain gradient representations. Then compare the results to the standard form of the Taylor series.

Lesson 6

Strain Gradient Representation of Discrete Structures

Purpose of Lesson   To present procedures for identifying and utilizing linearly independent sets of strain gradient quantities that a two-dimensional truss is capable of representing as a way to highlight the role of individual strain states as generalized coordinates.

The most natural coordinate system for describing the movement or displacements of a truss is to identify the change in the location of the individual nodes of the truss. In this lesson, strain gradient notation is used to provide an alternative set of independent coordinates for describing altered configurations of the truss. The nodal movements are expressed in terms of rigid body motions and various types of deformations or strain states. In other words, we use the strain gradient quantities as generalized coordinates.

The strain gradient coefficients are used to extract equivalent continuum parameters from truss structures. These parameters are used to represent the trusses as continua. In a word, we *reverse* the finite element procedure. We represent structures composed of discrete elements as continua instead of representing a continuum as a collection of discrete entities. This concrete application is designed to provide a context where the meaning of the individual strain gradient coefficients are directly related to physical systems. These capabilities are extended to develop an alternative approach for forming finite element stiffness matrices in Part III.

■ ■ ■

In this lesson, the direct relationship between strain gradient coefficients and the displacements of physical systems is demonstrated using truss structures. Since the joints of a truss correspond to the nodes of a finite element, the capabilities developed in this lesson apply directly to the formulation of finite element stiffness matrices (see Note 1).

Before beginning the formal development, the relationship between the strain gradient coefficients and the deformations of a square, four-node truss is presented to illustrate the type of results produced in this lesson. A four-node configuration is capable of representing the five deformations shown in Fig. 1. These deformations correspond to the following five strain states, which were developed in Lesson 5 for a four-node square:

Strain States

$$\varepsilon_x \quad \varepsilon_y \quad \gamma_{xy}$$

$$\varepsilon_{x,y} \quad \varepsilon_{y,x}$$

(6.1)

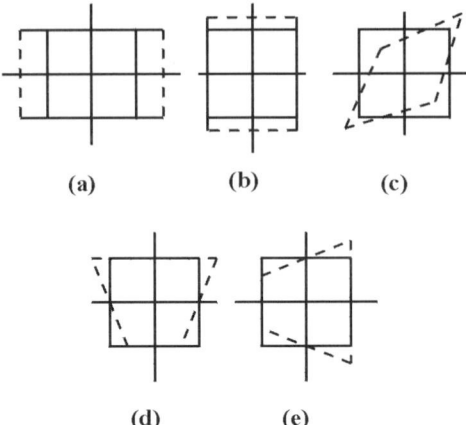

Figure 1. Deformation patterns for four-node configurations.

The strain state $(\varepsilon_x)_0$ is shown in Fig. 1a. This independent deformation mode is represented by having the two horizontal bars deform the same amount. Since the two bars have the same length, the two bars have the same amount of strain. Thus, this is a state of constant strain in the x direction. The vertical bars are undeformed.

The state of constant strain shown in Fig 1a is contrasted with the higher order strain state $(\varepsilon_{x,y})_0$ shown in Fig. 1d. In this figure, we see that the two horizontal bars have different deformations. The top bar is in tension and the bottom bar is in compression. Since these two bars with different values of strain are separated by a distance in the y direction, there is a rate of change in $(\varepsilon_x)_0$ with respect to the y direction. That is to say, the strain state existing in the overall structure can be denoted as $(\varepsilon_{x,y})_0$.

This comparison has provided a concrete example of the direct relationship between a discrete physical system and two strain gradient coefficients. An analogous situation is seen when $(\varepsilon_y)_0$ and $(\varepsilon_{y,x})_0$, as shown in Figs. 1b and 1e, are compared. A state of constant shear is shown in Fig. 1c. It is discussed in detail later when the procedures are developed for formally identifying the deformation patterns shown in Fig. 1.

Equivalent Continuum Representation

The close relationship between the physical behavior and the strain gradient coefficients enables us to represent a truss as an equivalent beam, as shown in Fig. 2. As mentioned earlier, the strain gradient coefficient $(\varepsilon_{x,y})_0$ corresponds to the strain distribution found in an Euler-Bernoulli beam. If we transform the strain energy for the truss from nodal coordinates to strain gradient coordinates, we can extract a parameter that corresponds to $(EI)_{\text{eq}}$, the flexural stiffness of a beam. This parameter is used to represent the truss as a continuous beam, as described next.

In Lesson 2, the governing differential equation for a prismatic Euler-Bernoulli beam was found as

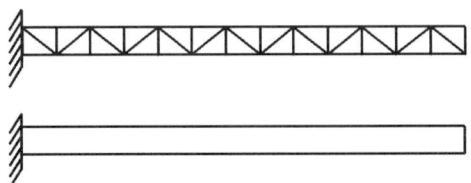

Figure 2. A truss represented as a continuous beam.

Governing Equation for an Euler-Bernoulli Beam

$$EIv_{,xxxx} - P(x) = 0 \qquad (6.2)$$

When the value for the flexural stiffness, $(EI)_{\text{eq}}$, found for the truss is substituted into Eq. 6.2, we have a continuous representation of the truss. In later examples, this continuous representation is found to provide results that accurately represent the truss being modeled.

This representation of truss structures as continuous elements can be viewed as an inversion of the finite element method. Instead of representing the continuum as a collection of discrete elements, this approach treats a structure composed of discrete elements as a continuum.

The availability of equivalent continuum properties enables a sophisticated analysis technique to be applied to complex structures during the early stages of design when there is no solid design. Because of the uncertainty of the design, there usually are limited resources and little time for screening any one of the several candidate designs being considered. The equivalent continuum approach is ideal for such a situation. The obviously good and bad candidates can be identified early in the design process with little effort by extracting the equivalent continuum parameters from each of the candidate designs. Strain gradient notation was originally formulated as a way of developing the preliminary design tool just discussed.

The deformation patterns shown in Fig. 1 apply equally well to a four-node finite element. This means that the capabilities developed in this lesson for truss structures apply directly to the formulation of finite element stiffness matrices. Furthermore, the use of the physically based notation enables the reformulation of the *finite difference* method so that it can represent the same complex geometries as the finite element method. The extended capabilities of the finite difference method in Part IV are used to develop error analysis procedures for the finite element method in Part V. Thus, the procedure for extracting the equivalent continuum parameters from truss structures provides the basis for reformulating the finite element and finite difference methods.

Extraction Procedure for Equivalent Continuum Parameters

The idea behind the analysis of a discrete structure as continuous elements is illustrated in Fig. 3 with a beamlike structure built up from repeated elements. In this figure, a repeated

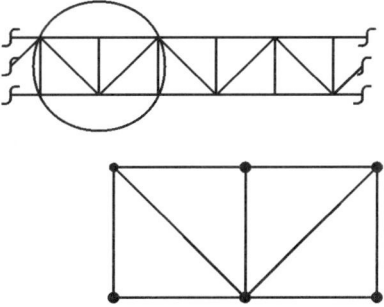

Figure 3. A repeated truss element.

element extracted from the overall structure is shown in the blowup. This building block element is treated as a differential volume. The continuum characteristics of the truss are extracted from this repeated element.

In brief, the equivalent continuum properties for a discrete structure are found as follows. After the repeated element is isolated from the discrete structure, the independent strain states that it can represent are identified. This information is used to form strain energy expressions in terms of the independent strain gradient quantities. These expressions are then compared to the strain energy expressions for the continuum representations to identify the equivalent continuum properties. The specific steps are outlined in Table 1.

Table 1. Steps for forming equivalent continuum parameters.

---

1. Isolate a repeated element.
2. Identify the strain gradient quantities associated with the structure.
3. Form the transformation from nodal to strain gradient quantities.
4. Transform the strain energy expression to a function of strain gradient quantities.
5. Identify the equivalent continuum parameters.

The steps of particular interest for use in later developments are operations 2 through 4. In step 2, the linearly independent strain gradient quantities required to represent a repeated element of a given configuration are identified. For a two-dimensional truss, the polynomials required to represent the structure are found by simply aligning the nodal configuration with Pascal's triangle. For trusses with a nonrepetitive nodal pattern, the process of identifying a linearly independent set of strain gradient coefficients is not completely obvious, and procedures from linear algebra are used to identify the set of independent quantities. These techniques are discussed in the references.

In step 3, the quantities identified in step 2 are used to form a coordinate transformation from nodal coordinates to strain gradient quantities. This is a straightforward process, but it is at the heart of the procedure for finding equivalent continuum parameters and, subsequently, finite element stiffness matrices and finite difference templates. In the formulation of finite element and finite difference quantities, this transformation must be inverted. Therefore, a linearly independent set of quantities must be identified.

In step 4, the strain energy of the truss expressed in nodal coordinates is transformed to a function of strain gradient quantities. The resulting expression is equated to the strain energy expression for a continuous element. This enables the equivalent continuum coefficients to be identified and extracted. A similar procedure is used to compute the strain energy contained in a portion of the continuum as a result of the deformations due to a specific strain state. This ability to compute the strain energy content of an arbitrary region under a well-defined strain state is the key to the alternative finite element formulation procedure presented in the first lesson of Part III.

Before presenting a complete example of the process of extracting the equivalent continuum parameters from a truss, the approach for identifying the linearly independent set of strain gradient quantities associated with a specific truss is presented. The transformation from nodal displacements to strain gradient quantities is then found.

Identification of Polynomial Representations

The set of linearly independent strain gradient quantities associated with the seven truss configurations shown in Fig. 4 are identified and interpreted in this section. The nodal

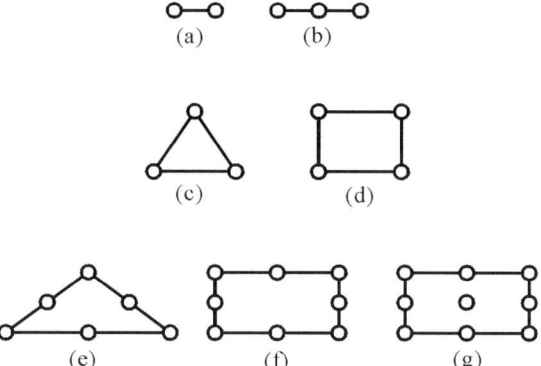

Figure 4. Sample truss configurations.

patterns or topology of these trusses match the shape and nodal locations of finite elements that are developed in Part III. In addition to relating to finite elements, the nine-node configuration matches the topology of the nine-node central difference template that is studied in later developments.

The algebraic terms required for the polynomial representation of each of the trusses shown in Fig. 4 are identified by relating the nodal patterns of the trusses to Pascal's triangle. Figure 5a shows the standard form of Pascal's triangle up to the fourth order. In Fig. 5b, Pascal's triangle is rotated so the terms in x are located along the x axis and the terms in y are along the y axis. The second form of Pascal's triangle provides a convenient tool with which to identify the polynomial terms needed to represent the nodal

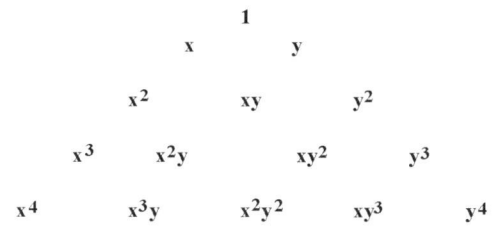

Figure 5. Pascal's triangle.

displacements of the trusses shown in Fig. 4. This is done by superimposing the nodal pattern of the trusses on the second form of Pascal's triangle. The trusses shown in Fig. 4 are now analyzed one at a time.

Single-Bar Element

The polynomial required to represent the single-bar element is now identified. In Fig. 6, the single, two-node bar element is superimposed on Fig. 5b. The first node is aligned with the constant term at the origin of the xy coordinate system. This location is dictated by the completeness requirements on the polynomial representations. Simply put, the polynomials must contain the lowest possible order of algebraic terms consistent with a linearly independent set. The second node is superimposed on the x term because the x axis was arbitrarily chosen as the local coordinate system. Thus, the local coordinates of the bar element can be represented with a linear polynomial of the following form:

$$u = a_1 + a_2x \qquad (6.3)$$

When the arbitrary coefficients of Eq. 6.3 are replaced by the strain gradient quantities contained in Table 1 or Appendix A of Lesson 5, the polynomial representation of a single bar in the local x coordinate system is

$$u = (u_{rb})_0 + (\varepsilon_x)_0 x \qquad (6.4)$$

Thus, this simplest truss, a two-degree-of-freedom bar, can represent two independent strain states, namely,

---

Independent Strain States

$$(u_{rb})_0 \quad (\varepsilon_x)_0 \qquad (6.5)$$

The quantities contained in Eq. 6.5 represent the rigid body motion and the normal strain in the x direction, respectively.

The meaning of these terms is illustrated in Fig. 7. In Fig. 7a, the bar is given a rigid body motion. The bar moves, with no internal strain, to the right with a positive rigid body displacement. The constant normal strain condition is shown in Fig. 7b. Thus, any possible motion of the bar along the x axis can be represented as a linear combination of these two independent strain states. In this capacity, the strain gradient quantities serve as

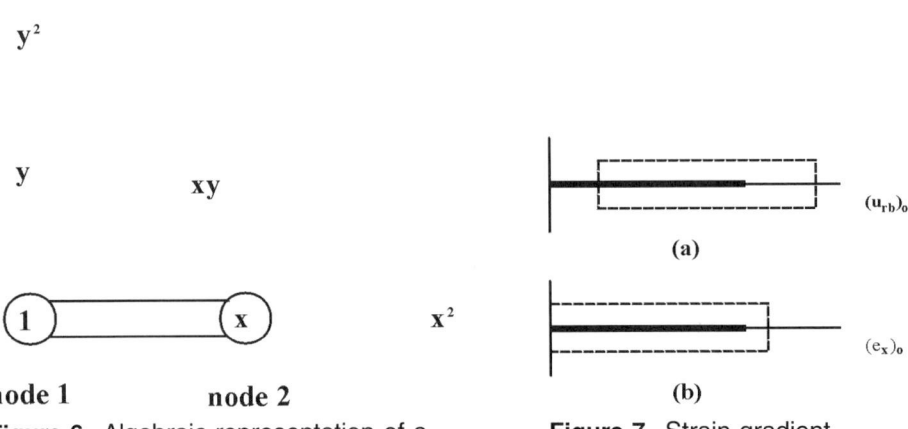

Figure 6. Algebraic representation of a bar element.

Figure 7. Strain gradient displacement patterns.

generalized coordinates. The operation just completed constitutes step 2 of the process for determining the equivalent continuum parameters shown in Table 1.

Equation 6.4 is used to form a coordinate transformation from the nodal displacements to the strain gradient quantities that constitute step 3 of the process of extracting the equivalent continuum parameters. This is accomplished by substituting the nodal locations of the two nodes of the bar shown in Fig. 6 into Eq. 6.4. When the nodal locations are substituted into Eq. 6.4, the results are

$$
\begin{aligned}
u(0) &= (u_{rb})_0 \\
u(L) &= (u_{rb})_0 + L(\varepsilon_x)_0
\end{aligned}
\tag{6.6}
$$

where $u(0)$ is the displacement of node 1, which is denoted as u_1, and $u(L)$ is the displacement of node 2, which is denoted as u_2

Equation 6.6 can be written in matrix form as

Nodal Displacement to Strain Gradient Transformation

$$
\left\{ \begin{array}{c} u_1 \\ u_2 \end{array} \right\} = \left[\begin{array}{cc} 1 & 0 \\ 1 & L \end{array} \right] \left\{ \begin{array}{c} (u_{rb})_0 \\ (\varepsilon_x)_0 \end{array} \right\}
\tag{6.7}
$$

Equation 6.7 constitutes the transformation from nodal displacement coordinates to the generalized coordinates given by the strain gradient coefficients. This transformation is used to transform the strain energy expression of the truss from nodal quantities to strain gradient quantities, which enables us to extract the equivalent continuum properties.

Later, when we form finite element stiffness matrices, this transformation is inverted and used to transform a strain energy expression from strain gradient quantities to nodal coordinates. The need to invert this transformation while at the same time using the lowest order possible polynomial terms explains the use of Pascal's triangle. For these simple configurations, the polynomials identified using Pascal's triangle satisfy the completeness requirements and form transformations that can be inverted.

When the finite element stiffness matrices are computed, we often find it more convenient to use the centroid of the element as the origin of the local reference system. When the displacement polynomials for the single bar element are evaluated with the origin at the center, the displacement of the two nodes as a function of the strain gradient quantities becomes

---

Alternative Coordinate Transformation

$$
\left\{ \begin{array}{c} u_1 \\ u_2 \end{array} \right\} = \left[\begin{array}{cc} 1 & -L/2 \\ 1 & L/2 \end{array} \right] \left\{ \begin{array}{c} (u_{rb})_0 \\ (\varepsilon_x)_0 \end{array} \right\}
\tag{6.8}
$$

In this case, the two independent displacements are shown in Fig. 8. The two displacement patterns are similar to those shown in Fig. 7. The rigid body displacement is obviously identical. The strain state $(\varepsilon_x)_0$ is shown in Fig. 8b as it exists in the new coordinate system. The constant strain condition is produced by having the two nodes move apart.

The transformation matrices in Eqs. 6.7 and 6.8 are called $[\Phi]$ matrices. The general form of Eqs. 6.7 and 6.8 can be written as

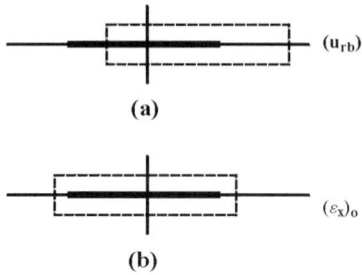

(a)

(b)

Figure 8. Strain gradient displacement patterns.

General Transformation Equation

$$\{d\} = [\Phi]\{\varepsilon_,\}$$

where $\{d\}$ are nodal displacements.

$[\Phi]$ is the transformation matrix.　　　　(6.9)

$\{\varepsilon_,\}$ is the vector of strain gradient coefficients.

Each of the columns of the transformation matrix can be interpreted as a displacement pattern associated with an individual strain gradient quantity. For the transformation matrices given in Eqs. 6.7 and 6.8, the first column represents the rigid body displacement of the two-node bar. The second column represents the constant strain states. The second columns differ in Eqs. 6.7 and 6.8 because the two cases have a different origin. These vectors play a role similar to eigenvectors in modal analyses, although these vectors are not orthogonal. That is to say, the strain gradient coefficients are generalized coordinates.

Two-Bar Element

The polynomial required to represent the simple three-degree-of-freedom structure composed of two constant strain bars as shown in Fig. 4b is now identified. The independent strain gradient quantities associated with the representation are then extracted. The transformation from nodal coordinates to strain gradient notation, the $[\Phi]$ matrix, is formed using this information.

In Fig. 9, the two-bar element is superimposed on Fig. 5b. As can be seen, this superposition shows that the nodal displacements of this structure can be represented by the following quadratic polynomial:

$$u = a_1 + a_2x + a_4x^2 \qquad (6.10)$$

The coefficient of the x^2 term is denoted by a_4 so that the notation of Eq. 6.10 corresponds to the notation of Lesson 5. When the strain gradient quantities from Lesson 5 are used to replace the arbitrary coefficients of Eq. 6.10, the result is

$$u = (u_{rb})_0 + (\varepsilon_x)_0 x + [(\varepsilon_{x,x})_0/2]x^2 \qquad (6.11)$$

The two-bar structure is represented by the following three independent strain states:

Independent Strain States

$$(u_{rb})_0 \quad (\varepsilon_x)_0 \quad (\varepsilon_{x,x})_0 \tag{6.12}$$

Before we interpret these generalized coordinates, the $[\Phi]$ matrix associated with this configuration is formed for the case where the local origin is located on the center node, as shown in Fig. 10. The individual equations contained in this coordinate transformation are formed by substituting the nodal coordinates of the three nodes into Eq. 6.11, one at a time. The result of this operation is

$$u(-L) = (u_{rb})_0 - L(\varepsilon_x)_0 + [L^2/2](\varepsilon_{x,x})_0$$
$$u(0) = (u_{rb})_0 \tag{6.13}$$
$$u(+L) = (u_{rb})_0 + L(\varepsilon_x)_0 + [L^2/2](\varepsilon_{x,x})_0$$

When the displacement at $x = -L$ is denoted as u_1, etc., Eq. 6.13 can be written in matrix notation as

Nodal Displacement to Strain Gradient Transformation

$$\begin{Bmatrix} u_1 \\ u_2 \\ u_3 \end{Bmatrix} = \begin{bmatrix} 1 & -L & L^2/2 \\ 1 & 0 & 0 \\ 1 & L & L^2/2 \end{bmatrix} \begin{Bmatrix} (u_{rb})_0 \\ (\varepsilon_x)_0 \\ (\varepsilon_{x,x})_0 \end{Bmatrix} \tag{6.14}$$

In this case, the $[\Phi]$ matrix, the transformation from nodal displacements to strain gradient quantities, is given as

Coordinate Transformation Matrix

$$[\Phi] = \begin{bmatrix} 1 & -L & L^2/2 \\ 1 & 0 & 0 \\ 1 & L & L^2/2 \end{bmatrix} \tag{6.15}$$

$\mathbf{y}^2$

$\mathbf{y}$ $\qquad\qquad \mathbf{xy}$

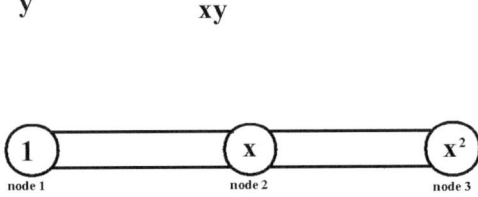

Figure 9. Algebraic representation of a two-bar element.

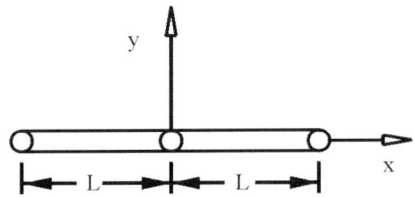

Figure 10. Three-node bar with coordinate system.

The meaning to the three independent strain states given in Eqs. 6.12 and 6.14 is now illustrated. The displacement pattern associated with $(u_{rb})_0$ is contained in column 1 of Eq. 6.15. It is denoted as $\{\Phi_1\}$ and shown in Fig. 11a. As can be seen, this represents the expected rigid body motion. The displacement pattern associated with $(\varepsilon_x)_0$ is contained in column 2 of Eq. 6.15. This deformation pattern is shown in Fig. 11b. In this case, the center node does not move and the two outside nodes move away from the center by the same amount. This causes both bars to contain the same level of constant strain. Thus, the strain state coefficient that denotes a constant strain condition in the x direction produces constant strain in all of the elements of the truss.

The displacement pattern for the strain state $(\varepsilon_{x,x})_0$ is shown in Fig. 11c. In this case, the center node does not move. The node on the right side of the element moves to the right. This puts the right-hand bar in tension. The node on the left side of the element moves to the right and puts the bar in compression. Thus, the two bars, one in compression and one in tension, are exhibiting different values of constant strain. This means that $(\varepsilon_x)_0$ changes from compression to tension as we move in the positive x direction. This shows that the strain gradient coefficient $(\varepsilon_{x,x})_0$ produces a variation in the strain $(\varepsilon_x)_0$ in the x direction in the truss. This demonstrates the meaning of $(\varepsilon_{x,x})_0$ for the case of a truss.

The study of these two examples of line elements was designed to introduce the use of Pascal's triangle in identifying the algebraic terms required to represent a structure. The identification of the polynomial representations enables the associated strain gradient quantities to be ascertained. Knowledge of these two characteristics enables us to form the $[\Phi]$ matrix, the transformation from nodal coordinates to strain gradient quantities. The column vectors of the $[\Phi]$ matrix, which contains the independent displacement patterns, were related to the individual strain gradient quantities to reinforce the recognition of the generalized coordinate nature of these quantities.

The discussion of the two deformation modes of the two-bar truss has highlighted the meaning of the higher order strain gradient quantity $(\varepsilon_{x,x})_0$. The rate of change of $(\varepsilon_x)_0$ in the x direction could easily be seen by the distinctive change in $(\varepsilon_x)_0$ in the two bars that constitute the second truss. In retrospect, we can see why the strain state $(\varepsilon_{x,x})_0$ could not exist in a structure consisting of a single constant strain bar. Consequently, the meaning of $(\varepsilon_{x,xx})_0$ in a three-bar line element should be clear. There would be an increasing or decreasing value of $(\varepsilon_x)_0$ in each of the three-bar elements.

Now that we have studied the one-dimensional case, we examine the two-dimensional case.

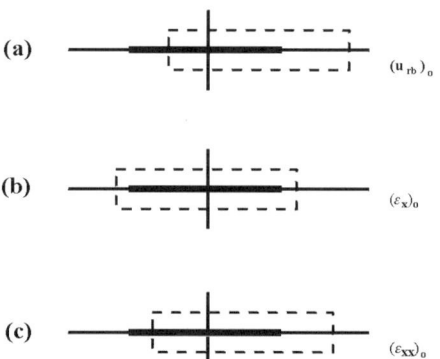

Figure 11. Strain gradient displacement patterns.

Three-Node Triangular Truss

In Fig. 12, a three-bar triangular truss is shown superimposed on the rotated form of Pascal's triangle. Although this structure has the same number of nodes as the two-bar element, the two-dimensional nature of this case is clear. This structure contains six degrees of freedom, three displacements in the x direction, and three displacements in the y direction. Thus, the displacements in the two coordinate directions can be represented by the following linear or first-order polynomials in x and y:

$$u = a_1 + a_2x + a_3y$$
$$v = b_1 + b_2x + b_3y$$

(6.16)

Although each of these representations contains three terms, note that the linear terms in y present in Eq. 6.16 replace the quadratic term in x contained in Eqs. 6.10 and 6.11 because of the topology, or layout, of the nodes.

When the strain gradient quantities from Lesson 5 are used to replace the arbitrary coefficients of Eq. 6.16, the result is

$$u = (u_{rb})_0 + (\varepsilon_x)_0 x + (\gamma_{xy}/2 - r_{rb})_0 y$$
$$v = (v_{rb})_0 + (\gamma_{xy}/2 + r_{rb})_0 x + (\varepsilon_y)_0 y$$

(6.17)

This three-node, six-degree-of-freedom triangular structure is represented by the following six independent strain states contained in Eq. 6.17:

Independent Strain States

$$(u_{rb})_0 \quad (v_{rb})_0 \quad (r_{rb})_0$$
$$(\varepsilon_x)_0 \quad (\varepsilon_y)_0 \quad (\gamma_{xy})_0$$

(6.18)

The first three terms are rigid body motions and the second three terms are constant strain states.

The transformation from nodal displacements to strain gradient quantities is formed by substituting the location of the nodes into Eq. 6.17. For the configuration shown in Fig. 12 with the node on the x axis having the coordinates $(a,0)$ and the leg on the y axis having the coordinates $(0,b)$, the transformation $[\Phi]$ is

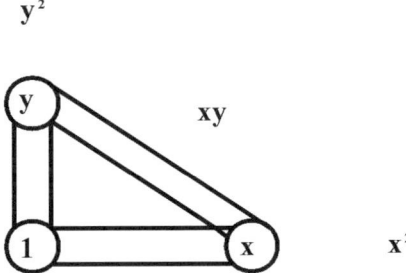

Figure 12. Algebraic representation.

Nodal Displacement to Strain Gradient Transformation

$$
\begin{Bmatrix} u_1 \\ u_2 \\ u_3 \\ v_1 \\ v_2 \\ v_3 \end{Bmatrix}
=
\begin{bmatrix}
1 & 0 & 0 & 0 & 0 & 0 \\
1 & 0 & 0 & a & 0 & 0 \\
1 & 0 & -b & 0 & 0 & b/2 \\
0 & 1 & 0 & 0 & 0 & 0 \\
0 & 1 & a & 0 & 0 & a/2 \\
0 & 1 & 0 & 0 & b & 0
\end{bmatrix}
\begin{Bmatrix} (u_{rb})_0 \\ (v_{rb})_0 \\ (r_{rb})_0 \\ (\varepsilon_x)_0 \\ (\varepsilon_y)_0 \\ (\gamma_{xy})_0 \end{Bmatrix}
\tag{6.19}
$$

The displacements associated with each of the generalized strain gradient coordinates are shown in Fig. 13. The rigid body motions are shown in Figs. 13a–c. They are similar to those shown in Fig. 1 of Lesson 5, as expected.

The constant normal strain in the x direction is shown in Fig. 13d. In this representation of $(\varepsilon_x)_0$, only the horizontal bar is deformed. It exhibits constant strain. If this were a continuous element, the movement of the hypotenuse of the triangle would induce an equal strain at every point in the region. A similar discussion holds for the constant normal strain in the y direction shown in Fig. 13e.

The most significant difference between this two-dimensional example and the one-dimensional examples previously discussed is the existence of the shear strain shown in Fig. 13f. Note that it consists of a "closing" of the right angle at the origin. In terms of the bar elements, this imposes a compression in the angled bar in the truss.

The relationship between the rigid body rotation and the shear stress can be seen by comparing Figs. 13c and 13e. In the rigid body rotation, both arms of the truss also change angles but the amounts are in the same direction and of the same magnitude so no strain is induced.

The similarities between the rigid body rotation and the shear strain can be seen when we compare the equations from linear elasticity for the two quantities. They are reproduced here for convenience as

Rigid Body Rotation–Shear Strain Comparison

$$
(r_{rb})_0 = \frac{1}{2}\left(\frac{\partial u}{\partial y} - \frac{\partial v}{\partial x}\right)
$$
$$
(\gamma_{xy})_0 = \left(\frac{\partial u}{\partial y} + \frac{\partial v}{\partial x}\right)
\tag{6.20}
$$

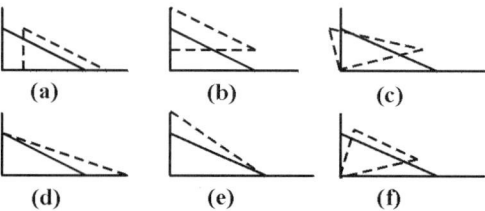

(a) (b) (c)

(d) (e) (f)

Figure 13. Strain gradient displacement patterns.

Both quantities contain the same two terms as a linear combination. The term $\partial v/\partial x$ can be seen to represent the rotation of the element along the x axis. That is to say, the displacement in the y direction of the line on the x axis increases linearly with x. The other term, $\partial u/\partial y$, represents the rotation of the element along the y axis. When the two arms rotate in the same direction, no deformation occurs, and we have the rigid body rotation. When the two bars rotate in the opposite direction, a deformation is produced and we have shear strain.

This example has introduced an approach for finding the algebraic terms required to represent the displacements in the two coordinate directions for a planar structure. The need for the linear terms in this three-node planar element instead of the quadratic representation in x that was required for the three-node bar was illustrated. Also, this example highlights the close relationship between the rigid body rotation and the constant strain expression. A slightly more complex two-dimensional case is now considered.

Four-Node Rectangular Truss

In Fig. 14, a four-node rectangular configuration is shown superimposed on the rotated Pascal's triangle. This configuration contains eight degrees of freedom so the polynomial representations in each of the coordinate directions must contain four terms. The first three terms in each direction are the constant terms and the two linear terms in x and y. We have three possible candidates for the fourth term, namely, x^2, xy, and y^2. The proper choice is the xy term. If we had the alternate truss configurations, as shown in Fig. 15, the other two algebraic terms, x^2 and y^2, would be the obvious candidates for the two cases shown. It could also be noted the choice of xy makes the polynomial symmetric. The significance of this is discussed when we form the finite element stiffness matrices in Part III. Nodal topologies similar to those shown in Fig. 15 are utilized in the reformulation of the finite difference method presented in Part IV.

The strain gradient representation of these polynomials is as follows:

$$u = (u_{rb})_0 + (\varepsilon_x)_0 x + (\gamma_{xy}/2 - r_{rb})_0 y + (\varepsilon_{x,y})_0 xy$$
$$v = (v_{rb})_0 + (\gamma_{xy}/2 + r_{rb})_0 x + (\varepsilon_y)_0 y + (\varepsilon_{y,x})_0 xy$$

(6.21)

These representations contain the following eight independent strain states:

Independent Strain States

$$(u_{rb})_0 \quad (v_{rb})_0 \quad (r_{rb})_0$$
$$(\varepsilon_x)_0 \quad (\varepsilon_y)_0 \quad (\gamma_{xy})_0$$
$$(\varepsilon_{x,y})_0 \quad (\varepsilon_{y,x})_0$$

(6.22)

The transformation from nodal displacements to strain gradient quantities is formed for the square with an edge length of $2a$ to give

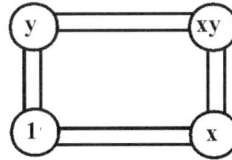

Figure 14. Algebraic representation of four-node configuration.

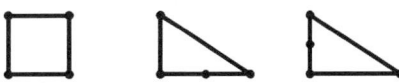

Figure 15. Alternate four-node configurations.

Nodal Displacement to Strain Gradient Transformation

$$
\begin{Bmatrix} u_1 \\ u_2 \\ u_3 \\ u_4 \\ v_1 \\ v_2 \\ v_3 \\ v_4 \end{Bmatrix} =
\begin{bmatrix}
1 & 0 & a & -a & 0 & -a/2 & a^2 & 0 \\
1 & 0 & a & a & 0 & -a/2 & -a^2 & 0 \\
1 & 0 & -a & a & 0 & a/2 & a^2 & 0 \\
1 & 0 & -a & -a & 0 & a/2 & -a^2 & 0 \\
0 & 1 & -a & 0 & -a & -a/2 & 0 & a^2 \\
0 & 1 & a & 0 & -a & a/2 & 0 & -a^2 \\
0 & 1 & a & 0 & a & a/2 & 0 & a^2 \\
0 & 1 & -a & 0 & a & -a/2 & 0 & a^2
\end{bmatrix}
\begin{Bmatrix} (u_{rb})_0 \\ (v_{rb})_0 \\ (r_{rb})_0 \\ (\varepsilon_x)_0 \\ (\varepsilon_y)_0 \\ (\gamma_{xy})_0 \\ (\varepsilon_{x,y})_0 \\ (\varepsilon_{y,x})_0 \end{Bmatrix}
\qquad (6.23)
$$

The rigid body motions for this case were discussed in Lesson 5. The constant strain states are shown in Fig. 16. The two constant normal strains shown in Figs. 16a and 16b are as expected. In the case of a positive value for ε_x, the two bars aligned along the x axis are elongated as shown. The normal strain ε_y is similarly distorted in the y direction.

The shear deformation shown in Fig. 16c exhibits the following characteristics. The bars on the edges are not deformed. However, any bars across the diagonals would either be compressed or stretched. The three possible ways to provide shear resistance can be seen in the figure. There could be a single bar across either of the diagonals or there could be bars across both diagonals.

The two higher order strain states, $\varepsilon_{x,y}$ and $\varepsilon_{y,x}$, are shown in Fig. 17. These two deformation patterns give a clear meaning to these higher order strain gradient terms. In Fig. 17a, we can see that the top bar is in tension, i.e., it is exhibiting a positive normal strain ε_x. The bottom bar is in compression, i.e., it is exhibiting a negative normal strain ε_x. This can be interpreted to mean that the two bars are representing two different levels of normal strain ε_x. They are separated by a distance in the y direction, so there is a rate of change in ε_x in the y direction. The quantity $\varepsilon_{x,y}$ has a value.

A similar argument could be made using Fig. 17b to explain the meaning of the term $\varepsilon_{y,x}$.

In addition to providing further demonstrations of the use of strain gradient notation, this example was primarily presented to show the treatment of nodal configurations that are described by incomplete polynomials and to give meaning to higher order strain gradient terms. The discussion of these terms and of $\varepsilon_{x,x}$ in the previous example should equip the reader to interpret the other higher order terms.

Higher Order Configurations

The previous examples were designed as much to highlight the characteristics of strain gradient notation as to identify the displacement polynomials required to represent the

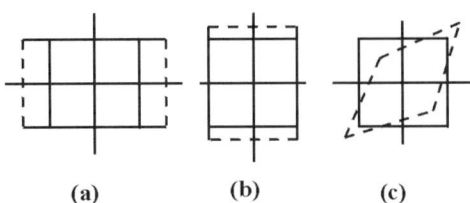

(a) (b) (c)

Figure 16. Displacement patterns for constant strains.

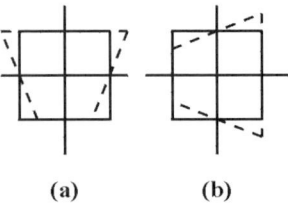

(a) (b)

Figure 17. Deflection patterns for flexure strains.

displacements of the configurations discussed. This section simply identifies the polynomials required to represent the six-node, the nine-node, and the eight-node configurations. The deformations of the new strain gradient states present in these higher order terms are not discussed in detail. In the preceding list, the eight-node configuration follows the nine-node case. This is *not* a typing error. By analyzing the nine-node case first, the approach for the eight-node case becomes much clearer. The three topologies are superimposed on Pascal's triangle in Fig. 18. They are discussed in turn.

In Fig. 18a, we can see that the displacements for the six-node triangle can be represented by the following polynomials:

$$u = a_1 + a_2x + a_3y + a_4x^2 + a_5xy + a_6y^2$$
$$v = b_1 + b_2x + b_3y + b_4x^2 + b_5xy + b_6y^2$$

$$(6.24)$$

These are complete second-order polynomials. All of the quadratic terms are present. In the next lesson, the significance of complete polynomial representations is discussed in detail. Briefly, we see that the use of incomplete displacement polynomials introduces modeling errors into individual finite elements. The strain representations resulting from incomplete displacement polynomials are also incomplete.

These incomplete polynomials introduce an error called parasitic shear that cannot always be removed using current practices. In Lesson 9 parasitic shear is shown to be a false representation of shear strain when an element based on incomplete polynomials represents a flexure stress. Incomplete polynomials also produce strain representations that are not invariant with rotation. This lack of invariance means that the strains in elements based on incomplete polynomials can depend on the orientation of the element. This modeling defect is demonstrated and discussed in the next lesson.

In Fig. 18b, the polynomials required to represent the displacements in the nine-node configuration are identified. The polynomial terms consist of the complete second-order terms, the two third-order terms, x^2y and xy^2, and the fourth order term, x^2y^2. The strain gradient components are identified after the eight-node case is discussed.

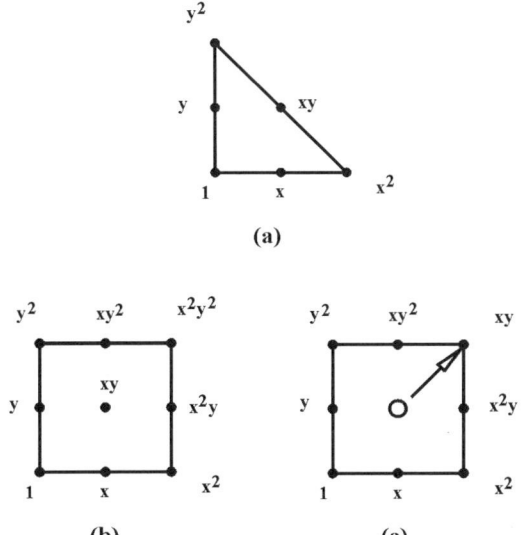

Figure 18. Algebraic representations of higher order elements.

In Fig. 18c, we can see that in the eight-node configuration no node is associated with the second-order term xy. In the eight-node element, the xy term must be included because of completeness requirements. The single fourth-order term is eliminated to make room for the second-order term. By retaining the two third-order terms, symmetry is retained in the polynomial. As mentioned earlier, the place of symmetry in the polynomial representations is discussed in Lesson 7.

The strain gradient quantities contained in the representation of the six-, eight- and nine-node elements are now identified. The strain gradient terms associated with the displacement polynomials for these three elements are found using the displacement representations developed in Lesson 5 to be

---

Six-Node Displacement Polynomial

$$u = (u_{rb})_0 + (\varepsilon_x)_0 x + \left(\frac{\gamma_{xy}}{2} - r_{rb}\right)_0 y + \frac{1}{2}(\varepsilon_{x,x})_0 x^2 + (\varepsilon_{x,y})_0 xy + \frac{1}{2}(\gamma_{xy,y} - \varepsilon_{y,x})_0 y^2$$

$$v = (v_{rb})_0 + \left(\frac{\gamma_{xy}}{2} + r_{rb}\right)_0 x + (\varepsilon_y)_0 y + \frac{1}{2}(\gamma_{xy,x} - \varepsilon_{x,y})_0 x^2 + (\varepsilon_{y,x})_0 xy + \frac{1}{2}(\varepsilon_{y,y})_0 y^2$$

(6.25)

Eight-Node Displacement Polynomials

$$u = \text{Six-Node Terms} + \frac{1}{2}(\varepsilon_{x,xy})_0 x^2 y + \frac{1}{2}(\varepsilon_{x,yy})_0 xy^2$$

$$v = \text{Six-Node Terms} + \frac{1}{2}(\varepsilon_{y,xx})_0 x^2 y + \frac{1}{2}(\varepsilon_{y,xy})_0 xy^2$$

(6.26)

Nine-Node Displacement Polynomials

$$u = \text{Eight-Node Terms} + \frac{1}{4}(\varepsilon_{x,xyy})_0 x^2 y^2$$

$$v = \text{Eight-Node Terms} + \frac{1}{4}(\varepsilon_{y,xxy})_0 x^2 y^2$$

(6.27)

The strain gradient polynomial representations of the six-, eight-, and nine-node configurations have been identified. We can now extract the strain gradient states that these polynomials are capable of representing. The six-node configuration can represent the following 12 strain states:

Six Nodes—Independent Strain States

$$(u_{rb})_0 \quad (v_{rb})_0 \quad (r_{rb})_0$$
$$(\varepsilon_x)_0 \quad (\varepsilon_y)_0 \quad (\gamma_{xy})_0$$
$$(\varepsilon_{x,x})_0 \quad (\varepsilon_{x,y})_0$$
$$(\varepsilon_{y,x})_0 \quad (\varepsilon_{y,y})_0 \quad\quad\quad (6.28)$$
$$(\gamma_{xy,x})_0 \quad (\gamma_{xy,y})_0$$

The displacement polynomials for the six-node configuration are of complete second order. This twelve-degree-of-freedom representation contains the three rigid body motions, the three constant strain states, and the six linear variations of the three strains as the 12 strain gradient components. This strain model consists of the complete first-order strain representations. The differences between the order of the displacement polynomials and the strain representations have important ramifications with respect to the strain extraction and error analysis procedures. The effects of these differences are discussed in later lessons.

The strain representation of the eight-node configuration contains the complete second-order polynomial terms plus the following third-order terms:

Eight Node—Additional Independent Strain States

$$(\varepsilon_{x,xy})_0 \quad (\varepsilon_{x,yy})_0$$
$$(\varepsilon_{y,xx})_0 \quad (\varepsilon_{y,xy})_0 \quad\quad\quad (6.29)$$

A ten-node triangle is required to produce a complete second-order representation of the strains. The nine-node element contains two more terms than the eight-node element, but the additional two terms are third-order strain or fourth-order displacement terms. This is required because of the topology of the nodal pattern. This idea is shown in Fig. 18 and is included as part of the problems at the end of this lesson. The additional terms are

Nine Node—Additional Independent Strain States

$$(\varepsilon_{x,xyy})_0 \quad (\varepsilon_{y,xxy})_0 \quad\quad\quad (6.30)$$

The strain states associated with the various nodal two-dimensional configurations just discussed are used throughout the remaining lessons in the discussion of the formulation and modeling characteristics of finite elements. The nine-node configuration is also used to formulate nine-node central difference templates that serve as the basis for the finite difference method. In other words, *the polynomial representations just formed for the displacements in terms of the strain gradient quantities provide the basis for the remainder of the developments contained in this book.*

Formulation of Equivalent Continuum Properties

In the previous section, strain gradient notation was used to developed coordinate transformations [Φ] (see Eqs. 6.7, 6.14, 6.19, and 6.23) from nodal displacement coordinates to generalized coordinates expressed as strain gradient quantities. Up to this point, these transformations have not been applied. To demonstrate the use of these transformations and to give a quantitative meaning to the strain gradient quantities as generalized coordinates, the equivalent continuum properties of the truss shown in Fig. 19 is now computed. The continuum parameters are extracted from the repeated element shown. Then these properties are applied to compute the overall deflection of the truss. The results are compared to the actual results found by analyzing the discrete structure.

The first three steps outlined in Table 1 for extracting the equivalent continuum parameters from a repeated element have been developed and applied in the previous section. Step 4, the transformation of the strain energy expressions of a repeated element from nodal to strain gradient coordinates is now performed. Then the equivalent continuum parameters are identified, as indicated in step 5 of Table 1.

Strain Energy Transformation

The strain energy for a repeated element as a function of nodal displacements is given as

---

Strain Energy in a Truss

$$U = \frac{1}{2}\{d\}^T[K]\{d\}$$

where $\{d\}$ is the vector of nodal displacements, u and v.

$[K]$ is the stiffness matrix for the repeated element.

(6.31)

The transformation from nodal displacements to strain gradient quantities was generalized in Eq. 6.9. This equation is repeated here for convenience as:

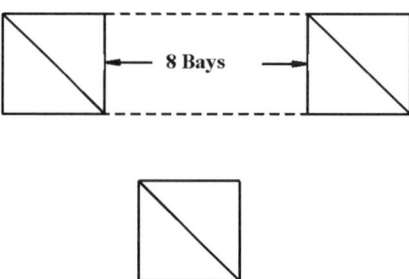

Figure 19. Truss structure with repeated element.

General Transformation Equation

$$\{d\} = [\Phi]\{\varepsilon_,\}$$

where $\{d\}$ are nodal displacements.   (6.32)

$[\Phi]$ is the transformation matrix.

$\{\varepsilon_,\}$ is the vector of strain gradient
coefficients.

The vector, $\{\varepsilon_,\}$ corresponds to vectors of the type on the right-hand sides of Eqs. 6.7, 6.14, 6.19, and 6.23. When Eq. 6.32 is substituted into Eq. 6.31, the result is

---

Strain Energy in Terms of Strain Gradient Quantities

$$U = \frac{1}{2}\{\varepsilon_,\}^T[\Phi]^T[K][\Phi]\{\varepsilon_,\}$$ (6.33)

The stiffness matrix expressed for the strain gradient quantities is extracted from Eq. 6.33 as

---

Stiffness Matrix for Strain Gradient Coordinates

$$[K]_{sg} = [\Phi]^T[K][\Phi]$$ (6.34)

The components of this matrix contain the strain energy quantities corresponding to the various strain states and the coupling between these strain states. For example, the diagonal element of $[K]_{sg}$ corresponding to the $(\varepsilon_x)_0$ strain gradient term contains the equivalent continuum parameters for longitudinal deformations. Similarly, the diagonal element of $[K]_{sg}$ corresponding to $(\varepsilon_{x,y})_0$ contains the equivalent continuum information for the Euler-Bernoulli beam behavior. In the next section, these two parameters are extracted for the example problem shown in Fig. 19. In Ref. 3, the meaning of the off-diagonal terms in Eq. 6.34 are discussed in detail and applied to the problems of assessing damage to a prospective space station structure.

Extraction of Equivalent Continuum Properties

The procedures for finding the equivalent continuum parameters for a truss are operationally defined by Eqs. 6.31–6.34. These equations provide step-by-step instructions for computing the equivalent continuum parameters. However, the application of this step-by-step procedure does not give much insight into the meaning of strain gradient quantities as generalized coordinates.

This insight is provided by applying Eq. 6.33 to determine the equivalent continuum parameter associated with a single strain gradient quantity. This is done for two different deformation modes. The first example determines the equivalent continuum parameter for the case of the truss acting as a longitudinal bar. The second example determines the equivalent continuum parameter for the truss acting as an Euler-Bernoulli beam.

Longitudinal Bar Example

As the first application, let us find the equivalent continuum property corresponding to the extension or longitudinal deformation of the truss shown in Fig. 19. Since we are only interested in one strain state, we use a reduced form of Eq. 6.32 containing only one column in the transformation matrix. The nodal displacements for the repeated element due to an extensional deformation are given as

---

Specialized Transformation Equation

$$\{d\} = \{\phi_4\}(\varepsilon_x)_0 \tag{6.35}$$

This equation gives the displacements of the nodes of the repeated element due to the strain state $(\varepsilon_x)_0$. The subscript 4 corresponds to the notation of Eq. 6.23 where the four-node configuration is considered. Thus $\{\phi_4\}$ is the fourth column of the $[\Phi]$ matrix in Eq. 6.23. In this role, the strain gradient term serves as a generalized coordinate. When Eq. 6.35 is substituted into Eq. 6.31, the strain energy expression reduces to the following scalar quantity:

Strain Energy for Extension

$$U = \frac{1}{2}(\varepsilon_x)_0\{\phi_4\}^T[K]\{\phi_4\}(\varepsilon_x)_0 \tag{6.36}$$

Equation 6.36 is a one-dimensional version of Eq. 6.33 so the role of $(\varepsilon_x)_0$ is clear. This expression represents the strain energy in the truss when it is experiencing a state of constant strain, which is represented by $(\varepsilon_x)_0$. The variable in this equation is the strain gradient quantity $(\varepsilon_x)_0$ so the magnitude of the strain energy depends on the size of $(\varepsilon_x)_0$. That is to say, the rest of the expression represents the strain energy contained in the repeated element when a unit normal strain is imposed on the stiffness model. We equate this strain energy expression to the analogous representation of strain energy in a continuous bar. This comparison identifies the equivalent continuum stiffness parameter we are seeking.

From Lesson 1, Eq. 1.13, we know that the strain energy of a continuous bar can be written as

---

Strain Energy in a Continuous Bar

$$U = \frac{AE}{2}\int_0^L \varepsilon_x^2 \, dL \tag{6.37}$$

For the case being discussed, only the state of constant longitudinal strain is being considered, so the strain is represented as

Reduced Strain Representation

$$\varepsilon_x = (\varepsilon_x)_0 \tag{6.38}$$

This equation is a reduced version of Eq. 5.23, which is reproduced here for convenience:

Strain Models with Strain Gradient Coefficients

$$
\begin{aligned}
\varepsilon_x &= (\varepsilon_x)_0 + (\varepsilon_{x,y})_0 y \\
\varepsilon_y &= (\varepsilon_y)_0 + (\varepsilon_{y,x})_0 x \\
\gamma_{xy} &= (\gamma_{xy})_0 + (\varepsilon_{x,y})_0 x + (\varepsilon_{y,x})_0 y
\end{aligned}
\tag{6.39}
$$

Equation 6.38 is formed by setting all of the strain gradient coefficients in Eq. 6.39 to zero with the exception of $(\varepsilon_x)_0$. Note that the existence of the normal strain terms in the shear strain expression of Eq. 6.39 is not a typographical error. As mentioned in Lesson 5, these terms result when incomplete polynomials are used in a displacement interpolation polynomial. These erroneous terms cause an error called parasitic shear in finite elements, which is discussed in detail in Part IV.

Equation 6.38 defines the total strain in the truss as being due to the constant strain component in the x direction. When Eq. 6.38 is substituted into Eq. 6.37, the strain gradient coefficient $(\varepsilon_x)_0$ can be factored out of the integral to give

Strain Energy Due to Longitudinal Deformation

$$U = \frac{AE}{2}(\varepsilon_x)_0^2 \int_0^L dL \tag{6.40}$$

When the integration is performed, the result is

Strain Energy in the Continuous Bar

$$U = \frac{AEL}{2}(\varepsilon_x)_0^2 \tag{6.41}$$

The strain energy quantity given by Eq. 6.41 represents the strain energy in a continuous bar with a constant cross section experiencing a state of constant strain. Equation 6.36 represents the strain energy in the truss when it is experiencing the same state of strain. When we equate these two expressions and simplify, the result is given as

Equivalent Continuum Parameter

$$(AE)_{\text{eq}} = \frac{\{\phi_4\}^T [K] \{\phi_4\}}{L} \tag{6.42}$$

The subscript eq is added to $(AE)_{eq}$ of Eq. 6.42 to identify this quantity as the equivalent stiffness parameter for the truss. This parameter is substituted into the differential equation for a bar to represent the deformations in the truss. When we inspect Eq. 6.42 closely, we see that the equivalent continuum property depends on the stiffness matrix of the truss, the deformation mode associated with the constant strain state, and the length of the repeated truss element. The numerator of Eq. 6.42 is identical to element K_{44} of Eq. 6.34.

The differential equation for the longitudinal deformation of a bar is given as Eq. 2.18. It is reproduced here for convenience as

---

Governing Equation for a Bar

$$(AE)_{eq} u_{,xx} - P(x) = 0 \qquad (6.43)$$

When the equivalent stiffness parameter given by Eq. 6.42 is used in the governing differential equation to find the displacement for the continuous model of the truss when it is loaded longitudinally, the exact numerical result for the truss is 99.99% of the approximate continuum result (0.10860/0.10861). The problem just solved highlights the meaning of one of the constant strain states.

Euler-Bernoulli Beam Example

This example determines the equivalent continuum parameter corresponding to a continuous beam for the truss shown in Fig. 19. In this case, the truss acts as an Euler-Bernoulli beam instead of a Timoshenko beam because the shear effects are small as a result of its high length-to-thickness ratio. This example was chosen to further clarify the nature of the flexure deformations represented by $(\varepsilon_{x,y})_0$ and shown in Fig. 17.

For the case of the flexure deformation, the transformation from nodal displacements to this generalized coordinate is

---

Specialized Transformation Equation

$$\{d\} = \{\phi_7\}(\varepsilon_{x,y})_0 \qquad (6.44)$$

This representation gives the displacements of the nodes of the repeated element of the truss due to a flexural deformation. The subscript 7 corresponds to the notation of Eq. 6.23 where the four-node configuration is considered. The seventh element in the vector of strain gradient quantities is the flexure term. Thus, $\{\phi_7\}$ corresponds to the seventh column of the $[\Phi]$ matrix in Eq. 6.23. In this role, the strain gradient term serves as a generalized coordinate. When Eq. 6.44 is substituted into Eq. 6.31, the result is

---

Strain Energy for Flexure

$$U = \frac{1}{2}(\varepsilon_{x,y})_0 \{\phi_7\}^T [K] \{\phi_7\}(\varepsilon_{x,y})_0 \qquad (6.45)$$

This expression represents the strain energy in the truss when it is deformed as a beam. The variable in this equation is the strain gradient quantity $(\varepsilon_{x,y})_0$. As a result, the magnitude of the strain energy depends on the size of this quantity. That is to say, the rest of the expression represents the strain energy contained in the repeated element when a unit ''flexure'' strain is imposed on the repeated element of the truss model. We equate this strain energy expression for the truss to the analogous representation of strain energy in an Euler-Bernoulli beam. This comparison identifies the equivalent continuum stiffness parameter we are seeking.

The strain energy of a continuous Euler-Bernoulli beam can be written as

Strain Energy in an Euler-Bernoulli Beam

$$U = \frac{E}{2} \int_0^L \int_\Omega \varepsilon_x^2 \, dA \, dL \tag{6.46}$$

Note that in Eq. 6.46, the strain expression is the total strain. It is not a Taylor series term. For the case being discussed, the strain is represented as

Reduced Strain Representation

$$\varepsilon_x = (\varepsilon_{x,y})_0 y \tag{6.47}$$

When Eq. 6.47 is substituted into Eq. 6.46, the Taylor series coefficient $(\varepsilon_{x,y})_0$ can be factored out of the integral to give

Strain Energy Due to Flexural Deformation

$$U = \frac{E}{2} (\varepsilon_{x,y})_0^2 \int_0^L \int_A y^2 \, dA \, dL \tag{6.48}$$

The area integral with the y^2 term is recognized as the second moment of the cross-sectional area. When both integrations are performed, the result is

Strain Energy in an Euler-Bernoulli Beam

$$U = \frac{EIL}{2} (\varepsilon_{x,y})_0^2 \tag{6.49}$$

This strain energy quantity represents the strain energy in an Euler-Bernoulli beam when it is experiencing a flexure deformation. Equation 6.45 represents the strain energy in the repeated element of the truss when it is undergoing flexure. When we equate the two expressions, the result is given as

Equivalent Continuum Parameter

$$(EI)_{\text{eq}} = \frac{\{\phi_7\}^T [K] \{\phi_7\}}{L} \tag{6.50}$$

The quantity defined by Eq. 6.50 is the equivalent flexural stiffness parameter for the truss. This equivalent continuum parameter for flexure differs from the parameter given by Eq. 6.42 for tension only by the shape of the deformation produced in the nodes by the strain states. In other words, the different bars may be deformed in different ways in the two strain states. The numerator is identical to element K_{77} of Eq. 6.34.

When this parameter is substituted into the governing differential equation for an Euler-Bernoulli beam given by Eq. 6.2, the continuous representation of the truss as a beam is produced. When the truss is deformed with a transverse load on the end, the exact numerical result is 100.07% of the approximate continuum result (15.376/15.273).

Closure

This lesson has presented and applied procedures for identifying the strain gradient quantities required to represent the displacements of simple nodal configurations. The result was used to formulate transformation matrices from nodal displacements to strain gradient coordinates. The $[\Phi]$ matrices for the seven configurations shown in Fig. 4 are used in Part III to develop finite element stiffness matrices.

These transformations were applied to compute the equivalent continuum parameters of a truss. These examples were designed to highlight the generalized coordinate nature of the strain gradient quantities.

The equations that compute the strain energy in the discrete structures, Eq. 6.33 and its specializations, Eqs. 6.36 and 6.45, hint at an alternative procedure for formulating finite element stiffness matrices. By directly computing the left-hand side of Eq. 6.34, which is only possible with strain gradient notation, the finite element stiffness matrix can be computed by rearranging Eq. 6.34 to produce $[K]$, the finite element stiffness matrix.

We see later that the evaluation of the left-hand side of Eq. 6.34 requires the computation of only a few simple integrals. This fact makes the new approach computationally competitive with the isoparametric method for computing finite element stiffness matrices. The formulation of finite element stiffness matrices by the new method is developed and applied in Part III. In summary, this lesson has provided the background for the new approach for developing finite element stiffness matrices presented in Part III.

Notes

1. One capability required to form finite element stiffness models that is not presented in this book is the assembly of global stiffness matrices from elemental stiffness matrices. This procedure is presented with a good set of examples in *Matrix Structural Analysis* by McGuire and Gallagher (Ref. 4).

References and Other Reading

1. Dow, J. O., Bodley, C. S., and Feng, C. C. "An Equivalent Continuum Representation of Structures Composed of Repeated Elements," *Proceedings of the 24th AIAA/ASME/ASCE/AHS Structures, Structural Dynamics, and Materials Conference*, Lake Tahoe, Nev., May 2–4, 1983, pp. 630–640. This is the first presentation of strain gradient notation. It is applied to three-dimensional beamlike structures.

2. Dow, J. O., Su, Z. W., Bodley, C. S., and Feng, C. C. "An Equivalent Continuum Representation of Structures Composed of Repeated Elements," *AIAA Journal*, Vol. 23, No. 10, Oct. 1985, pp. 1564–1569. The work of Ref. 1 is extended to platelike structures.

3. Dow, J. O., and Huyer, S. "Continuum Models of Space Station Structures," *ASCE Journal of Aerospace Engineering*, Vol. 2, No. 4, Oct. 1989, pp. 212–230.

4. McGuire, W., and Gallagher, R. H. *Matrix Structural Analysis*, John Wiley and Sons, New York, 1979. Chapter 3 contains a good presentation of element assembly with examples. Chapter 11 gives a brief presentation of various solution algorithms.

Lesson 6   Problems

The first five problems deal with the idea of refining finite element meshes in regions of high strain. The objective is to get smaller finite elements in areas of high rates of change in strains so the low-order polynomials in the finite element representations provide a better approximation of the actual result.

1. Assume that you have a rectangular region that is broken into four-node squares or rectangles. Assume that the right-hand end must represent high strains. Draw two more meshes with smaller square elements at the right side than exist in your original mesh. *Hint*: Without violating any constraints, i.e., each node must connect to a node, you will have most likely arrived at a uniform refinement over the whole region (see Note 1).

2. Now represent the region with triangular elements of approximately equal shape. Now draw two successive refinements of your original triangular mesh where there are smaller elements at the right-hand side. *Hint*: Note that the triangular elements can be graduated in size without violating the nodal constraints.

3. How could such a reduction be accomplished with rectangular elements? *Hint*: A special type of element is required. See Fig. 6.15 and Problem 4.

4. What polynomials are required to represent a five-node rectangular or square element with the fifth node subdividing one edge? *Hint*: Such elements are known as transition elements.

5. What strain gradient terms can the polynomials of this five-node element represent?

6. What type of element seems to be better for general meshes, rectangular or triangular? *Hint*: Give reasons relating to the number of nodes and the completeness of the strain representations.

Lesson 7
Strain Transformations

Purpose of Lesson   To develop the standard second-order tensor transformations for representing strains in rotated coordinate systems and to extend these transformations to higher order strain gradient quantities to provide a better understanding of the strain modeling characteristics of individual finite elements.

The availability of these transformations in a strain gradient context allows the strain modeling characteristics of individual finite elements to be identified with an *a priori* analysis during element formulation. This means that the characteristics of the strain representations can be found without performing a series of test problems in an attempt to identify the modeling deficiencies in an element.

We will see that the strain models inherent in elements based on incomplete polynomials do not possess rotational invariance. That is to say, if the same problem is modeled with a different orientation with respect to the global coordinate axes, the displacements for the two representations will be the *same* but the strain results may *differ*. The invariance of the displacement results implies that accurate strain results can be extracted from a finite element result even if the strains in the elements used in the model are not as good as they could be. This capability is developed and used to form error analysis procedures in Part V.

■   ■   ■

In Lesson 6, procedures were developed to identify the strain states that specific displacement approximation polynomials are capable of representing. In this lesson, the study of these polynomials continues by establishing procedures for analyzing their strain modeling characteristics. This analysis examines the transformational characteristics of the strain gradient terms present in the polynomials representing the strains under a coordinate rotation, as shown in Fig. 1. The transformational characteristics of the strains are examined by first formulating the standard strain transformations for rotated coordinates. Then these transformations are extended to higher order strain gradient terms. Specifically, strain modeling characteristics are generalized by examining the transformation characteristics of the flexure term $(\varepsilon_{x,y})_0$.

Scope of the Lesson

The four- and six-node configurations are specifically studied. The four-node configuration is chosen because it is the simplest topology based on incomplete polynomials. The six-node configuration is studied because it is the first configuration that is both complete and contains higher order strain gradient terms. The general conclusion reached from studying these two configurations is that the higher order strain representations contained in elements formed from incomplete displacement polynomials are not rotationally invariant. This means that the contribution of the higher order strain states depend on the orientation of the element. We will find that this characteristic does

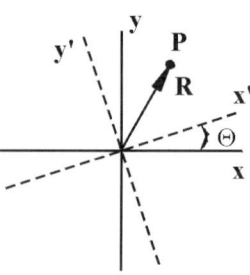

$$R = xi + yj$$
$$R = x'i' + y'j'$$

Figure 2. A point located in rotated and nonrotated coordinate systems.

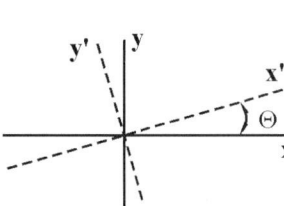

Figure 1. A rotated coordinate system.

not affect the displacement results. The invariance of the displacements and the variable nature of the strains open the door for alternative strain extraction procedures, which are discussed in later lessons.

The development proceeds as follows. The coordinate transformation for the position vector of a point in a rotated coordinate system serves as the starting point. This transformation is extended to the displacement vector. Then the definitions of the strains are applied in the rotated coordinate system. An application of the chain rule for differentiation and the use of the original coordinate transformation produce the standard strain transformation equations. Then, a similar process is applied to get the transformational characteristics of the higher order strain states. These transformations are used to analyze the strain modeling characteristics of finite elements. The results of this *a priori* analysis are validated with numerical results using three-, four-, six-, and nine-node finite element models.

Coordinate Transformations

Let us assume that we have a point P located with respect to the two-coordinate systems oriented with respect to each other, as shown in Fig. 2. The radius vector R from the origin to the point P must be the same vector regardless of the coordinate system in which it is expressed. This enables us to relate the two vector representations of R to each other. When this is done, the coordinates of a general point P expressed in terms of its location in the xy system can be written in terms of the rotated $x'y'$ system as:

Coordinate Transformation

$$x = x' \cos \theta - y' \sin \theta$$
$$y = x' \sin \theta + y' \cos \theta$$

(7.1)

Since these coordinates are orthogonal, the transpose of the transformation matrix is equal to the inverse of the transformation matrix. As a result, the inverse transformation is given as

Inverse Coordinate Transformation

$$x' = x\cos\theta + y\sin\theta$$
$$y' = -x\sin\theta + y\cos\theta$$

(7.2)

In addition to serving as the model for the transformation of the displacement vector in the next section, these transformations are used in the formulation of the strain transformations.

Strain Transformations

The expressions relating the strains in the two-coordinate systems are now formed. The strain in the rotated system $x'y'$ is found in terms of the strains in the nonrotated system xy. The starting point for the derivation is the transformation for the displacement components u' and v' in the $x'y'$ system as a function of the displacement components u and v in the xy system. This transformation has the same form as Eq. 7.2 and is given as

Displacement Transformation

$$u' = u\cos\theta + v\sin\theta$$
$$v' = -u\sin\theta + v\cos\theta$$

(7.3)

The expression for the normal strain with respect to the rotated x' axis, $\varepsilon_{x'}$, as a function of the strains in the xy coordinate system is now formed. This is accomplished by applying the definition of the normal strain in the x' direction to u', as given in Eq. 7.3. When this derivative is taken, the result is

Strain Definition

$$\varepsilon_{x'} = \frac{\partial u'}{\partial x'} = \frac{\partial u}{\partial x'}\cos\theta + \frac{\partial v}{\partial x'}\sin\theta$$

(7.4)

The two derivatives on the right-hand side of Eq. 7.4 can be evaluated explicitly in terms of derivatives in x and y. This is accomplished by applying the chain rule to the two derivative terms. When this is done, the explicit forms of the two derivatives become

Displacement Derivatives

$$\frac{\partial u}{\partial x'} = \frac{\partial u}{\partial x}\frac{\partial x}{\partial x'} + \frac{\partial u}{\partial y}\frac{\partial y}{\partial x'}$$
$$\frac{\partial v}{\partial x'} = \frac{\partial v}{\partial x}\frac{\partial x}{\partial x'} + \frac{\partial v}{\partial y}\frac{\partial y}{\partial x'}$$

(7.5)

The second term in each component on the right-hand side of Eq. 7.5 is available to us from Eq. 7.1. When the appropriate derivatives of Eq. 7.1 are taken, the result is

Derivatives of Coordinate Transformation

$$\frac{\partial x}{\partial x'} = \cos\theta$$

$$\frac{\partial y}{\partial x'} = \sin\theta$$

(7.6)

When these expressions are substituted into Eq. 7.5, the result is

Displacement Derivatives Simplified

$$\frac{\partial u}{\partial x'} = \frac{\partial u}{\partial x}\cos\theta + \frac{\partial u}{\partial y}\sin\theta$$

$$\frac{\partial v}{\partial x'} = \frac{\partial v}{\partial x}\cos\theta + \frac{\partial v}{\partial y}\sin\theta$$

(7.7)

When the expanded form of these derivatives are substituted into Eq. 7.4, the result is

Strain Transformation

$$\varepsilon_{x'} = \left(\frac{\partial u}{\partial x}\cos\theta + \frac{\partial u}{\partial y}\sin\theta\right)\cos\theta + \left(\frac{\partial v}{\partial x}\cos\theta + \frac{\partial v}{\partial y}\sin\theta\right)\sin\theta$$

(7.8)

Equation 7.8 can be simplified by introducing the definitions for the strain components in the xy coordinate system to give

Strain Transformation

$$\varepsilon_{x'} = \varepsilon_x \cos^2\theta + \varepsilon_y \sin^2\theta + \gamma_{xy}\sin\theta\cos\theta$$

(7.9)

A similar process can be applied to transform the other two strain components, $\varepsilon_{y'}$ and $\gamma_{x'y'}$. When this is done, the transformations of the three strain components are

Strain Transformations

$$\varepsilon_{x'} = \varepsilon_x \cos^2\theta + \varepsilon_y \sin^2\theta + \gamma_{xy}\sin\theta\cos\theta$$

$$\varepsilon_{y'} = \varepsilon_x \sin^2\theta + \varepsilon_y \cos^2\theta - \gamma_{xy}\sin\theta\cos\theta$$

$$\gamma_{x'y'} = 2(\varepsilon_y - \varepsilon_x)\sin\theta\cos\theta + \gamma_{xy}(\cos^2\theta - \sin^2\theta)$$

(7.10)

The strain transformations presented in Eq. 7.10 are, of course, the expected result. The transformed strain components are functions of the three strain components in the unrotated coordinate system and the angle of rotation. These equations provide an insight into the modeling characteristics of finite elements.

This equation shows that if a finite element is capable of representing the three constant strain components; the element is capable of representing constant strains with respect to any axis system. That is to say, if an element is capable of representing all three constant strains, the constant strains transform invariantly under a coordinate rotation. Since the capability of representing the three constant strain components is a requirement of the convergence criteria, most finite elements can represent constant strain regardless of the orientation of the finite element. Later, we see that if the higher order strain states do not transform invariantly, the constant strain components can be contaminated.

By comparing Eq. 7.3 to Eq. 7.10, a general observation concerning transformations under rotations can be made. Equation 7.3 contains the trigonometric functions to the first power. This is a characteristic of the transformation of vectors. When a derivative of this vector is taken with respect to one of the coordinate directions, the new quantity has different transformational characteristics. This is seen in Eq. 7.9 where the derivative of u has a transformation containing the trigonometric quantities to the second power. Similarly, when we take derivatives of the strain quantities in the next section so we can study the transformational characteristics of the higher order strain gradient terms, we see that the trigonometric functions are contained in the transformation to the third power.

This increase in order of the trigonometric functions in a transformation when a derivative is taken is a standard characteristic of tensors. That is to say, when a derivative of a tensor quantity is taken with respect to one of the coordinate directions, a new quantity is formed that is a tensor of one order higher. In this example, we have seen that when a derivative of the displacement is taken, the result is a second-order tensor, namely, strain. This is as expected since the displacement is a vector, which is also a first-order tensor. By extension, the higher order strain gradient quantities are third-order and higher order tensors. In summary, the order of a tensor is identified by the power to which the trigonometric quantities are raised in a transformation due to a rotation.

Higher Order Strain Transformations

In previous lessons, we developed and discussed higher order strain gradient quantities. They are the derivatives of the strain terms. We defined the rigid body motions to be zeroth-order strain states, the constant strains as first-order strain states, and the derivatives of the strain terms as the higher order strain states. In this section, we extend the transformations just developed to these higher order strain states.

This is done by focusing on the transformation of the second-order strain gradient quantity, the generalized flexure term $(\varepsilon_{x',y'})_0$. This specific strain state was chosen to demonstrate the development of the higher order strain transformations because the conclusions drawn from this transformation can easily be tested and verified. Also, the development of the transformation for this term can be generalized to the other second-order strain states and to the transformations of even higher order strain states.

Generalized flexure in the plane is defined as the derivative of the normal strain with respect to the coordinate perpendicular to the normal strain. This definition is the generalization of the deformation pattern in an Euler-Bernoulli beam. In a beam, flexure around the z axis and along the x axis is given as $d\varepsilon_x/dy$ or as $\varepsilon_{x,y}$.

Figure 3 shows a beam element deformed with a constant curvature. A loading such as this is used to demonstrate the accuracy of the conclusions drawn from the developments made here. The half of the element above the neutral axis is in tension. The lower half is in compression. That is to say, the normal strain ε_x varies linearly from top to bottom. The beam is experiencing flexure. The strain pattern is defined as $(\varepsilon_x)_y$, or as $\varepsilon_{x,y}$, which corresponds to the definition of flexure. The definition of generalized flexure is given in terms of the rotated coordinates as

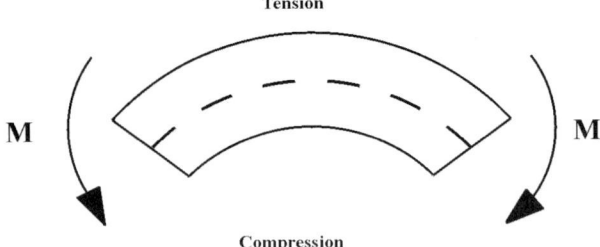

Figure 3. Beam flexure.

Definition of Generalized Flexure

$$\frac{\partial \varepsilon_{x'}}{\partial y'}$$ (7.11)

Formulation of a Generalized Flexure Expression

The generalized flexure expression is formed by taking the derivative of Eq. 7.9 with respect to y' in accordance with Eq. 7.11. When this is done, the result is

---

Derivative of Strain Transformation

$$\frac{\partial \varepsilon_{x'}}{\partial y'} = \frac{\partial \varepsilon_x}{\partial y'}\cos^2\theta + \frac{\partial \varepsilon_y}{\partial y'}\sin^2\theta + \frac{\partial \gamma_{xy}}{\partial y'}\sin\theta\cos\theta$$ (7.12)

The three derivatives of the strain components on the right-hand side of Eq. 7.12 taken with respect to y' are now explicitly evaluated. Since the strain terms are defined in the xy coordinate system, the chain rule is applied to each of these terms to give the following results

---

Strain Derivatives

$$\frac{\partial \varepsilon_x}{\partial y'} = \frac{\partial \varepsilon_x}{\partial x}\frac{\partial x}{\partial y'} + \frac{\partial \varepsilon_x}{\partial y}\frac{\partial y}{\partial y'}$$

$$\frac{\partial \varepsilon_y}{\partial y'} = \frac{\partial \varepsilon_y}{\partial x}\frac{\partial x}{\partial y'} + \frac{\partial \varepsilon_y}{\partial y}\frac{\partial y}{\partial y'}$$ (7.13)

$$\frac{\partial \gamma_{xy}}{\partial y'} = \frac{\partial \gamma_{xy}}{\partial x}\frac{\partial x}{\partial y'} + \frac{\partial \gamma_{xy}}{\partial y}\frac{\partial y}{\partial y'}$$

The first term in each of the components of the right-hand side of these equations is simply a derivative of a strain component in a nonrotated coordinate system. Thus, these terms can be interpreted as strain gradient quantities. The second term in each component of the equations is available to us by taking the derivatives of Eq. 7.1, the coordinate transformation, to give:

Derivatives of Coordinate Transformation

$$\frac{\partial x}{\partial y'} = -\sin\theta$$

$$\frac{\partial y}{\partial y'} = \cos\theta \tag{7.14}$$

When Eq. 7.13 is simplified with Eq. 7.14 and the derivatives of the strains are expressed in strain gradient notation, the result is

Strain Derivatives Simplified

$$\frac{\partial \varepsilon_x}{\partial y'} = -\varepsilon_{x,x}\sin\theta + \varepsilon_{x,y}\cos\theta$$

$$\frac{\partial \varepsilon_y}{\partial y'} = -\varepsilon_{y,x}\sin\theta + \varepsilon_{y,y}\cos\theta \tag{7.15}$$

$$\frac{\partial \gamma_{xy}}{\partial y'} = -\gamma_{xy,x}\sin\theta + \gamma_{xy,y}\cos\theta$$

The availability of Eq. 7.15 enables Eq. 7.12 to be expressed in quantities defined in the unrotated coordinate system as

Generalized Flexure Expression

$$\frac{\partial \varepsilon_{x'}}{\partial y'} = \varepsilon_{x,y}\cos^3\theta + (\gamma_{xy,y} - \varepsilon_{x,x})\cos^2\theta\sin\theta$$

$$+ (\varepsilon_{y,y} - \gamma_{xy,x})\cos\theta\sin^2\theta - \varepsilon_{y,x}\sin^3\theta \tag{7.16}$$

The presence of the sine and cosine terms to the third power show that Eq. 7.16 is indeed the transformation of a third-order tensor quantity. The fact that this result corresponds to our previous speculation means that we do not have to form the transformations for the higher order strain gradient quantities to understand their behavior under a rotation. Furthermore, we can see that this transformation contains all six possible second-order strain gradient terms, namely, the first derivatives of the strain terms, which are given as

Second-Order Strain Gradient Terms

$$\begin{array}{cc} (\varepsilon_{x,x})_0 & (\varepsilon_{x,y})_0 \\ (\varepsilon_{y,x})_0 & (\varepsilon_{y,y})_0 \\ (\gamma_{xy,x})_0 & (\gamma_{xy,y})_0 \end{array} \tag{7.17}$$

Equations 7.16 and 7.17 show that for any element to accurately represent flexure around an arbitrary axis, the element must be capable of representing all six of these second-order strain gradient terms. In Lesson 6, we saw in Eq. 6.28 that the six-node triangle is the simplest configuration capable of representing all of the second-order strain states. Thus, it can be concluded that the six-node element can represent flexure invariantly under rotation.

The four-node configuration does not represent the complete set of the second-order strain gradient terms. Thus, the four-node element cannot represent flexure strains invariantly under rotation. The effect of this modeling deficiency is demonstrated in the examples that follow.

We now present the displacements and strain results for a series of problems solved at different orientations with several types of elements to demonstrate the significance of Eq. 7.16. These results also motivate the study of the various elements developed and analyzed in Part III.

Flexural Modeling Capability of the Four-Node Quadrilateral

In the analysis of the four-node configuration in Lesson 6, it was found that the approximate displacement polynomials of this element are capable of modeling the following strain gradient terms:

Strain States Represented by a Four-Node Quadrilateral

$$
\begin{array}{ccc}
(u_{rb})_0 & (v_{rb})_0 & (r_{rb})_0 \\
(\varepsilon_x)_0 & (\varepsilon_y)_0 & (\gamma_{xy})_0 \\
(\varepsilon_{x,y})_0 & (\varepsilon_{y,x})_0 &
\end{array}
\tag{7.18}
$$

The first two rows are the rigid body motions and the constant strain terms. The third row of Eq. 7.18 contains the second-order strain gradient terms that the four-node quadrilateral element is capable of representing. We can see that this element represents only two of the second-order strain gradient terms, i.e., flexure along the x and y axes, respectively. The four-node quadrilateral element is not capable of representing the following four second-order strain gradient terms:

Second-Order Strain Gradient Terms Not Modeled by the Four-Node Quadrilateral

$$
(\varepsilon_{x,x})_0 \quad (\varepsilon_{y,y})_0 \quad (\gamma_{xy,x})_0 \quad (\gamma_{xy,y})_0
\tag{7.19}
$$

Thus, we can see that the flexure terms in the four-node configuration do not transform invariantly under rotation because four of the second-order strain gradient quantities are missing. Furthermore, the element cannot accurately represent flexure along any axis because the $(\varepsilon_{y,y})_0$ term is missing. The presence of $(\varepsilon_{x,y})_0$ "activates" the $(\varepsilon_{y,y})_0$ strain gradient term as a result of the Poisson effect. If Poisson's ratio, ν, were zero the absence of $(\varepsilon_{y,y})_0$ would be of no consequence.

At this point, one might be tempted to conclude that the two extra degrees of freedom associated with the four-node quadrilateral element are "wasted" because of the modeling deficiencies just noted. However, before these deficiencies are considered fatal to the usefulness of this element and it is eliminated from the library of recommended elements, let us study the solution to a set of problems designed to evaluate the modeling characteristics of this element.

Experimental Validation of Generalized Flexure Analysis

The contents of this section are designed to validate the *a priori* analysis of the strain modeling characteristics of the four-node element that have been presented. Specifically, the problems solved in this section support the conclusions that the four-node element cannot invariantly represent flexure strains under rotation. Furthermore, the solutions to these problems demonstrate the effects of this modeling deficiency on the displacements and strain representations. The results from the four-node models are compared to models constructed from other elements. These comparisons further extend our understanding of the capabilities of the *a priori* analysis of individual elements.

The behavior of the four-node element is studied by applying three different loads to the cantilever beam structure shown in Fig. 4. This model contains 120 degrees of freedom. The three different load cases applied to the model are also shown in Fig. 4. These loadings produce progressively more complex strain distributions in the beam. The first load case consists of a constant tension. In an accurate solution, this loading produces a state of constant strain in the beam. The second load case is a couple. It produces a constant moment in the beam and, hence, a state of constant flexure. The third load case produces a linearly varying moment in the beam. This produces a strain state that is a linearly varying flexure. Each of the load cases produces a well-defined strain distribution in the beam. This enables us to easily analyze the accuracy of the strain distributions produced by the finite element models because we know what to expect.

This sequence of problems was chosen to demonstrate the modeling characteristics of the four-node element under a full range of conditions. In the first load case, the *a priori* analysis predicts that the four-node element can exactly represent constant strain in the beam regardless of the orientation of the beam.

In the second load case, the *a priori* analysis predicts that the four-node element can represent the state of constant flexure correctly if the element is oriented along the line of constant flexure. It further predicts that the element cannot represent this state of strain if the orientation of the beam is changed. Finally, the *a priori* analysis implicitly shows that the four-node element is too stiff as it represents a state of constant flexure because the second-order strain state that would allow the Poisson effect to be represented under a

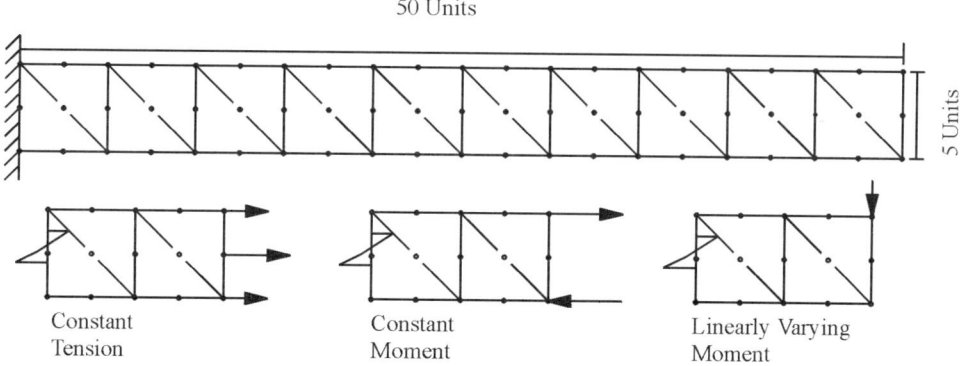

Figure 4. Cantilever beam problem with three load cases.

state of constant flexure is not present in the four-node element; namely, the strain state $(\varepsilon_{x,x})_0$ is absent from the four-node element. This is equivalent to constraining this deformation mode to zero so the four-node element is too stiff when it represents constant flexure. When we extract the strain gradient components from the strain representations produced by the four-node elements, we find how this modeling deficiency affects the strain model.

In the third load case, the four-node element is asked to represent a linearly varying state of flexure. That is to say, the strain state induced by the linearly varying moment is $(\varepsilon_{x,yx})$. The four-node element does not contain this strain state or the strain state needed to represent the companion Poisson effect. This strain state can only be approximated with a model consisting of four-node elements by a highly refined finite element mesh.

Each four-node model is solved for three different orientations with respect to the global reference frame, namely, 0, 15, and 30 degrees, to demonstrate the strain transformation characteristics. As mentioned earlier, we see that the relative displacement results are independent of the orientation of the elements. However, the strain representations contained in elements based on incomplete polynomials are not the same for different orientations. In other words, the displacements are invariant with rotation but the strain representations may vary with orientation in elements based on incomplete polynomials.

In addition to solving these problems with the four-node element, we also solve the problems with three-, six-, and nine-node elements. This provides more insight into the capabilities of *a priori* analysis of individual elements.

The displacement results are discussed first. This is followed by an analysis of the strain distributions in terms of strain gradient components and the strain transformations previously developed.

Displacement Results

The displacement results for the set of problems just defined are presented in this section. The maximum tip displacements for the three load cases are given in Table 1. The displacement for the constant tension case is an extension in the longitudinal direction. The displacement for the other cases is the lateral displacement of the tip.

Table 1 presents the displacements produced by the analysis of the problems shown in Fig. 4 with the following five element types: (1) three-node constant strain triangle, (2) four-node bilinear element uncorrected for parasitic shear, (3) four-node bilinear element corrected for parasitic shear, (4) six-node linear strain triangle, and (5) nine-node element corrected for parasitic shear. The concept of parasitic shear was introduced in application

Table 1. Maximum tip displacements.

Element type	Maximum displacement		
	Constant tension	Constant moment	Varying moment
Three node	0.1324	10.72	7.22
Four node	0.1320	17.81	11.94
Four node[a]	0.1319	19.48	13.05
Six node	0.1331	19.91	13.31
Nine node[a]	0.1331	19.96	13.41

[a] Parasitic shear has been removed from this model.

4 of Lesson 5. It was shown that parasitic shear consisted of the presence of erroneous normal strain terms in the shear strain expressions in elements formed with incomplete displacement polynomials.

Later, we see that this error is easily corrected in the four-node element. However, the modeling deficiency cannot be removed from the eight- and nine-node elements without the use of strain gradient notation. If an attempt is made to remove the error, either too few terms are removed and a residual of the error remains or too many terms are removed and a new error is introduced. This new modeling deficiency consists of the presence of deformation states that have no energy content associated with them. These spurious zero energy modes introduce instabilities into the displacement results. In Part III we find that with the use of strain gradient notation that this error need never exist.

These problems were also solved with the beam rotated 15 and 30 degrees with respect to the global coordinates, as shown in Fig. 5. That is to say, the elements are formulated in a local coordinate system aligned with the global coordinate system. By changing the orientation of the strains produced by the loading condition with respect to the elemental coordinate system (and the global coordinate system, in this case), the effect of a rotation on the elemental strain representation can be studied. The relative displacement results are identical for these cases. The effect on the strain gradient components is discussed in the next section.

Constant Tension

The results for the constant tension case show that all of the elements give essentially the same results. This is as expected since all of the elements are capable of representing the constant strain states induced by this loading condition. The displacements shown are the average of the displacements in the end nodes of the model. The small differences in the displacements are due to the mesh layout and the boundary conditions at the support. The four-node element with and without parasitic shear produced nearly identical results. This is to be expected since the sources of the errors, the flexure terms, are not active in this load case. Identical results were found for the three different orientations, 0, 15, and 30 degrees.

Constant Moment

For the loading case of a constant moment, the maximum lateral displacement of the free end of the finite element models varies widely. The constant strain triangle cannot accurately represent this problem because this simple element is incapable of representing the linear strain variation produced by this loading condition. To better represent this problem with three node elements, the mesh would have to be heavily

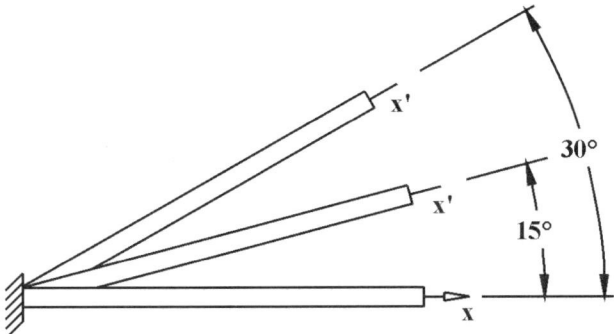

Figure 5. Rotated cantilever beam problem.

refined. That is to say, the capacity to represent this strain distribution with constant strain triangles would have to be provided by refining the mesh layout, since it cannot be done by the individual elements.

The necessary mesh refinements to produce accurate results can be accomplished automatically with an adaptive refinement procedure. The error estimation component of an adaptive refinement capability that is developed in Part V depends on the jumps in the interelement strains and misrepresentation of the boundary stresses in the finite element models to identify the regions of error. These modeling errors are discussed in the next section.

The four-node quadrilateral, with and without parasitic shear, produces a better result than does the constant strain triangle. This is as expected because the four-node element is capable of representing higher order strain states. The stiffening effect of parasitic shear terms can be seen by comparing the results for the four-node elements with and without parasitic shear. The results improve markedly when the parasitic shear terms are removed. In fact, the results for the corrected element nearly match the results for the higher order elements. The reason for this is discussed in the next section when the strain modeling characteristics are discussed.

The linear strain triangle is the lowest order element that can be expected to accurately represent the constant moment problem with its linear strain distribution. This is the case since the element is capable of representing all six linear strain terms. Similarly, the nine-node element can be expected to represent this problem accurately, and it does. The displacements are the same for all of the finite element models when the problem is rotated 15 and 30 degrees.

Linearly Varying Moment

The representations of the problem loaded with a linearly varying moment produce somewhat surprising results. The three-node element and the four-node element with parasitic shear give the expected inaccurate results. However, the corrected four-node element and the six-node element give results that are more accurate than might be expected. The maximum displacements are very close to the results for the nine-node element. In fact, when the mesh is uniformly refined by subdividing each of the nine-node elements into four elements, the nine-node results improve by only 2%. Thus, the corrected four-node and the six-node elements give very good results. The reasons for this are discussed in terms of the strain gradient components in the next section when the strain representations are discussed.

Analysis of Strains

The analysis of the finite element strain models presented in this section is made possible by the use of strain gradient notation. This notation enables the strain representation in the entire finite element to be captured with a minimum of information in an understandable format. That is to say, the strain gradient coefficients totally define the strain distribution since they are the coefficients of the Taylor series representation. No plots are needed to describe the strain distribution completely. This form of presentation enables the strains at any point in the element to be computed.

As a result of this compact representation, any discontinuity or jump between adjacent elements is easily computed. The compatibility of strains between elements is used in later discussions to assess the accuracy of the finite element results. Furthermore, the points of maximum strain and of maximum principal strains can be identified exactly for use in failure models. The strain gradient form of presentation differs from the usual approach, where the strains are given at a number of points on the element. With the standard pointwise form of presentation, it is difficult to ascertain the behavior of individual elements.

The strain modeling behavior of the various element types is now analyzed for the three load cases.

Constant Tension

The first load case is designed to produce constant normal strains in the model. Table 2 contains the nonzero strain gradient components for the finite elements when the problem is aligned with the x axis, i.e., the rotation is zero. The higher order terms in the four-, six-, and nine-node elements are zero.

These results show that the normal strain ε_x is constant in each element. The Poisson effect produces a constant value for ε_y of 0.30 of the longitudinal strain ε_x. The shear strain is zero. In other words, all of the element types provide the same results.

Later when we discuss error analysis, we see that the accuracy of the results is associated with jumps in the strains between elements and with the accurate representation of the boundary stresses. We have an accurate result if the interelement jumps are zero and the boundary stresses are satisfied. In this case, the strains are the same in all of the elements so there are no interelement jumps in the strains. Similarly, the boundary stresses are satisfied, namely, $\tau_{xy} = 0$ and $\sigma_y = 0$. The shear stress is zero because the shear strain is zero. The fact that the normal stress σ_y is zero can be seen by substituting the appropriate strains into the constitutive equation $(\varepsilon_y - \nu\varepsilon_x = 0)$. Thus, we can conclude that the result is accurate. This subject is discussed further in the lessons concerning error analysis.

Table 3 contains the strain results for this problem when it is rotated 15 degrees. As can be seen, the strains contained in the individual elements differ for this problem orientation. It should be remembered that the reference axes for the elements used in this model are still aligned with the global axes. When the transformation for strains, as given by Eq. 7.10, are used to transform the strains to a coordinate system aligned with the beam, the results match Table 1. Thus, the strains are invariant for this constant strain case.

Table 2. Strain gradient components for constant tension (0 degrees rotation).

	Element type			
	Three node	**Four node**	**Six node**	**Nine node**
$(\varepsilon_x)_0$	2.667	2.667	2.667	2.667
$(\varepsilon_y)_0$	-0.800	-0.800	-0.800	-0.800
$(\gamma_{xy})_0$	0.000	0.000	0.000	0.000

Table 3. Strain gradient components for constant tension (15 degrees rotation).

	Element type			
	Three node	**Four node**	**Six node**	**Nine node**
$(\varepsilon_x)_0$	2.434	2.434	2.434	2.434
$(\varepsilon_y)_0$	-0.567	-0.567	-0.567	-0.567
$(\gamma_{xy})_0$	1.734	1.734	1.734	1.734

This invariance is as we would expect because all of these elements represent the complete set of constant strain terms. The results for the 30-degree rotation case is not shown, but these results parallel those for the 15-degree rotation.

Constant Moment

The second load case, a constant moment along the length of the beam, is designed to produce a linear variation in the normal strains with respect to y and a shearing strain of zero when the beam is aligned with the x axis. Table 4 contains the constant and linear strain gradient components for the elements being studied. The higher order terms in the nine-node element are zero because this strain distribution can be completely captured by the linear variations. The four-node elements used in this analysis are corrected for parasitic shear. The results for the nine-node element corrected for parasitic shear and for the uncorrected element are virtually identical for this case because the terms that produce parasitic shear are not active for this loading condition.

The displacement results for the six- and nine-node elements were virtually identical for this loading condition. An inspection of the active strain gradient terms in Table 4 shows that the strain distributions in the two elements are identical. As expected, the normal strains vary with y. All of the other terms are zero. The component in the y direction exhibits the expected Poisson effect of 0.3.

As mentioned earlier, the accuracy of the result can be assessed by the jumps in the strains between the elements and the accuracy in satisfying the boundary stresses. In the six- and nine-node representations, there are no jumps in the strains between elements. This is easily seen because each element has the same value for the strain in the y direction and there is no variation in the x direction. The boundary conditions are satisfied because the shear strains are zero and the Poisson effect is satisfied as in the previous problem. Thus, the result can be judged as accurate. The small differences between the displacements for the two models can be assumed to be largely due to differences in the representations at the fixed and loaded boundaries. Boundary modeling is discussed further in the lessons dealing with error analysis.

One of the objectives of this preliminary study of the strain modeling characteristics of elements is to observe how an element responds when it is incapable of representing the strain state it is attempting to model. In the case of the four-node configuration, the

Table 4. Strain gradient components for constant moment (0 degrees rotation).

	Element type			
	Three node	**Four node**	**Six node**	**Nine node**
$(\varepsilon_x)_0$	-21.48^a	0.0	0.0	0.0
$(\varepsilon_y)_0$	3.24^a	5.36	0.0	0.0
$(\gamma_{xy})_0$	10.75^a	17.85^a	0.0	0.0
$(\varepsilon_{x,x})_0$	$-^b$	—	0.0	0.0
$(\varepsilon_{x,y})_0$	—	14.28	16.00	16.00
$(\varepsilon_{y,x})_0$	—	0.0	0.0	0.0
$(\varepsilon_{y,y})_0$	—	—	-4.80	-4.80
$(\gamma_{xy,x})_0$	—	—	0.0	0.0
$(\gamma_{xy,y})_0$	—	—	0.0	0.0

[a]These entries vary in the elements by sign or with zero.
[b]The dash (—) indicates that the element is not capable of representing this strain state.

element does not possess the capability of representing the variation of ε_y in the y direction as do the two higher order elements.

The four-node element compensates for the inability to represent the linear variation in ε_y by engaging components of constant strain. This is seen in Table 4, where the six- and nine-node elements both have a nonzero value for $(\varepsilon_{y,y})_0$. The four-node element attempts to model this linearly varying distribution of ε_y with a constant representation and a contribution by γ_{xy} in some of the four-node elements.

As mentioned earlier, the accuracy of the displacement result produced by the four-node element was close to that of the higher order elements. The reason for this can be seen by comparing the terms that largely contributed to the displacement, namely, the flexure terms $\varepsilon_{x,y}$. The displacement of the four-node element is 98% of the result produced by the higher order element and the value of the flexure term in the four-node element is 98% of that of the higher order term.

As was inevitable, the three-node element represents the strain distribution with constant strains. In this case, the model approximates the flexure with a pattern of constant values of ε_x that alternate between zero and the value shown in Table 4. As an aside, significant interelement jumps exist between the elements so the lack of accuracy can be recognized by studying the strains.

This problem was also solved with the beam rotated 15 and 30 degrees to the global coordinates, as shown in Fig. 5. As mentioned earlier, the displacements in the rotated coordinates are identical to those found for the nonrotated problem. Thus, the displacements are invariant.

The strain gradient coefficients found for the four-, six-, and nine-node elements when the problem is rotated 15 degrees are shown in Table 5. The three-node element results are not presented because they are invariant with rotation and can be computed from Table 4 with Eq. 7.10. Two entries are given for the four-node results because the strain representations in the upper and lower elements (as noted in the footnotes to the table) are different. The first column contains the results for the upper element. The second column contains the results for the lower elements.

When the higher order strain gradient terms for the six- and nine-node elements are transformed as third-order tensors parallel to the coordinate axes, the results match those for the nonrotated case. Thus, the strains for these two elements are invariant as was expected. This problem was also solved for a rotation of 30 degrees and the same invariance was found.

Table 5. Strain gradient components for constant moment (15 degrees rotation).

	Element type			
	Four node[a]	Four node[b]	Six node	Nine node
$(\varepsilon_x)_0$	8.087	− 2.222	0.0	0.0
$(\varepsilon_y)_0$	7.579	− 2.718	0.0	0.0
$(\gamma_{xy})_0$	9.032	− 33.20	0.0	0.0
$(\varepsilon_{x,x})_0$	—	—	− 3.785	− 3.779
$(\varepsilon_{x,y})_0$	15.92	15.92	14.113	14.110
$(\varepsilon_{y,x})_0$	4.286	4.280	0.8836	0.8824
$(\varepsilon_{y,y})_0$	—	—	− 3.282	− 3.285
$(\gamma_{xy,x})_0$	—	—	− 2.694	− 2.687
$(\gamma_{xy,y})_0$	—	—	10.045	10.050

[a] These entries are for the upper layer of elements.

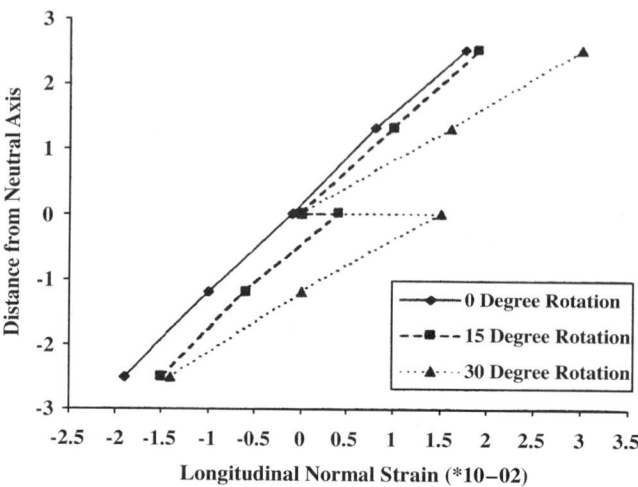

Figure 6. Strain profiles for a four-node element.

The strain profiles for the four-node element for the three orientations are shown in Fig. 6. As can be seen, the strain models produced for the three orientations are different. The result for the original orientation is smooth. There are no interelement jumps in the strains between the elements. However, for the two rotated cases, there is a jump in the strains between elements. Thus, we can conclude that the strain models for elements based on incomplete polynomials are not invariant. This is the case because all of the terms needed for the transformation are not present. A similar result is seen in the next load case for the nine-node element because it does not posses the ability to represent all of the strain states needed to model the strain distributions produced by the linearly varying moment.

Linearly Varying Moment

The third load case produces a flexural distribution that varies linearly along the length of the beam and a parabolic shear strain distribution. Table 6 contains the strain gradient

Table 6. Strain gradient components for linearly varying moment (0 degrees rotation).

	Element type	
	Six node	Nine node
$(\varepsilon_x)_0$	0.660	0.0
$(\varepsilon_y)_0$	-0.415	0.0
$(\gamma_{xy})_0$	-1.929	-1.733
$(\varepsilon_{x,x})_0$	5.280^a	0.0
$(\varepsilon_{x,y})_0$	8.00	8.00
$(\varepsilon_{y,x})_0$	0.2480^a	0.0
$(\varepsilon_{y,y})_0$	-2.404	-2.400
$(\gamma_{xy,x})_0$	0.0	0.0
$(\gamma_{xy,y})_0$	0.2480^a	0.0
$(\varepsilon_{x,xy})_0$	—	-0.465
$(\varepsilon_{y,xy})_0$	—	-0.0957

[a] The values alternate from plus to minus from element to element.

components for the six- and nine-node elements resulting from solving the problem shown in Fig. 4. The nine-node element is corrected for parasitic shear.

The nine-node element represents ε_x as a combination of the linear term $(\varepsilon_{x,y})_0$ and the quadratic term $(\varepsilon_{x,xy})_0$. The negative sign on the quadratic term indicates that the strain in the element is higher on the side closest to the support. The representation of ε_y consists of the single quadratic term $(\varepsilon_{y,xy})_0$. These two normal strain components satisfy interelement strain compatibility. That is to say, there are no jumps in the interelement strains. Also, σ_y satisfies the boundary conditions on the top and bottom surfaces. Thus, the normal strain distributions are represented exactly.

The nine-node element represents the shear strain as a constant across the section. All of the shear strain components are zero except for the constant term. The expected result is a parabolic distribution. Although interelement compatibility is satisfied for the shear strains between elements, i.e., there are no jumps in the shear strain between elements, the shear strain boundary conditions on the top and bottom surfaces are not satisfied. To represent the parabolic distribution with one element, the shear strain model would need to contain $(\gamma_{xy,yy})_0$. The nine-node element does not contain this term. Thus, it cannot accurately represent the strain distribution at the element level. The lowest order element that contains this term would be a ten-node triangle. In the next section, the mesh is refined to improve the shear model.

The six-node element is not capable of representing the quadratic normal strain components required to exactly represent this problem. Instead, the six-node element represents the normal strains as a combination of a constant and a linear term. The alternating signs on the linear terms indicate that interelement discontinuities exist in the normal strains. Also, the boundary conditions are not satisfied on the top and bottom surfaces.

The six-node elements attempt to represent the parabolic shear with a constant component and two linear components. The distribution does not satisfy interelement compatibility or the boundary conditions.

This problem has demonstrated the behavior of the six-node element when it is incapable of exactly representing the strain distribution being modeled. As expected, the inability to represent the exact distribution has resulted in interelement incompatibilities in the strains and a failure to satisfy the boundary conditions.

This problem was also solved for rotations of 15 and 30 degrees. The strains in the six- and nine-node elements behaved as expected. The six-node representations transformed invariantly, albeit incorrectly, since the element is based on complete polynomials. The strains for the nine-node element did not transform invariantly. This is shown in Fig. 6 for ε_x across the section. The fact that these strain distributions do not coincide means that the strains do not transform invariantly since the element is based on incomplete polynomials.

Model Improvement through Mesh Refinement

As discussed in the previous section, the nine-node model represented the normal strain distributions exactly. However, the expected parabolic shear strain distribution was represented as a constant across the section, as shown in Fig. 7. To improve the shear strain representation, the mesh for the nine-node model was uniformly refined by subdividing each element into four elements.

The lateral displacements of the end were nearly unchanged after the mesh refinement. The maximum displacement went from 13.37 to 13.40 resulting in a 0.22% change. The distributions of the two normal strain components were unchanged. This validates the idea that the normal strain representations for this model are exact and that the lateral displacements are largely due to flexure.

The shear strain representation for the refined mesh is shown in Fig. 7. As can be

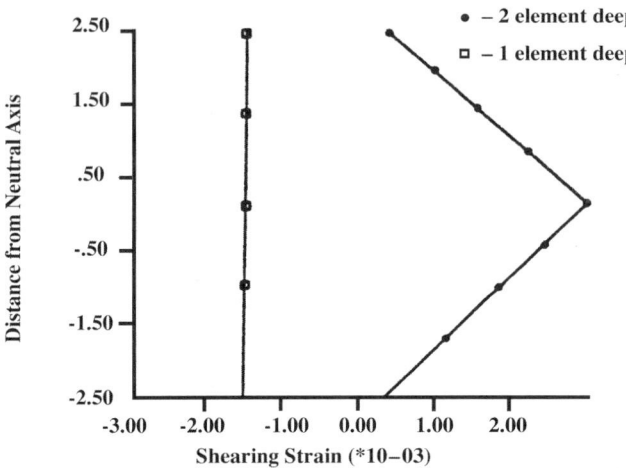

Figure 7. Strain profiles.

seen, the refined representation of the shear strains is closer to the expected parabolic distribution. The improved representation does not exactly satisfy the boundary conditions for shear. To satisfy this condition, the shear strain must be zero on the boundary. However, the refined model is closer to satisfying this condition than the original model. In this model, there are no jumps in the shear strains between elements. That is to say, strain compatibility between elements is satisfied. Since the shear strain model consists of linear strain elements it can be expected to transform invariantly.

Closure

This lesson has developed and validated an approach for identifying the strain modeling characteristics of finite elements in terms of the displacement approximation polynomials. As a result, the behavior of elements can be partially analyzed at the formulation stage. The strain modeling characteristics do not have to be totally deduced from an *ad hoc* set of test problems.

The development of the transformational characteristics of the second-order strain gradient terms provides a mechanism for explaining why the strain models for elements based on incomplete polynomials are not always invariant. For example, a four-node element contains a complete set of constant strain terms and an incomplete set of linear strain gradient terms. Thus, if this element is representing constant strain, the strain representations are rotationally invariant. However, if the second-order strain gradient terms are active, the strain representation is not invariant.

Displacement Invariance

The example problems that validated the conclusions concerning the strain modeling deficiencies of elements based on incomplete displacement polynomials provided evidence that these elements need not be condemned. The displacement results produced by the four-node elements corrected for parasitic shear are accurate. Furthermore, the displacement results are invariant with rotations even though the strain models are not. These two observations mean that the strain energy content in each element is constant regardless of the strain representation associated with a given orientation. These characteristics imply that the development of alternate strain extraction procedures would improve the usefulness of these elements.

The usefulness of the four-node element is further enhanced by the fact that as a mesh is refined, the coefficients of the constant strain terms increase in size relative to the higher order terms as convergence is approached. Since the constant strains transform invariantly, the effect of the invariance of the second-order terms is submerged as the model is improved. A problem at the end of the lesson demonstrates the increase in participation of the constant strain terms as a model is refined.

Parasitic Shear

The need to correct elements for parasitic shear was demonstrated by the results of the example problems. A comparison of the displacement results for the corrected and uncorrected four-node elements given in Table 1 showed that the displacements were significantly reduced when the higher order strain states in an element with parasitic shear were active. The statement in the previous section concerning the increased participation of the constant strain states when a problem is refined implies that the effect of parasitic shear is reduced under mesh refinement. However, more elements are required to produce a given level of accuracy if parasitic shear effects are not removed from the element. This reduces the efficiency of the model since the problem size must be increased to achieve the same level of accuracy if uncorrected elements are used.

We noted in this lesson that parasitic shear could not be eliminated from eight- and nine-node elements without the guidance of strain gradient notation. We see in the next set of lessons that the current numerical approach of attempting to remedy the modeling deficiency is not successful for eight- and nine-node elements.

In the current approach for attempting to correct these elements, either the parasitic shear effects are not completely removed or another error is introduced. This new error introduces modes that have a deformation but contain no strain energy. These modes are known as spurious zero-energy modes. They behave somewhat like rigid body motions but can introduce instabilities into solutions. The effects of this error can be more serious than the reduced displacement caused by the presence of parasitic shear. Both parasitic shear and spurious zero-energy modes can be simply and definitively removed from eight- and nine-node elements with the alternative element formulation procedure developed in the next lesson. The numerical correction of parasitic shear is successful for four-node elements and some plate elements.

Error Analysis and Mesh Refinement

The ability to evaluate the accuracy of the results of a finite element analysis by examining the jumps in the interelement strains and the accuracy of the boundary stress representations was introduced. The ability to identify regions of high error enables the finite element to be improved in these often critical areas. This capability, in combination with automatic mesh refinement, is known collectively as adaptive mesh refinement.

When adaptive refinement reaches general use, the overall accuracy of finite element results will be improved. Furthermore, adaptive mesh refinement is a missing link in the development of computer-aided design procedures. In later lessons on error analysis, procedures for guiding the automatic improvements of finite element meshes in critical areas is developed and applied. This work suggests alternative strain extraction procedures.

Finally, the improvement in the strain representations with an improved mesh was shown in the final example. Later we demonstrate that the very act of increasing the number of elements in an inaccurate model does not necessarily improve the result. The additional elements must be in the correct location. The effect of adding elements in areas where they are not needed can be likened to adding unsymmetric trial functions to a Rayleigh-Ritz solution to a problem with a symmetric result. The desire to develop more efficient finite element representations further motivates the use of adaptive mesh refinement procedures.

Future Developments

This lesson, in addition to developing a powerful tool for evaluating the modeling characteristics of individual elements, was designed to provide an overview of the characteristics of the finite element method in general. The remaining lessons address all of the issues raised here in more detail. The next part develops a procedure based on the use of strain gradient notation to formulate finite element stiffness matrices and apply the procedure to develop the stiffness matrices for several configurations. In this process, the source of parasitic shear and spurious zero-energy modes is identified. As a result of this knowledge, these modeling errors can be definitively removed from these elements.

We see that this alternative form of element formulation is computationally competitive with existing methods. Furthermore, elements with prescribed characteristics can be developed if one should so choose. As part of the element development presentation, the source of another modeling deficiency is identified. It is well known that high-aspect-ratio elements do not provide good results. We will see that this error is caused by the same mechanism that produces a modeling error known as shear locking. Shear locking exists in some out-of-plane bending elements. This error is characterized by an unnatural stiffening in elements that are thin with respect to their dimensions in the plane. This error is caused by the inherent nature of the polynomials used and cannot be directly eliminated. Its effects can be controlled by correcting the stiffness matrix with factors that depend on element geometry or by placing limitations on element geometry.

References and Other Reading

1. Dow, J. O., and Byrd, D. E. "The Identification and Elimination of Artificial Stiffening Errors in Finite Elements," *International Journal for Numerical Methods in Engineering*, Vol. 26, Mar. 1988, pp. 743–762.

2. Dow, J. O., Harwood, S. A., Jones, M. S., and Stevenson, I. "Validation of a Finite Element Error Estimator," *AIAA Journal*, Vol. 29, No. 10, Oct. 1991, pp. 1736–1742.

3. Dow, J. O., and Byrd, D. E. "Error Estimation Procedures for Plate Bending Elements," *AIAA Journal*, Vol. 28, No. 4, Apr. 1990, pp. 685–693.

4. Dow, J. O., Cabiness, H. D., and Ho, T. H. "A Generalized Finite Element Evaluation Procedure," *ASCE Journal of Structural Engineering*, Vol. 111, No. 2, Feb. 1985, pp. 435–452.

Lesson 7 Problems

1. Derive the transformation for shear strain equivalent to Eq. 7.9. *Hint*: This problem is designed to solidify the understanding of the strain transformation and the meaning of the results.

2. Derive the transformation for $\gamma_{xy,x}$ equivalent to Eq. 7.16. *Hint*: This problem extends the transformation capabilities to third-order tensors. It is used to evaluate the modeling capabilities of higher order elements.

3. Derive the transformation expression for the fourth-order tensor quantity $\varepsilon_{x,yx}$. Note: This equation is used to evaluate the modeling capabilities of 8- and 9-node quadrilateral elements and the 10-node triangle. This result is not a *pro forma* exercise because the higher order compatibility equation may come into play.

4. Compute the principal strains and their directions for the point halfway between the neutral axis and the top of the beam ($y = 1.25$ and the local origin is at the centroid of the element) for the nine-node model of the constant moment case for 0 and 15 degrees (see Tables 4 and 5). Give the orientations with respect to both the beam axis and the global coordinate axes. *Hint*: The objective of this problem is to clearly demonstrate the Taylor

series nature of the strain representations and the orientation of the element coordinates with respect to the global system.

5. A single nine-node element exactly represents the strain profile for the constant moment case (see Table 4.). Take the origin of the elemental coordinate systems to be at the centroid of the element. (a) What would the strain gradient components be in the individual elements if the mesh were subdivided once and then again? (b) How does the strain energy composition change with respect to the change in the composition of the strain states? (c) What conclusion can be drawn with respect to the distribution of the strain energy as the model is further refined? *Hint*: This problem is designed to provide a better understanding of element behavior.

Part III

The Strain Gradient Reformulation of the Finite Element Method

Introduction

Part III develops and applies a new procedure for formulating finite element stiffness matrices that is based on strain gradient notation. The approach is used to form the stiffness matrices for several elements. The modeling characteristics of the finite elements created using the new approach are evaluated during the formulation process. This *a priori* evaluation of the modeling characteristics of the individual elements is possible because of the transparent nature of the strain gradient notation.

The new approach for forming finite element stiffness matrices is put in context with respect to finite element analysis as a whole in this introduction. This context is provided by giving an overview of the finite element analysis process, by discussing the modeling requirements of individual finite elements, and by presenting the advantages of the strain gradient formulation process.

The finite element method is a form of the Rayleigh-Ritz solution technique. The finite element method overcomes the primary limitation[1] of the original form of the Rayleigh-Ritz method by subdividing complex shapes into a finite number of small regions with relatively simple boundary conditions, as shown in Fig. 1. By subdividing the domain of a larger problem into subregions, the finite element method essentially reduces the intractable problem of finding trial functions for complex domains into two separate but solvable problems, namely, (1) the formulation of accurate finite element stiffness matrices and (2) the development practical error analysis procedures.

An Overview of the Finite Element Method

The finite element version of the Rayleigh-Ritz method consists of the following *two separate but related problems*. The first problem requires the formulation of individual finite element stiffness matrices that adequately represent the behavior of the subregions. The second problem is the identification of a mesh that subdivides the problem in such a way that the results of the finite element analysis accurately represents the actual solution

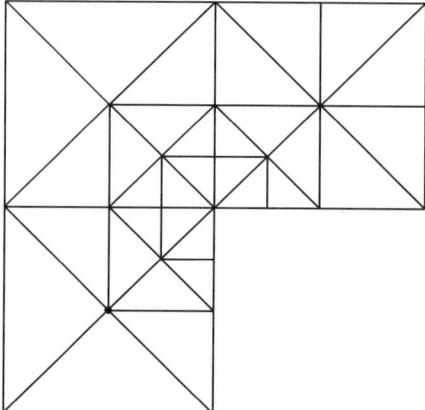

Figure 1. Finite element subdivisions.

[1]The primary limitation of the standard form of the Rayleigh-Ritz method is the need to define trial functions that satisfy boundary conditions on the whole domain. This limitation is discussed in Lesson 4 of Part I.

of the continuous problem. The use of strain gradient notation provides powerful new approaches for solving both of these problems.

Problem 1 — Element Formulation

The new procedure for formulating finite element stiffness matrices presented in Lesson 8 uses the capabilities developed in Lessons 6 and 7 for determining the equivalent continuum parameters for trusses to form finite element stiffness matrices. That is to say, the procedures contained in Lesson 6 and 7 are inverted to create a new approach for formulating finite element stiffness matrices. The crux of the method is the identification of the strain gradient quantities that provide the basis functions for the finite element. The process for identifying these quantities is formalized by relating the nodal locations of the element being formed to the structure of Pascal's triangle. Once the basis functions are identified, the strain energy is computed for the element due to each of the deformation patterns given by the strain gradient quantities. These strain energy quantities are then transformed to displacement coordinates to finish the formulation process.

The new formulation procedure enables modeling errors to be identified and removed from individual finite element stiffness matrices in ways that were not before possible. This capability exists because the modeling capabilities are clearly identified by the physically based notation. Furthermore, the number of integrations that must be performed is reduced. This eliminates the need to use Gaussian quadrature integration, so the errors introduced by the transformations required by the isoparametric formulation procedure are eliminated. These errors are discussed in detail in the lessons of Part III. The isoparametric approach is presented in Appendix A of Lesson 8 so the reader can conveniently compare the new approach to the standard approach for forming finite element stiffness matrices.

Problem 2 — Mesh Refinement

The use of strain gradient notation provides the basis for developing new methods for evaluating the accuracy of finite element results. Procedures for estimating local errors in finite element results are developed and applied in Part V. The ability to compute local error estimates provides a way to ensure the accuracy of these approximate solutions. If the solution is not sufficiently accurate in a given region, the mesh can be refined to improve the representation.

The error analysis procedures developed in Part V evaluate the modeling accuracy of the results produced by individual elements that make up the total finite element solution. The errors in individual elements are estimated using either aggregate or pointwise error measures. The aggregate or elemental errors are estimated by evaluating the strain energy content of individual elements. The pointwise measures estimate the errors in the stresses and/or strains at individual points in each element. We see in Part V that the resolution of the pointwise error measures is much greater than that for the aggregate error measures based on strain energy. This higher resolution recommends the pointwise measures for general use.

Both types of error measures are computed by comparing the finite element result to a result based on the finite difference method. That is to say, the theoretical basis for the error measures comes from comparing the results of two different approaches for solving the problem. In this case, an approximation based on the variational approach is compared to an approximation based on the differential approach.

The practical basis for these error measures results from the fact that both the finite element and the finite difference methods are developed from the same polynomial functions. As mentioned earlier, an approximation of the finite difference result for the current nodal pattern is formed from the finite element displacements. The finite difference method is formulated using strain gradient notation in Part IV.

An Overview of the Strain Gradient Finite Element Formulation Process

The first step in a finite element analysis is the subdivision of the continuum into a finite number of regions. The use of subdivisions enables low-order polynomial interpolation functions to be written that approximate the displacements over each of the geometrically simple finite elements. The identification of the interpolation polynomials that adequately represent the rigid body motions and the strains in the finite element is of primary importance in producing accurate finite element representations. The ability of a finite element model to accurately represent the behavior of the continuum depends on how well the strain approximations represent the actual deformations. The procedures developed in Lesson 6 for rationally identifying the interpolation polynomials needed to form equivalent continuum representations of discrete structures is used in Lesson 8 to identify the interpolation polynomials needed for a finite element with a given nodal pattern.

Once these interpolation functions are identified, they are substituted into the linear elasticity strain-displacement relations to produce the strain models. The modeling capabilities of these approximate strain representations are evaluated at this point by comparing the strain models to the Taylor series expansions for strains. The approximation polynomials are corrected if modeling deficiencies are detected. The complete Taylor series expansions of the strains in strain gradient notation serve as the templates for the ideal strain representations.

The strain models and the interpolation functions are used to reduce (or transform) functional representations of the potential energy for each finite element into a polynomial with a finite number of variables, or degrees of freedom. The principle of minimum potential energy is applied and the resulting equilibrium equations for the individual elements are put in matrix form as $[K]\{D\} = \{F\}$. The components of this equation are the finite element stiffness matrix $[K]$, the vector of unknown nodal displacements $\{D\}$, and the applied load vector $\{F\}$. The individual or elemental stiffness matrices and load vectors are then assembled into the global stiffness matrix $[K_G]$ to approximate the total problem as shown in Fig. 2 (see Note 1, Lesson 6).

An eight-step procedure for formulating the individual finite element stiffness matrices is presented in Lesson 8. The general characteristics of individual finite elements and overall finite element models are now discussed. This overview provides the background for the detailed development contained in the lessons that follow.

General Modeling Requirements

As mentioned earlier, the transparent nature of strain gradient notation enables the modeling characteristics of individual elements to be seen during the formulation procedure. Before we can assess the modeling characteristics, we need to know the modeling characteristics that are required for a successful element. As we will see, all elements must contain certain basic modeling characteristics if they are to produce accurate results. These requirements are now identified. This knowledge assists us in identifying the approximation polynomials needed to form successful finite elements.

Intuitive physical arguments are now given to show that all finite element stiffness matrices must be able to represent rigid body motions and constant strain states if the results are to converge to the exact result as the model is refined. The physical meaning of

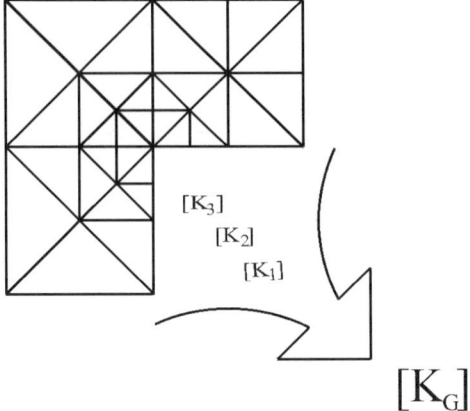

Figure 2. Global stiffness matrix formulation.

a rigid body motion is obvious from its name, i.e., if a body experiences rigid body motion, it does not deform. The need for a finite element to be able to represent rigid body motions can be seen by considering the following problem. Let us assume that a finite element model is loaded as shown in Fig. 3.

The loaded section of the model deforms. The unrestrained portion of the model beyond the loaded segment contains elements that may not deform. These elements experience only rigid body rotations and rigid body displacements. If these individual finite elements are not capable of representing rigid body motions, they must deform. This would overly constrain the model and the results would not accurately represent the actual solution. Thus, we can conclude that all finite elements must be capable of representing rigid body motion if they are to produce accurate results.

An intuitive physical argument can be made to demonstrate that a finite element must also be capable of representing constant strains. Let us assume that a continuum is loaded so that it experiences deformations due to constant strains. If the individual finite elements cannot represent constant strains, the finite element model will represent the actual deformations as rigid body motions or as a more complex deformation. Since the continuum can experience constant strain, we can conclude that individual finite elements

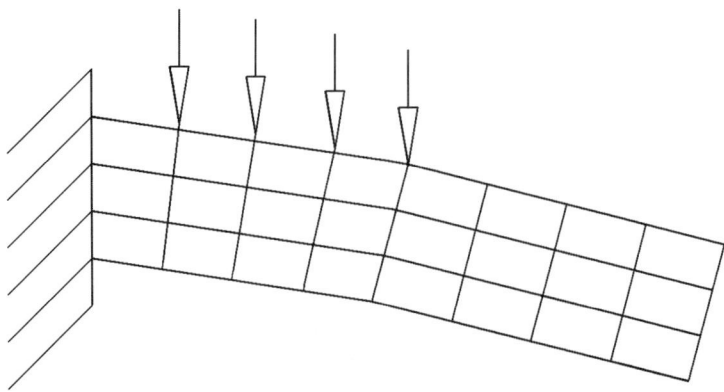

Figure 3. A finite element model with elements experiencing rigid body motion.

must be capable of representing constant strain states if accurate results are to be obtained. The requirement for all elements to be able to represent constant strains is discussed in the context of convergence criteria in the next section.

Arguments similar to those just presented could be made for each higher order strain state, until, in the limit, the approximate solution would be required to contain an infinite number of strain states. This would be equivalent to finding the exact solution to the problem and we would have strayed from the finite element method. However, this verbal argument shows that the complete representation of the strains serves as the ideal template for evaluating the strain modeling capabilities of a finite element. The ability to represent complex strain distributions is intimately tied to the issues of error analysis in individual elements and mesh refinements.

Convergence Criteria

A primary question of interest concerning the modeling characteristics of an element is the identification of the characteristics that a finite element must possess if the approximate solution is to converge to the actual result as the mesh is refined. The discussion concerning rigid body motion illustrated by Fig. 3 showed that all finite elements must be capable of representing rigid body motion if they are to converge to the exact result. We now more formally demonstrate that each finite element must also be capable of representing constant strains if a finite element solution is to converge to the exact result.

Let us imagine that the model shown in Fig. 3 is refined again and again until in the limit the finite element model consists of an infinite number of infinitely small elements. As was seen in the development of strain gradient notation, the strains can be expanded as Taylor series, e.g., $\varepsilon_x = (\varepsilon_x)_0 + (\varepsilon_{x,x})_0\, x + \cdots$. In the limit, as the dimensions of the element approach zero, all but the constant strain term in the Taylor series expansion go to zero. Thus in the limit, each element must be capable of representing constant strain if the finite element model is to converge to the exact result.

Finite element stiffness matrices can be tested for the ability to model rigid body motions and constant strain states with a procedure known in the literature as the "patch test" (see Ref. 1). In the patch test, a group or patch of elements is assembled and loaded in such a way as to produce constant strain in the model. The individual elements with various orientations must reproduce the expected rigid body and constant strain states or they fail the test. A failure to pass the patch test indicates that an element does not possess the capacity to represent rigid body motion and/or constant strain states. This means that a model utilizing such elements cannot converge to the exact result. Elements that do not pass the patch test are used at the peril of the analyst.

Finite Element Interpolation Functions

The discussions of the general modeling requirements and the convergence criteria have shown that successful finite elements must be capable of representing rigid body motions and the constant strain states. These modeling requirements define the first few terms that must be contained in the interpolation functions for successful finite elements. The required terms can be seen by studying the Taylor series expansions for the displacements. The displacement expressions derived in Lesson 5 are partially reproduced here for convenience as

---

Displacement Polynomials

$$u(x, y) = (u_{rb})_0 + (\varepsilon_x)_0 x + [(\gamma_{xy}/2) - r_{rb}]_0 \frac{1}{2}(\varepsilon_{x,x})_0 x^2$$
$$+ (\varepsilon_{x,y})_0 xy + \cdots$$
$$v(x, y) = (v_{rb})_0 + [(\gamma_{xy}/2) + r_{rb}]_0 x + (\varepsilon_y)_0 y$$
$$+ \frac{1}{2}(\gamma_{xy,x} - \varepsilon_{x,y})_0 x^2 + (\varepsilon_{y,x})_0 xy + \cdots$$

As can be seen in this equation, the rigid body displacements, $(u_{rb})_0$ and $(v_{rb})_0$, are directly related to the constant, or zeroth-order, terms in these polynomials. The rigid body rotation, $(r_{rb})_0$, and the constant strain states, $(\varepsilon_x)_0$, $(\varepsilon_y)_0$, and $(\gamma_{xy})_0$ are related to the first-order, or linear, terms. Thus, the need to represent both rigid body motions and constant strain states requires that the constant and first-order terms be present in the approximation polynomials for the displacements.

The remainder of the polynomial terms required for a successful element are identified using the procedures developed in Lessons 6 and 7. For the relatively simple geometry utilized in finite elements, the correct polynomial is identified by relating the geometry of the element to the structure of Pascal's triangle. This approach identifies linearly independent low-order polynomial terms that correspond to strain gradient quantities. The identification of the required approximation polynomials for several configurations is presented in Lesson 6.

Discretization Errors in Finite Element Models

Errors are introduced into the finite element method by the very process of subdividing the problem into subregions. These errors are produced because the subregions are not capable of representing the full range of behavior of the continuum. That is to say, errors will exist if the simple polynomial interpolation functions of the individual finite elements cannot capture the actual strain behavior in the continuum. This error is known as *discretization error*.

Discretization errors can be reduced under the control of the analyst through the use of adaptive mesh refinement procedures. In this process, the mesh refinement is guided by procedures that determine the level of error contained in each of the individual elements. The elements with the largest errors are further subdivided. The discretization error can be reduced to any predetermined level by this approach.

Physically, the mesh refinement can be interpreted as placing more elements in regions of high strain gradient. Thus each element is required to represent a simpler deformation and the model more accurately represents the continuum. The effect of such a refinement can be seen for a one-dimensional case in Fig. 4. A complex displacement is approximated by a relatively coarse mesh in Fig. 4a. The mesh is refined in Fig. 4b by dividing each element in half. In this representation, each of the two smaller elements is required to approximate a simpler deformation than did the original element. Adaptive mesh refinement procedures improve finite element solutions by automatically identifying and refining regions with high levels of error. These procedures are discussed in Part V.

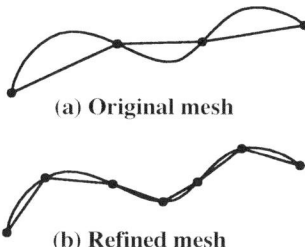

(a) Original mesh

(b) Refined mesh

Figure 4. Finite element mesh refinement.

Modeling Errors in Individual Finite Elements

Another type of error can exist in finite element results. This error is due to modeling deficiencies in individual elements. These errors are introduced when the displacement interpolation polynomials do not accurately represent the behavior of the continuum. These errors are called *elemental errors*. The use of strain gradient notation enables these errors to be identified and eliminated or controlled during the formulation of the individual elements.

Three well-known types of elemental errors exist. They are called parasitic shear, spurious zero-energy modes, and shear locking. In some discussions of elemental errors, parasitic shear and shear locking are not differentiated. This aggregation of errors usually occurs in the discussion of out-of-plane bending elements since both sources of elemental error may exist concurrently in these elements. Although both errors are caused by flaws in the strain models, analyses using strain gradient notation have shown that the causes of these two errors are distinctly different (see Ref. 2). The methods developed for eliminating or controlling these errors are based on identifying the causes of the three errors.

Parasitic Shear

Parasitic shear is an elemental error caused by the presence of erroneous normal strains in the shear strain expressions. In the case of plane stress elements, these erroneous terms are produced when the shear strain expressions are formed from incomplete polynomials. A comparison of the geometries of widely used elements with Pascal's triangle shows that quadrilateral elements are formed with incomplete polynomials. Thus, all quadrilateral plane elasticity elements must be corrected for parasitic shear.

Parasitic shear terms can be identified and removed during the formulation process in elements generated with the strain gradient approach. However, in the standard formulation procedures, it is difficult or impossible to remove the error during the formulation process. Instead, an attempt is made to remove the error during the numerical integration of the strain energy function using a procedure called reduced-order integration (see Note 1). This procedure is successful for the four-node element but it fails for the eight- and nine-node plane elasticity elements. Note that the use of selective reduced-order integration does eliminate parasitic shear in eight- and nine-node *plate-bending elements*. The source and removal of parasitic shear is discussed in detail in the next set of lessons.

In Lesson 12, it is shown that parasitic shear introduces a *qualitative error* in laminated composite plate elements that contain coupling between the in-plane and out-of-plane deformations when they are modeled using the Mindlin representation. A qualitative error is an error that is *not removed* by mesh refinement. In this element, the qualitative error takes the form of a coupling mode of deformation going in the wrong direction. For example, when a laminated plate element is given an axial load, the plate bends upward instead of downward because of the existence of the parasitic shear term. This is the first case that has been identified where an element must be formed using strain gradient. The isoparametric form of the element cannot be corrected with existing methods.

Spurious Zero-Energy Modes

The second source of elemental error, spurious zero-energy modes (see Ref. 2), is due to a mistake that is made while trying to correct for parasitic shear in eight- and nine-node elements. In the case of these elements, the uncritical application of reduced-order integration removes strain-modeling terms that belong in the strain energy expression of the finite element model. When these necessary and legitimate sources of strain energy are removed, the modes of deformation associated with these strain states do not contain any strain energy. The effect on the element is the creation of what appears to be extra rigid body modes. Even though functional-analysis-based techniques exist for eliminating the effects of this error from the finite element solutions, these procedures need never be utilized because parasitic shear can be removed during the formulation stage so extra zero-energy modes are not introduced into the model by mistake.

Shear Locking

A third type of elemental error known as *shear locking* is contained in out-of-plane bending elements with independent shear strain models. This error is due to improper modeling of the out-of-plane shear strains by the displacement polynomials. This error begins to dominate the problem by making the individual element overly stiff when the thickness becomes small with respect to the other dimensions of the element. This error is found in Timoshenko beam elements and Mindlin plate elements unless corrections are made. The cause of this modeling error was identified through the use of strain gradient notation [see Ref. 2].

An approach based on St. Venant's principle has been developed for controlling this modeling deficiency through insights provided by the use of strain gradient notation. The St. Venant correction factor is similar to *ad hoc* penalty methods previously developed to control shear locking. However, the strain gradient approach enables these correction factors to be developed using physical arguments.

Although this book focuses on plane stress problems, a Timoshenko beam is developed in Lesson 12 to identify the source of shear locking and to demonstrate that this error is indeed different from parasitic shear. The Timoshenko beam element is corrected using a St. Venant factor to control this error. Strain-gradient-based Mindlin plate elements containing the St. Venant correction factor are developed and demonstrated in Ref. 3. The mechanism that causes shear locking is also shown to cause plane elasticity elements with high length-to-width ratios to be overly stiff. This modeling deficiency is usually controlled by disallowing the use of high-aspect-ratio elements. The presentation is first made from the point of view of the out-of-plane bending elements because the development is more direct. Then the error due to aspect ratio stiffening is considered.

Isoparametric Finite Elements

A large number of the finite elements in current use are computed with the isoparametric formulation procedure. In the discussions that follow, this approach is referred to as the standard finite element formulation procedure unless otherwise specified. In the isoparametric approach, the element displacement approximations are written directly in terms of nodal displacements of the finite element. This eliminates the need to perform the computationally intensive operation of inverting a matrix. These displacement approximations are known as Lagrange interpolation functions.

The computational effort is further reduced by mapping the actual geometry of the finite element onto a standard shape or "parent" element. Typical isoparametric mappings are shown in Fig. 5. Similar mappings are available for triangular elements. The mapping of every finite element onto the same shape simplifies the numerical integration, which is performed with Gaussian quadrature (see Note 1). The form of this mapping gives the isoparametric elements their name. The mapping has the same (iso-) form as the interpolation functions. However, the variables (parameters) in the mapping function are the geometric location of the nodal points instead of the nodal displacements, as is the case in the interpolation functions. Thus finite elements formed by this process are called "isoparametric" elements (see Refs. 4–9).

Errors in the modeling characteristics of isoparametric finite elements with curved edges have been identified using analyses based on strain gradient notation (see Ref. 10). These errors consist of incorrect values in the strain energy when the element is deformed by higher order strain states. These modeling deficiencies are produced by the nonlinearities contained in the isoparametric mappings. There are no errors in the rigid body motions or constant strain states so the elements containing this modeling deficiency pass the patch test. Elements that do not contain these errors have been developed using the strain gradient approach (see Ref. 11).

Advantages of Self-Referential Strain Gradient Notation

Virtually all of the developments contained in this part of the book are made possible through the use of strain gradient notation. As we saw in Lessons 5, 6, and 7, this notation directly relates the displacement interpolation polynomials to the sources of displacement in the continuum, namely, the rigid body motions and the strains. The close connection between the notation and the source of the displacements enables the mathematical models developed with this notation to be directly related to the physical problem being analyzed. This enables the elemental errors to be identified and corrected during the formulation stage using physical arguments.

The *a priori* element evaluation capabilities provided by strain gradient notation differs from the way elemental modeling errors are usually identified. In the usual scenario, the errors are identified when anomalies are detected in finite element results. The errors are identified in this hit-or-miss approach because the use of polynomials with

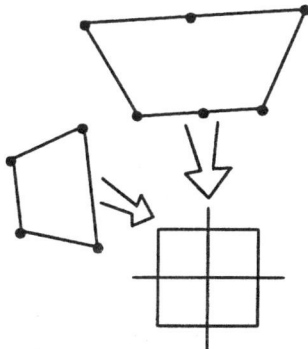

Figure 5. Isoparametric mappings.

arbitrary coefficients does not enable the finite element model to be directly compared to the problem being represented. As a result, no formal procedure for evaluating element performance exists that is universally used.

Contents of Part III

The lessons that make up Part III are designed to highlight the value of the strain gradient approach in the formulation of the finite elements. This notation has the capacity to identify and eliminate elemental errors with direct physical arguments that have not existed before. In addition, this approach substantially reduces the number of integrals that must be evaluated during the computation of an elemental stiffness matrix. Furthermore, the integrals have a simpler form than do the integrals that must be evaluated in the isoparametric formulation method. This eliminates the need for the isoparametric mappings. It is not clear whether the strain gradient approach is always computationally more efficient than the standard approach for the same elements. However, experience to date has shown that an efficiently executed strain gradient approach is always competitive with the isoparametric approach. Furthermore, elements with properties that were not possible before can now be developed.

Lesson Outlines

Lesson 8 develops the strain gradient formulation procedure and demonstrates it for the three-node constant strain triangle. The significance of this approach is seen to be the ability to directly relate the modeling capabilities of the elements to the continuum as the element is formed.

Lesson 9 utilizes the development of Lesson 8 to identify the presence of parasitic shear in the process of formulating the stiffness matrix for the four-node quadrilateral element. The structure of the strain energy expression enables the terms that cause parasitic shear to be removed before the expression is integrated. This eliminates the need to correct the element through the use of a reduced-order Gaussian quadrature integration formula.

The six-node linear strain triangle is developed in Lesson 10. This element is developed from complete polynomials and, hence, contains no parasitic shear terms. This validates the contention in Lesson 9 that the source of parasitic shear in plane strain and plane stress elements is due to the use of incomplete polynomials. The lesson also shows that the isoparametric formulation procedure develops elements with the incorrect amount of strain energy when the original shape of the element contains curved edges. This flaw is corrected with the strain gradient formulation procedure.

In Lesson 11, the eight- and nine-node elements are developed. This development identifies the source of spurious zero-energy modes in finite elements. The cause of this modeling deficiency is shown to be due to the improper use of reduced-order Gaussian quadrature integration to correct for parasitic shear. Furthermore, we see that parasitic shear cannot be correctly removed from isoparametric elements through the use of reduced-order integration.

Lesson 12 develops the stiffness matrices for two Timoshenko beam elements with different numbers of nodes. This is done to show that shear locking and parasitic shear are independent modeling errors in out-of-plane bending elements. We see that shear locking is contained in the polynomial representation itself and cannot be removed by simply removing a polynomial term. A correction factor must be included in the strain model used in the element to control the effects of shear locking in the element. The effect of parasitic shear as the source of qualitative errors in laminated composite plate elements is also presented. The cause of aspect ratio stiffening is also shown to be due to the same mechanism as shear locking.

Notes

1. The mapping of elements of general shape onto a regular shape in the isoparametric formulation procedure enables the strain energy integrals to be computed with fixed integration limits. This mapping is designed to accommodate the approximate Gaussian quadrature method. In this method, the integrand is evaluated at a small number of points at specified locations. The weighted sum of these values estimates the integral. For example, if a 3×3 pattern is used, an integral with a complete fifth-order polynomial integrand is evaluated exactly. An integral with a higher order integrand is evaluated approximately. If a 2×2 pattern is used, a complete third-order polynomial is evaluated exactly. In comparison to the 3×3 case, the 2×2 case is referred to as reduced-order integration. Furthermore, if a 2×3 or a 3×2 pattern is used, a polynomial with fifth-order terms in one variable and third-order terms in the other variable are evaluated exactly. In comparison to the 3×3 case, these patterns produce selectively reduced-order integration. Expositions of the Gaussian quadrature method are given in most finite element textbooks (e.g., Ref. 4).

References and Other Readings

1. Irons, B., and Ahmad, S. *Techniques of Finite Elements*, Ellis Horwood Limited, Chichester, England, 1980.

2. Dow, J. O., and Byrd, D. E. ''The Identification and Elimination of Artificial Stiffening Errors in Finite Elements,'' *International Journal of Numerical Methods in Engineering*, Vol. 26, May 1988, pp. 743–762.

3. Dow, J. O., and Byrd, D. E. ''Error Estimation Procedures for Plate Bending Elements,'' *AIAA Journal*, Vol. 28, Apr. 1990, pp. 685–693.

4. Zienkiewicz, O. C. *The Finite Element Method in Structural and Continuum Mechanics*, McGraw–Hill Publishing Co. Ltd., London, 1967.

5. Zienkiewicz, O. C. *The Finite Element Method*, 3rd ed. McGraw–Hill Publishing Co. Ltd., London, 1977.

6. Zienkiewicz, O. C., and Taylor, R. L. *The Finite Element Method*, 4th ed. McGraw–Hill Publishing Co. Ltd., London, 1989.

7. Cook, R. D. *Concepts and Applications of Finite Element Analysis*, John Wiley and Sons, Inc., New York, 1974.

8. Cook, R. D., Malkus, D. S., and Plesha, M. E. *Concepts and Applications of Finite Element Analysis*, 3rd ed. John Wiley and Sons, Inc., New York, 1989.

9. Segerlind, L. J. *Applied Finite Element Analysis*, John Wiley and Sons, Inc., New York, 1984.

10. Dow, J. O., Ho, T. H., and Cabiness, H. D. ''A Generalized Finite Element Evaluation Procedure,'' *ASCE Journal of Structural Engineering*, Vol. 111, No. 2, Feb. 1985, pp. 435–452.

11. Dow, J. O., Cabiness, H. D., and Ho, T. H. ''A Linear Strain Element with Curved Edges,'' *ASCE Journal of Structural Engineering*, Vol. 112, No. 4, Apr. 1986, pp. 692–708.

Lesson **8**

The Development of Strain-Gradient-Based Finite Elements

Purpose of Lesson   To develop and demonstrate a procedure for formulating finite element stiffness matrices using strain gradient notation.

The use of strain gradient notation enables a direct connection to be made between the mathematical representation of a physical problem and the actual physical problem. The physical nature of the notation enables the modeling characteristics of finite element stiffness matrices to be assessed and corrected during their development. This contrasts with the current approach in which the modeling deficiencies are inferred from errors found in the solutions to problems.

The strain gradient approach provides an alternative to existing finite element formulation procedures. In addition to enabling several types of elemental errors to be removed using physical arguments, the strain gradient approach possesses certain computational advantages over other approaches. The formulation procedure using strain gradient notation requires the evaluation of fewer integrals than does the standard approach. Furthermore, these integrals are of simpler form than the integrals found in other approaches. Developments in later lessons show that the strain gradient approach enables new types of finite elements to be formulated. The strain gradient formulation procedure is demonstrated in this lesson with the development of the stiffness matrix for a three-node constant strain triangle. The isoparametric approach for formulating finite element stiffness matrices is presented in Appendix A.

■   ■   ■

The finite element form of the Rayleigh-Ritz solution technique subdivides a larger problem into a finite number of small regions with relatively simple boundary conditions. A complex region is typically divided into rectangular and triangular subregions, as shown in Fig. 1. The displacements over each subregion or finite element are approximated with low-order polynomial trial functions having n degrees of freedom. A strain energy expression is formed in terms of these displacement approximations for each of the elements and the principle of minimum potential energy is applied. The resulting n equilibrium equations are put in the form of an $n \times n$ stiffness matrix and a load vector with n rows. The individual matrix representations for each subregion are assembled to form a mathematical model that approximates the overall problem, as symbolized in Fig. 1.

In this lesson, a procedure for formulating finite element stiffness matrices based on the use of self-referential strain gradient notation is developed. Strain gradient notation directly relates the displacement approximation polynomials used in the finite element formulation to the physical causes of the displacements in the continuum, namely, the rigid body motions and the strain quantities. That is to say, the coefficients of the displacement approximation polynomials are written in terms of rigid body motions and

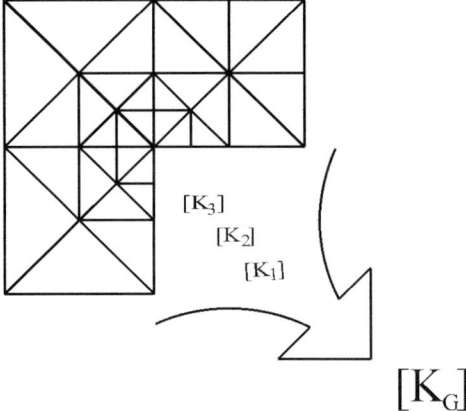

$[K_3]$

$[K_2]$

$[K_1]$

$[K_G]$

Figure 1. Finite element model for a typical continuum.

strain gradient terms. This notation is derived in Lesson 5. It is discussed and applied at length in Lessons 6 and 7.

The close relationship between the notation and the source of the displacements enables the mathematical model to be directly compared to the physical problem being solved. This enables the modeling characteristics of the individual elements to be assessed and, if necessary, corrected as the element is formulated. During the development of this procedure, we see that the strain gradient approach possesses certain computational advantages over the isoparametric approach. The isoparametric formulation procedure is presented in Appendix A of this lesson.

Finite Element Formulation Procedure

The strain-gradient-based procedure for formulating finite element stiffness matrices is an eight-step process, as shown in Table 1. Step 1 identifies the strain gradient quantities that serve as generalized coordinates for the displacement interpolation polynomials for a finite element of a given configuration. These coefficients identify the independent strain states that a finite element is capable of representing. Step 2 forms and evaluates the strain models that result from the displacement polynomials formed in step 1. As one part of this evaluation, the strain models are compared to the Taylor series expansion of the strain components. Step 3 forms the strain energy expressions for the subregion being

Table 1. Strain-gradient-based finite element formulation procedure.

1. Polynomial identification.
2. Strain model evaluation.
3. Formulation of strain energy expressions in strain gradient coordinates.
4. Integration of strain energy expressions.
5. Formulation of strain gradient to nodal coordinate transformation.
6. Transformation of strain energy expression.
7. Formulation of applied load work function.
8. Application of principle of minimum potential energy to produce the force-displacement relation.

modeled. Because these expressions have a simplicity that does not exist in other methods used to form finite element stiffness matrices, the integrations of these strain energy expressions are considered as a separate operation in step 4. The results of step 4 are strain energy quantities expressed in terms of strain gradient coordinates. To get the stiffness matrix in terms of the desired nodal coordinates, the strain energy expression formed in step 4 must be transformed to these coordinates. The required transformation is formed in step 5 and the strain energy expression is transformed to nodal coordinates in step 6. The work function that introduces the applied loads into the potential energy expressions is formed in step 7. In step 8, the principle of minimum potential energy is applied to form the force-displacement relation that contains the finite element stiffness matrix.

The eight steps required to form a finite element stiffness matrix are illustrated one at a time in the next eight sections for the three-node, six-degree-of-freedom triangle.

Step 1 — Polynomial Identification

The first step in the strain gradient formulation procedure is the identification of the approximation polynomials that correspond to the number of degrees of freedom and geometry of the element being formulated. These polynomials must be of the lowest order that will provide a linearly independent basis set for the element. This step was discussed at length in Lesson 6 but it is reiterated here in condensed form so that this lesson can stand alone.

The requirement for the low-order displacement approximation polynomials was discussed in the introduction to Part III preceding this lesson. The constant terms are needed to represent the rigid body displacements. The linear terms are required to represent the rigid body rotation and constant strains. The quadratic terms are the next possible candidates for addition to the approximation polynomials because of completeness requirements. If the geometry of the element requires more terms in the displacement models, the cubic terms are the first candidates etc.

If a choice exists concerning which of these higher order terms to use in the approximation polynomial, symmetric terms are often chosen so the element will be as invariant as possible with respect to its orientation. For example, in the case of a four-node element, a choice exists between taking x^2, xy, or y^2 as the fourth term in the displacement approximations. The xy term is usually chosen because of the symmetry argument. The need for the requirement that the strain gradient terms be linearly independent is discussed in step 5 when the transformation from strain gradient coordinates to nodal displacement coordinates is developed.

A simple mechanism for identifying the polynomials that satisfy the requirements just outlined exists for the nodal configurations encountered in finite elements. The terms in the polynomials are identified by relating the nodal locations of the element to Pascal's triangle, as was done in Lesson 6. The strain gradient quantities associated with the coefficients of the approximation polynomials are taken from Table 1 of Lesson 5. This table contains the strain gradient form of the coefficients that correspond to the algebraic terms identified with Pascal's triangle. This process identifies the strain gradient quantities or strain states that the element is capable of representing.

The three-node triangular element shown in Fig. 2 contains six degrees of freedom. When the geometry of the element is compared to Pascal's triangle, the lowest order polynomials with three unknown coefficients that can represent the u and v displacements contains a constant term, an x term, and a y term. The strain gradient representation is completed by taking the coefficients of the appropriate terms from Table 1 of Lesson 5. The subscript 0 on the strain gradient coefficients is a reminder that the displacement

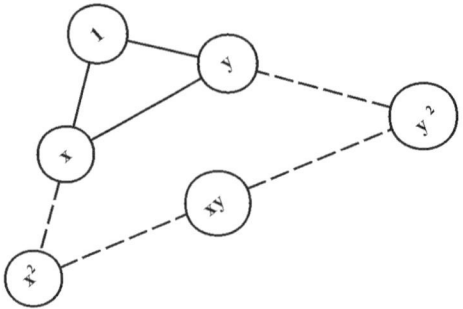

Figure 2. A triangular finite element region.

approximations are Taylor series expansions around a local origin. The approximation polynomial for this case is

---

Strain Gradient Displacement Approximations

$$u(x, y) = (u_{rb})_0 + (\varepsilon_x)_0 x + [(\gamma_{xy}/2) - r_{rb}]_0 y$$
$$v(x, y) = (v_{rb})_0 + [(\gamma_{xy}/2) + r_{rb}]_0 x + (\varepsilon_y)_0 y$$

$$(8.1)$$

The strain gradient terms that an element can represent are contained in the displacement polynomials found by the process just described. Since this triangular element has six degrees of freedom, it is capable of representing six linearly independent strain states. When these terms are extracted from Eq. 8.1, they are found to be

---

Strain States Represented

$$(u_{rb})_0 \quad (v_{rb})_0 \quad (r_{rb})_0$$
$$(\varepsilon_x)_0 \quad (\varepsilon_y)_0 \quad (\gamma_{xy})_0$$

$$(8.2)$$

The six strain states are the three planar rigid body motions and the three planar constant strain quantities. Thus, the three-node element satisfies the minimum modeling requirements of an element that is capable of converging to the exact result, as discussed in the introduction to Part III. This element is called the constant strain triangle (CST) because it is only capable of representing constant strain states.

This step in the formulation process identifies the interpolation functions for an element and, consequently, the strain states that the element is capable of representing. As such, the result of this step provides the basis for the remainder of the formulation process.

Step 2 — Strain Approximation and Model Evaluation

The second step in the formulation process is the development and evaluation of the strain polynomials for the element. The approximate strain representations are formed by substituting the displacement polynomials from the previous step into the linear elasticity definitions of strains. The resulting strain models are evaluated by comparing them to the Taylor series expansion for the strains on a term-by-term basis. If the coefficients in the approximations match those of the Taylor series, they are judged to be correct.

Two types of errors exist: one of commission and one of omission. Both types of error exist in elements formed from incomplete polynomials. It is formally demonstrated in Lesson 9 that extra terms exist in the shear strain expression in elements formed with incomplete polynomials. These terms produce a modeling deficiency called parasitic shear, which was shown informally in Lesson 7 to reduce the expected displacements in finite element models. In other words, the comparison to the Taylor series identified the terms that overstiffened the individual finite elements. This comparison also identifies the terms that are truncated from the strain representation. In Lesson 7, we saw that the lack of these terms caused the strain models to vary with rotation.

Such a comparison shows that the three-node triangle can represent no terms beyond the constant strain coefficients. The definition of the three strains that are active in plane stress problems are

---

Strain-Displacement Relations

$$\varepsilon_x = \frac{\partial u}{\partial x} \qquad \varepsilon_y = \frac{\partial v}{\partial y} \qquad \gamma_{xy} = \frac{\partial u}{\partial y} + \frac{\partial v}{\partial x} \tag{8.3}$$

When the components of Eq. 8.1 are substituted into Eq. 8.3, the result is

---

Strain Representations

$$\varepsilon_x = (\varepsilon_x)_0 \qquad \varepsilon_y = (\varepsilon_y)_0 \qquad \gamma_{xy} = (\gamma_{xy})_0 \tag{8.4}$$

The components of Eq. 8.4 are the strain models contained in the three-node finite element being developed. These one-term polynomials consist of the constant term in the Taylor series expansion for each of the strains. Since these approximations match the Taylor series representation, no elemental errors are introduced through the strain approximations. This is to say, the strain models do not contain any erroneous terms that differ from the Taylor series representation. However, the element can only represent constant strain.

The remainder of the derivation of the element stiffness matrix is simplified if the strain approximations are put in matrix form. When this is done, the result is

---

Strain Representation in Matrix Form

$$\{\varepsilon\} = [T]\{\varepsilon_,\}$$

where

$$\{\varepsilon\}^T = [\varepsilon_x \ \varepsilon_y \ \gamma_{xy}] \tag{8.5}$$

$$\{\varepsilon_,\}^T = [(u_{rb})_0 \ (v_{rb})_0 \ (r_{rb})_0 \ (\varepsilon_x)_0 \ (\varepsilon_y)_0 \ (\gamma_{xy})_0]$$

$$[T] = [[T_0]|[T_\varepsilon]] = \begin{bmatrix} \begin{bmatrix} 0 & 0 & 0 \\ 0 & 0 & 0 \\ 0 & 0 & 0 \end{bmatrix} \begin{bmatrix} 1 & 0 & 0 \\ 0 & 1 & 0 \\ 0 & 0 & 1 \end{bmatrix} \end{bmatrix}$$

Equation 8.5 represents the strains in the continuum with a discrete model in terms of rigid body motions and the constant terms of the Taylor series expansion for strains. The rigid body terms are introduced into this equation by including the null partition $[T_0]$ in the transformation matrix $[T]$. It might not be obvious at this point why the rigid body motions are included in the strain transformation because, by definition, rigid body motions do not produce strains. In a subsequent step of the finite element formulation process, the strain energy expression is transformed to nodal displacements and the final result must include the rigid body modes.

This step has developed the strain approximations that are used to form the strain energy expression in terms of the strain gradient variables. The use of this physically based notation enables the strain approximations to be compared to the full Taylor series expansion to determine if any erroneous terms are introduced into the strain model. In this simple case, we saw that no errors were introduced. However, in the next lesson, we see that erroneous terms are introduced into the shear strain model for a four-node element.

Step 3 — Formulation of the Strain Energy Expression

The third step in the formulation process is the development of a strain energy function expressed in terms of strain gradient quantities. We will see that this function contains fewer integrals that must be evaluated than does the strain energy expression produced by the commonly used formulation procedures. Furthermore, these integrals have a form that makes their evaluation simple. The use of strain gradient coordinates eliminates the need to map the element onto a simple shape, as is done in the isoparametric procedure so that the integrals can be evaluated numerically using Gaussian quadrature (see Note 1).

The strain energy function for the plane stress problem written in terms of strains was developed in Lesson 3. It is reproduced here for convenience as

---

Strain Energy Expression

$$U = \frac{1}{2} \int_{\Omega} \left(\frac{E}{1 - \nu^2} \right) \left(\varepsilon_x^2 + 2\nu\varepsilon_x\varepsilon_y + \varepsilon_y^2 + \left(\frac{1 - \nu}{2} \right) \gamma_{xy}^2 \right) d\Omega \qquad (8.6)$$

This expression can be written in matrix form as

Matrix Form of the Strain Energy Expression for Plane Stress

$$U = \frac{1}{2} \int_{\Omega} \{\varepsilon\}^T [E] \{\varepsilon\} \, d\Omega$$

where

$$\{\varepsilon\}^T = [\varepsilon_x \; \varepsilon_y \; \gamma_{xy}]$$

$$[E] = \frac{E}{(1 - \nu^2)} \begin{bmatrix} 1 & \nu & 0 \\ \nu & 1 & 0 \\ 0 & 0 & \alpha \end{bmatrix}$$

$$\alpha = \frac{1 - \nu}{2}$$

(8.7)

At this point in the Rayleigh-Ritz procedure, the finite degree of freedom approximations are introduced into the strain energy expression for the continuum. This is accomplished in the finite element method by substituting the strain approximations given by Eq. 8.5 into Eq. 8.7 to give

Matrix Form of the Discrete Strain Energy Expression

$$U = \frac{1}{2}\{\varepsilon_,\}^T \left[\int_\Omega [T]^T [E][T] \, d\Omega \right] \{\varepsilon_,\}$$

$$U = \frac{1}{2}\{\varepsilon_,\}^T \overline{U} \{\varepsilon_,\} \qquad (8.8)$$

where

$$\overline{U} = \int_\Omega [T]^T [E][T] \, d\Omega.$$

Equation 8.8 is a discrete approximation of the strain energy with strain gradient quantities as the independent variables. The strain gradient components can be viewed as the ''generalized coordinates'' for this Rayleigh-Ritz model. At this point, the differences between the standard finite element formulation procedure and the strain gradient approach can be observed.

The integrals that must be evaluated for this form of the strain energy expression are simpler than those that must be evaluated in the standard formulation procedure. Because of the significance of this difference, the evaluation of the integrals is treated as a separate step. After the integrals are evaluated, the finite element stiffness matrix is formed by transforming the strain energy expression from strain gradient coordinates to nodal displacements.

Step 4 — Integration of the Strain Energy Terms

The integrals contained in Eq. 8.8 are now evaluated. They are contained in $\overline{U}$, which is defined in Eq. 8.8. The matrix $\overline{U}$ can also be considered as the stiffness matrix for the finite element in the generalized coordinates $\{\varepsilon_,\}$. The definition of $\overline{U}$ is repeated here for convenience as

Strain Energy Integrals

$$\overline{U} = \int_\Omega [T]^T [E][T] \, d\Omega \qquad (8.9)$$

Equation 8.9 is expanded using Eq. 8.5 to give

Strain Energy Integrals

$$\overline{U} = \int_\Omega \begin{bmatrix} [T_0]^T[E][T_0] & [T_0]^T[E][T_\varepsilon] \\ [T_\varepsilon]^T[E][T_0] & [T_\varepsilon]^T[E][T_\varepsilon] \end{bmatrix} d\Omega \qquad (8.10)$$

Since the transformation $[T_0]$ is a null matrix, any term containing it is zero. Thus, the lower right partition $\overline{U}_{22}$ of Eq. 8.10 contains the only nonzero elements in this expression. When this partition is expanded, it has the following form:

Expanded Form of the Strain Energy Integrals

$$\overline{U}_{22} = \frac{E}{(1-\nu^2)} \int_\Omega \begin{bmatrix} 1 & \nu & 0 \\ \nu & 1 & 0 \\ 0 & 0 & \alpha \end{bmatrix} d\Omega \tag{8.11}$$

As we can see, all of the terms under the integral are constants, so only one integral needs to be evaluated to form the strain energy expression for the three-node triangle. The common integral has the following form:

Finite Element Area

$$A = \int_\Omega d\Omega \tag{8.12}$$

This integral simply computes the area of the triangle. The area is contained in each of the terms of the strain energy expression given by Eq. 8.11. The evaluation of this one integral contrasts to the 21 integrals that must be evaluated in the isoparametric procedure for formulating the constant strain triangle. When Eq. 8.12 is substituted into Eq. 8.11, the final form of the nonzero strain energy terms for the constant strain triangle is

---

Expanded Form of the Strain Energy Integrals

$$\overline{U}_{22} = \frac{EA}{(1-\nu^2)} \begin{bmatrix} 1 & \nu & 0 \\ \nu & 1 & 0 \\ 0 & 0 & \alpha \end{bmatrix} \tag{8.13}$$

When higher order elements are formed in later lessons, more than 1 integral must be integrated. For example, in the case of the six-node linear strain triangle, the area integral, the two first moments of area, and the three second moments of the area must be evaluated. That is to say, only 6 different integrals must be evaluated. This contrasts to the 78 integrals that must be evaluated during the standard formulation of the six-node triangle.

We now interpret the meaning of the individual terms in Eq. 8.13. The diagonal components correspond to the strain energy contained in the element when it is deformed by a single strain gradient term. For example, the first diagonal element of Eq. 8.11 models the strain energy contained in the element when it is representing the constant strain state $(\varepsilon_x)_0$. Similarly, the other diagonal terms contain the strain energy due to the other constant strain states. For example, the strain energy content for the strain state of constant shear was computed in Application 5 of Lesson 5.

The off-diagonal terms contain the strain energy contained in the element as a result of the coupling between the strain states. The only nonzero coupling term present in a constant strain triangle corresponds to the Poisson effect, which couples ε_x and ε_y. The remaining off-diagonal terms are zero. This indicates that no coupling exists between the shear and normal strains.

The strain energy expression just developed is expressed in terms of strain gradient quantities (see Note 2). In this form, fewer integrals with simpler integrands must be evaluated (see Note 1). The presence of fewer and simpler integrals has computational implications that are discussed later. The ability to form the strain energy expressions in terms of strain gradient quantities enables elements to be developed that are guaranteed to contain exact representations of the strain energy content. The significance of being able to form finite elements that contain the correct amount of strain energy is discussed when the six-node linear strain triangle is developed in a Lesson 10.

Step 5 — Formulation of Coordinate Transformations

The finite element stiffness matrix emerges when the principle of minimum potential energy is applied. Before this principle can be applied, the strain energy function given in Eq. 8.8 must be expressed in terms of nodal displacements. This change of variables is accomplished by developing a transformation from strain gradient quantities to nodal displacements. The required transformation is formed by inverting the transformation from nodal displacements to strain gradient quantities, which we now construct. This transformation was developed and applied in Lesson 6.

The development of the transformation from nodal displacements to strain gradient quantities is based on the fact that the displacement components at any point in the continuum are represented as a linear combination of the displacements produced by the individual strain gradient terms, as given in Eq. 8.1. The displacements at each of the element nodes are found by substituting the nodal coordinates into Eq. 8.1. This produces one equation for each degree of freedom in the element. The rows of the transformation matrix are most easily developed by putting Eq. 8.1 in matrix form as

---

Matrix Form of the Displacement Approximations

$$u = [A_x]\{\varepsilon_,\} \qquad v = [A_y]\{\varepsilon_,\}$$

where

$$[A_x] = [1 \ 0 \ -y \ x \ 0 \ y/2]$$
$$[A_y] = [0 \ 1 \ x \ 0 \ y \ x/2]$$

$\{\varepsilon_,\}$ is defined in Eq. 8.5.

(8.14)

Before forming the full transformation, let us illustrate the relationship between one of the strain gradient quantities and the nodal displacements. As an example, let us find the vector of nodal displacements associated with the shear strain $(\gamma_{xy})_0$. This implies that we are assuming the other strain gradient terms are not active. In other words, the nodal displacements of the finite element are due solely to the shear strain. The displacements given by Eq. 8.14 or Eq. 8.1 for this single strain gradient term are the following:

Displacements Due to Shear Strain

$$u = \frac{y}{2}(\gamma_{xy})_0 \qquad v = \frac{x}{2}(\gamma_{xy})_0$$

(8.15)

The vector of nodal displacements associated with this strain state is found by substituting the coordinate location of each of the nodes into Eq. 8.15 (see Note 2). When this is done, the result is

---

Nodal Displacements Due to Shear Strain

$$\{d\} = \{\phi_6\}(\gamma_{xy})_0$$

where

$$\{d\}^T = [u_1 \ u_2 \ u_3 \ v_1 \ v_2 \ v_3] \tag{8.16}$$

$$\{\phi_6\}^T = \frac{1}{2}[y_1 \ y_2 \ y_3 \ x_1 \ x_2 \ x_3]$$

In this equation, $\{d\}$ is the vector of nodal displacements produced by the shear strains. The vector $\{\phi_6\}$ provides the pattern of the nodal displacements that produce shear strain in the element. The subscript 6 denotes the fact that this displacement is associated with $(\gamma_{xy})_0$ and this term is the sixth term in the $\{\varepsilon_,\}$ vector (see Eq. 8.5). The multiplier $(\gamma_{xy})_0$ identifies the level of participation of the shear strain. If this quantity is zero, then there is no displacement due to shear strain. If $(\gamma_{xy})_0$ has a larger value than the other strain gradient terms, the shear strain will dominate the overall deformation of the element. Similar vectors can be formed for each of the other five strain gradient quantities associated with the constant strain triangle. The total transformation results when a linear combination of these displacement vectors is formed as follows:

Nodal Displacement to Strain Gradient Transformation

$$\{d\} = [\Phi]\{\varepsilon_,\}$$

where

$$[\Phi] = [\phi_1 \ \phi_2 \ \phi_3 \ \phi_4 \ \phi_5 \ \phi_6] \tag{8.17}$$

In this transformation, ϕ_1 is associated with $(u_{rb})_0$, ϕ_2 is associated with $(v_{rb})_0$, etc. The $[\phi]$ matrix for the constant strain triangle has the following form:

Transformation Matrix for the CST

$$[\Phi] = \begin{bmatrix} 1 & 0 & -y_1 & x_1 & 0 & y_1/2 \\ 1 & 0 & -y_2 & x_2 & 0 & y_2/2 \\ 1 & 0 & -y_3 & x_3 & 0 & y_3/2 \\ 0 & 1 & x_1 & 0 & y_1 & x_1/2 \\ 0 & 1 & x_2 & 0 & y_2 & x_2/2 \\ 0 & 1 & x_3 & 0 & y_3 & x_3/2 \end{bmatrix} \tag{8.18}$$

This matrix can be interpreted as follows. The first column is the vector of nodal displacements associated with the rigid body motion in the x direction, $(u_{rb})_0$ etc. The sixth column is identical to the vector contained in Eq. 8.16 for the nodal displacements associated with the shear strain $(\gamma_{xy})_0$.

The desired transformation from strain gradient quantities to nodal displacements is formed by finding the inverse of the matrix given in Eq. 8.18. When this is done, the transformation can be written as

---

Strain Gradient to Nodal Displacement Transformation

$$\{\varepsilon_.\} = [\Phi]^{-1}\{d\}$$

(8.19)

The need to form the inverse of the $[\Phi]$ matrix to form the desired transformation is the source of the requirement that the strain gradient vectors be linearly independent. The use of Pascal's triangle ensures that the vectors representing the strain states meet this criterion.

Orthogonality Characteristics of the $[\Phi]$ Matrix

Lest the reader conclude that the need to form the inverse of the $[\Phi]$ matrix renders the strain gradient approach computationally noncompetitive, the orthogonality character-istics of the $[\Phi]$ matrix are examined for a triangle. We see that several of the vectors of the $[\Phi]$ matrix are orthogonal to each other. We then see that the number of orthogonal vectors is increased if the centroid is chosen as the local origin and/or if the local coordinate axes correspond to the principal axes.

The orthogonality characteristics are studied by forming the product $[\Phi]^T [\Phi]$ for four examples. The number of zero off-diagonal terms in the product indicates the orthogonality of the vector pairs. Thus, the larger the number of zero off-diagonal terms, the easier it is to form the inverse of the $[\Phi]$.

The first two cases examined are shown in Fig. 3. Figure 3a shows a triangle with the origin at node 1. The $[\Phi]$ matrix for this element is shown as Fig 4a. Figure 3b shows the

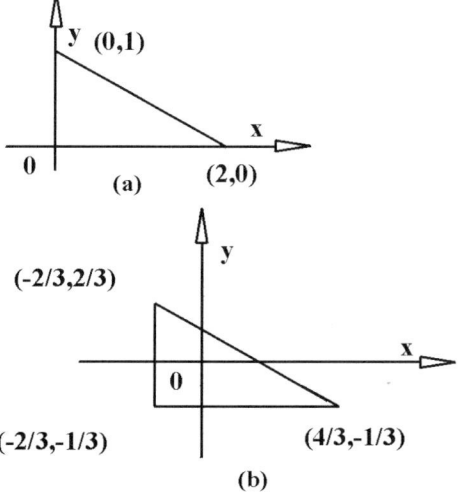

Figure 3. Two identical triangles with local coordinate systems.

$$\begin{bmatrix} 1 & 0 & 0 & 0 & 0 & 0 \\ 1 & 0 & 0 & 2 & 0 & 0 \\ 1 & 0 & -1 & 0 & 0 & 1/2 \\ 0 & 1 & 0 & 0 & 0 & 0 \\ 0 & 1 & 2 & 0 & 0 & 1 \\ 0 & 1 & 0 & 0 & 1 & 0 \end{bmatrix} \quad \begin{bmatrix} 1 & 0 & 1/3 & -1 & 0 & -1/3 \\ 1 & 0 & 1/3 & 1 & 0 & -1/3 \\ 1 & 0 & -2/3 & 1 & 0 & 2/3 \\ 0 & 1 & -1 & 0 & -1/3 & -1 \\ 0 & 1 & 1 & 0 & -1/3 & 1 \\ 0 & 1 & 0 & 0 & 2/3 & 0 \end{bmatrix}$$

$$\text{(a)} \qquad\qquad\qquad\qquad\qquad \text{(b)}$$

Figure 4. The [Φ] matrices for the triangles shown in Fig. 3.

same triangle with the origin at the centroid. The [Φ] matrix for this case is shown in Fig. 4b.

The orthogonality characteristics of these two [Φ] matrices are shown in Fig. 5. The matrices show the upper triangular elements for the product $[\Phi]^T [\Phi]$ for the two cases. The X's and 0's indicate whether or not the vectors associated with the off-diagonal term are orthogonal. The D's indicate diagonal elements. If the indicated value is zero, the vector pairs are orthogonal. If an X is given, the vectors are not orthogonal. For example, the (1,2) term in Fig. 5a is zero. This indicates that the vectors ϕ_1 and ϕ_2 for the triangle shown in Fig. 3a are orthogonal. Inversely, the fact that the (1,3) term is nonzero indicates that the vectors ϕ_1 and ϕ_3 are not orthogonal.

When we count the nonzero elements contained in Fig. 5a, we see that 7 of the 15 off-diagonal terms are zero. This means that over half of the possible orthogonality conditions are satisfied. The effort involved in finding the inverse of this [Φ] matrix is substantially less than the effort required to find the inverse for a case where none of the vectors are orthogonal.

For the matrix shown in Fig. 5b, we see that only five of the off-diagonal terms are nonzero. Thus, the effort to find the inverse for this case is reduced even further.

To illustrate the effect of locating the axes coincident with the principal axes, the orthogonality conditions for the triangles shown in Fig. 6 are evaluated. In Fig. 6a, the origin of the local system is located at node 1. The triangle shown in Fig. 6b has the origin located at the centroid and the axes are aligned with the principal axes.

The orthogonality characteristics of the [Φ] matrices for these two triangles are shown in Fig. 7. The number of nonzero terms is 11 for the case with the origin located at node 1. There is only 1 nonzero off-diagonal term for the case where the origin is located

$$\begin{bmatrix} D & 0 & X & X & 0 & X \\ & D & X & 0 & X & X \\ & & D & 0 & 0 & X \\ & & & D & 0 & 0 \\ & & & & D & 0 \\ & & & & & D \end{bmatrix} \quad \begin{bmatrix} D & 0 & 0 & 0 & 0 & 0 \\ & D & 0 & 0 & 0 & 0 \\ & & D & X & X & X \\ & & & D & 0 & X \\ & & & & D & X \\ & & & & & D \end{bmatrix}$$

$$\text{(a)} \qquad\qquad\qquad\qquad \text{(b)}$$

Figure 5. Structure of the upper triangle of $[\Phi]^T [\Phi]$.

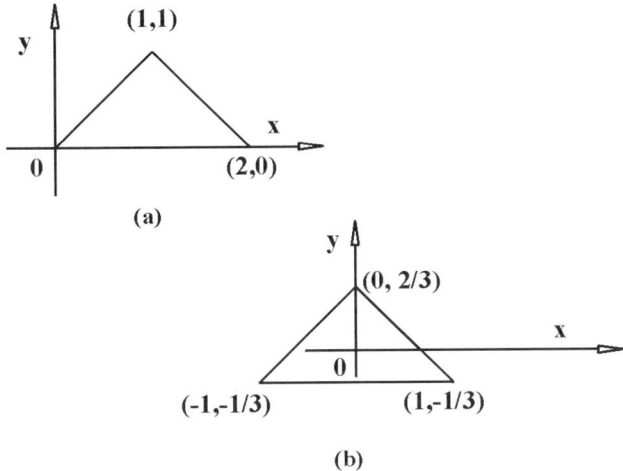

Figure 6. Two identical triangles with local coordinate systems.

$$\begin{bmatrix} D & 0 & X & X & 0 & X \\ & D & X & 0 & X & X \\ & & D & X & X & X \\ & & & D & 0 & X \\ & & & & D & X \\ & & & & & D \end{bmatrix} \qquad \begin{bmatrix} D & 0 & 0 & 0 & 0 & 0 \\ & D & 0 & 0 & 0 & 0 \\ & & D & 0 & 0 & X \\ & & & D & 0 & 0 \\ & & & & D & 0 \\ & & & & & D \end{bmatrix}$$

(a)   (b)

Figure 7. Structure of the upper triangle of $[\Phi]^T[\Phi]$.

at the centroid and the local axes are principal axes. These examples illustrate that the strain gradient approach can be made computationally competitive with the isoparametric approach. The same type of orthogonality conditions are exhibited by other element types.

The inversion of the $[\Phi]$ matrix is the most computationally intensive step in the formulation of a strain-gradient-based finite element. The reduction in effort resulting from the simplified form of the integrals that must be evaluated balances with the inversion to make the strain gradient formulation as computationally efficient as the isoparametric element formulation procedure. In this section, we have seen ways to take advantage of the special structure of the transformation matrix to reduce the effort required for this inversion. When this is done, the strain gradient element formulation procedure is even more efficient than the isoparametric approach (see Appendix A).

Step 6 — Transformation to Nodal Coordinates

We are now in a position to transform the strain energy to nodal displacements. This is accomplished by substituting Eq. 8.19 into Eq. 8.8. When this is done, the result is

Strain Energy Expression in Terms of Nodal Displacements

$$U = \frac{1}{2}\{d\}^T \left[\int_\Omega [\Phi]^{-T}[T]^T[E][T][\Phi]^{-1} \, d\Omega \right] \{d\} \tag{8.20}$$

Equation 8.20 can be written in terms of the finite element stiffness matrix as

Definition of a Strain-Gradient-Based Finite Element Stiffness Matrix

$$U = 1/2\{d\}^T[K]\{d\}$$

where

$$[K] = \left[\int_\Omega [\Phi]^{-T}[T]^T[E][T][\Phi]^{-1} \, d\Omega \right]. \tag{8.21}$$

The general definition of the finite element stiffness matrix is given in Eq. 8.21. That this is indeed the stiffness matrix is validated in step 8 with the application of the principle of minimum potential energy. The stiffness matrix for any finite element can be derived using this equation. This equation is used in subsequent sections to derive other types of elements.

Step 7 — Formulation of the Load Vector

The load vector associated with a finite element is derived from the work function expressed in terms of nodal displacements. Finite elements can be loaded with both distributed loads and point loads. In this derivation, the point loads are applied at the nodes of the element. The work function for a finite element can be written as

Finite Element Work Function

$$W = \int_\Omega f_x u \, d\Omega + \int_\Omega f_y v \, d\Omega + \sum [(F_x)_i u_i + (F_y)_i v_i]$$

where f_x and f_y are distributed loads in the x and y directions.

$(F_x)_i$ and $(F_y)_i$ are concentrated loads at the ith node.

(8.22)

The components of the work function must all be put in terms of the nodal displacements. This is accomplished by first transforming the continuous displacements to discrete functions of the strain gradient variables using the approximation polynomials (see Eq. 8.14). The strain gradient variables are then transformed to nodal displacement with Eq. 8.19. When this is done, the result is

Finite Element Work Function

$$W = \int_\Omega f_x [A_x][\Phi]^{-1}\{d\}\, d\Omega + \int_\Omega f_y [A_y][\Phi]^{-1}\{d\}\, d\Omega$$
$$+ \sum [(F_x)_i u_i + (F_y)_i v_i] \tag{8.23}$$

Equation 8.23 can be written more compactly as

Finite Element Work Function

$$W = \{d\}^T \{F\}$$
$$\text{where } \{F\}^T = [[F_{x_i}]|[F_{y_i}]]$$
$$F_{x_i} = \frac{\partial W}{\partial u_i} \qquad F_{y_i} = \frac{\partial W}{\partial v_i} \tag{8.24}$$

The equivalent nodal loads are extracted from this expression when the principle of minimum potential energy is applied to this function.

Step 8 — Formulation of the Finite Element Force-Displacement Relation

The final form of the finite element force-displacement relation is found by applying the principle of minimum potential energy to the potential energy function for the finite element and its system of loads. The potential energy function is formed from Eqs. 8.21 and 8.24 as

Potential Energy Function

$$V = \frac{1}{2}\{d\}^T [K]\{d\} - \{d\}^T \{F\} \tag{8.25}$$

When the principle of minimum potential energy is applied the result is

Finite Element Force-Displacement Relation

$$[K]\{d\} = \{F\} \tag{8.26}$$

As can be seen by comparing Eq. 8.26 to Eq. 8.25, the principle of minimum potential energy need not be explicitly applied in the future to identify the stiffness matrix of the load vector for individual elements. We need only form the potential energy function and extract the stiffness matrix and the load vector. This is the procedure that is followed in subsequent lessons to form stiffness matrices.

Modeling Characteristics for a Constant Strain Triangle

The development just presented enables us to discuss some of the overall modeling characteristics of the constant strain triangle. Since this element is incapable of representing higher order strain states, it is not able to exactly represent the behavior of strain fields with high gradients. When constant strain elements attempt to model high-gradient strain fields, their inability to represent this behavior can be identified by the high levels of interelement discontinuities in the strains. These discontinuities can be reduced by refining the model.

These ideas are illustrated in Fig. 8. In Fig. 8a, a high-strain-gradient region is modeled with three elements. Each of these elements is attempting to represent a strain field that is not constant. Since it is impossible for these elements to represent anything but constant strain, interelement strain discontinuities exist. In Fig. 8b, each of the elements is divided in half. As can be seen, the representation is closer to the desired result and the interelement discontinuities are reduced.

This simple figure has illustrated the way in which a mesh refinement of constant strain elements produces a more accurate result. The modeling characteristics of higher order elements are discussed in subsequent lessons.

In later developments, error analysis procedures that utilize interelement discontinuities to guide the automatic refinement of meshes to produce more accurate results are developed. In these procedures, a smoothed strain field (one with no discontinuities) is developed for a problem. This smoothed solution is then compared to the finite element result. The level of error in the individual elements is estimated as the difference between these two solutions. The elements with high error are then subdivided. This procedure has been found to improve the finite element model in regions of high error.

Closure

A procedure based on the use of strain gradient notation has been developed for formulating finite element stiffness matrices. The procedure can be condensed down to the six steps outlined in the following box. The principle of minimum potential energy was applied in step 8 to formally identify the stiffness matrix. As we saw in step 6, the stiffness matrix is available directly from the strain energy expression. This means that after this one application of the principle of minimum potential energy, we do not have to formally apply it again. Similarly, the development of the load vector was included here

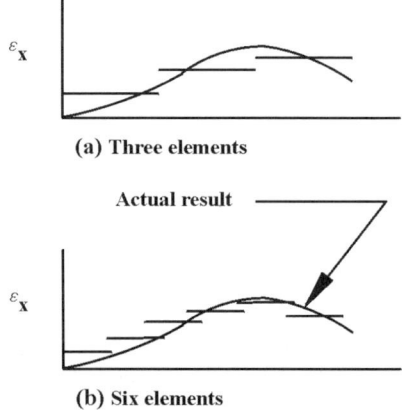

(a) Three elements

(b) Six elements

Figure 8. Representation of interelement discontinuities.

for completeness. Since the load vector formulation is identical for the strain gradient approach as it is for the standard finite element approach, the formulation of the load vectors is not considered further in this book.

Steps to Form a Strain-Gradient-Based Finite Element Stiffness Matrix

1. Identify the displacement approximation polynomials that correspond to the finite element geometry and the strain gradient quantities associated with these polynomials.

2. Form the approximate strain models from the displacement polynomials and evaluate the accuracy of their representation.

3. Develop the strain energy expression in strain gradient coordinates.

4. Identify the unique integrals contained in the strain energy expression just formed and evaluate them.

5. Form the strain gradient-to-nodal coordinate transformation matrix.

6. Transform the strain energy expression to a function of nodal displacements to form the finite element stiffness matrix.

The close relationship between the notation and the sources of displacements in the continuum enables sources of elemental errors in individual elements to be identified. This capability is demonstrated in the lessons that follow as several types of error are identified in finite element stiffness matrices.

Notes

1. A major motivation for the use of isoparametric elements is to reduce the computational effort required in evaluating the strain energy integrals (see Appendices A and B). The small number and simple form of the integrals required for the strain gradient formulation (see Eq. 8.8) reduces the motivation for the use of isoparametric transformations. The integrals of Eq. 8.8 for several types of elements were evaluated using Green's Theorem (see Ref. 1). This requires a simple program for the evaluation of these terms. Note that the nonlinear nature of the isoparametric transformations produces the errors in the strain energy discussed in Note 2 (see Refs. 2, 3, and 4).

2. Strain gradient notation enables the accuracy of the strain energy models in finite element matrices to be evaluated in a way not possible before. Equation 8.1 provides a link between the nodal displacements that produce a given strain state and the strain model. Through the use of Eq. 8.1 the displacements that produce a given strain state can be found. This is the essence of Eqs. 8.16 and 8.17. The strain energy can be computed for this strain state using Eq. 8.8. An element to be analyzed is given the nodal displacements specified by Eq. 8.17 and the strain energy is obtained as $1/2\{d\}^T[K]\{d\}$. The strain energy is then computed using Eq. 8.8. The two strain energy computations are compared. If there is a difference, the finite element is not modeling the strain correctly. This procedure was used to show that distorted isoparametric elements incorrectly represent the strain energy for the higher order strain states. This is discussed in Refs. 2, 3, and 4. An example in Lesson 6 computed the strain energy due to a specific strain gradient term.

Appendix A

The Isoparametric Formulation Procedure

Introduction

In current applications of the finite element method, the isoparametric formulation procedure must be considered the standard approach for computing finite element stiffness matrices. The basic objective of the isoparametric approach is to reduce the computational requirements for forming the stiffness matrices for any element shape, such as those shown in Fig. A.1. The isoparametric approach was developed when computing capabilities were much more limited than they are today. The ways in which the computational requirements are reduced are noted during the development of the procedure.

A price is paid, however, for this computational efficiency in the form of errors in many of the resulting isoparametric elements. In some cases, the errors are inherent in the isoparametric approach itself and, in others, the errors are disguised during the formulation by the abstract nature of the notation used in the development. That is to say, the notation is not physically interpretable and cannot be directly related to the problem being solved. In general, both types of errors are exhibited as mistakes in the strain field representations of the individual elements. These errors are outlined in the next section. They are discussed in detail in the lessons of this part, where the various strain gradient elements are developed and compared to their isoparametric counterparts.

The objective of this appendix is to outline the isoparametric formulation procedure and to identify the causes of the errors *inherent* to the isoparametric approach. Errors are introduced by the isoparametric formulation process when the Jacobian of the isoparametric transformation is not a constant. The Jacobian is a function of position within the element when the element is distorted, as shown in Figs. A.1c–e. A four-sided element is said to be distorted when it has a nonconstant Jacobian. This occurs when it is not a parallelogram. A triangular element has a nonconstant Jacobian when it has one or more curved edges.

The development of the isoparametric formulation procedure and the identification of the modeling errors due to the isoparametric process are presented in the context of a three-node bar. This element was chosen for this demonstration because it is simple enough that the equations involved are not overly complex and the causes of the modeling errors can be clearly seen.

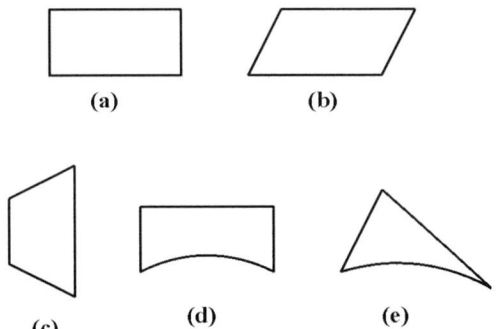

Figure A.1. Typical element shapes.

Alternative Element Formulation Procedure and Elemental Errors

An alternative approach for formulating and computing finite element stiffness matrices is the strain gradient approach. This procedure is developed in Lesson 8 and applied in Lessons 9–12. This alternative formulation procedure has the following advantages: (1) it utilizes a physically interpretable notation that provides *a priori* insights into the modeling characteristics of individual elements, (2) it has none of the modeling deficiencies inherent to the isoparametric elements, and (3) it is computationally competitive with the isoparametric scheme. The modeling errors shown to exist in isoparametric elements are identified using the physically interpretable notation that is at the heart of the strain gradient approach. This notation provides a direct relationship between the nodal displacements in an element and the strain representation that should exist in the element. This capability enables the modeling characteristics of individual finite elements to be evaluated to a depth that was never before possible.

In Lesson 10, we see that isoparametric six-node elements with curved edges misrepresent the strain distributions for the linear strain states. Examples demonstrate the magnitude of the pointwise strain errors and the errors in the strain energy errors in moderately curved elements. No similar errors exist in the corresponding strain gradient elements. The rigid body motions and the constant strain states are accurately represented regardless of the configuration of the element for both isoparametric and strain gradient elements.

In Lesson 9, we see that the four-node isoparametric element misrepresents the limited set of linear strain states that it is capable of representing when an element has a nonparallelogram shape. This modeling deficiency further stiffens the element in flexure. It is also shown that the four-node element contains parasitic shear unless it is corrected for this modeling error. In the case of the *four-node element*, parasitic shear can be removed from isoparametric elements by integrating the shear strain energy expression with a one-point Gaussian quadrature rule (see Appendix B). The process of integrating the shear strain energy expression with a lower order Gaussian quadrature rule than is required to capture all of the terms in the strain energy expression is called reduced-order integration. If a different order rule is used in the x and y directions, the process is called "selectively reduced-order" integration. Note that the use of any form of reduced-order integration *does not work* for eight- and nine-node isoparametric plane stress or plane strain elements. As is discussed later, the use of reduced-order Gaussian quadrature integration successfully corrects eight- and nine-node isoparametric *plate* elements for parasitic shear.

In Lesson 11, we see that the eight- and nine-node plane elasticity elements contain parasitic shear. However, as mentioned, this defect cannot be removed from isoparametric elements using reduced-order Gaussian quadrature integration. If a 2×2 rule is used to integrate the shear strain energy expression, not all of the erroneous terms are removed. If a 1×1 rule is used, two components of the basis set are removed and a new error is introduced. The removal of the two elements of the basis set introduces two extra zero-energy modes into the element. These zero-energy modes are introduced in addition to the three rigid body modes. These erroneous modes are called spurious zero-energy modes. The effects of the spurious zero-energy modes are demonstrated in Lesson 11.

There are ways of ensuring that these erroneous modes are not activated and of removing their effects if they are present but the need to consider this error adds a pseudo-complexity to the finite element method that is not needed. The parasitic shear terms can be removed from strain gradient elements and spurious zero-energy modes never appear.

As was the case for the four-node element, the eight- and nine-node isoparametric elements contain strain modeling errors if the element has a nonparallelogram shape.

Nonparallelogram shapes include elements that are essentially parallelograms but have one or more curved edges. These elements do not represent the linear and higher order strain states accurately. The magnitudes of these errors are discussed in Lesson 11.

Cause of Strain Modeling Errors in Distorted Isoparametric Elements

In all cases where the linear and higher order strain states are erroneously represented, the misrepresentations are caused by the presence of a nonconstant Jacobian in the isoparametric transformation. When the Jacobian is not a constant, it is a function of the location in the region where it is being computed. A Jacobian is nonconstant if the finite element is distorted. A rectangular region is distorted if it is not a parallelogram or if it has one or more curved edges. A triangular element is distorted if it has one or more curved edges. The error caused by a nonconstant Jacobian can be exacerbated by the inability of the Gaussian quadrature integration rules used in the element formulation to accurately integrate the strain energy expressions. The problem with the Gaussian quadrature procedure is outlined at the end of this appendix and discussed in detail in Appendix B.

The effects of all of the errors due to strain distortions contained in isoparametric elements can be removed by refining the finite element model sufficiently. These errors are completely absent when each of the elements represents only constant strains. This, however, eliminates any advantage of using higher order elements. The errors are reduced under less extreme refinement as the amount of strain energy contained in the individual elements is "bound" in the constant strain states. Note, however, that specialized elements have been identified where refinement does not eliminate "qualitative" errors due to parasitic shear. This is discussed next.

Errors in Out-of-Plane Bending Elements

In Lesson 12, out-of-plane bending elements are discussed briefly. One of the primary purposes of Lesson 12 is to identify the cause of a modeling error called shear locking and to show its relation to parasitic shear. As mentioned earlier, it is noted in Lesson 12 that isoparametric eight- and nine-node *plate-bending elements* with *isotropic* properties can be corrected for parasitic shear using selectively reduced-order Gaussian quadrature integration procedures.

Another purpose of Lesson 12 is to show that errors exist in specialized isoparametric elements that cannot be removed either by reduced-order Gaussian quadrature integration or by mesh refinement. Laminated composite plate-bending elements have been identified in which the coupling between, say, bending and torsion, is qualitatively misrepresented because of parasitic shear terms. That is to say, when the element is bent downward, the torsional rotation is in the wrong direction. These errors cannot be corrected by means of mesh refinement or controlled in other ways. Since parasitic shear terms cannot be correctly removed from higher order isoparametric elements, the isoparametric version of these specialized elements qualitatively misrepresent the problem. This error can be removed from strain gradient elements. These are the first elements that have been identified that *must* be formulated using strain gradient notation to provide accurate representation of the physical problem.

Let us now present the isoparametric formulation procedure via an example. The example chosen for this demonstration is the three-node bar element. This element is simple enough to easily see the details of the isoparametric formulation and complex enough to highlight the causes of the strain modeling deficiencies in distorted isoparametric elements. These errors are demonstrated with sample calculations.

An Outline of the Isoparametric Formulation Procedure

In Fig. A.2a, a three-node bar element is shown that has an actual physical length of one unit. In this case, node 1 is located anywhere along the x axis and node 3 is located one unit to the right of node 1. The interior node is denoted as node 2. The location of this node is retained as a parameter in this development. That is to say, we are not going to fix the position of node 2 in this development. Its position will be carried along as a variable in the example problems presented to illustrate the modeling errors inherent in isoparametric elements. The variable location of node 2 is used to demonstrate the effect of element distortion of the results produced by the isoparametric formulation procedure. That is to say, the noncentered location of the internal node is analogous to a nonparallelogram shape and to any element with one or more curved edges. It is understood that a three-node bar with a noncentered internal node is rarely used.

In Fig. A.2b, the canonical or standard locations of the nodes of any bar are shown in natural or dimensionless coordinates. The left-hand end of the bar in physical coordinates (node 1) is mapped onto the end of the canonical bar with a location of $\zeta = -1$. Similarly, the right end of the bar in physical coordinates is mapped onto the right end of the canonical bar with a location of $\zeta = +1$. Thus, the canonical bar has a length in non-dimensional coordinates of two.

A key element of the isoparametric method is seen in the location of node 2 in the natural coordinates. The interior node is mapped onto the origin of the natural coordinates regardless of its location in the physical coordinate system. This feature causes the Jacobian to be a function of ζ.

Later, we find that physical considerations limit the location of the interior node. In the case of the three-node bar, the interior node can be located anywhere between the quarter points of the bar. For a bar of unit length, this means that the interior node can be located between $1/4 < x < 3/4$. The two limiting cases for the location of the interior node are shown in Fig. A.3. One of the purposes of this development is to show that the Jacobian of the transformation from the physical coordinates to the dimensionless coordinates is a function of ζ unless node 2 is located in the center of the element. We also see that no computational difficulties arise if the node is centered at $x = 1/2$ because the Jacobian is a constant.

The mapping of the physical problem onto the canonical locations of the nodes in dimensionless coordinates enables the strain energy expressions in the natural coordinate system to be integrated using Gaussian quadrature integration. The efficiency of Gaussian quadrature integration contributes to the overall efficiency of the computation of the stiffness matrices of isoparametric elements. We will see that the use of Gaussian quadrature contributes to the errors present in the strain energy for the isoparametric elements. Gaussian quadrature integration is discussed in Appendix B.

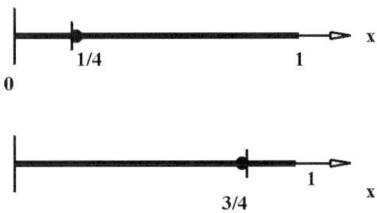

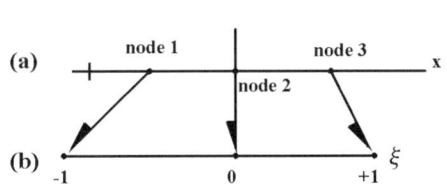

Figure A.2. The isoparametric mapping.

Figure A.3. The limit cases of the isoparametric mapping.

Step 1 — Identification of the Displacement Interpolation Function

In an isoparametric element, the stiffness matrix is computed in the dimensionless natural coordinate system. This means that the displacement interpolation function must be expressed in the dimensionless ζ coordinate. Lagrangian interpolation functions are used in the isoparametric method to formulate in-plane elements. The use of Lagrangian interpolation functions contributes to the computational efficiency of the isoparametric element because no transformation matrix must be inverted as part of the computation. Lagrangian interpolation functions are discussed in most standard finite element books (see Refs. 7–12). The Lagrangian interpolation function for a three-node bar element is

---

Displacement Interpolation Function

$$u = N_1 u_1 + N_2 u_2 + N_3 u_3$$

where

$$N_1 = \frac{\zeta}{2}(\zeta - 1) \qquad N_2 = (1 + \zeta)(1 - \zeta) \qquad N_3 = \frac{\zeta}{2}(1 + \zeta)$$

(A.1)

The N's in this interpolation function are known as shape functions. They indicate the three shapes that the bar can take when each of the nodes is independently displaced. Any displacement is a linear combination of these shape functions. Note that the interpolation functions are expressed in terms of the nodal displacements and that we have no knowledge of the strain states or strain distributions that this element can represent. This is in contrast to the strain gradient approach, where we have intimate knowledge of the strain modeling characteristics of an element.

Step 2 — Strain Approximation and Model Evaluation

The displacement interpolation function is not used directly in the isoparametric formulation procedure to transform to nodal coordinates. In the isoparametric approach, the displacement function is used only to form the strain model in the manner that follows. The strain expression for ε_x is formed by taking the derivative of u with respect to x to give

---

Strain Model

$$\varepsilon_x = \frac{du}{dx} = \frac{dN_1}{dx} u_1 + \frac{dN_2}{dx} u_2 + \frac{dN_3}{dx} u_3$$

(A.2)

However, a problem appears when an attempt is made to evaluate Eq. A.2. We cannot directly differentiate the shape functions with respect to x because they are functions of ζ, not of x. Thus, we must apply the chain rule to form the derivatives contained in Eq. A.2. When this is done, the result is

Strain Model Revisited

$$\varepsilon_x = \frac{dN_1}{d\zeta}\frac{d\zeta}{dx}u_1 + \frac{dN_2}{d\zeta}\frac{d\zeta}{dx}u_2 + \frac{dN_3}{d\zeta}\frac{d\zeta}{dx}u_3 \tag{A.3}$$

Another obstacle presents itself in Eq. A.3. We do not currently have a coordinate transformation available to us with which to compute the derivative term $d\zeta/dx$. In the isoparametric formulation procedure, the coordinate transformation that relates ζ and x has the following form:

Coordinate Transformation from x to ζ

$$x = N_1 x_1 + N_2 x_2 + N_3 x_3 \tag{A.4}$$

As can be seen, Eq. A.4 has the same form as Eq. A.1. The shape functions used in the interpolation polynomial given by Eq. A.1 are also used in the coordinate transformation given by Eq. A.4. Thus, the coordinate transformation has the *same* form as the interpolation function and it is expressed in the *same* parameter ζ. This is the characteristic that gives the *iso*parametric element its name. In Greek, the prefix *iso-* means the *same*.

This coordinate transformation maps node 1 in the physical x coordinate system onto the point $\zeta = -1$ in the ζ system. Node 2, regardless of its location in the x coordinate system, is mapped onto the point, $\zeta = 0$. Finally, node 3 is mapped onto the point $\zeta = +1$.

Let us revisit Eq. A.3 now that we have a function relating x and ζ. The quantity we must compute is $d\zeta/dx$. We cannot directly compute this quantity from Eq. A.4 because we do not have ζ as a function of x.

Equation A.4 gives us the inverse form of this transformation. The transformation given by Eq. A.4 is desired for the isoparametric formulation because we want to compute the stiffness matrix in the natural ζ coordinates. This enables the required integrations to be performed using the Gaussian quadrature procedure.

This means that we must extract the desired quantity, namely, $d\zeta/dx$, from Eq. A.4 indirectly. This is done by first forming the derivative for $dx/d\zeta$ as

Formulation of the Jacobian

$$\frac{dx}{d\zeta} = \frac{dN_1}{d\zeta}x_1 + \frac{dN_2}{d\zeta}x_2 + \frac{dN_3}{d\zeta}x_3$$

where

$$\frac{dN_1}{d\zeta} = \frac{d}{d\zeta}\left[\left(\frac{\zeta}{2}\right)(\zeta - 1)\right] = \zeta - \frac{1}{2} \tag{A.5}$$

$$\frac{dN_2}{d\zeta} = \frac{d}{d\zeta}[(1 + \zeta)(1 - \zeta)] = -2\zeta$$

$$\frac{dN_3}{d\zeta} = \frac{d}{d\zeta}\left[\left(\frac{\zeta}{2}\right)(1 + \zeta)\right] = \zeta + \frac{1}{2}$$

The expression for $dx/d\zeta$ is known as the Jacobian of the transformation. In the case of a one-dimensional problem, the Jacobian is a single term. In the case of multiple dimensions, the Jacobian is a matrix. The determinant of the Jacobian is the ratio of the differential volumes for the two-coordinate systems. In the one-dimensional case, the Jacobian and the determinant of the Jacobian are the same quantity. Thus, we can express $dx/d\zeta$ as $|J|$ in this one-dimensional case to indicate the determinant of the Jacobian.

We can form the $d\zeta/dx$ expression needed to compute the strain expression by finding the reciprocal of Eq. A.5. When this is done, the result is

---

Formation of $1/|J|$

$$\frac{d\zeta}{dx} = \frac{1}{|J|} = \frac{1}{(\zeta - 1/2)x_1 - 2\zeta x_2 + (\zeta + 1/2)x_3} \tag{A.6}$$

The fact that the determinant of the Jacobian is a function of ζ has important implications. This fact limits the geometry of the element, adversely affects the strain representations, and causes the Gaussian quadrature integration rules to integrate the strain energy expressions incorrectly. These effects are discussed in the context of the three-node bar element in the next subsection.

We are now in a position to complete the strain model so it can be used to compute the stiffness matrix for a specific problem. When Eq. A.6 and the derivatives of the shape functions from Eq. A.5 are substituted into Eq. A.3, the strain model can be written in matrix notation as

---

Final Strain Model

$$
\begin{aligned}
\varepsilon_x &= \frac{1}{|J|} \begin{bmatrix} \dfrac{dN_1}{d\zeta} & \dfrac{dN_2}{d\zeta} & \dfrac{dN_3}{d\zeta} \end{bmatrix} \begin{Bmatrix} u_1 \\ u_2 \\ u_3 \end{Bmatrix} \\[2ex]
&= \frac{1}{|J|} [(\zeta - 1/2)\ (-2\zeta)\ (\zeta + 1/2)] \begin{Bmatrix} u_1 \\ u_2 \\ u_3 \end{Bmatrix} \\[2ex]
&= \frac{[(\zeta - 1/2)\ (-2\zeta)\ (\zeta + 1/2)]}{[(\zeta - 1/2)x_1 - 2\zeta x_2 + (\zeta + 1/2)x_3]} \begin{Bmatrix} u_1 \\ u_2 \\ u_3 \end{Bmatrix} \\[2ex]
&= [B]\{u\}
\end{aligned}
\tag{A.7}
$$

This expression for the strain is a function of ζ. It is now ready for substitution into the strain energy expression so the stiffness matrix can be computed using Gaussian quadrature integration. In many publications and texts, the matrix that accomplishes the strain to nodal displacement transformation is called the $[B]$ matrix. The $[B]$ matrix does not exist explicitly in the strain gradient approach because the transformation takes place in two steps. The $[B]$ matrix is equivalent to the product $[T][\Phi]^{-1}$ used in the strain gradient formulation.

Geometric Restrictions

As mentioned earlier, the structure of the determinant of the Jacobian, as given in Eq. A.6, restricts the location of the interior node in the three-node bar element. This restriction is built into the interpretation of the determinant of the Jacobian as the ratio of the two differential volumes. This ratio cannot be less than or equal to zero. The physical meaning of a negative Jacobian is easy to see in the case of a two-dimensional element. As shown in Fig. A.4, a six-node element is progressively distorted until a portion of the element overlaps itself. The determinant of the Jacobian in the overlapped region is negative. Similarly, the ratio of the differential volumes cannot be zero. If it were zero, a singularity would exist in the strain expression given by Eq. A.7 because of the presence of a zero in the determinant.

We now identify the restrictions put on the geometry of the three-node bar by the requirement that the Jacobian be positive. This is done for the example problem shown in Fig. A.1. In this case, node 1 is located at $x = 0$ and node 3 is located at $x = 1$. When these nodal locations are substituted into Eq. A.6, the reciprocal of the Jacobian becomes

---

$1/|J|$ for the Demonstration Problem

$$\frac{d\zeta}{dx} = \frac{1}{|J|} = \frac{1}{1/2 + \zeta(1 - 2x_2)} \tag{A.8}$$

As can be seen in Eq. A.8, the effect of the denominator is the greatest at the boundary values of the ζ. Let us look at the limit put on the location of the interior node when $\zeta = -1$. This is done by setting the denominator of Eq. A.8 to zero and solving for x_2:

Evaluation of $|J|$ for the Demonstration Problem

$$1/2 + \zeta(1 - 2x_2) = 0$$
$$1/2 - 1 + 2x_2 = 0 \tag{A.9}$$
$$x_2 = 1/4$$

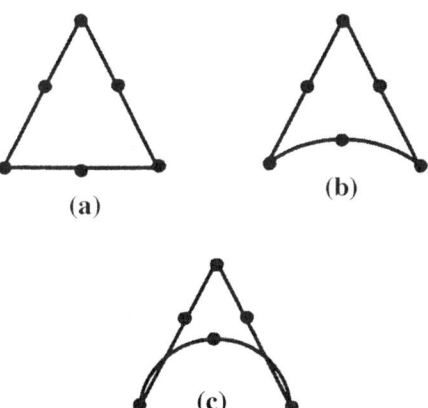

Figure A.4. The limit cases of the isoparametric mapping.

When the same procedure is applied to compute the limit location of $x_2 = 0$ when $\zeta = +1$, the result is $x_2 = 3/4$. Thus, the position of the interior node is restricted to a location between the quarter points $(1/4 < x_2 < 3/4)$. If the interior node is located outside of this region, the inverse of the Jacobian is negative or equal to ∞. In either case, the strain representation is given by Eq. A.7.

The determinant of the Jacobian is not a function of ζ if the coefficient of the ζ term, namely, $(1 - 2x_2)$, is equal to zero. When this coefficient is rearranged, it is found to equal zero when $x_2 = 1/2$. That is to say, the determinant of the Jacobian is constant when the interior node is in the center of the bar.

This has an easy physical interpretation. The ratio of the lengths in the physical coordinate system and the natural coordinate system on the two sides of the bar are the same. Thus, the ratio of the differential lengths is constant throughout the bar.

Strain Model Evaluation

The use of strain gradient notation enables the three strain states in a three-node bar to be identified. This is discussed in detail in Part II. The three-node bar is capable of representing the following three strain states: (1) a rigid body displacement, (2) a constant strain state, and (3) a linear variation of strain. In strain gradient notation, these three strain states are denoted as $(u_{rb})_0$, $(\varepsilon_x)_0$ and $(\varepsilon_{x,x})_0$. The subscript zero indicates that these coefficients are Taylor series terms. The displacement patterns produced by these three individual strain states are expressed analytically as

---

Strain Distributions in the Three-Node Bar

$$u = (u_{rb})_0$$
$$u = x(\varepsilon_x)_0$$
$$u = \frac{x^2}{2}(\varepsilon_{x,x})_0 \tag{A.10}$$

Let us substitute the displacement pattern associated with the three strain states, one at a time, into the strain representation given by Eq. A.7. The displacement pattern for the rigid body displacement requires that each node have the same displacement. When the displacement pattern for the rigid body motion is substituted into Eq. A.7, the strain in the element is found to be

---

Strain Representation for Rigid Body Motion

$$\varepsilon_x = \frac{[(\zeta - 1/2)\,(-2\zeta)\,(\zeta + 1/2)]}{[(\zeta - 1/2)x_1 - 2\zeta x_2 + (\zeta + 1/2)x_3]} \begin{Bmatrix} u_1 \\ u_2 \\ u_3 \end{Bmatrix}$$

$$= \frac{[(\zeta - 1/2)\,(-2\zeta)\,(\zeta + 1/2)]}{[(\zeta - 1/2)x_1 - 2\zeta x_2 + (\zeta + 1/2)x_3]} \begin{Bmatrix} 1 \\ 1 \\ 1 \end{Bmatrix} \{(u_{rb})_0\} \tag{A.11}$$

$$= \frac{[0]\{(u_{rb})_0\}}{[(\zeta - 1/2)x_1 - 2\zeta x_2 + (\zeta + 1/2)x_3]}$$

$$= 0$$

As can be seen, the numerator for this expression is zero. Thus, the strain contained in a three-node bar with any geometrically acceptable nodal location undergoing rigid body motion contains the correct strain representation, namely, a strain value of zero.

When the displacement pattern for a constant strain state, as given by the second expression in Eq. A.10, is substituted into the expression for strain given by Eq. A.7, the strain in the element is found to be

---

Strain Representation for Constant Strain State

$$\varepsilon_x = \frac{[(\zeta - 1/2)\ (-2\zeta)\ (\zeta + 1/2)]}{[(\zeta - 1/2)x_1 - 2\zeta x_2 + (\zeta + 1/2)x_3]} \begin{Bmatrix} u_1 \\ u_2 \\ u_3 \end{Bmatrix}$$

$$= \frac{[(\zeta - 1/2)\ (-2\zeta)\ (\zeta + 1/2)]}{[(\zeta - 1/2)x_1 - 2\zeta x_2 + (\zeta + 1/2)x_3]} \begin{Bmatrix} x_1 \\ x_2 \\ x_3 \end{Bmatrix} \{(\varepsilon_x)_0\} \qquad (\text{A.12})$$

$$= \frac{[(\zeta - 1/2)x_1 - 2\zeta x_2 + (\zeta + 1/2)x_3]}{[(\zeta - 1/2)x_1 - 2\zeta x_2 + (\zeta + 1/2)x_3]} \{(\varepsilon_x)_0\}$$

$$= \{(\varepsilon_x)_0\}$$

As can be seen, the numerator and the denominator are identical. When they are canceled, the resulting strain representation is the expected result, a constant strain.

When the displacement pattern for a linear strain state is substituted into the strain expression, the strain representation is given as

---

Strain Representation for a Linear Strain State

$$\varepsilon_x = \frac{[(\zeta - 1/2)\ (-2\zeta)\ (\zeta + 1/2)]}{[(\zeta - 1/2)x_1 - 2\zeta x_2 + (\zeta + 1/2)x_3]} \begin{Bmatrix} u_1 \\ u_2 \\ u_3 \end{Bmatrix}$$

$$= \frac{[(\zeta - 1/2)\ (-2\zeta)\ (\zeta + 1/2)]}{[(\zeta - 1/2)x_1 - 2\zeta x_2 + (\zeta + 1/2)x_3]} \begin{Bmatrix} x_1^2/2 \\ x_2^2/2 \\ x_3^2/2 \end{Bmatrix} \{(\varepsilon_{x,x})_0\} \qquad (\text{A.13})$$

$$= \frac{[(\zeta - 1/2)x_1^2 - 2\zeta x_2^2 + (\zeta + 1/2)x_3^2]}{2[(\zeta - 1/2)x_1 - 2\zeta x_2 + (\zeta + 1/2)x_3]} \{(\varepsilon_{x,x})_0\}$$

There is no immediate reduction in the form of the strain representation when the displacement pattern for the linear strain distribution is substituted into Eq. A.7. When the interior node is located in the center, the denominator of Eq. A.13 becomes a constant and the strain representation is the expected linear variation. However, if the interior node is not centered, the strain representation is not a linear representation.

The effect of the location of the internal node on the strain distribution will be illustrated for the bar shown in Fig. A.3. When Eq. A.13 is evaluated with $x_1 = 0$ and $x_3 = 1$, the resulting expression for ε_x is a function of the location of the interior node, x_2:

Strain Representation for a Linear Strain State

$$\varepsilon_x = \frac{-2\zeta x_2^2 + (\zeta + 1/2)}{2[-2\zeta x_2 + (\zeta + 1/2)]} \{(\varepsilon_{x,x})_0\} \tag{A.14}$$

When Eq. A.14 is evaluated for a sequence of interior node locations, the strain distributions are as shown in the columns of Table A.1. The distribution for the case where the internal node is located in the center of the bar produces the expected linear result. Any deviation from these values indicates an error. As can be seen, the strain distribution errors increase as the interior point moves away from the center of the beam and toward the limit point.

This modeling error is caused by the presence of a function of ζ in the denominators of Eqs. A.13 and A.14. This determinant is a function of ζ if the interior node is not centered in the bar. This means that the Jacobian of the isoparametric transformation is not a constant. If the center node is near one of the limit points, i.e., $x_2 = 1/4$ or $x_2 = 3/4$, the Jacobian approaches zero when $\zeta = \pm 1$. This means that the strain becomes unbounded. This is seen in Table A.1.

This subsection has shown the source of one strain modeling error in isoparametric elements. The rigid body and constant strain states are represented accurately regardless of the configuration of the element. The higher order strain state is misrepresented in distorted elements because the Jacobian is not constant. This pattern holds for planar and three-dimensional elements. The results of this for planar elements is shown in detail in the lessons contained in this part.

This presentation also explains why the distorted elements converge to the correct solution. In most cases as the model is refined, the strain energy content in the model is contained in the constant strain states, which are modeled correctly in isoparametric elements. This error is a result of the isoparametric approach. It is not due to the finite element method itself. For example, the strain gradient approach to formulating stiffness matrices does not introduce this strain modeling deficiency.

Step 3 — Formulation of the Strain Energy Expression

In Lesson 1, we developed the strain energy expression for a bar as

Strain Energy Expression for a Bar

$$U = \frac{1}{2} \int_0^L EA\varepsilon_x^2 \, dx \tag{A.15}$$

The strain energy expression is transformed to a function of nodal displacements to form the finite element stiffness matrix. This is accomplished by substituting Eq. A.7 into Eq. A.14 to give

Strain Energy Expression for a Bar

$$U = \frac{1}{2} \int_0^L EA\{u\}^T [B]^T [B]\{u\} \, dx \tag{A.16}$$

Table A.1. Strain distribution vs interior node location.

ζ	x_2 Location										
	0.25 +	0.30	0.35	0.40	0.45	0.50	0.55	0.60	0.65	0.70	0.75 −
−1.0	≈ −∞	−1.60	−0.64	−0.30	−0.12	0.0	0.09	0.16	0.22	0.27	0.31
−0.8	−1.00	−0.43	−0.20	−0.06	0.03	0.10	0.16	0.21	0.25	0.30	0.33
−0.6	−0.06	0.02	0.07	0.12	0.16	0.20	0.23	0.27	0.30	0.33	0.36
−0.4	0.25	0.25	0.26	0.27	0.28	0.30	0.32	0.33	0.35	0.37	0.39
−0.2	0.41	0.40	0.40	0.40	0.40	0.40	0.40	0.41	0.42	0.43	0.44
0.0	0.50	0.50	0.50	0.50	0.50	0.50	0.50	0.50	0.50	0.50	0.50
0.2	0.56	0.57	0.58	0.59	0.60	0.60	0.60	0.60	0.60	0.60	0.59
0.4	0.61	0.63	0.65	0.67	0.68	0.70	0.72	0.73	0.74	0.75	0.75
0.6	0.64	0.67	0.70	0.73	0.77	0.80	0.84	0.88	0.93	0.98	1.06
0.8	0.67	0.70	0.75	0.79	0.84	0.90	0.97	1.06	1.20	1.43	2.00
1.0	0.69	0.73	0.78	0.84	0.91	1.00	1.12	1.30	1.64	2.60	≈ +∞

ε_x

In this expression, the $[B]$ matrix is a function of ζ and the variable of integration is x. The integral is expressed in terms of ζ by changing the integration variable from x to ζ by applying Eq. A.5 or Eq. A.6 to give

---

Strain Energy Expression for a Bar

$$U = \frac{1}{2}\int_{-1}^{+1} EA\{u\}^T[B]^T[B]\{u\}|J|\,d\zeta$$

$$= \frac{1}{2}\{u\}^T[K]\{u\}$$

(A.17)

When the stiffness matrix is explicitly expressed in terms of the functions of ζ, the result is

---

Stiffness Matrix for a Bar

$$[K] = \int_{-1}^{+1} \frac{EA}{|J|^2}[B]^T[B]|J|\,d\zeta$$

$$= \int_{-1}^{+1} \frac{EA}{|J|}[B]^T[B]\,d\zeta$$

(A.18)

As can be seen, the determinant of the Jacobian is present in the denominator of the final form of the stiffness matrix. As was the case in the strain representation, the presence of |J| in the determinant has no effect on the stiffness properties when the element is representing the rigid body displacement or the constant strain state. However, the representation of the linear strain state is affected if the interior node is not located in the center of the element. The effect is less pronounced on the strain energy because the effects of the errors are averaged. This effect is discussed in detail in Appendix B and demonstrated at the end of this appendix.

Let us now form the fully expanded expression for the stiffness matrix in preparation for studying the effect of the determinate of the Jacobian in the denominator of the stiffness matrix. When Eq. A.18 is expanded, the result is

---

Stiffness Matrix for a Bar

$$[K] = \int_{-1}^{+1} EA\frac{\begin{Bmatrix}(\zeta-1/2)\\(-2\zeta)\\(\zeta+1/2)\end{Bmatrix}[(\zeta-1/2)\;(-2\zeta)\;(\zeta+1/2)]}{[(\zeta-1/2)x_1 - 2\zeta x_2 + (\zeta+1/2)x_3]}\,d\zeta$$

$$= \int_{-1}^{+1} EA\frac{\begin{bmatrix}(\zeta-1/2)(\zeta-1/2) & (\zeta-1/2)(-2\zeta) & (\zeta-1/2)(\zeta+1/2)\\(-2\zeta)(\zeta-1/2) & (-2\zeta)(-2\zeta) & (-2\zeta)(\zeta+1/2)\\(\zeta+1/2)(\zeta-1/2) & (\zeta+1/2)(-2\zeta) & (\zeta+1/2)(\zeta+1/2)\end{bmatrix}}{[(\zeta-1/2)x_1 - 2\zeta x_2 + (\zeta+1/2)x_3]}\,d\zeta$$

(A.19)

As can be seen, the evaluation of the stiffness matrix requires the computation of six relatively complex integrals. The equivalent strain gradient element requires the computation of three integrals, namely, the length, the first moment of the length, and the second moment of the length. For the three-node bar, there is not a significant reduction in the number of integrals that must be evaluated if the strain gradient approach is used. However, in the evaluation of two-dimensional elements, the savings is considerable. When a six-node linear strain element is evaluated using the strain gradient approach, only 6 simple integrals must be evaluated instead of the 78 relatively complex integrals that must be computed when the isoparametric approach is used. This is discussed in the context of the individual elements in the lessons of this part.

Let us specialize Eq. A.19 for the unit length bar being used here as an example. The location of the interior node continues as a parameter in the expression for the stiffness matrix. When this is done, the result is

Stiffness Matrix for a Bar

$$[K] = \int_{-1}^{+1} 2EA \frac{\begin{bmatrix} (\zeta - 1/2)(\zeta - 1/2) & (\zeta - 1/2)(-2\zeta) & (\zeta - 1/2)(\zeta + 1/2) \\ (-2\zeta)(\zeta - 1/2) & (-2\zeta)(-2\zeta) & (-2\zeta)(\zeta + 1/2) \\ (\zeta + 1/2)(\zeta - 1/2) & (\zeta + 1/2)(-2\zeta) & (\zeta + 1/2)(\zeta + 1/2) \end{bmatrix}}{[1 + 2(1 - 2x_2)\zeta]} d\zeta$$

(A.20)

When element k_{11} is extracted from Eq. A.20, the explicit expression for this single element of the stiffness matrix is

Sample Stiffness Element

$$k_{11} = \int_{-1}^{+1} \frac{2AE(\zeta - 1/2)^2}{[1 + 2(1 - 2x_2)\zeta]} d\zeta$$

(A.21)

This is the sample stiffness element that is used to demonstrate the properties of the Gaussian quadrature integration procedure in Appendix B. In brief, we see in Appendix B that the presence of a denominator in Eq. A.21 that is a function of ζ raises the order of the strain energy polynomial. Since the Gaussian quadrature rule used to evaluate the integral is usually of low order, the integral is often evaluated incorrectly.

This is seen in Table A.2, where the location of the interior node is varied and the result of integrating the function with different Gaussian quadrature rules is shown. As can be seen, when the interior point is centered at $x = 0.50$, the result is evaluated correctly with a two-point rule. However, when the interior node is not centered, the approximation of the integral changes with different order Gaussian quadrature rules. This indicates that the order of the integrand is changed by the presence of the nonconstant Jacobian.

Closure

This appendix has demonstrated the isoparametric approach for formulating stiffness matrices and identified the causes of modeling errors in distorted isoparametric elements. This was accomplished with the example of a three-node bar element. This problem is

Table A.2. Integral value vs interior node location for the two-, three-, and four-point rules for Eq. A.20.

Location	Two points	Three points	Four points
0.25	5.500	8.500	10.750
0.30	4.322	5.242	5.487
0.35	3.561	3.874	3.910
0.40	3.028	3.124	3.128
0.45	2.635	2.652	2.653
0.50	2.334	2.334	2.334
0.55	2.095	2.106	2.107
0.60	1.901	1.944	1.946
0.65	1.742	1.834	1.844
0.70	1.610	1.779	1.824
0.75	1.500	1.8334	2.084

both simple enough that each step of the formulation is shown in full detail and complex enough that the source of strain modeling errors in isoparametric elements can be identified.

We have seen that the sources of strain modeling errors result from the use of the Gaussian quadrature integration procedure. This integration procedure requires that all finite elements, regardless of shape, be mapped onto a standard region so that Gaussian quadrature integration can be applied. This mapping introduces an error into the strain representation when an element has a distorted shape.

Then, the very use of Gaussian quadrature integration introduces a further error into the finite element model. The strain energy is evaluated incorrectly because the Gaussian quadrature rules typically used are not of a high enough order to capture all of the additional terms introduced by the mapping into the strain energy expression. Note that the isoparametric approach was initially introduced as a way to reduce the amount of computation required to form the finite element stiffness matrices when computing capacity was much less than it is today.

The results of the isoparametric formulation are contrasted to the results produced by the strain gradient approach. This comparison indicates that the strain gradient approach contains none of the errors present in isoparametric elements. Furthermore, the strain gradient approach is computationally competitive with the isoparametric approach. Finally, all strain gradient elements can be corrected for parasitic shear without introducing other errors. This is not the case for isoparametric elements of eight nodes and higher order that represent plane stress or plane strain.

The isoparametric version of these higher order elements cannot be corrected for parasitic shear using reduced-order Gaussian quadrature without introducing spurious zero-energy modes. Finally, the use of strain gradient notation has identified the source of shear locking in out-of-plane bending elements and aspect ratio stiffening in plane elasticity elements. This knowledge enables these modeling deficiencies to be controlled using rational, physically based criteria.

Note that all of the strain modeling deficiencies just outlined, with the exceptions of spurious zero-energy modes and the qualitative errors in laminated composite plate elements, are ameliorated as the mesh is refined. This negates the use of higher order elements.

In summary, we can conclude that the use of isoparametric elements in current finite element packages can produce accurate results, with some exceptions, if the finite element models are sufficiently refined. This adds a computational inefficiency that is counter to the original reason for using the isoparametric approach in the first place. Although it is not mandatory to replace isoparametric elements with strain gradient elements in finite element packages, it seems logical to use the strain gradient approach to teach the finite element method and in research. The clarity of presentation that accompanies the use of strain gradient notation provides insights that are not otherwise available.

Appendix B

Gaussian Quadrature Integration

The Gaussian quadrature integration procedure is basic to the isoparametric form of the finite element method. In the isoparametric approach, the stiffness matrices are created by mapping the individual elements onto a standard element geometry in a dimensionless or natural coordinate system so the strain energy expressions can be numerically integrated with the Gaussian quadrature integration procedure. Numerical integration with the Gaussian quadrature approach is so computationally efficient that the overhead of the transformation to natural coordinates is acceptable in the isoparametric element formulation because of the large number of integrals that must be computed. For example, a six-node isoparametric element requires the integration of 78 distinct expressions to form the stiffness matrix.

This appendix outlines the Gaussian quadrature procedure in one and two dimensions, demonstrates its use, and presents some of its characteristics. We see that the application of the Gaussian quadrature integration scheme in higher dimensions follows directly from the one-dimensional case.

One-Dimensional Application

In its standard form, the one-dimensional Gaussian quadrature integration procedure approximates the value of an integral on the range -1 to $+1$ by evaluating the function being integrated at a specified number of sampling points, multiplying these quantities by weighting factors, and summing the products. The procedure is exact for polynomials of order $(2n-1)$, where n is the number of sampling points. In equation form, the approximation of the integral is computed as

---

One-Dimensional Quadrature Formula

$$I = \int_{-1}^{+1} f(x)\,dx \approx \sum_{i=1}^{n} w_i\, f(a_i)$$

where   (B.1)

    w_i are weighting factors.

    a_i are sampling points.

    n is the number of sampling points.

The sampling points or Gaussian points for the one-, two-, and three-point Gaussian quadrature rules are shown in Fig. B.1. The function being integrated is evaluated at these points on the x axis. The Gaussian points for the one-, two-, and three-point rules are 0.0, -0.57735 and $+0.57735$, and $-0.7746, 0.0$, and $+0.7746$, respectively. The weighting factors for the three cases are 2.0; 1.0 and 1.0; and 5/9, 8/9, and 5/9, respectively. The weighting factors must sum to a value of 2.00. This is an obvious condition when it is considered that each rule must be capable of integrating a constant function exactly. That is to say, the constant value found at each sampling point must, in one way or another, be multiplied by the width of the interval, which is 2 in the standard one-dimensional form of the Gaussian quadrature procedure (see Note 1).

For example, the three-point rule approximates the integral of the function f as

Three-Point Rule

$$I_{3pt} = 5/9f(-0.7746) + 8/9f(0.0) + 5/9f(+0.7746) \qquad \text{(B.2)}$$

The three-point rule given by Eq. B.2 approximates the integral of the function f over the range -1 to $+1$ by evaluating the function being integrated at the three sampling points -0.7746, 0.0, and $+0.7746$, multiplying the quantities found in the first step by the designated weighting factors, and summing the three products.

The procedure for finding the Gaussian points and the weighting factors is outlined in Note 1. These quantities are found by forcing the approximation to evaluate polynomials of the order $(2n - 1)$ exactly, as mentioned earlier. Thus, the one-point rule evaluates first-order, i.e., $2 * 1 - 1 = 1$, polynomials exactly; the two point rule evaluates third-order, i.e., $2 * 2 - 1 = 3$, polynomials exactly; etc.

The result of evaluating higher order polynomials with a low-order Gaussian quadrature rule can be seen with the following examples. In Fig. B.2, an odd function and an even function with no constant term are shown. That is to say, both the odd and even functions are zero at the origin. As we can see, the odd terms are antisymmetric. This means that when these terms are integrated over a symmetric interval, the integrals will be zero. All of the Gaussian quadrature integration rules evaluate the odd terms correctly as zero because of the symmetry of the sampling points and weighting factors.

However, when even terms that are of an order higher than $(2n - 1)$ are evaluated, the results produced by the Gaussian quadrature method are not exact. The one-point rule evaluates all of the higher order terms as zero because the sampling point is located at zero. Thus, the one-point rule can be viewed as truncating a function at the linear term. The higher order Gauss quadrature rules cannot be viewed as truncating the polynomials.

The behavior of the two- and three-point Gaussian quadrature rules with the higher order even terms is demonstrated by the results contained in Table B.1. In this table, the even terms from 4th to 10th order are integrated with the two- and three-point rules. As is seen in the first row, the three-point rule integrates the 4th-order term exactly. This is as expected since $(2n - 1)$ for the three-point rule is five or 5th order. However, the two-point rule produces a result that is neither zero nor correct. This phenomenon is seen in the remainder of the polynomial terms contained in Table B.1, where neither the two- nor

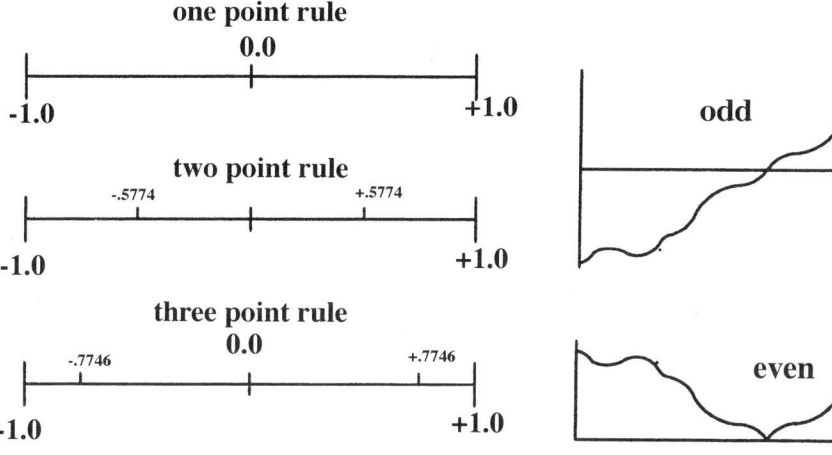

Figure B.1. Gaussian quadrature sampling points.

Figure B.2. Odd and even functions.

Table B.1. Integration of even terms.

Term	Two point	Three point	Exact
x^4	0.222	0.400	0.400
x^6	0.074	0.244	0.286
x^8	0.025	0.144	0.222
x^{10}	0.008	0.086	0.182

the three-point rules produces results that are zero or correct for these even terms. Thus, the Gaussian quadrature rules do not in general truncate the polynomial terms beyond the limit of accuracy given by the expression $(2n - 1)$.

The extension of the Gaussian quadrature integration to two dimensions is presented next. This numerical integration process also applies to triangles. The triangular case is not discussed; however, see Note 2.

Two-Dimensional Application

In its standard form, the two-dimensional form of the Gaussian quadrature procedure approximates the value of an integral on the square region bounded by $(-1 < x < +1)$ and $(-1 < y < +1)$ in a manner analogous to the one-dimensional approach. The function is evaluated at a number of sampling points, these quantities are multiplied by weighting factors, and the products are summed to form the approximation as

Two-Dimensional Quadrature Formula

$$I = \int_{-1}^{+1} \int_{-1}^{+1} f(x, y)\, dx\, dy \approx \sum_{i=1}^{n} \sum_{j=1}^{m} w_{ij}\, f(a_i, b_j)$$

where

w_{ij} are weighting factors.

a_i, b_j are sampling points in the x and y locations.

n, m are the numbers of sampling points in the

x and y directions.

(B.3)

The sampling points and weighting factors are formed from the values of the one-dimensional cases. The Gaussian points for the 2×2, the 3×3, and the 2×3 Gaussian quadrature integration procedure are shown in Fig. B.3. As we can see, the locations of the sampling points are defined by the locations of the Gaussian points in the one-dimensional case. In the 2×2 case, the x coordinates of both rows are the same as the x values of the one-dimensional case. The y values are the identical locations in the y direction. The same is true in the 3×3 case.

The only novelty seen in Fig. B.3 is the 2×3 case. In this application of Gaussian quadrature method, the x coordinates of the sampling points are given by the three-point rule and the y coordinates are given by the two-point rule. As expected, a 2×3 rule exists and the Gaussian points are located analogously. In the finite element method, the use of 1×2, 2×1, 2×3, and 3×2 rules are referred to as selectively reduced

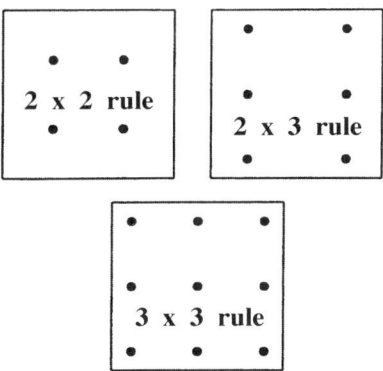

Figure B.3. Two-dimensional Gaussian point locations.

integration. Selectively reduced order integration is used successfully to remove an error called parasitic shear from eight- and nine-node *plate* elements. Selectively reduced-order integration *does not* accurately correct eight- and nine-node plane stress or strain elements for parasitic shear. In fact, the attempt to correct for parasitic shear with a lower order Gaussian quadrature rule can introduce another error called spurious zero-energy modes. This phenomenon is discussed in detail in Lesson 11.

The weighting factors for the two-dimensional rules are all developed from the weighting factors for the one dimensional rules. For example, the weight factors for the 3×2 and 3×3 rules are formed using the following vector products:

Two-Dimensional Weighting Factors, 2×3 and 3×3 Rules

$$w_{2 \times 3} = \begin{Bmatrix} 1 \\ 1 \end{Bmatrix} [5/9 \ 8/9 \ 5/9] = \begin{bmatrix} 5/9 & 8/9 & 5/9 \\ 5/9 & 8/9 & 5/9 \end{bmatrix}$$

$$w_{3 \times 3} = \begin{Bmatrix} 5/9 \\ 8/9 \\ 5/9 \end{Bmatrix} [5/9 \ 8/9 \ 5/9] = \begin{bmatrix} 0.31\ldots & 0.49\ldots & 0.31\ldots \\ 0.49\ldots & 0.79\ldots & 0.49\ldots \\ 0.31\ldots & 0.49\ldots & 0.31\ldots \end{bmatrix} \quad (B.4)$$

These weighting factors sum to a total of four, as expected since the integral is being evaluated over a square with side dimensions of 2.

The polynomial terms that can be evaluated exactly are also directly related to the one-dimensional case. For example, the 1×1 rule (i.e., a point at the origin with a weight factor of 4.0) evaluates the following four-term, bilinear polynomial exactly: $z = f(x, y) = c_0 + c_1 x + c_2 y + c_3 xy$. This rule does not evaluate the x^2, y^2, or any higher order terms. The 1×1 rule evaluates the second-order xy term because it is linear in x and linear in y. As was the case in the one-dimensional application of the one-point rule, the 1×1 rule truncates any polynomial to the four-term, bilinear polynomial given earlier.

As expected, the behavior of the two-dimensional rules is similar to that of the one-dimensional rules. All of the odd terms are evaluated correctly as zero and the higher order even terms are evaluated incorrectly. This is seen in Table B.2 for the five fourth-order polynomial terms evaluated by the 2×2 rule. In this case, the 3×3 rule produces exact results because the value of $(2n - 1)$ for the three-point rule is 5 so fifth-order terms will be evaluated exactly. The 2×2 rule evaluates the two odd terms xy^3 and $x^3 y$ as 0. The $x^2 y^2$ term is evaluated correctly as expected because it contains both x and y to only

Table B.2. Integration of fourth-order terms with 2 × 2 and 3 × 3 rules.

Term	2 × 2	3 × 3	Exact
x^4	0.444	0.800	0.800
x^3y	0.0	0.0	0.0
x^2y^2	0.444	0.444	0.444
xy^3	0.0	0.0	0.0
y^4	0.444	0.800	0.800

the second power. The two fourth-order terms in x and y are evaluated by the 2 × 2 rule at a value less than the exact result, while the 3 × 3 rule reproduced the exact result. Thus, the two-dimensional rules behave in a manner similar to the one-dimensional cases.

Applications to Isoparametric Finite Elements

The previous sections have outlined the general characteristics of the integration of polynomials with the Gaussian quadrature approach. The method finds its major application in the finite element method in the evaluation of individual terms of isoparametric element stiffness matrices. This section demonstrates the behavior of the Gaussian quadrature method in this application with the case of a typical term from the stiffness matrix of a one-dimensional bar with three nodes.

A typical stiffness element for the three-node bar expressed in the natural coordinate ξ is the k_{11} element. This equation is given as Eq. A.21 in Appendix A of this lesson. This element contains the following expression:

Typical One-Dimensional Quadratic Stiffness Element

$$K' = \int_{-1}^{+1} \left[\frac{2(\xi - 1/2)}{2(1 + c\xi)} \right]^2 2(1 - c\xi) \, d\xi$$

$$K' = \int_{-1}^{+1} \frac{2(\xi - 1/2)^2}{(1 + c\xi)} \, d\xi \tag{B.5}$$

where

$$c = 2(1 - 2x_2)$$

The term in brackets in the first line is part of a strain energy density expression. The second term containing ξ in the integrand of the first line is the Jacobian of the transformation from the physical coordinate x to the natural coordinate ξ. Remember that the Jacobian is the ratio of the differential area or length in the physical coordinates to the differential area or length in the natural coordinates. As we can see, the Jacobian appears as a square in the denominator of the strain energy density expression (the term in brackets). One of these terms is canceled by the Jacobian contained in the differential to give the reduced form presented as the second line in Eq. B.5.

The parameter c in Eq. B.5 depends on the location of the interior node of the three-node element. The geometry of the bar being used in this example is shown in Fig. B.4. The bar has a total length of unity. The center node can be located on the region from

$1/4 < x < 3/4$. If the node is located at either $1/4$ or $3/4$, the Jacobian can take on the value of zero at some point on the bar. If the interior node is located outside of the interval, the Jacobian can take on negative values on the bar. These are not physically possible cases. The ratio of the differential areas must be positive everywhere. Note that the parameter c becomes zero when the interior node is located at the center of the element.

The meaning of a zero and a negative Jacobian is easily seen in the case of a six-node triangle. A triangle that is not distorted is shown in Fig. B.5a. The Jacobian in this case is a constant at every point in the region. However, when the element has a curved edge, the Jacobian becomes a function of the natural coordinates in the integrand. In Fig. B.5b, an element with an initially curved edge is shown. In this case, the Jacobian varies over the region, but it is everywhere positive. In Fig. B.5c, the element has a physically impossible initial curvature. There is an overlap. This is a physically impossible case, but it is not mathematically impossible. The Jacobian is negative on the portion of the region with the overlap. It is zero where the edges cross.

We now demonstrate the effect of the location of the interior node on the Gauss quadrature integration of the stiffness term given in Eq. B.5. When the integrand of Eq. B.5 is written as a partial fraction and the quotient and the remainders are found, the result is

---

Integrand of Eq. B.5 in Partial Fraction Form

$$I = \frac{2\xi^2 - 2\xi + 1/2}{c\xi + 1}$$

$$= (c\xi + 1)\left[\frac{2}{c}\xi - \frac{2(c+1)}{c^2}\right] + \frac{r}{(c\xi + 1)} \tag{B.6}$$

where

$$r = \frac{1}{2} + \frac{2(c+1)}{c^2}$$

When the interior node is located at the center of the bar, the parameter c is zero and, as seen in the first line of Eq. B.6, the integrand is a quadratic function in ξ. As a result, this integral will be evaluated exactly by the two-point rule. This conclusion is validated in

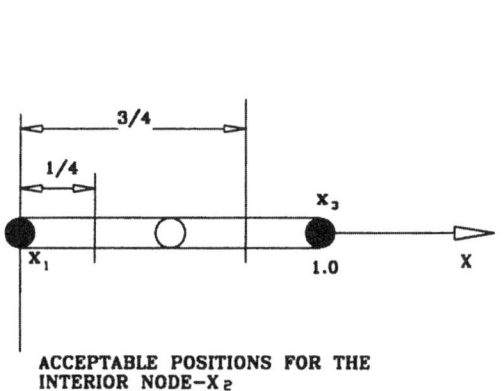

ACCEPTABLE POSITIONS FOR THE
INTERIOR NODE—X_2

Figure B.4. Geometry of a three-node bar element.

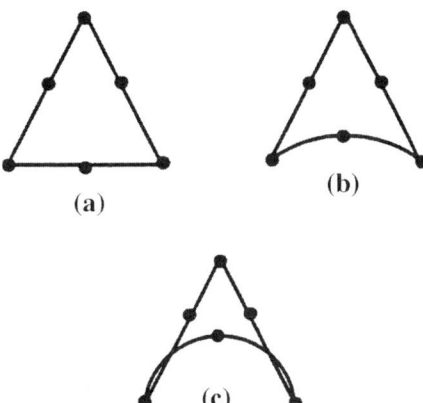

Figure B.5. Six-node triangle geometries.

Table B.3 for x_2 equal to 0.5. The same value for the integral is given for the three different rules.

However, when the interior point is not located in the center of the bar, the parameter c is not zero and the remainder term in the partial fraction exists. This essentially raises the polynomial order of the integrand. Evidence of this increase in order is seen in Table B.3 for each case where the interior node is not located at the center of the bar. Each rule gives a different result, which implies that higher order terms are being evaluated differently by the different rules. The difference increases as the interior node moves away from the center of the bar. Furthermore, note that the three Gaussian quadrature rules produce a result at 0.25 and 0.75 when the integral contains a singularity. A sharp rise in the approximations produced by the Gaussian quadrature rules for the three- and four-point rules hints at this singularity.

The changes seen in the values of the integrals with the change in the location of the interior node seen in Table B.3 are due to the transformation used to map the arbitrary region onto the standard region used in the Gaussian quadrature integration method. The transformation introduces nonlinearities into the element, which makes them overly stiff when the elements are initially distorted. The effects of this nonlinearity are seen in the discussions of the four-, six-, eight-, and nine-node elements presented in Lessons 8, 10, and 11.

Closure

This presentation has outlined the use of the Gaussian quadrature integration procedure in one and two dimensions. The behavior of this approximate integration method was shown when it was applied to polynomial expressions of different orders. The approach is exact when applied to polynomials of order $(2n - 1)$ or lower. Higher order odd terms are evaluated correctly as zero. The higher order even terms are not evaluated correctly.

The behavior of the method when it is applied to a typical stiffness element of an isoparametric element was demonstrated by integrating a term from a three-node bar. We saw that the term increased in apparent polynomial order when the element contained an ''initial distortion'' by having the interior node located at points other than the center. This means that the Gaussian quadrature rule that will integrate the term correctly in the

Table B.3. Integral value vs interior node location for the two-, three-, and four-point rules for Eq. B.6.

Location	Two points	Three points	Four points
0.25	5.500	8.500	10.750
0.30	4.322	5.242	5.487
0.35	3.561	3.874	3.910
0.40	3.028	3.124	3.128
0.45	2.635	2.652	2.653
0.50	2.334	2.334	2.334
0.55	2.095	2.106	2.107
0.60	1.901	1.944	1.946
0.65	1.742	1.834	1.844
0.70	1.610	1.779	1.824
0.75	1.500	1.8334	2.084

undistorted element will not exactly integrate the element when it is initially distorted. The same result is seen in two-dimensional elements and is discussed in the main text.

Note that the stiffness matrices for elements developed with the strain gradient approach are not normally integrated with the Gaussian quadrature approach. They are typically integrated with Green's theorem. The reason for this is twofold: (1) there are very few integrals that must be evaluated in the strain gradient approach, e.g., 6 terms vs 78 terms for the six-node element, and (2) the integrands that appear in the strain gradient approach are simple expressions. The elimination of the use of the Gaussian quadrature procedure removes the need for the mapping to the standard region and the consequent nonlinearities introduced into initially distorted elements by this mapping.

Notes

1. The procedures for finding the Gaussian points and the weighting factors are presented in Refs. 10 and 11. In brief, these quantities are found by minimizing the error in the computation of a $(2n - 1)$ polynomial. This is done by forming an error function and forcing it to a minimum. In the case of this polynomial, the Gaussian points and weighting factors evaluate the integrals exactly.

2. A Gaussian quadrature procedure exists for triangular regions. The function being evaluated is integrated over a standard equilateral triangle centered at the origin. The Gaussian points and weighting functions for triangles are given in Ref. 11.

References and Other Readings

1. Harwood, S. A. "Finite Element Error Analysis Using the Finite Difference Method," M.S. Thesis, University of Colorado, 1989.

2. Dow, J. O., Ho, T. H., and Cabiness, H. D. "A Generalized Finite Element Evaluation Procedure," *ASCE Journal of Structural Engineering*, Vol. 111, No. 2, Feb. 1985, pp. 435–52.

3. Cabiness, H. D. "A Strain-Gradient Based Finite Element Evaluation Technique," M.S. Thesis, University of Colorado, 1985.

4. Ho, T. H. "A Generalized Finite Element Evaluation Procedure," Ph.D. Thesis, University of Colorado, 1984.

5. Dow, J. O., and Byrd, D. E. "The Identification and Elimination Artificial Stiffening Errors in Finite Elements," *International Journal for Numerical Methods in Engineering*, Vol. 26, 1988, pp. 743–762.

6. Byrd, D. E. "Identification and Elimination of Errors in Finite Element Analysis", Ph.D. Dissertation, University of Colorado, 1988.

7. Zienkiewicz, O. C. *The Finite Element Method in Structural and Continuum Mechanics*, McGraw-Hill Publishing Co. Ltd., London, 1967.

8. Zienkiewicz, O. C. *The Finite Element Method*, 3rd ed. McGraw-Hill Publishing Co. Ltd., London, 1977.

9. Zienkiewicz, O. C., and Taylor, R. L. *The Finite Element Method*, 4th ed. McGraw-Hill Publishing Co. Ltd., London, 1989.

10. Cook, R. D. *Concepts and Applications of Finite Element Analysis*, John Wiley and Sons, Inc., New York, 1974.

11. Cook, R. D., Malkus, D. S., and Plesha, M. E. *Concepts and Applications of Finite Element Analysis*, 3rd ed. John Wiley and Sons, Inc., New York, 1989.

12. Segerlind, L. J. *Applied Finite Element Analysis*, John Wiley and Sons, Inc., New York, 1984.

Lesson 8 Problems

1. Form the stiffness matrix for the two-node bar element. Have the x axis along the length of the bar with the origin at the left end of the bar. What strain states is this element capable of representing? How many integrals must be taken during the formulation process? *Hint*: This is a warm-up for the strain gradient approach for formulating finite element stiffness matrices.

2. Re-do Problem 1 with the origin at the center of the bar element. Compare the results with those of Problem 1 and comment on the simplification that results from having the origin at the centroid of the region.

3. Form the stiffness matrix for an equilateral constant strain triangle. Does the size of the element effect the stiffness matrix? Use a unit thickness, $E = 10$, and a Poisson ratio of 0.3.

4. Form the interpolation polynomials for a rectangular four-node element. *Hint*: They are done for you in Lesson 9. However, try it on your own first.

5. Form the stiffness matrix for a three-node bar with the nodes located at (0.0, 1.0, 3.0). *Hint*: Compare this stiffness matrix to that produced by the isoparametric approach. See Appendix A.

6. Form the equivalent nodal loads for the two-node bar element of Problem 1 when it is loaded with a uniform load of w. *Hint*: The nodal loads found in Problems 6 and 7 are the only exercises concerning nodal loads.

7. Form the equivalent nodal load for the three-node bar element of Problem 5 when it is loaded with a uniform load w. *Hint*: Compare these results to those of Problem 6.

Lesson 9

Four-Node Quadrilateral Element

Purpose of Lesson   To formulate the stiffness matrix and evaluate the modeling characteristics of a four-node quadrilateral finite element.

The strain gradient formulation procedure developed in Lesson 8 is applied to the four-node quadrilateral. Evaluation of the strain models reveals that the incomplete second-order displacement approximation polynomials used in the development introduce an error commonly known as *parasitic shear* into the element. In the strain gradient approach, this error can be eliminated directly during the formulation process. This differs from the isoparametric approach, which indirectly eliminates parasitic shear during the computation of the stiffness matrix.

Furthermore, we see that the incomplete displacement approximations used to formulate the four-node quadrilateral do not enable the element to uniformly represent higher order strain states along general axes. This modeling deficiency cannot be corrected without changing to a higher order element.

■   ■   ■

The stiffness matrix for the four-node quadrilateral element shown in Fig. 1 is formulated using the procedure developed in Lesson 8. This element has displacements u and v in the x and y directions at each of the four nodes, so the element has eight degrees of freedom. The load vector is not formed because the procedure for introducing applied loads into the model is well defined in Lesson 8. The modeling characteristics are discussed in detail during the formulation steps.

Step 1 — Polynomial Identification

The geometries of several four-node quadrilateral elements are related to Pascal's triangle in Fig. 2. As expected, the approximation polynomials for both the u and v displacements contain the following terms: 1, x, y, and xy.

This representation is an incomplete second-order polynomial because it does not contain the x^2 and y^2 terms. We saw in Lesson 7 that the strain models do not possess rotational invariance because the approximate displacement polynomials are incomplete.

The strain gradient representation of the displacements is completed by identifying the strain gradient coefficients of the displacement approximations. This is accomplished by taking the coefficients of the appropriate terms from Table 1 of Lesson 5. The approximate displacement representations for the four-node quadrilateral element are:

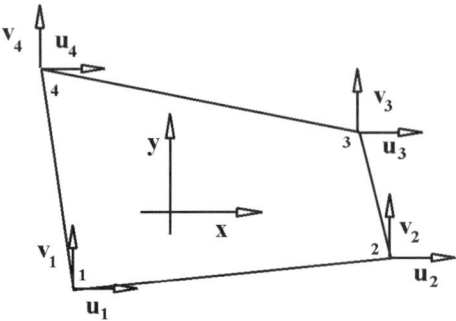

Figure 1. Four-node quadrilateral with coordinate definitions.

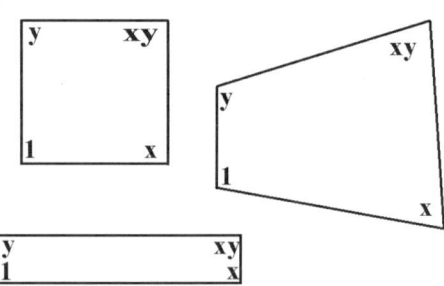

Figure 2. Polynomial identification with Pascal's triangle.

Strain Gradient Displacement Approximations

$$u(x, y) = (u_{rb})_0 + (\varepsilon_x)_0 x + [(\gamma_{xy}/2) - r_{rb}]_0 y$$
$$+ (\varepsilon_{x,y})_0 xy$$
$$v(x, y) = (v_{rb})_0 + [(\gamma_{xy}/2) + r_{rb}]_0 x + (\varepsilon_y)_0 y$$
$$+ (\varepsilon_{y,x})_0 xy \tag{9.1}$$

These polynomials contain the eight linearly independent strain states that this eight-degree-of-freedom element is capable of representing. When these linearly independent variables are extracted from Eq. 9.1, they are

Strain States Represented

$$(u_{rb})_0 \quad (v_{rb})_0 \quad (r_{rb})_0$$
$$(\varepsilon_x)_0 \quad (\varepsilon_y)_0 \quad (\gamma_{xy})_0 \tag{9.2}$$
$$(\varepsilon_{x,y})_0 \quad (\varepsilon_{y,x})_0$$

The first six terms show that this element is capable of representing the three rigid body modes and the three constant strain states in the plane. Therefore, the four-node quadrilateral satisfies the convergence criteria for a finite element. This means that a problem modeled with this element will converge to the actual result in the limit as each element approaches zero area.

The displacements associated with the three rigid body motions are shown in Fig. 3. The deformations associated with the three constant strain states and the two higher order strain states are shown in Figs. 4 and 5, respectively. Each of these eight deformation patterns represent the displacements of the finite element due to a single strain gradient term. Inversely, these are the displacements that are produced by one strain state when the other seven terms are zero. These displacement patterns are given relative to an elemental coordinate system with the origin at the centroid of the element.

The rigid body modes can be interpreted as zeroth-order strain terms because they produce no strain in the element. The three strain gradient terms that produce constant strain can be considered as first-order strain states. The two higher order strain gradient quantities $(\varepsilon_{x,y})_0$ and $(\varepsilon_{y,x})_0$ shown in Fig. 5 are second-order strain states. They can be

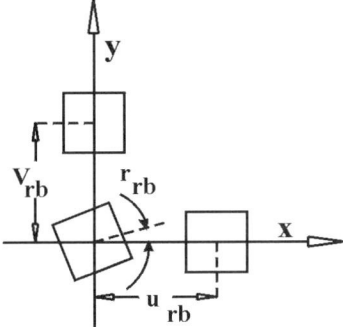

Figure 3. Displacements for rigid body motions.

interpreted physically in terms of the displacement patterns they produce in elements. The term $(\varepsilon_{x,y})_0$ represents the linear variation of $(\varepsilon_x)_0$ in the y direction. The portion of the element above the neutral axis is in tension. The lower portion is in compression. That is to say, the value of ε_x varies along the y axis. This figure provides a physical interpretation of the meaning of $(\varepsilon_{x,y})_0$.

This deformation pattern is similar to the displacement exhibited by a beam element deformed by a constant moment. Thus, the term $(\varepsilon_{x,y})_0$ can be interpreted as a flexure term. A similar interpretation applies to the term $(\varepsilon_{y,x})_0$, which represents flexure around the z axis along the y axis. The concept of flexure around a general axis was used in Lesson 7 as an example of a third-order tensor transformation. This transformation was used to explain why the strain representations in elements formed with incomplete displacement polynomials are not rotationally invariant.

Step 2 — Strain Approximation and Model Evaluation

When the displacement approximations for the four-node quadrilateral are substituted into the definitions of strain, the strain representations contained in this element are found to be

---

Four-Node Quadrilateral Strain Representations

$$\varepsilon_x = (\varepsilon_x)_0 + (\varepsilon_{x,y})_0 y$$
$$\varepsilon_y = (\varepsilon_y)_0 + (\varepsilon_{y,x})_0 x \qquad (9.3)$$
$$\gamma_{xy} = (\gamma_{xy})_0 + (\varepsilon_{x,y})_0 x + (\varepsilon_{y,x})_0 y$$

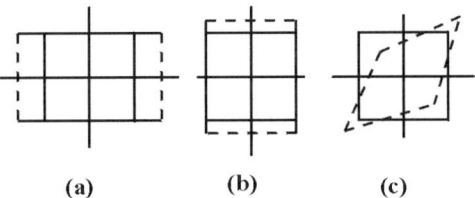

(a)   (b)   (c)

Figure 4. Deformations for constant strain states.

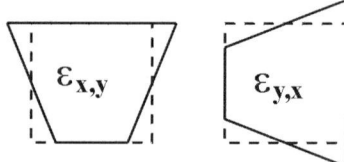

Figure 5. Flexure deformations.

The highest order term contained in each of these strain representations is a linear term. This means that the four-node quadrilateral cannot represent any quadratic or higher order strain states. However, we can identify the modeling capabilities of the four-node quadrilateral by comparing Eq. 9.3 to the Taylor series expansion for the strains. Equation 9.3 need be compared only to the Taylor series representation truncated at the linear terms because no higher order terms exist in the finite element model. To facilitate this comparison, the complete second-order Taylor series strain representations developed in Lesson 5 are reproduced here as

---

Complete Second-Order Taylor Series Strain Expansions

$$\varepsilon_x = (\varepsilon_x)_0 + (\varepsilon_{x,x})_0 x + (\varepsilon_{x,y})_0 y$$
$$\varepsilon_y = (\varepsilon_y)_0 + (\varepsilon_{y,x})_0 x + (\varepsilon_{y,y})_0 y \qquad (9.4)$$
$$\gamma_{xy} = (\gamma_{xy})_0 + (\gamma_{xy,x})_0 x + (\gamma_{xy,y})_0 y$$

Incomplete Normal Strain Models

The normal strain ε_x is represented in Eq. 9.3 by the constant term $(\varepsilon_x)_0$ and the linear term $(\varepsilon_{x,y})_0$ that varies with y. These terms enable the element to independently represent a constant value of ε_x and flexure. When we compare this approximation to the complete Taylor series expansion given in Eq. 9.4, we see that the element is not capable of representing the other second-order strain term in the ε_x representation, namely, $(\varepsilon_{x,x})_0$.

A similar comparison of the normal strain ε_y with the Taylor series expansion shows that the element can represent a constant value of ε_y and flexure. However, the element is not capable of representing $(\varepsilon_{y,y})_0$. The four-node quadrilateral element is sometimes referred to as the bilinear element because it is capable of representing only two of the four linear strain gradient terms.

As seen in Lesson 7, the absence of $(\varepsilon_{x,x})_0$ and $(\varepsilon_{y,y})_0$ does not enable this element to accurately represent the Poisson effects due to the flexure strains given by $(\varepsilon_{y,x})_0$ and $(\varepsilon_{x,y})_0$. As a result, the element is overly stiff in flexure. The lack of these two terms was also seen in Lesson 7 to contribute to the inability of the four-node quadrilateral to represent strain invariance under rotation.

Shear Strain and Parasitic Shear

The shear strain γ_{xy} is represented in Eq. 9.3 by a constant term and two linear terms. However, when we compare this expression to the equivalent Taylor series in Eq. 9.4, we see that the two linear terms are incorrect. The two normal strain terms $(\varepsilon_{x,y})_0$ and $(\varepsilon_{y,x})_0$ should not be present in the shear strain representation. The two linear terms in the shear strain representation should be $(\gamma_{xy,x})_0$ and $(\gamma_{xy,y})_0$. Physically, this is obvious because the shear strains are independent of the normal strains. Thus, in its corrected form, the shear strain representation in a four-node element can model only constant shear strain.

The presence of the two normal strain terms in the shear expression causes a modeling error that is known as parasitic shear (see Notes 1 and 2). The effect of this error can be described as follows. When the element is attempting to represent flexure via either $(\varepsilon_{x,y})_0$ or $(\varepsilon_{y,x})_0$, an erroneous shear strain is developed. This tends to reduce the deformation produced by a given loading because part of the work done by the load is absorbed as strain energy in the erroneous shear strains. Thus, the deformations produced by this element when it is attempting to represent general flexure are smaller than they should be because of this modeling defect. This effect was clearly demonstrated in the examples presented in Lesson 7.

Elimination of Parasitic Shear

Parasitic shear can be removed from elements formulated with the strain gradient approach by simply removing the erroneous terms from the shear strain expression before the shear strain energy expression is evaluated. This operation is demonstrated in the next step of the element development.

We should understand that parasitic shear was recognized as a modeling deficiency long before strain gradient notation clearly identified its cause. It was identified by the overly stiff nature of the four-node quadrilateral when it represents flexure. The error is corrected in the isoparametric formulation of the four-node quadrilateral by evaluating the shear energy integral with a lower order Gaussian quadrature formula. This truncates the polynomial contained in the integrand and eliminates the effect of the parasitic shear terms.

When we develop the eight- and nine-node quadrilateral elements, we will see that these elements contain parasitic shear because they are developed from incomplete polynomial displacement representations. As we saw in Lesson 6, these elements contain complete representations of the linear terms. However, the higher order terms are not complete and they contain parasitic shear terms. In the attempt to correct this modeling deficiency in isoparametric elements using reduced-order or selectively reduced-order numerical integration, either too few or two many terms are removed.

When not all of the erroneous terms are removed from the shear strain expression, a residual of the overstiffening due to parasitic shear remains. When too many terms are removed, sources of strain energy are incorrectly eliminated from the strain energy expression. This means that deformation patterns that should contain strain energy do not produce any strain energy. Instead, these deformation patterns act as rigid body modes. That is to say, these deformation patterns do not contain any strain energy so it appears that these elements contain more than the three required rigid body modes. An element containing these extra zero-energy modes is said to contain spurious zero-energy modes (see Ref. 1). These modes introduce an instability known as hourglassing into the deformations. This effect is shown in Fig. 3 of Lesson 11 when procedures for accurately eliminating parasitic shear in eight- and nine-node elements are identified.

Higher Order Modeling Deficiencies

Although parasitic shear can be removed, the use of incomplete polynomials in the displacement representations causes a modeling error that cannot be corrected in an element. The nature of this error can be seen by studying the expression for flexure for an arbitrary system of axes in the four-node quadrilateral. The generalized expression for flexure is derived in Lesson 7 and repeated here for convenience as

Generalized Flexure Expression

$$\frac{\partial \varepsilon_{x'}}{\partial y'} = \varepsilon_{x,y} \cos^3 \theta - \varepsilon_{y,x} \sin^3 \theta$$
$$+ (\gamma_{xy,y} - \varepsilon_{x,x}) \cos^2 \theta \sin \theta$$
$$+ (\varepsilon_{y,y} - \gamma_{xy,x}) \cos \theta \sin^2 \theta \qquad (9.5)$$

This expression is a generalization of the two higher order strain gradient terms $\varepsilon_{x,y}$ and $\varepsilon_{y,x}$ represented by the four-node quadrilateral. These two terms define flexure along the element coordinate axes. The generalized expression defines flexure along any axis. Flexure along the coordinate x axis is shown in Fig. 6a. Generalized flexure along the arbitrary x' axis is shown in Fig. 6b. In Fig. 6a, the flexure along the x axis is represented by $\varepsilon_{x,y}$. In Fig. 6b, the flexure is represented by the expression given by Eq. 9.5.

The generalized flexure expression given in Eq. 9.5 can be seen to contain all six possible second-order strain gradient terms, namely, the first derivatives of the three strain components. Thus, any element that is capable of accurately representing flexure around any axis must be capable of representing all six of these second-order strain gradient terms. However, the four-node quadrilateral is not capable of representing all six of these second-order terms. This element can only represent $(\varepsilon_{x,y})_0$ and $(\varepsilon_{y,x})_0$. It is not capable of representing $(\varepsilon_{x,x})_0$, $(\varepsilon_{y,y})_0$, $(\gamma_{xy,x})_0$, and $(\gamma_{xy,y})_0$.

Thus, we can conclude that the four-node quadrilateral is not capable of accurately representing flexure except with respect to the x and y coordinate axes. Even in these cases, the four-node element contains a deficiency. The element cannot represent the Poisson effect that accompanies the two flexure terms, i.e., the element cannot represent $(\varepsilon_{x,x})_0$ and $(\varepsilon_{y,y})_0$. This means that the six-node linear strain triangle is the simplest element that is capable of representing flexure around any axis. These conclusions are supported by results presented in Lesson 7.

Step 3 — Formulation of the Strain Energy Expression

In the third step of the formulation process, a strain energy function expressed in terms of strain gradient quantities is formed using the strain representations given in Eq. 9.3. The development transforms the strain energy expression for plane stress in the continuum to

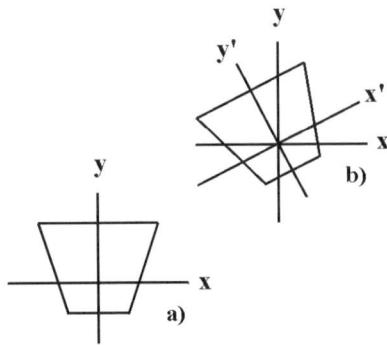

Figure 6. Flexure.

a function of the eight strain gradient variables contained in a four-node element. The matrix form of the expression for the strain energy contained in the continuum given in Lesson 8 is reproduced here for convenience:

Strain Energy Expression for the Continuous Plane Stress Problem

$$U = \frac{1}{2} \int_{\Omega} \{\varepsilon\}^T [E] \{\varepsilon\} \, d\Omega$$

where

$$\{\varepsilon\}^T = [\varepsilon_x \; \varepsilon_y \; \gamma_{xy}]$$

$$[E] = \frac{E}{(1-\nu^2)} \begin{bmatrix} 1 & \nu & 0 \\ \nu & 1 & 0 \\ 0 & 0 & \alpha \end{bmatrix}$$

$$\alpha = \frac{1-\nu}{2}$$

(9.6)

The strain energy expression given in Eq. 9.6 is transformed to a function of the strain gradient variables by replacing the strain terms with the polynomial representations of the strain components given by Eq. 9.3. Equation 9.3 can be written in matrix form as

---

Strain Representation in Matrix Form

$$\{\varepsilon\} = [T]\{\varepsilon_,\}$$

where $\quad \{\varepsilon\}^T = [\varepsilon_x \; \varepsilon_y \; \gamma_{xy}]$

$$\{\varepsilon_,\}^T = [(u_{rb})_0 \; (v_{rb})_0 \; (r_{rb})_0 \; (\varepsilon_x)_0 \; (\varepsilon_y)_0 \; (\gamma_{xy})_0 \; (\varepsilon_{x,y})_0 \; (\varepsilon_{y,x})_0]$$

$$[T] = [[T_0]|[T_\varepsilon]]$$

$$[T_0] = \begin{bmatrix} 0 & 0 & 0 \\ 0 & 0 & 0 \\ 0 & 0 & 0 \end{bmatrix}$$

(9.7)

$$[T_\varepsilon] = \begin{bmatrix} 1 & 0 & 0 & y & 0 \\ 0 & 1 & 0 & 0 & x \\ 0 & 0 & 1 & \underline{x} & \underline{y} \end{bmatrix}$$

Parasitic shear is caused by the two erroneous flexure terms in the shear strain expression as shown in Eq. 9.3. These erroneous terms are contained in the shear strain component of Eq. 9.7 by the x and y terms that are underlined in the third row of $[T_\varepsilon]$. Thus, if the transformation given in Eq. 9.7 is used to develop the strain energy expression for the four-node quadrilateral, the element will contain parasitic shear. Parasitic shear can be removed from the element by removing these two erroneous terms from the shear strain component, which is the third row of the transformation matrix.

When the strain transformation given by Eq. 9.7 is substituted into the strain energy expression given by Eq. 9.6, the result is

---

Matrix Form of the Discrete Strain Energy Expression

$$U = \frac{1}{2}\{\varepsilon_{,}\}^T \left[\int_\Omega [T]^T [E][T]\, d\Omega \right] \{\varepsilon_{,}\}$$

$$U = \frac{1}{2}\{\varepsilon_{,}\}^T \bar{U}\{\varepsilon_{,}\}$$

(9.8)

This equation is a discrete approximation of the strain energy written with strain gradient quantities as the independent variables. To highlight the simple form of the integrals that result when Eq. 9.8 is expanded, they are formed separately in the next step.

Step 4 — Integration of the Strain Energy Terms

The integrals contained in the $\bar{U}$ matrix of Eq. 9.8 are now evaluated. The $\bar{U}$ matrix can be interpreted as the finite element stiffness matrix expressed in terms of the generalized strain gradient coordinates. The $\bar{U}$ matrix is given as

---

Strain Energy Integrals

$$\bar{U} = \int_\Omega [T]^T [E][T]\, d\Omega$$

(9.9)

Equation 9.9 can be expanded using Eq. 9.7 to give

Strain Energy Integrals

$$\bar{U} = \int_\Omega \begin{bmatrix} [T_0]^T[E][T_0] & [T_0]^T[E][T_\varepsilon] \\ [T_\varepsilon]^T[E][T_0] & [T_\varepsilon]^T[E][T_\varepsilon] \end{bmatrix} d\Omega$$

(9.10)

Only the lower right partition $\bar{U}_{22}$ is nonzero in this strain energy expression since the rigid body portion of the transformation $[T_0]$ is the null matrix. When this partition is expanded and integrated over an area of constant thickness t, the result is

Expanded Form of the Strain Energy Matrix

$$\overline{U}_{22} = \frac{tE}{(1-\nu^2)} \begin{bmatrix} I_1 & \nu I_1 & 0 & I_3 & \nu I_2 \\ \nu I_1 & I_1 & 0 & \nu I_3 & I_2 \\ 0 & 0 & \alpha I_1 & \alpha I_2 & \alpha I_3 \\ I_3 & \nu I_3 & \alpha I_2 & (I_6 + \alpha I_4) & (\nu + \alpha)I_5 \\ \nu I_2 & I_2 & \alpha I_3 & (\nu + \alpha)I_5 & (I_4 + \alpha I_6) \end{bmatrix}$$

$$\text{where } I_1 = \int_\Omega d\Omega \qquad I_4 = \int_\Omega x^2 \, d\Omega \qquad (9.11)$$

$$I_2 = \int_\Omega x \, d\Omega \qquad I_5 = \int_\Omega xy \, d\Omega$$

$$I_3 = \int_\Omega y \, d\Omega \qquad I_6 = \int_\Omega y^2 \, d\Omega$$

The erroneous terms contained in the strain energy matrix due to parasitic shear are the following three terms: αI_4, αI_5, and αI_6. When these terms are removed from Eq. 9.11, the strain energy matrix becomes

Corrected Strain Energy Matrix

$$\overline{U}_{22} = \frac{tE}{(1-\nu^2)} \begin{bmatrix} I_1 & \nu I_2 & 0 & I_3 & \nu I_2 \\ \nu I_2 & I_1 & 0 & \nu I_3 & I_2 \\ 0 & 0 & \alpha I_1 & \alpha I_2 & \alpha I_3 \\ I_3 & \nu I_3 & \alpha I_2 & I_6 & \nu I_5 \\ \nu I_2 & I_2 & \alpha I_3 & \nu I_5 & I_4 \end{bmatrix} \qquad (9.12)$$

The significant feature of Eqs. 9.11 and 9.12 is that only six integrals must be evaluated to formulate the stiffness matrix for the four-node quadrilateral. The small number and simple integrands contained in these integrals mean that it is not necessary to transform to a coordinate system that makes the integrals easier to evaluate as is done in the isoparametric method. These integrals can be evaluated using brute force numerical integration, Green's theorem, or symbol manipulation (see Appendix A). The last two approaches have proven the most effective to date (see Ref. 2).

The six integrals defined in Eq. 9.11 can be recognized as the area, the two first moments of area and the three second moments of area. This recognition can be used to further simplify the evaluation of the stiffness matrix. If the centroid is chosen as the origin of the local or elemental coordinate system, some of the off-diagonal terms of Eq. 9.12 become zero. With the centroid as the origin, the integrals identified as I_2 and I_3 are zero by definition. This reduces Eq. 9.12 to

Corrected Strain Energy Matrix at the Centroid

$$\overline{U}_{22} = \frac{tE}{(1-v^2)} \begin{bmatrix} I_1 & 0 & 0 & 0 & 0 \\ 0 & I_1 & 0 & 0 & 0 \\ 0 & 0 & \alpha I_1 & 0 & 0 \\ 0 & 0 & 0 & I_6 & v I_5 \\ 0 & 0 & 0 & v I_5 & I_4 \end{bmatrix} \tag{9.13}$$

If the elemental axes are also principal axes for the area of the element, the product of inertia I_5 would also be zero. Equation 9.13 would then be further simplified. This possibility has yet to be exploited, but it is discussed further in the transformation step discussed next.

Step 5 — Formulation of Coordinate Transformations

The transformation required to put the strain energy expressions in terms of nodal coordinates is now formed. This coordinate transformation is formed by using the interpolation functions given in Eq. 9.1. A transformation from nodal displacements to strain gradient variables is first formed. This process was developed in detail and applied to the three-, four-, six-, eight-, and nine-node configurations in Lesson 6. The inverse transformation is now developed.

The transformation from nodal displacements to strain gradient variables can be viewed as a set of boundary conditions applied to the interpolation functions. That is to say, eight linearly independent equations are formed for the four-node element by specifying the displacements at the nodal locations of the finite element. If we look at a general nodal point i, the nodal displacements are u_i and v_i. The coordinates of this node are x_i and y_i. Thus, two transformation equations are available at each node from Eq. 9.1 as the following:

Generic Boundary Equation

$$\begin{Bmatrix} u_i \\ v_i \end{Bmatrix} = \begin{bmatrix} 1 & 0 & -y_i & x_i & 0 & y_i/2 & x_i y_i & 0 \\ 0 & 1 & x_i & 0 & y_i & x_i/2 & 0 & x_i y_i \end{bmatrix} \{\varepsilon_,\} \tag{9.14}$$

The transformation for all four nodes can be written as

Nodal Displacement to Strain Gradient Transformation for the Four-Node Quadrilateral

$$\begin{Bmatrix} u_1 \\ u_2 \\ u_3 \\ u_4 \\ v_1 \\ v_2 \\ v_3 \\ v_4 \end{Bmatrix} = \begin{bmatrix} 1 & 0 & -y_1 & x_1 & 0 & y_1/2 & x_1 y_1 & 0 \\ 1 & 0 & -y_2 & x_2 & 0 & y_2/2 & x_2 y_2 & 0 \\ 1 & 0 & -y_3 & x_3 & 0 & y_3/2 & x_3 y_3 & 0 \\ 1 & 0 & -y_4 & x_4 & 0 & y_4/2 & x_4 y_4 & 0 \\ 0 & 1 & x_1 & 0 & y_1 & x_1/2 & 0 & x_1 y_1 \\ 0 & 1 & x_2 & 0 & y_2 & x_2/2 & 0 & x_2 y_2 \\ 0 & 1 & x_3 & 0 & y_3 & x_3/2 & 0 & x_3 y_3 \\ 0 & 1 & x_4 & 0 & y_4 & x_4/2 & 0 & x_4 y_4 \end{bmatrix} \begin{Bmatrix} (u_{rb})_0 \\ (v_{rb})_0 \\ (r_{rb})_0 \\ (\varepsilon_x)_0 \\ (\varepsilon_y)_0 \\ (\gamma_{xy})_0 \\ (\varepsilon_{x,y})_0 \\ (\varepsilon_{y,x})_0 \end{Bmatrix} \tag{9.15}$$

Equation 9.15 can be written in symbolic form as

Transformation Matrix

$$\{d\} = [\Phi]\{\varepsilon_{,}\} \tag{9.16}$$

This transformation was given in Lesson 6 as Eq. 6.23 and in Lesson 8 as Eq. 8.17. Again, the columns of $[\Phi]$ are the vectors of the displacements produced by the individual strain states. These displacement patterns are shown graphically in Figs. 3, 4, and 5. The inverse of this transformation is required to complete the formulation of the finite element stiffness matrix. The inverse is written as

---

Strain Gradient to Nodal Displacement Transformation

$$\{\varepsilon_{,}\} = [\Phi]^{-1}\{d\} \tag{9.17}$$

As discussed earlier, this inverse need not necessarily be computed numerically by direct inversion. The inverse can be formed symbolically or through the use of the Gram-Schmidt orthogonalization process. These approaches have been used successfully.

The choice of the elemental coordinate system origin as the centroid simplifies the formulation of the inverse. This is the case because the transformation given by Eq. 9.15 is made "more orthogonal." The minimum effect is to make the rigid body rotation vector orthogonal to additional strain gradient vectors. In some cases, other vectors became orthogonal to each other. When a transformation contains a larger number of orthogonal vectors, a smaller effort is required to compute a fully orthogonal transformation.

Step 6 — Formulation of Stiffness Matrix

With the availability of the strain energy expression in terms of strain gradient variables and the transformation from strain gradient variables to nodal displacements, we can now form the strain energy expression in terms of nodal displacements by substituting Eq. 9.17 into Eq. 9.8 to give

---

Transformed Strain Energy Expression

$$U = \frac{1}{2}\{d\}^T[\Phi]^{-T}\overline{U}[\Phi]^{-1}\{d\} \tag{9.18}$$

The finite element stiffness matrix can be extracted from Eq. 9.18 according to the principle of minimum potential energy as (see Lesson 8, Eqs. 8.25 and 8.26)

Four-Node Quadrilateral Finite Element Stiffness Matrix

$$[K] = [\Phi]^{-T}\overline{U}[\Phi]^{-1}$$

where   (9.19)

$$\overline{U} = \int_\Omega [T]^T[E][T]\,d\Omega$$

This result is identical to the general definition of the strain-gradient-based finite element stiffness matrix given in Lesson 8, Eq. 8.21.

Closure

The general procedure for formulating strain-gradient-based finite element stiffness matrices developed in Lesson 8 has been applied to the four-node quadrilateral element. The eight strain states represented by the element are capable of representing rigid body motions, constant strain states, and flexures with respect to the two elemental coordinate axes.

The idea of generalized flexure developed in Lesson 7 was used to show that the element is incapable of representing flexure around general axes. This deficiency is caused by the fact that the element is not capable of representing all of the second order strain gradient terms. The development of this analytic approach to the *a priori* analysis of element performance suggests well-defined numerical tests to validate the conclusions drawn from the analysis. Furthermore, the tests suggested for evaluating individual element performance can be extended to compare the performance of competing elements of different numbers of degrees of freedom.

The use of strain gradient notation has identified the source of parasitic shear in the four-node quadrilateral as being due to the use of incomplete polynomials in the interpolation functions. By identifying the presence of parasitic shear during the formulation stage, this modeling error can be removed during the formulation by simply removing the incorrect terms. This was demonstrated in the transition from Eq. 9.11 to Eq. 9.12.

Furthermore, the use of strain gradient notation provides guidance in simplifying the computation of the stiffness matrix. For example, by choosing the centroid of the area as the origin of the elemental coordinate system, the number of integrations that must be evaluated is reduced. In addition, the number of orthogonal strain gradient modes in the $[\Phi]$ matrix is increased by this choice of origin. The presence of additional orthogonality conditions reduces the effort required to form the inverse of the matrix.

Notes

1. The use of strain gradient notation identifies the source of parasitic shear as the two erroneous normal strain terms in a way that is not possible if arbitrary coefficients are utilized in the development of the four-node quadrilateral. We can see this by considering the following representation of the displacement approximation based on arbitrary coefficients: $u = a_1 + a_2 x + a_3 y + a_4 xy$ and $v = b_1 + b_2 x + b_3 y + b_4 xy$. When the shear strain is found using this notation, the result is $\gamma_{xy} = (a_3 + b_2) + a_4 x + b_4 y$. In this notation, the erroneous terms identified using strain gradient notation could be wrongly interpreted as the two linear terms contained in the Taylor series expansion of the shear strain, as given in Eq. 9.4.

2. In the next lesson, the six-node, 12-degree-of-freedom linear strain triangle is formulated using complete second-order displacement polynomials. We see that this element is capable of representing the two linear strain gradient terms contained in the Taylor series expansion for each of the three strain components. In the development of the linear strain triangle, we see that the erroneous terms present in the four-node quadrilateral are cancelled by the presence of other terms in the displacement representations. Thus, we can clearly see that the source of parasitic shear is due to the use of incomplete polynomials in the displacement representations.

Appendix A
Strain Energy Evaluation Using Green's Theorem

The following six integrals emerge when the stiffness matrix for a four- or a six-node finite element is formed using the strain gradient approach:

Strain Energy Integrals

$$I_1 = \int_\Omega d\Omega \quad I_4 = \int_\Omega x^2 \, d\Omega$$

$$I_2 = \int_\Omega x \, d\Omega \quad I_5 = \int_\Omega xy \, d\Omega \tag{A.1}$$

$$I_3 = \int_\Omega y \, d\Omega \quad I_6 = \int_\Omega y^2 \, d\Omega$$

This is a small number of integrals to evaluate when compared with the 36 or 78 integrals that must be evaluated when the four- or six-node elements are formed using the isoparametric approach.

The need to evaluate this large number of integrals prompts the use of the Gaussian quadrature procedure, which, in turn, leads to the isoparametric approach. As we have seen, the isoparametric mapping can lead to errors in the evaluation of these integrals when distorted elements are formed, i.e., elements with nonconstant Jacobians. With the reduced number of integrals that must be evaluated in the strain gradient approach, alternative approaches for evaluating these integrals should be considered. In this appendix, Green's theorem is presented as an alternative to the Gauss quadrature approach (see Lesson 3, Appendix A). The use of Green's theorem to compute strain energy quantities is demonstrated with the example of a general quadrilateral.

Green's Theorem

Green's theorem can be stated as follows:

Let C be a piecewise smooth simple closed curve and let Ω be the region consisting of C and its interior. If M and N are functions that are continuous and have continuous first partial derivatives throughout the open region D containing Ω then

---

Green's Theorem

$$\int_\Omega \left(\frac{\partial N}{\partial x} - \frac{\partial M}{\partial y} \right) d\Omega = \int_c M \, dx + \int_c N \, dy \tag{A.2}$$

As we can see, the left-hand side of this equation consists of an area integral over the region Ω. The right-hand side consists of two line integrals around the closed curve C. Thus, Green's theorem reduces an integration over an area to two line integrations around a closed curve. That is to say, Green's theorem is a two-dimensional analog of integration by parts.

A general closed curve is shown in Fig. A.1a. A general quadrilateral element is shown in Fig. A.1b. The boundary of the quadrilateral consists of a piecewise simple

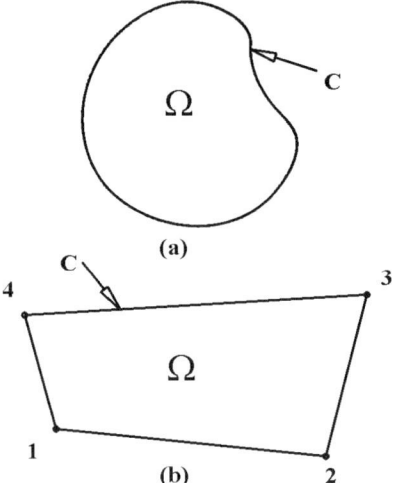

Figure A.1. Piecewise simple closed curves.

curve made up of the four sides of the quadrilateral. As a consequence, we see that each of the line integrals on the right-hand side of Eq. A.2 can be broken into four components for ease of evaluation.

Evaluation Using Green's Theorem

We now demonstrate the process of applying Green's theorem to the integrals we want to evaluate with an example. Let us choose the integral I_5 from Eq. A.1 as the example. This integral is given as

---

Example Integral

$$I_5 = \int_\Omega xy \, d\Omega \qquad (A.3)$$

When we compare Eq. A.3 to Green's theorem, we see that this integral corresponds to the left-hand side of Eq. A.2. As we can see, both of these integrals are evaluated over the area of the region of interest, which in this case is the area of the finite element being formulated.

Thus, our task is to choose the functions N and M so that the integral on the left-hand side of Eq. A.2 is equal to the integral of Eq. A3. There are no other constraints on the functions used, so the choice of these functions is not unique. This means that we can choose the functions N and M with the goal of making the integrals as easy to evaluate as possible.

For a quadrilateral, there are four separate sides, as shown in Fig A.1b. This implies that each of the two line integrals on the right-hand side of Eq. A.2 can be broken into four parts. Thus, there is a total of eight line integrals to be evaluated.

We can reduce this number of line integrals by a factor of 2 by choosing either M or N to be zero. This choice is perfectly acceptable as long as we can find a single expression for the remaining N or M that produces the desired integrand on the left-hand side of Eq. A.2. In this example, N is chosen to be zero.

This means that we want a function that will equal xy after a derivative with respect to y is taken. The function that satisfies this condition is $(1/2)xy^2$. As a result, we will choose N and M for this example to be the following:

Arbitrary Functions

$$M = -\frac{1}{2}xy^2$$
$$N = 0$$

(A.4)

When we introduce these functions into the left-hand side of Eq. A.2, we obtain

Left-Hand Side of Green's Theorem

$$\int_\Omega \left[\frac{\partial(0)}{\partial x} - \frac{\partial(-1/2xy^2)}{\partial y} \right] d\Omega = \int_\Omega xy\, d\Omega$$

(A.5)

When we compare Eq. A.5 to Eq. A.3, we see that our choices of M and N have produced the area integral that we want to evaluate. When Green's theorem is applied to these functions for M and N, the result is

Green's Theorem Applied

$$\int_\Omega xy\, d\Omega = \int_c -1/2xy^2\, dx$$

(A.6)

Thus, we have achieved our goal. The task of evaluating an area integral has been replaced with the task of evaluating a line integral. Furthermore, the choice of $N = 0$ has reduced the number of line integrals that must be evaluated by a factor of 2. The use of Green's theorem provides us with an approach for evaluating the integrals contained in Eq. A.1 that is different from the standard approach of applying an isoparametric mapping and using the Gaussian quadrature technique to evaluate these integrals.

Application to a Quadrilateral

We now demonstrate the use of Green's theorem in this application by evaluating the line integral on the right-hand side of Eq. A.6 for an arbitrary four-node quadrilateral similar to the one shown in Fig. A.1b. To evaluate this integral, it is necessary to express y as a function of x so that the line integral can be integrated in closed form.

The desired expression of y as a function of x is simply the equation of one of the lines bounding the region Ω. In this case, the line C consists of four straight line segments. The equation that describes each of the straight line segments is given as

Equation of a Boundary Segment

$$y = mx + b$$

(A.7)

The parameters m and b are the slope and the y intercept of the line. These quantities are available in terms of the nodal coordinates of the quadrilateral being evaluated and are given after Green's theorem is expressed in terms of Eq. A.7.

When Eq. A.7 is substituted into the right-hand side of Eq. A.6 and the expressions are integrated in terms of the four line segments that bound the region, the result is the following:

Green's Theorem Evaluated

$$\int_c -1/2xy^2 \, dx = -\frac{1}{2}\left(\frac{m_1^2 x^4}{4} + \frac{2b_1 m_1^2 x^3}{3} + \frac{b_1^2 x^2}{2}\right)\Bigg|_{x_1}^{x_2}$$

$$= -\frac{1}{2}\left(\frac{m_2^2 x^4}{4} + \frac{2b_2 m_2^2 x^3}{3} + \frac{b_2^2 x^2}{2}\right)\Bigg|_{x_2}^{x_3}$$

$$= -\frac{1}{2}\left(\frac{m_3^2 x^4}{4} + \frac{2b_3 m_3^2 x^3}{3} + \frac{b_3^2 x^2}{2}\right)\Bigg|_{x_3}^{x_4}$$

$$= -\frac{1}{2}\left(\frac{m_4^2 x^4}{4} + \frac{2b_4 m_4^2 x^3}{3} + \frac{b_4^2 x^2}{2}\right)\Bigg|_{x_4}^{x_1}$$

(A.8)

The limits on the evaluated integrals are the x coordinates of the nodes with the numbering, as shown in Fig. A.1b. The slope m_1 and the y intercept b_1 correspond to the line connecting node 1 and node 2, etc.

The slopes of the straight lines can be computed in terms of the geometry of the element by substituting the nodal coordinates of the two end points of the line into Eq. A.7. This is equivalent to using the two point form of the equation of a straight line. When the slope is evaluated, the result is

---

Slope of Line Segment

$$m_i = \frac{y_j - y_i}{x_j - x_i}$$

(A.9)

---

where the subscripts i and j correspond to the lower and upper bounds on the evaluated integrals in Eq. A.8.

The y intercept can be found from Eq. A.7 once the slopes are known as

---

y Intercept

$$b_i = y_i - m_i x_i$$

(A.10)

We are now in the position of being able to compute the value of the integral using Green's theorem, as given in Eq. A.8.

Numerical Evaluation

The procedure just developed is now applied to the specific quadrilateral shown in Fig. A.2. The first step is to calculate the slopes for the four lines that constitute the closed boundary of the element. When this is done, the four slopes are found to be the following:

Boundary Slopes

$$m_1 = \frac{1-1}{6-1} = 0$$
$$m_2 = \frac{5-1}{10-6} = 1$$
$$m_3 = \frac{4-5}{1-10} = \frac{1}{9}$$
$$m_4 = \frac{4-1}{1-1} = \text{undefined}$$

(A.11)

The next step is to compute the y intercept for the individual lines. When this is done, the results are as follows:

---

y Intercepts

$$b_1 = 1 - 0 * 1 = 1$$
$$b_2 = 1 - 1 * 6 = -5$$
$$b_3 = 5 - \frac{1}{9} * 10 = \frac{35}{9}$$
$$b_4 = \text{undefined}$$

(A.12)

When the four line integrals contained in Eq. A.8 are computed, one at a time, the results are as follows:

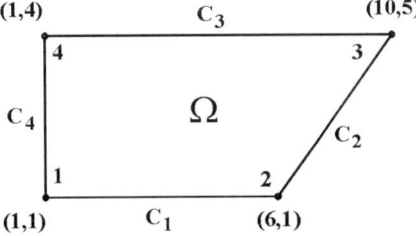

Figure A.2. A specific quadrilateral.

Line c1

$$\int_{c1} -1/2xy^2 \, dx = -\frac{1}{2}\left(\frac{m_1^2 x^4}{4} + \frac{2b_1 m_1^2 x^3}{3} + \frac{b_1^2 x^2}{2}\right)\Big|_{x_1}^{x_2}$$

$$= -\frac{1}{2}\left(\frac{0^2 * x^4}{4} + \frac{2*1*0*x^3}{3} + \frac{1^2*x^2}{2}\right)\Big|_1^6 \quad\quad \text{(A.13)}$$

$$= -8.750$$

Line c2

$$\int_{c2} -1/2xy^2 \, dx = -\frac{1}{2}\left(\frac{m_2^2 x^4}{4} + \frac{2b_2 m_2^2 x^3}{3} + \frac{b_2^2 x^2}{2}\right)\Big|_{x_2}^{x_3}$$

$$= -\frac{1}{2}\left(\frac{1^2 * x^4}{4} + \frac{2*-5*1*x^3}{3} + \frac{15^2*x^2}{2}\right)\Big|_6^{10} \quad\quad \text{(A.14)}$$

$$= -181.33333$$

Line c3

$$\int_{c3} -1/2xy^2 \, dx = -\frac{1}{2}\left(\frac{m_3^2 x^4}{4} + \frac{2b_3 m_3^2 x^3}{3} + \frac{b_3^2 x^2}{2}\right)\Big|_{x_3}^{x_4}$$

$$= -\frac{1}{2}\left(\frac{(1/9)^2 * x^4}{4} + \frac{2*(35/9)*(1/9)*x^3}{3} + \frac{(35/9)^2*x^2}{2}\right)\Big|_{10}^1$$

$$= 533.6250$$

$$\text{(A.15)}$$

Line c4

$$\int_{c4} -1/2xy^2 \, dx = -\frac{1}{2}\left(\frac{m_4^2 x^4}{4} + \frac{2b_4 m_4^2 x^3}{3} + \frac{b_4^2 x^2}{2}\right)\Big|_{x_4}^{x_1}$$

$$= -\frac{1}{2}[\ldots]\Big|_1^1 \quad\quad \text{(A.16)}$$

$$= 0.0$$

When the four integrations are summed, the result is the following:

Evaluated Integral

$$\int_\Omega xy \, d\Omega = \int_c -1/2xy^2 \, dx$$

$$= 343.5417 \quad\quad \text{(A.17)}$$

This result corresponds to the area computed by directly integrating the area integral given as Eq. A.3.

Closure

We have just demonstrated an alternative approach to the isoparametric method for evaluating the six integrals involved in forming the stiffness matrix for a four-node quadrilateral element. The other five integrals can be evaluated in a similar manner. All that must be done is to form appropriate expressions for the M and N functions to reproduce the integrands of Eq. A.1 in the left-hand side of Green's theorem given by Eq. A.2. A set of functions that can be used to evaluate all six integrals is given as

---

M **and** N **Functions for Use in Green's Theorem**

$$I_1: M = -y \qquad\qquad N = 0$$
$$I_2: M = -xy \qquad\qquad N = 0$$
$$I_3: M = -1/2y^2 \qquad\qquad N = 0$$
$$I_4: M = -1/2xy^2 \qquad\qquad N = 0 \qquad\qquad\text{(A.18)}$$
$$I_5: M = -1/2x^2y \qquad\qquad N = 0$$
$$I_6: M = -1/3y^3 \qquad\qquad N = 0$$

As mentioned, the integrals given by Eq. A.1 are the same integrals that must be evaluated in the process of formulating the stiffness matrix for a six-node finite element. If the sides are straight, the process is identical to that just presented for the four-node quadrilateral. If the sides are curved, an expression for y that is a quadratic function of x must be formed. The coefficients of this expression are evaluated by substituting the nodal coordinates into the equation for the quadratic curve. It is difficult to envision many circumstances where more than one edge would be curved, so the effort required to evaluate these integrals for a six-node element is not much more than is required for a four-node element. Once the coefficients for the three edges of the six-node element are found, the process for evaluating the six integrals is identical to the one just demonstrated for the four-node element. This process is taken to completion in Appendix B of Ref. 3.

References and Other Readings

1. Dow, J. O., and Byrd, D. E. "The Identification and Elimination of Artificial Stiffening Errors in Finite Elements," *International Journal for Numerical Methods in Engineering*, Vol. 26, 1988, pp. 743–762.

2. Harwood, S. A. "Finite Element Error Analysis Using the Finite Difference Method," M.S. Thesis, University of Colorado, 1989.

3. Ho. T. H. "A Generalized Finite Element Evaluation Procedure," Ph.D. Dissertation, University of Colorado, 1984.

Lesson 9   Problems

1. Form the interpolation polynomials for a six-, eight-, *or* nine-node element. *Hint*: Form the equivalent of Eq. 9.14 for another element. These matrices are formed in later lesson. This problem is to be done as a warm-up.

2. Form the strain approximations derived from the displacement functions used in Problem 1 and compare them to the complete Taylor series strain representations. What are the limitations of the finite element strain models? *Hint*: Form the equivalent of Eq. 9.3 for the element studied in Problem 1. The result is available in later lessons.

3. Put the strain representations in matrix form. Remove any parasitic shear terms. *Hint*: This task is similar in content to Eq. 9.7. The results are available in later lessons. The six-node element will not have any parasitic shear terms.

4. Form the discrete strain energy expression in terms of strain gradient quantities. Identify the number of different integrals that must be evaluated in the general case. *Hint*: This problem is designed to reinforce the idea that stiffness matrices generated using strain gradient notation are computationally efficient to form.

5. Form the $[\Phi]$ matrix for the element used in Problem 1. *Hint*: Form the equivalent of Eq. 9.15. Draw a figure of the two second-order shear strain states, i.e., the derivatives with respect to x and y. Describe how the figure represents the strain state. (*Hint*: Put the origin at the center of the x coordinate of the element.) Compare the shapes found here to the element shapes in Fig. 3 of Lesson 11.

6. Form an expression for the stiffness matrix using compact matrix notation, e.g., use $[\Phi]$ etc. instead of the full matrices.

7. In a sentence or six, explain how to assemble two finite elements into a single structure. See Problem 3 of Lesson 1. *Hint*: This is designed to demonstrate how easy element assembly is in theory. Sometimes the bookkeeping is a chore.

8. Find the area of a rectangle using Green's theorem. *Hint*: See Appendix A.

9. Find the area of a right triangle using Green's theorem.

10. Find the centroid of a rectangle using Green's theorem.

11. Find the centroid of a right triangle using Green's theorem.

Lesson **10**

Six-Node Linear Strain Element

Purpose of Lesson   To formulate the stiffness matrix and evaluate the modeling characteristics of a six-node linear strain finite element.

The strain gradient formulation procedure developed in Lesson 8 is applied to form the stiffness matrix for the six-node finite element using displacement representations that are complete second-order polynomials. This polynomial basis enables the six-node element to represent all six of the linear variations in the strain components. The six-node element is also known as the linear strain triangle (LST) because of this capability. Since the element is based on complete polynomials it does not contain any parasitic shear terms.

The six-node element is the simplest element that is capable of representing curved boundaries. The stiffness matrices for the six-node strain gradient and iso-parametric representations with *curved* edges are not identical. This occurs because the six-node isoparametric element misrepresents the linear strain distributions. These modeling errors produce an element that is overly stiff. In contrast, the strain gradient element properly represents the strain distribution, so it contains the proper amount of strain energy. Both types of elements satisfy the convergence criteria for finite elements.

The effects of the modeling errors on the performance of the six-node isoparametric element are demonstrated by evaluating single elements and with an example problem. These results highlight the differences in the modeling character-istics of the two approaches.

■   ■   ■

The stiffness matrix for the general six-node element is formulated here using the procedure developed in Lesson 8. The six-node element has displacements u and v in the x and y directions at each of the nodes, so the element has 12 degrees of freedom. Two configurations of six-node elements are shown in Fig. 1. A triangle is shown in Fig. 1a and a more general six-node element is shown in Fig. 1b. The curved edge is represented by the second-order displacement polynomial as a parabola. The standard nodal numbering is used.

The modeling characteristics of the linear strain element are discussed in detail as the development proceeds. The inherent modeling capabilities are deduced from the displacement and strain approximation polynomials. The behavior under deformation for different element configurations is shown by presenting the results of tests of single elements. These tests show some of the effects of curved edges on the stiffness characteristics of the elements. The applied load vector for the six-node element is not formed because the procedure for introducing the loading conditions into the model is well defined in Lesson 8.

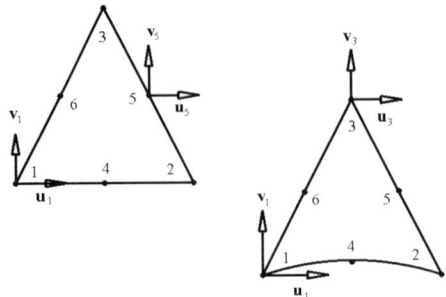

Figure 1. Six-node elements with coordinate definitions.

Step 1 — Polynomial Identification

The geometries of several six-node linear strain elements are related to Pascal's triangle in Fig. 2. As expected, the approximation polynomials for both the u and v displacements contain the following terms: 1, x, y, x^2, xy, and y^2. This representation is a complete second-order polynomial. As discussed in previous lessons, plane stress elements based on complete displacement polynomials also possess strain representations that are complete polynomials. Since the strain representations are complete, parasitic shear is not present in the element and the strains are rotationally invariant.

The strain gradient representation of the displacements is completed by identifying the strain gradient coefficients of the displacement approximations. This is accomplished by taking the coefficients of the appropriate terms from Table 1 of Lesson 5. The approximate displacement representations for the six-node linear strain element are

---

Strain Gradient Displacement Approximations

$$u(x,y) = (u_{rb})_0 + (\varepsilon_x)_0 x + [(\gamma_{xy}/2) - r_{rb}]_0 y + (\varepsilon_{x,x}/2)_0 x^2 + (\varepsilon_{x,y})_0 xy$$
$$+ [(\gamma_{xy,y} - \varepsilon_{y,x})/2]_0 y^2$$

$$v(x,y) = (v_{rb})_0 + [(\gamma_{xy}/2) + r_{rb}]_0 x + (\varepsilon_y)_0 y + [(\gamma_{xy,x} - \varepsilon_{x,y})/2]_0 x^2$$
$$+ (\varepsilon_{y,x})_0 xy + [(\varepsilon_{y,y})/2]_0 y^2$$

(10.1)

These polynomials contain the 12 linearly independent strain states that this 12-degree-of-freedom element is capable of representing. When these linearly independent variables are extracted from Eq. 10.1, they are

---

Strain States Represented

$$
\begin{array}{ccc}
(u_{rb})_0 & (v_{rb})_0 & (r_{rb})_0 \\
(\varepsilon_x)_0 & (\varepsilon_y)_0 & (\gamma_{xy})_0 \\
(\varepsilon_{x,x})_0 & (\varepsilon_{y,x})_0 & (\gamma_{xy,x})_0 \\
(\varepsilon_{x,y})_0 & (\varepsilon_{y,y})_0 & (\gamma_{xy,y})_0
\end{array}
$$

(10.2)

The first six terms show that this element is capable of representing rigid body modes and the constant strain states in the plane. Therefore, the six-node linear strain element satisfies the modeling requirements guaranteeing that a problem modeled with this

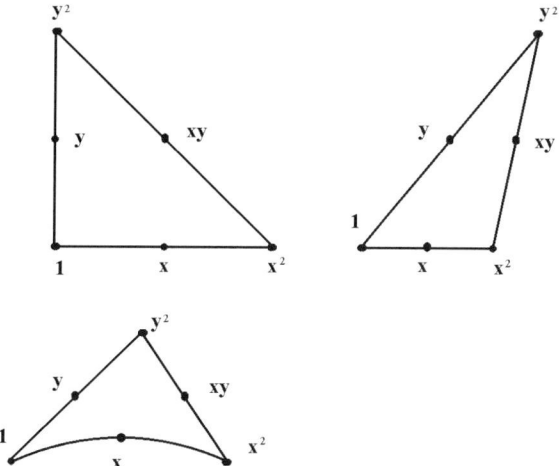

Figure 2. Polynomial identification with Pascal's triangle.

element will converge to the actual result in the limit as each element approaches zero area. The displacements associated with the three rigid body motions are shown in Fig. 3. The deformations associated with the three constant strain states are shown in Fig. 4. These displacement patterns are given relative to an elemental coordinate system with the origin shown in the figure. In practice, the local origin is usually taken as the centroid of the element. This choice reduces the number of nonzero integrals that must be computed in the process of formulating the element stiffness matrix. Furthermore, the computation of the inverse of the $[\Phi]$ matrix can be simplified.

The displacements associated with the higher order strain gradient quantities representing the linear variations in x of the three strain components are shown in Fig. 5. The deformation pattern associated with $(\varepsilon_{x,x})_0$ is shown in Fig. 5a. As we can see, the strain to the right of the line connecting nodes 4 and 5 is higher than the strain to the left of this line. Thus, the level of strain in the x direction varies across the element in the x direction.

The displacement pattern associated with $(\varepsilon_{y,x})_0$ is shown in Fig. 5b. As we can see, the strain in the line connecting nodes 1 and 3 is constant and smaller than the strain in the line connecting nodes 4 and 5. Thus, the level of strain in the y direction varies as x gets larger.

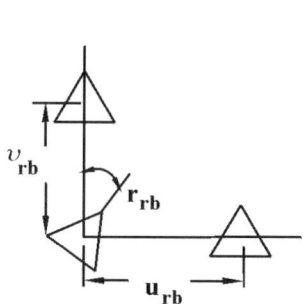

Figure 3. Displacements for rigid body motions.

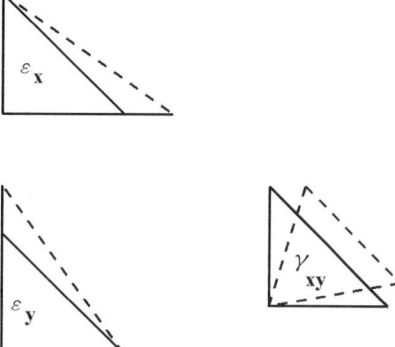

Figure 4. Deformations for constant strain states.

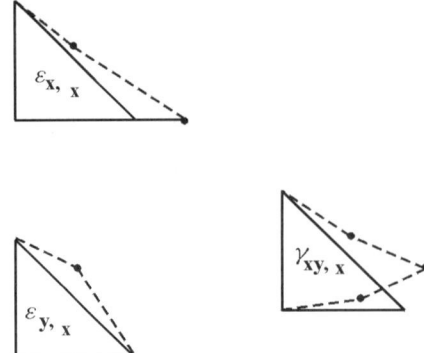

Figure 5. Deformations due to linear strain variations.

The deformation pattern due to $(\gamma_{xy,x})_0$ is shown in Fig. 5c. As we can see, the rotations of the line connecting nodes 1 and 4 and nodes 4 and 2 are different. The increase in the rotation with x is indicative of the change in the shear strain in the x direction.

Step 2 — Strain Approximation and Model Evaluation

When the displacement approximations for the six-node element are substituted into the definitions of strain, the strain representations contained in this element are found to be

---

Strain Representations in the Six-Node Element

$$\varepsilon_x = (\varepsilon_x)_0 + (\varepsilon_{x,x})_0 x + (\varepsilon_{x,y})_0 y$$
$$\varepsilon_y = (\varepsilon_y)_0 + (\varepsilon_{y,x})_0 x + (\varepsilon_{y,y})_0 y \qquad (10.3)$$
$$\gamma_{xy} = (\gamma_{xy})_0 + (\gamma_{xy,x})_0 x + (\gamma_{xy,y})_0 y$$

When Eq. 10.3 is compared to Eqs. 9.3 and 9.4 of Lesson 9, the strain modeling characteristics of the six-node element are made clear. In contrast to the normal strain representation for the four-node quadrilateral given by Eq. 9.3, the normal strain representations for the six-node element consist of complete first-order polynomials. Similarly, when we compare the shear strain representations for the two elements, we see that no parasitic shear terms exist in the six-node model and the shear strain representation is a complete first-order polynomial. When we compare Eq. 10.3 to Eq. 9.4 we see that Eq. 10.3 matches the actual truncated Taylor series representation. Thus, the strain modeling capability of the six-node element is as good as it can be for a linear representation. Since the three strain components are complete first-order polynomials, the six-node element is often referred to as the linear strain element.

The behavior of this strain model under rotation can be demonstrated by examining the transformation characteristics of one of the higher order terms. The transformation matrix for the flexure term around an arbitrarily rotated set of axes given by Eq. 9.5 is reproduced here for convenience as

Generalized Flexure Expression

$$\frac{\partial \varepsilon_{x'}}{\partial y'} = \varepsilon_{x,y} \cos^3 \theta - \varepsilon_{y,x} \sin^3 \theta$$
$$+ (\gamma_{xy,y} - \varepsilon_{x,x}) \cos^2 \theta \sin \theta \qquad (10.4)$$
$$+ (\varepsilon_{y,y} - \gamma_{xy,x}) \cos \theta \sin^2 \theta$$

As we can see, this transformation contains the same six linear terms that are present in the six-node element. Thus, this linear strain gradient quantity transforms invariantly with rotation. Since Eq. 10.4 is representative of the transformation equations of the other five linear strain gradient terms, we can conclude that all of the linear strain terms transform invariantly under rotation for the six-node element. Since the element represents all three constant strain terms, these strain components also transform correctly.

In summary, this analysis shows that the strain model for the six-node element contains no inherent modeling errors. The only limitation is the truncation of the displacement and strain representations that are inherent in the nature of the finite element method.

Step 3 — Formulation of the Strain Energy Expression

A strain energy function expressed in terms of strain gradient quantities is now formed using the strain representations given in Eq. 10.3. The development transforms the strain energy expression for plane stress in the continuum to a function with a finite number of strain gradient variables. The matrix form of the strain energy expression given in Lesson 8 is reproduced here for convenience:

Matrix Form of the Continuous Strain Energy Expression for Plane Stress

$$U = \frac{1}{2} \int_{\Omega} \{\varepsilon\}^T [E] \{\varepsilon\} \, d\Omega$$

$$\text{where} \quad \{\varepsilon\}^T = [\varepsilon_x \ \varepsilon_y \ \gamma_{xy}] \qquad (10.5)$$

$$[E] = \frac{E}{(1 - \nu^2)} \begin{bmatrix} 1 & \nu & 0 \\ \nu & 1 & 0 \\ 0 & 0 & \alpha \end{bmatrix}$$

$$\alpha = \frac{1 - \nu}{2}$$

The strain energy expression is transformed to a function of a finite number of variables by replacing the strain terms in Eq. 10.5 with the polynomial representation of Eq. 10.3. Equation 10.3 can be written in matrix form as

Strain Representation in Matrix Form

$$\{\varepsilon\} = [T]\{\varepsilon_,\}$$

where $\{\varepsilon\}^T = [\varepsilon_x \; \varepsilon_y \; \gamma_{xy}]$

$\{\varepsilon_,\}^T = [\text{See Eq. 10.2}]$

$[T] = [[T_0][T_\varepsilon]]$

$$[T_0] = \begin{bmatrix} 0 & 0 & 0 \\ 0 & 0 & 0 \\ 0 & 0 & 0 \end{bmatrix}$$

(10.6)

$$[T_\varepsilon] = \begin{bmatrix} 1 & 0 & 0 & x & 0 & 0 & y & 0 & 0 \\ 0 & 1 & 0 & 0 & x & 0 & 0 & y & 0 \\ 0 & 0 & 1 & 0 & 0 & x & 0 & 0 & y \end{bmatrix}$$

The vector $\{\varepsilon_,\}$ contains the 12 strain gradient components in the same order as they are shown in Eq. 10.2. That is to say, the $\{\varepsilon_,\}$ vector in Eq. 10.6 is analogous to the $\{\varepsilon_,\}$ vector contained in Eq. 9.7 of Lesson 9.

When the strain transformation given by Eq. 10.6 is substituted into the strain energy expression given by Eq. 10.5, the result is:

Matrix Form of the Discrete Strain Energy Expression

$$U = \frac{1}{2}\int_\Omega \{\varepsilon_,\}^T [T]^T [E][T]\{\varepsilon_,\} \, d\Omega$$

$$U = \frac{1}{2}\{\varepsilon_,\}^T \left[\int_\Omega [T]^T [E][T] \, d\Omega \right] \{\varepsilon_,\}$$

(10.7)

$$U = \frac{1}{2}\{\varepsilon_,\}^T \overline{U}\{\varepsilon_,\}$$

This equation is a discrete approximation of the strain energy written with strain gradient quantities as the independent variables. The expansion of Eq. 10.7 produces simple integrals that are formed separately in the next step.

Step 4 — Integration of the Strain Energy Terms

When the integrals that must be evaluated are extracted from Eq. 10.7, they are given as

Strain Energy Integrals

$$\overline{U} = \int_\Omega [T]^T [E][T] \, d\Omega$$

(10.8)

Equation 10.8 can be expanded using Eq. 10.6 to give

Strain Energy Integrals

$$\overline{U} = \int_\Omega \begin{bmatrix} [T_0]^T [E][T_0] & [T_0]^T [E][T_\varepsilon] \\ [T_\varepsilon]^T [E][T_0] & [T_\varepsilon]^T [E][T_\varepsilon] \end{bmatrix} d\Omega \tag{10.9}$$

Only the lower right 9×9 partition of the $\overline{U}$ matrix is nonzero in this strain energy expression since the rigid body portion of the transformation $[T_0]$ is the null matrix. When this partition is expanded and integrated, the result is

---

Expanded Form of the Strain Energy Matrix

$$\overline{U}_{22} = \frac{tE}{(1-\nu^2)} \begin{bmatrix}
I_1 & \nu I_1 & 0 & I_2 & \nu I_2 & 0 & I_3 & \nu I_3 & 0 \\
\nu I_1 & I_1 & 0 & \nu I_2 & I_2 & 0 & \nu I_3 & I_3 & 0 \\
0 & 0 & \alpha I_1 & 0 & 0 & \alpha I_2 & 0 & 0 & \alpha I_3 \\
I_2 & \nu I_2 & 0 & I_4 & \nu I_4 & 0 & I_5 & \nu I_5 & 0 \\
\nu I_2 & I_2 & 0 & \nu I_4 & I_4 & 0 & \nu I_5 & I_5 & 0 \\
0 & 0 & \alpha I_2 & 0 & 0 & \alpha I_4 & 0 & 0 & \alpha I_5 \\
I_3 & \nu I_3 & 0 & I_5 & \nu I_5 & 0 & I_6 & \nu I_6 & 0 \\
\nu I_3 & I_3 & 0 & \nu I_5 & I_5 & 0 & \nu I_6 & I_6 & 0 \\
0 & 0 & \alpha I_3 & 0 & 0 & \alpha I_5 & 0 & 0 & \alpha I_6
\end{bmatrix} \tag{10.10}$$

where

$$I_1 = \int_\Omega d\Omega \qquad I_4 = \int_\Omega x^2 \, d\Omega$$

$$I_2 = \int_\Omega x \, d\Omega \qquad I_5 = \int_\Omega xy \, d\Omega$$

$$I_3 = \int_\Omega y \, d\Omega \qquad I_6 = \int_\Omega y^2 \, d\Omega$$

The significant feature of Eq. 10.10 is that only six integrals with simple integrands must be evaluated to formulate the stiffness matrix for the six-node element. The simple integrands contained in these integrals mean that they can be computed with little effort. Thus, it is not necessary to make special efforts to simplify the evaluation of these integrals, as is done in the isoparametric method. The integrals contained in Eq. 10.10 have been efficiently integrated using Green's theorem or symbolic computation (see Ref. 2).

The six integrals defined in Eq. 10.10 can be recognized as the area, the two first moments of area and, the three second moments of area. This recognition can further simplify the evaluation of the stiffness matrix. If the centroid is chosen as the origin of the local or elemental coordinate system, the evaluation of Eq. 10.10 is further simplified. With the centroid as the origin, the integrals identified as I_2 and I_3 are zero by definition.

This reduces Eq. 10.10 to

Simplified Strain Energy Matrix at the Centroid

$$\overline{U}_{22} = \frac{tE}{(1-\nu^2)} \begin{bmatrix} I_1 & \nu I_1 & 0 & 0 & 0 & 0 & 0 & 0 & 0 \\ \nu I_1 & I_1 & 0 & 0 & 0 & 0 & 0 & 0 & 0 \\ 0 & 0 & \alpha I_1 & 0 & 0 & 0 & 0 & 0 & 0 \\ 0 & 0 & 0 & I_4 & \nu I_4 & 0 & I_5 & \nu I_5 & 0 \\ 0 & 0 & 0 & \nu I_4 & I_4 & 0 & \nu I_5 & I_5 & 0 \\ 0 & 0 & 0 & 0 & 0 & \alpha I_4 & 0 & 0 & \alpha I_5 \\ 0 & 0 & 0 & I_5 & \nu I_5 & 0 & I_6 & \nu I_6 & 0 \\ 0 & 0 & 0 & \nu I_5 & I_5 & 0 & \nu I_6 & I_6 & 0 \\ 0 & 0 & 0 & 0 & 0 & \alpha I_5 & 0 & 0 & \alpha I_6 \end{bmatrix} \qquad (10.11)$$

Step 5 — Formulation of Coordinate Transformation Matrices

The strain energy is now transformed from strain gradient variables to nodal displacement coordinates. This transformation produces the finite element stiffness matrix being sought. This change of variables is accomplished by using the interpolation functions given in Eq. 10.1 to form a transformation from nodal displacements to strain gradient quantities. This transformation is then inverted to form the desired transformation to nodal coordinates. Equation 10.1 can be written in terms of the coordinate location of the ith node as

---

Nodal Displacement – Strain Gradient Relation

$$\begin{Bmatrix} u_i \\ v_i \end{Bmatrix} = \begin{bmatrix} 1 & 0 & -y_i & x_i & 0 & y_i/2 & x_i^2/2 & -y_i^2/2 & 0 & x_i y_i & 0 & y_i^2/2 \\ 0 & 1 & x_i & 0 & y_i & x_i/2 & 0 & x_i y_i & x_i^2/2 & -x_i^2/2 & y_i^2/2 & 0 \end{bmatrix} \{\varepsilon_,\}$$

$$(10.12)$$

The transformation for all six nodes can be written as

Nodal Displacement to Strain Gradient Transformation

$$\{d\} = [\Phi]\{\varepsilon_,\} \qquad (10.13)$$

The full form of the transformation matrix $[\Phi]$ for the six-node element is

Transformation Matrix for the Six-node Element

$$[\Phi] = \begin{bmatrix} 1 & 0 & -y_1 & x_1 & 0 & y_1/2 & x_1^2/2 & -y_1^2/2 & 0 & x_1y_1 & 0 & y_1^2/2 \\ 1 & 0 & -y_2 & x_2 & 0 & y_2/2 & x_2^2/2 & -y_2^2/2 & 0 & x_2y_2 & 0 & y_2^2/2 \\ 1 & 0 & -y_3 & x_3 & 0 & y_3/2 & x_3^2/2 & -y_3^2/2 & 0 & x_3y_3 & 0 & y_3^2/2 \\ 1 & 0 & -y_4 & x_4 & 0 & y_4/2 & x_4^2/2 & -y_4^2/2 & 0 & x_4y_4 & 0 & y_4^2/2 \\ 1 & 0 & -y_5 & x_5 & 0 & y_5/2 & x_5^2/2 & -y_5^2/2 & 0 & x_5y_5 & 0 & y_5^2/2 \\ 1 & 0 & -y_6 & x_6 & 0 & y_6/2 & x_6^2/2 & -y_6^2/2 & 0 & x_6y_6 & 0 & y_6^2/2 \\ 0 & 1 & x_1 & 0 & y_1 & x_1/2 & 0 & x_1y_1 & x_1^2/2 & -x_1^2/2 & y_1^2/2 & 0 \\ 0 & 1 & x_2 & 0 & y_2 & x_2/2 & 0 & x_2y_2 & x_2^2/2 & -x_2^2/2 & y_2^2/2 & 0 \\ 0 & 1 & x_3 & 0 & y_3 & x_3/2 & 0 & x_3y_3 & x_3^2/2 & -x_3^2/2 & y_3^2/2 & 0 \\ 0 & 1 & x_4 & 0 & y_4 & x_4/2 & 0 & x_4y_4 & x_4^2/2 & -x_4^2/2 & y_4^2/2 & 0 \\ 0 & 1 & x_5 & 0 & y_5 & x_5/2 & 0 & x_5y_5 & x_5^2/2 & -x_5^2/2 & y_5^2/2 & 0 \\ 0 & 1 & x_6 & 0 & y_6 & x_6/2 & 0 & x_6y_6 & x_6^2/2 & -x_6^2/2 & y_6^2/2 & 0 \end{bmatrix} \tag{10.14}$$

This transformation from nodal displacements to strain gradient quantities was given in Lesson 8 as Eq. 8.17. Again, the columns of $[\Phi]$ are the vectors of the displacements produced by the individual strain states. The inverse of this transformation is required to complete the formulation of the finite element stiffness matrix. The inverse is written as

Strain Gradient to Nodal Displacement Transformation

$$\{\varepsilon_{,}\} = [\Phi]^{-1}\{d\} \tag{10.15}$$

As was discussed earlier, this inverse need not necessarily be computed numerically by direct inversion. The inverse can be formed symbolically or through the use of the Gram-Schmidt orthogonalization process. These approaches have been used successfully.

The choice of the elemental coordinate system origin as the centroid simplifies the formulation of the inverse because the transformation given by Eq. 10.15 is made "more orthogonal." The minimum effect of choosing the centroid as the origin is to make the rigid body rotation vector orthogonal to additional strain gradient vectors. In some cases, other vectors became orthogonal to each other. When a transformation contains a larger number of orthogonal vectors, a smaller effort is required to compute a fully orthogonal transformation, which can then be inverted by transposition.

Step 6 — Formulation of Stiffness Matrix

With the availability of the strain energy expression in terms of strain gradient variables and the transformation from strain gradient variables to nodal displacements, we can now form the strain energy expression in terms of nodal displacements by substituting Eq. 10.15 into Eq. 10.7 to give

Transformed Strain Energy Expression

$$U = \frac{1}{2}\{d\}^T[\Phi]^{-T}\overline{U}[\Phi]^{-1}\{d\} \tag{10.16}$$

The finite element stiffness matrix can be extracted from Eq. 10.16 according to the principle of minimum potential energy as (see Lesson 8, Eq. 8.25):

Six-Node Finite Element Stiffness Matrix

$$[K] = [\Phi]^{-T}\overline{U}[\Phi]^{-1}$$

where   (10.17)

$$\overline{U} = \int_{\Omega} [T]^{T}[E][T]\,d\Omega$$

This result is identical to the general definition of the strain gradient based finite element stiffness matrix given in Lesson 8, Eq. 8.21, and Lesson 9, Eq. 9.19.

Single-Element Evaluation Procedures

We now compare the modeling capabilities of the six-node strain-gradient-based elements to that of geometrically identical isoparametric elements. Such an evaluation is possible because strain gradient notation directly relates the strain distributions and the displacements in the continuum. This direct relationship between the displacements and the strains enables us to impose a set of nodal displacements on an element that would produce a known strain distribution in the continuum. We can then compare the strain energy content and the strain distributions produced in the element to the expected results as a way to evaluate the performance of the finite elements.

The evaluation process is demonstrated by comparing the ability of a sequence of six-node strain gradient and isoparametric elements with an initially curved edge to represent the strain state $(\varepsilon_{x,y})_0$. Figure 6 contains a sequence of six-node elements with progressively more curvature on one edge. The curvature is measured by the depth of the arc given as a proportion of the length of the initially straight edge. The element shown in Fig. 6a has no curvature. The elements shown in Figs. 6a and 6b have curvatures of 0.1 and 0.2. Curved elements are used later in an example.

The displacement polynomials that produce a known strain distribution in the continuum or in an element are formed by setting all of the strain gradient terms to zero in Eq. 10.1 except for the strain state being imposed on the element in the evaluation process. In this example, we impose the strain distribution $\varepsilon_{x,y}$ on the elements. Thus, we set all of the strain gradient quantities in Eq. 10.1 to zero except for $(\varepsilon_{x,y})_0$ to give

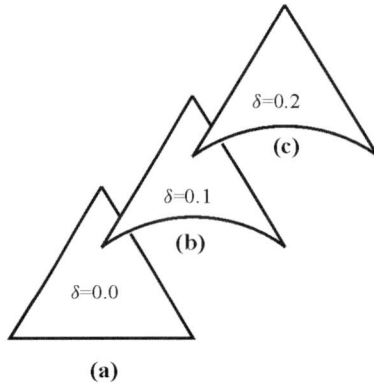

(a)

Figure 6. Elements with curved edges.

Displacements Due to the Single Strain Gradient Term $(\varepsilon_{x,y})_0$

$$u(x, y) = xy(\varepsilon_{x,y})_0$$

$$v(x, y) = -\frac{x^2}{2}(\varepsilon_{x,y})_0$$

$$(10.18)$$

The vector of nodal displacements that should produce this strain distribution in an element is formed by substituting the coordinates of the six element nodes into Eq. 10.18. This vector is contained in Eq. 10.14 as the tenth column of the $[\Phi]$ matrix. When this vector is extracted, the nodal displacements required to produce a strain distribution of $\varepsilon_{x,y}$ are given as:

Displacement Vector

$$\{d\} = \{\phi_{10}\}(\varepsilon_{x,y})_0$$

where

$$(10.19)$$

$$\{\phi_{10}\}^T = [(x_1 y_1) \cdots (x_6 y_6)(-x_1^2/2) \cdots (-x_6^2/2)]$$

We can now compute the strain energy actually contained in the finite element being evaluated using the following expression:

Finite Element Strain Energy

$$U = 1/2\{d\}^T[K]\{d\}$$

$$(10.20)$$

We now want to compare the strain energy computed using Eq. 10.20 to the strain energy that should actually be contained in the element (and the continuum). To compute the actual strain energy, we need expressions for the strain components due to the strain state being studied. The strain distribution produced by a given strain state is available to us from Eq. 10.3. In a process analogous to that used to find Eq. 10.18, we set all of the strain gradient quantities except for the desired strain state to zero in Eq. 10.3. When this is done for this case, the result is

Strain Distribution for a Single Strain State

$$\varepsilon_x = y(\varepsilon_{x,y})_0$$

$$\varepsilon_y = 0$$

$$\gamma_{xy} = 0$$

$$(10.21)$$

This equation defines the strain distribution in the continuum and the strain distribution that should be contained in the finite element due to a single strain state. The expected strain distribution is compared to the strain distribution actually existing in the element as a way of assessing the accuracy of the element.

In addition to comparing the strain distributions, we also compare the strain energy contained in the two types of element to that contained in an identical region of the continuum. The strain energy in the continuum is computed by using the results of Eq. 10.21 in Eq. 10.5 and integrating over the region represented by the element. The strain energy contained in an element is computed with Eq. 10.20. The results of this comparison for the sequence of elements shown in Fig. 6 are presented in the next section.

Example Results

The errors in the strain distributions for the sequence of elements shown in Fig. 6 are presented in Fig. 7. The plots show the difference between the strain distribution that should be contained in the element as given by Eq. 10.21 and the strains actually contained in the isoparametric elements. There are no errors in the representations of the strain gradient elements regardless of the initial curvature. This means that the elements developed using the strain gradient approach represent the linear strain distributions exactly for each configuration.

The same is not true for the isoparametric elements. As seen in Fig. 7a, the undistorted element represents the strain distribution exactly. However, as seen in Figs. 7b and 7c, the initially curved isoparametric elements contain substantial levels of error. The maximum error in the element with the initial curvature of 0.1 is -26.9%. The maximum error in the elements with an initial curvature of 0.2 is -77.3%. These errors are introduced by the transformation used in the isoparametric formulation to map the distorted element on an equilateral parent element for easy integration using Gaussian quadratures. This mapping is not used in the strain gradient formulation because only 6 simple integrals must be evaluated. This is very different from the isoparametric approach where 78 integrals must be evaluated.

In Fig. 7, the locations of the Gaussian points are indicated on the elements. The Gaussian points for the three-point rule are located on the boundary. The Gaussian points for the seven-point rule are located on the interior. The levels of error at these points can be estimated.

The errors in the strain distributions produce errors in the strain energy content of the elements. The errors in the strain energy content of the strain gradient elements and isoparametric elements evaluated with three- and seven-point rules are shown in Table 1. There are no errors in the strain energy content of the strain gradient elements regardless of the initial configuration.

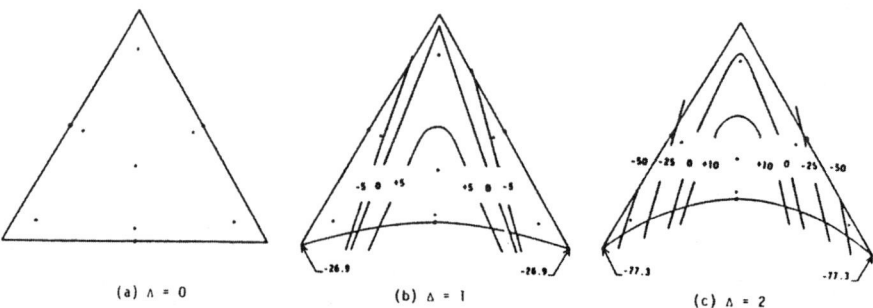

(a) ʌ = 0   (b) ʌ = 1   (c) ʌ = 2

Figure 7. Strain distribution errors in isoparametric elements with curved edges.

Table 1. Error in strain energy.

Curvature (δ)	Percent error in the strain energy		
	Strain gradient	Isoparametric, three points	Isoparametric, seven points
0.0	0.0	0.0	0.0
0.1	0.0	26.4	48.6
0.2	0.0	101.1	179.6

As expected because of the errors in the strain distributions, there are errors in the strain energy content for the initially curved isoparametric elements. As we can see, the level of error depends directly on the amount of distortion. The level of error also depends on the order of the Gaussian quadrature used to approximate the strain energy integrals. The seven-point rule was found to match a highly refined numerical integration of the elements, so the seven-point answer can be considered to be exact. The three-point rule is exact for the undistorted case, but it does not accurately integrate the strain energy integrals for the distorted cases.

The differences in the results produced by the three- and seven-point rules are as expected. The three-point rule produces the exact result for polynomials up to the fifth order. The polynomials in the strain energy expression for the undistorted elements are of the fourth order so an exact result is expected. However, the isoparametric mapping raises the order of the integrands contained in the strain energy integrals beyond the fifth order in elements with curved edges. Thus, the three-point rule cannot be expected to produce the same results as the seven-point rule for the elements with curved edges.

These examples have captured the strain modeling characteristics of strain gradient and isoparametric finite elements. Six-node strain gradient elements exactly reproduce the strain distribution and the strain energy for any configuration of element for any strain state. The same is not true for the isoparametric elements. Elements with initially straight edges represent the strain distributions and strain energies correctly but the initially distorted elements do not correctly represent these quantities for the higher order strain states. Since the curved isoparametric elements correctly represent the rigid body modes and the constant strain states, curved isoparametric elements satisfy the finite element convergence criteria. The effects of this modeling error are seen in an example contained in the next section.

Summary of General Element Evaluation Results

In Refs. 1 and 2, the results of testing 135 initial element configurations with all nine strain states are reported. Four types of initial configurations are studied, namely, (1) equilateral triangles with displaced vertices, (2) non-equilateral triangles with displaced vertices, (3) equilateral triangles with one or two edges curved, and (4) nonequilateral triangles with one or two edges curved. The following general conclusions summarize the results:

1. The strain energy and strain distribution representations for all strain gradient elements are correct regardless of initial configuration.

2. The strain energy and strain distribution representations for all isoparametric elements with straight edges are correct.

3. The strain energy and strain distribution representations in all isoparametric elements with curved edges are correct for the constant strain states.

4. The strain energy and strain distribution representations in all isoparametric elements with curved edges are *incorrect for the linear strain states*.

The effect of the strain modeling errors in the isoparametric elements is not as bad as might be expected. As a finite element model is refined, the percentage of the strain energy contained in the constant strain states increases. Since the isoparametric elements represent the constant strains exactly, the total error in a region drops because the errors contained in an individual element drop as effect of the linear strain states is reduced. Furthermore, the most common use of elements with curved edges is on boundaries so the use of curved edged elements is limited. However, care must be taken at critical points.

The effect of the modeling errors in curved isoparametric elements is demonstrated for the example of a circular hole in a plane stress problem loaded with an axial tension, as shown in Fig. 8. Due to symmetry only a quarter of the problem is analyzed. The problem is solved with three mesh refinements and with three different types of elements on the boundary. The boundary is modeled with straight edged elements, curved isoparametric elements, and curved strain gradient elements. The values of the stress concentration factors found at the critical point at the top of the hole are given in Table 2.

As we can see, the model with the curved isoparametric elements on the boundary produces the lowest levels of stress for every refinement. This result is consistent with the fact that the curved isoparametric elements are overly stiff. However, all three of the stresses are within 5% of the result of 3.00 that would be expected for a semi-infinite region. This example has shown that all of the finite element models produce acceptable results and has demonstrated the effect of the overstiffening in curved isoparametric elements.

Closure

The procedure for generating strain-gradient-based finite elements has been applied to develop the six-node linear strain element. The six-node element is based on complete second-order displacement polynomials. As a result, the element is capable of

Table 2. Stress concentration problem results.

Number of elements— quarter mesh	Straight-edged elements	Curved-edge elements	
		Isoparametric	Strain gradient
8	2.48	2.34	2.63
16	2.96	2.63	3.10
64	3.10	2.96	3.15

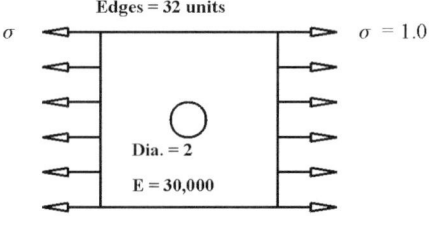

Figure 8. Stress concentration problem.

representing the six linear variations in the three strain components. Since all of the linear variations are represented, the strain representations are rotationally invariant.

The single-element evaluation procedure presented here provides a way to analyze the behavior of all types of elements. In fact, the strain gradient approach was first used in conjunction with the finite element method to study the behavior of single elements. It was only after the single-element evaluation procedure was developed that it was recognized that the process could be inverted to form error free elements. That is to say, if the process could identify elements, it could be used to eliminate the errors.

The single-element evaluation process can be viewed as an extension of the patch test, which is a standard approach for evaluating the convergence capabilities of elements. That is to say, the patch test determines the ability of different types of elements to represent rigid body motions and constant strain states.

The single-element evaluation test procedure was used to compare the performance of the strain-gradient-based elements and isoparametric elements. We found that strain gradient elements exactly represent the strain distributions and the strain energy content for the nine strain states that the linear strain elements are capable of representing for every element configuration. This contrasts to the isoparametric elements, which are overly stiff when they have curved edges, which means that the isoparametric elements with curved edges are more likely to underestimate the stresses and strains than strain-gradient-based elements. This last sentence cannot be made stronger because there is no upper bound on stresses and strains due to the Rayleigh-Ritz criterion.

References and Other Readings

1. Dow, J. O., and Byrd, D. E. "The Identification and Elimination Artificial Stiffening Errors in Finite Elements," *International Journal for Numerical Methods in Engineering*, Vol. 26, 1988, pp. 743–762.

2. Dow, J. O., Ho, T. H., and Cabiness, H. D. "A Generalized Finite Element Evaluation Procedure," *ASCE Journal of Structural Engineering*, Vol. 111, No. 2, Feb. 1985, pp. 435–452.

3. Cabiness, H. D. "A Strain-Gradient Based Finite Element Evaluation Technique," M.S. Thesis, University of Colorado, 1985.

4. Ho, T. H. "A Generalized Finite Element Evaluation Procedure," Ph.D. Thesis, University of Colorado, 1984.

5. Byrd, D. E. "Identification and Elimination of Errors in Finite Element Analysis," Ph.D. Dissertation, University of Colorado, 1988.

6. Harwood, S. A. "Finite Element Error Analysis Using the Finite Difference Method," M.S. Thesis, University of Colorado, 1989.

Lesson 10   Problems

1. Form the [Φ] matrix for the isosceles triangle shown in Fig. 1 of this lesson. *Hint*: This problem and those that follow are designed to provide more experience with the transformation from strain gradient quantities to nodal displacements.

2. Compare the result of Problem 1 to Eq. 11.14 of Lesson 11 and the [Φ] matrix following Eq. A.4 in Appendix A in Lesson 13. *Hint*: Look at the similarities in the results.

3. Form the inverse of the [Φ] matrix formed in Problem 1. Check the inverse by computing $[\Phi]^{-1}[\Phi]$ for the six-node element. *Hint*: The objective of this problem is to provide more experience with the meaning of the inverse, namely, the transformation

from nodal displacements to strain gradient quantities. In Part IV, we see that these inverse transformations are finite difference operators.

4. Given a nodal displacement pattern of $u_1 = 1.0$ and $v_1 = 2.0$ with all other displacements taken to be zero, find the strain gradient components contained in this displacement pattern. *Hint*: Solve the equation $\{d\} = [\Phi] \{\varepsilon_{,}\}$. Use the results from Problem 3.

Lesson 11
Eight- and Nine-Node Elements

Purpose of Lesson   To formulate the stiffness matrices and evaluate the modeling characteristics of eight- and nine-node finite elements using strain gradient notation.

Both the eight- and nine-node elements represent the complete set of linear strain terms plus incomplete representations of the quadratic strain terms. In addition, the nine-node element represents two of the cubic strain terms. As a result, the strain representations of these elements are not rotationally invariant. Both elements contain higher order parasitic shear terms that cannot all be removed by using reduced-order Gaussian quadrature integration without introducing another error known as spurious zero-energy modes. The use of strain gradient notation enables us to remove all of the parasitic shear terms from the elements without inducing spurious zero-energy modes.

The eight- and nine-node elements are capable of representing curved boundaries. When these elements are initially distorted, either with a curved edge or with a nonparallelogram shape, the isoparametric representations of the higher order strain states are distorted and the elements are overly stiff. In contrast, the strain gradient representations contain the correct strain representations and the proper amount of strain energy. Since both types of elements represent the rigid body modes and constant strain states exactly, they both satisfy the convergence criteria for the finite element method.

The effects of the modeling differences between the two types of elements are demonstrated by evaluating single elements and with an example problem. These results highlight the differences in the modeling characteristics of the two approaches.

■   ■   ■

The stiffness matrices for the eight- and nine-node elements are formulated here using the procedure developed in Lesson 8. These elements have displacements u and v in the x and y directions at each of the nodes, so they have 16 and 18 degrees of freedom, respectively, as shown in Fig. 1. Two generic eight- and nine-node elements are shown in Fig. 1. The difference between the two elements is that the nine-node element has a node on the interior, usually located at the center.

These three elements represent a nondistorted element, an element distorted away from a parallelogram shape, and a distorted element with a curved edge. The curved edge is represented by the displacement polynomials as a parabola. Later, we see that the two distorted elements do not represent the strains or the strain energy correctly in isoparametric elements. We also argue that the nine-node element should be preferred to the eight node element. The reasons are primarily tied to the more uniform spacing of the nodes in the nine-node elements, which proves useful when the error analysis procedures developed in Part V are applied.

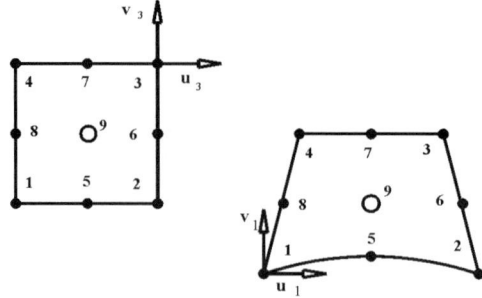

Figure 1. Eight- and nine-node elements with coordinate definitions.

The modeling characteristics of the two elements are discussed in detail as the development proceeds. The inherent modeling capabilities are deduced from the displacement and strain approximation polynomials. The behavior of various configurations of distorted elements are shown by presenting the results of tests of single elements. These tests show some of the effects on the stiffness and strain modeling characteristics of the elements when they deviate from a parallelogram shape. The applied load vectors are not formed because the procedure for introducing the loading conditions into the model is well defined in step 7 of Lesson 8.

Step 1 — Polynomial Identification

The polynomials used to formulate the eight- and nine-node elements are related to their geometries or topologies with Pascal's triangle in Fig. 2. As expected, the approximation polynomials for the nine-node element contain one more term than do the approximation polynomials for the eight-node element. The u and v displacement approximations for the nine-node element contain the additional term, $x^2 y^2$. The polynomials associated with the nine-node template were discussed originally in Lesson 6 in the context of a truss (see Fig. 18 of that lesson.) The resulting polynomial representations of the u and v displacements for the nine-node element are expressed in terms of arbitrary coefficients as

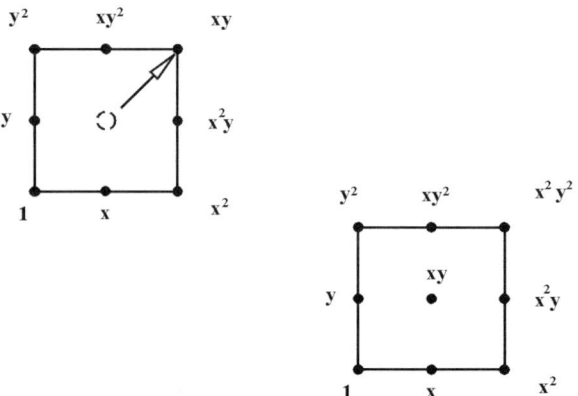

Figure 2. Polynomial identification with Pascal's triangle.

General Displacement Polynomial Representations

$$u = a_1 + a_2 x + a_3 y + a_4 x^2 + a_5 xy + a_6 y^2 + a_7 x^2 y + a_8 xy^2 + a_9 x^2 y^2$$
$$v = b_1 + b_2 x + b_3 y + b_4 x^2 + b_5 xy + b_6 y^2 + b_7 x^2 y + b_8 xy^2 + b_9 x^2 y^2 \tag{11.1}$$

The strain gradient representation of the displacements is completed by identifying the strain gradient coefficients of the displacement approximations. We accomplish this by taking the coefficients of the appropriate terms from Appendix A of Lesson 5. When the general coefficients are replaced by the strain gradient coefficients, the displacement approximations for the nine-node element are given as

Strain Gradient Displacement Representations

$$u = (u_{rb})_0 + (\varepsilon_x)_0 x + \left(\frac{\gamma_{xy}}{2} - r_{rb}\right)_0 y + \frac{1}{2}(\varepsilon_{x,x})_0 x^2 + (\varepsilon_{x,y})_0 xy + \frac{1}{2}(\gamma_{xy,y} - \varepsilon_{y,x})_0 y^2$$
$$+ \frac{1}{2}(\varepsilon_{x,xy})_0 x^2 y + \frac{1}{2}(\varepsilon_{x,yy})_0 xy^2 + \frac{1}{4}(\varepsilon_{x,xyy})_0 x^2 y^2$$
$$v = (v_{rb})_0 + \left(\frac{\gamma_{xy}}{2} + r_{rb}\right)_0 x + (\varepsilon_y)_0 y + \frac{1}{2}(\gamma_{xy,x} - \varepsilon_{x,y})_0 x^2 + (\varepsilon_{y,x})_0 xy + \frac{1}{2}(\varepsilon_{y,y})_0 y^2 \tag{11.2}$$
$$+ \frac{1}{2}(\varepsilon_{y,xx})_0 x^2 y + \frac{1}{2}(\varepsilon_{y,xy})_0 xy^2 + \frac{1}{4}(\varepsilon_{y,xxy})_0 x^2 y^2$$

This set of polynomials is capable of representing the following strain gradient quantities:

Nine-Node Independent Strain States

$$(u_{rb})_0 \quad (v_{rb})_0 \quad (r_{rb})_0$$
$$(\varepsilon_x)_0 \quad (\varepsilon_y)_0 \quad (\gamma_{xy})_0$$
$$(\varepsilon_{x,x})_0 \quad (\varepsilon_{x,y})_0$$
$$(\varepsilon_{y,x})_0 \quad (\varepsilon_{y,y})_0$$
$$(\gamma_{xy,x})_0 \quad (\gamma_{xy,y})_0 \tag{11.3}$$
$$(\varepsilon_{x,xy})_0 \quad (\varepsilon_{x,yy})_0$$
$$(\varepsilon_{y,xx})_0 \quad (\varepsilon_{y,xy})_0$$
$$(\varepsilon_{x,xyy})_0 \quad (\varepsilon_{y,xxy})_0$$

The displacement polynomials for the eight-node configuration contain the three rigid body motions, the three constant strain states, the six linear variations of the three strain components, and four of the eight quadratic terms. The nine-node element represents all of these terms and 2 of the 10 cubic terms. That is to say, the nine-node element represents the two additional strain states, $(\varepsilon_{x,xyy})_0$ and $(\varepsilon_{y,xxy})_0$.

The displacement polynomials given in Eq. 11.2 are presented in a convenient form in Table 1. The coefficients of the polynomial terms for u and v are given in the last two columns. For example, the multipliers of the x^2 terms for the columns labeled u and v are identical to the terms multiplied by x^2 in Eq. 11.2. Rows 17 and 18 of Table 1 are eliminated to give the displacement representations for the eight-node element.

Since the displacement models contain the complete set of second-order polynomial terms, the strain models contain complete linear strain representations. However, the

Table 1. The coefficients for the nine-node element displacement polynomials.

i	$\{\varepsilon_,\}$	u	v
1	$(u_{rb})_0$	1	0
2	$(v_{rb})_0$	0	1
3	$(r_{rb})_0$	$-y$	x
4	$(\varepsilon_x)_0$	x	0
5	$(\varepsilon_y)_0$	0	y
6	$(\gamma_{xy})_0$	$y/2$	$x/2$
7	$(\varepsilon_{x,x})_0$	$x^2/2$	0
8	$(\varepsilon_{y,x})_0$	$-y^2/2$	$xy/2$
9	$(\gamma_{xy,x})_0$	0	$x^2/2$
10	$(\varepsilon_{x,y})_0$	xy	$-x^2/2$
11	$(\varepsilon_{y,y})_0$	0	$y^2/2$
12	$(\gamma_{xy,y})_0$	$y^2/2$	0
13	$(\varepsilon_{x,xy})_0$	$x^2y/2$	0
14	$(\varepsilon_{x,yy})_0$	$xy^2/2$	0
15	$(\varepsilon_{y,xy})_0$	0	$xy^2/2$
16	$(\varepsilon_{y,xx})_0$	0	$x^2y/2$
17	$(\varepsilon_{x,xyy})_0$	$x^2y^2/4$	0
18	$(\varepsilon_{y,xxy})_0$	0	$x^2y^2/4$

higher order strain representations are not complete. As a result, the eight- and nine-node elements contain parasitic shear if they are not corrected. Furthermore, since these elements are formed from incomplete polynomials, the strain representations are not rotationally invariant.

These characteristics are discussed in detail in the next section. The deformation patterns of the individual strain states are not presented for these elements, as was done for the four- and six-node elements. The presentations contained in the previous lessons can be extended by the reader if this information is desired. The deformation patterns are not discussed because the higher order strain states contained here are simply gradients of strain states that have already been discussed. The displacement patterns for the eight- and nine-node elements are identical for the strain states that they have in common.

Step 2 — Strain Approximation and Model Evaluation

When we substitute the displacement approximations for the eight-node element into the definitions of strain, we find the strain representations contained in this element to be

Strain Representations in the Eight-Node Element

$$\varepsilon_x = (\varepsilon_x)_0 + (\varepsilon_{x,x})_0 x + (\varepsilon_{x,y})_0 y + (\varepsilon_{x,xy})_0 xy + 1/2(\varepsilon_{x,yy})_0 y^2$$
$$\varepsilon_y = (\varepsilon_y)_0 + (\varepsilon_{y,x})_0 x + (\varepsilon_{y,y})_0 y + 1/2(\varepsilon_{y,xx})_0 x^2 + (\varepsilon_{y,xy})_0 xy \qquad (11.4)$$
$$\gamma_{xy} = (\gamma_{xy})_0 + (\gamma_{xy,x})_0 x + (\gamma_{xy,y})_0 y + (\varepsilon_{y,xx} + \varepsilon_{x,yy})_0 xy + 1/2\left[(\varepsilon_{x,xy})_0 x^2 + (\varepsilon_{y,xy})_0 y^2\right]$$

The strain modeling characteristics of the eight-node element is discussed first since the only difference between the strain representations in the eight- and nine-node elements is the presence of two higher order strain gradient terms in the nine-node model. As a result, the modeling characteristics for the nine-node element are an extension of those for the eight-node element, so everything that is said about the eight-node element applies directly to the nine-node element.

The representations for the three strain components in the eight-node element contain the constant strain terms, the complete set of linear terms and certain quadratic terms. The two normal strain components contain two of the three quadratic Taylor series components. These quantities can be interpreted as higher order flexure modes. The components missing from the two normal strain models represent higher order extension modes, namely, $(\varepsilon_{x,xx})_0$ and $(\varepsilon_{y,yy})_0$. The missing coefficients cause the element to misrepresent the Poisson effects of the higher order flexure terms.

For example, when the flexure mode represented by $(\varepsilon_{x,yy})_0$ is active, there is no representation of $(\varepsilon_{y,yy})_0$ to model the Poisson effect that occurs in the physical system. This means that when the element represents the higher order flexure mode, no strain gradient coefficient exists to represent the corresponding extension exactly, so other strain gradient terms must be recruited in an attempt to represent this strain state. The lack of this term can be viewed as an error that overconstrains the element and reduces the deformation in line with Rayleigh's principle. This deficiency is analogous to the effect of the missing $(\varepsilon_{y,y})_0$ and $(\varepsilon_{x,x})_0$ coefficients in the strain representation of the four-node element.

Compatibility Equation

The shear strain representation given in Eq. 11.4 contains several interesting features. Let us first examine the coefficient of the xy term. This coefficient contains the two normal strain terms, $(\varepsilon_{x,yy})_0$ and $(\varepsilon_{y,xx})_0$. At first glance, this expression might appear to be a parasitic shear term since it contains normal strain quantities. However, this is not the case. As we now see, this term represents the correct Taylor series coefficient for the xy term in the shear strain representation. The expected Taylor series coefficient for the xy component is $(\gamma_{xy,xy})_0$. When we equate the expected coefficient of the xy term to the coefficient actually present in Eq. 11.4, the result is

---

Compatibility Equation

$$(\gamma_{xy,xy})_0 = (\varepsilon_{y,xx} + \varepsilon_{x,yy})_0 \tag{11.5}$$

We can recognize Eq. 11.5 as the compatibility equation for two-dimensional elasticity. This equation was derived in Lesson 5 as Eq. 5.18 when the arbitrary coefficients were evaluated in terms of strain gradient quantities. Thus, the coefficient of the xy term containing the normal strain quantities given by Eq. 11.4 is a legitimate coefficient because it embodies the compatibility relation for two-dimensional elasticity. Note that this coefficient would not have been recognized as an embodiment of the compatibility equation without the use of strain gradient notation. In contrast, the normal strain coefficients of the x^2 and y^2 terms in the shear strain expression of Eq. 11.4 are parasitic shear terms.

Parasitic Shear

As we see later, the two erroneous parasitic shear terms contained in the shear strain expression can be easily removed from the eight-node element when it is formulated in strain gradient notation. However, these erroneous terms cannot be removed from isoparametric elements without inducing another error. The reason for the inability to successfully remove the parasitic shear terms is now discussed in detail.

When the expression for the shear strain energy is formed by squaring Eq. 11.4 and multiplying by the shear modulus, the result has the following form:

Finite Element Shear Strain Energy

$$U_{\gamma_{xy}} = \frac{E}{2(1-\nu^2)} \int_\Omega (f_1 + f_2)^2 \, d\Omega$$

$$= \frac{E}{2(1-\nu^2)} \int_\Omega (f_1^2 + 2f_1 f_2 + f_2^2) \, d\Omega \qquad (11.6)$$

where

$$f_1 = f_1(1, x, y, xy) = \text{correct Taylor series terms.}$$
$$f_2 = f_2(x^2, y^2) = \text{parasitic shear terms.}$$

This shear strain energy expression contains three types of terms. The first term in the integrand consists of the square of the correct shear strain terms, f_1. Thus, the correct expression in the integrand consists of constant, linear, and second-order terms. The second and third terms contain parasitic shear contributions and are, therefore, sources of modeling errors. The lowest order terms of the erroneous contribution to the shear strain energy are of second order. A corrected element must not contain the effects of the last two terms. In the strain gradient approach, the parasitic shear terms contained in Eq. 11.4 are simply removed from the shear strain expression and the element formulation continues. This results in an element without parasitic shear.

Parasitic Shear in Isoparametric Elements

As we saw in Lesson 9, the parasitic shear terms can be successfully removed from four-node isoparametric elements by integrating the shear strain energy with a reduced-order Gaussian quadrature rule. That is to say, a one-by-one Gaussian quadrature rule is used to evaluate the shear strain expression instead of a two-by-two rule. The use of a Gaussian quadrature integration rule of lower order than is needed to capture an integral exactly is called *reduced-order integration*. The characteristics of Gaussian quadrature integration and the effect of using reduced-order integration are discussed in detail in Appendix B of Lesson 8.

The use of a reduced-order Gaussian quadrature rule *fails* when we apply it to the eight- or nine-node elements. The reason for this failure is now explained. To capture all of the terms in Eq. 11.6, we must use the three-by-three Gaussian quadrature rule. This would include the effect of all of the erroneous parasitic shear terms in the integral. An attempt to remove the erroneous terms through the use of reduced-order Gaussian quadrature integration produces the following results. If the shear strain energy is integrated with a two-by-two rule, some of the terms contained in the third set of terms in Eq. 11.6 are removed, while the effect of the first two terms of Eq. 11.6 are included without change. Thus, the use of the two-by-two rule removes the effect of some of the parasitic shear terms but it does not remove all of the parasitic shear terms.

To eliminate the effects of all of the parasitic shear terms, the shear strain integral given by Eq. 11.6 must be evaluated with a one-by-one rule. As discussed in the Gaussian quadrature appendix in Lesson 8, the one-by-one rule truncates the polynomial to the terms linear in x and y, i.e., the constant, x, y, and xy terms are included in the integral. As indicated in Eq. 11.6, all of the parasitic shear terms are at least functions of x^2 or y^2 so they are eliminated when the one-by-one rule is used. We now see that the use of the one-by-one rule also eliminates terms that belong in the shear strain representation and, hence, produces two types of error.

Spurious Zero-Energy Modes

The use of the one-by-one rule evaluates the constant and the linear terms correctly and eliminates the x^2, y^2, and higher order terms. This means that some of the correct shear strain energy terms are eliminated from the strain energy expression. This produces a modeling error because the element does not contain the correct strain energy content. However, the nature of the Gaussian quadrature integration introduces another error which can negatively effect the deformations produced by the finite element model.

Although the use of the one-by-one rule correctly evaluates the constant and linear components remaining in the first term of Eq. 11.6, the use of the one-by-one rule completely removes the effects of the linear shear terms, $(\gamma_{xy,x})_0$ and $(\gamma_{xy,y})$. This occurs since the use of the reduced-order rule eliminates the squared terms and the linear terms are evaluated as zero because they are odd functions and integrate to zero on symmetric regions. Thus, the only term included in the shear strain energy expression is the constant shear strain component.

From a linear algebra point of view, the removal of the strain gradient coefficients $(\gamma_{xy,x})$ and $(\gamma_{xy,y})_0$ eliminates two terms from the basis set of the eight-node element. This has the effect of adding two zero-energy deformation modes to the element. That is to say, an eight-node isoparametric element with the shear strain energy expression evaluated with a one-by-one rule contains *five* zero-energy modes instead of the three expected rigid body modes. These zero-energy modes consist of the three rigid body modes and two *spurious zero-energy modes*. The removal of the coefficient of the xy term in the shear strain expression does not introduce a zero energy mode because the two normal strain terms are contained in the strain energy expression for the normal strains, so strain energy is still produced when the element is deformed in these modes.

One effect of the spurious zero-energy modes is seen in Fig. 3. This figure shows a cantilever beam deformed by an end load. This loading activates the spurious zero-energy modes into a pattern that is called hourglassing. This name is derived from the shape produced by two adjacent elements.

We can see the cause of these erroneous deformations when the shape of the strain states that are eliminated by the one-by-one Gaussian quadrature rule are studied. The

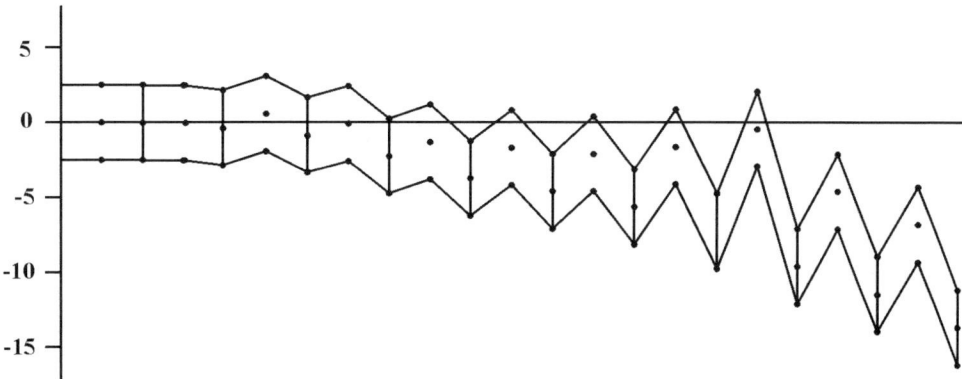

Figure 3. Spurious zero-energy mode induced hourglassing.

deformation patterns of the two shear strain states eliminated by the reduced-order integration are shown in Fig. 4. When we compare Figs. 3 and 4, we can see that the individual elements in Fig. 3 are exhibiting the deformation pattern associated with $(\gamma_{xy,x})_0$ shown in Fig. 4. The deformation pattern due to $(\gamma_{xy,y})_0$ is constrained because of the boundary conditions at the fixed end. The individual elements of the model simply cannot deform in the shape of this spurious zero-energy mode at the fixed end or when this boundary condition is applied. Thus, the use of strain gradient notation has demonstrated the cause of spurious zero-energy modes and explained one of its manifestations exactly.

This presentation also explains the rationale and the success of two methods currently used to eliminate the effects of spurious zero-energy modes. In one approach, the finite element model has constraints imposed on it so that the spurious zero-energy modes cannot be activated. We saw the efficacy of this approach in the example shown in Fig. 3. In this case, the spurious zero energy mode $(\gamma_{xy,y})_0$ is not present because of the fixed constraint on the left end of the beam model.

In the second approach for eliminating the effects of the spurious zero-energy modes from the finite element solution, the deformations produced by the spurious zero-energy modes are removed from a result to give a corrected displacement solution. For example, the displacements due to the spurious zero-energy modes contained in the displacements of Fig. 3 would be removed to give an improved result. A comparison of the mode shapes in Fig. 4 with the deformations of Fig. 3 shows that this approach can be successful.

Suggested Approaches

The best approach for eliminating the effects of parasitic shear and spurious zero energy modes is to eliminate them from the finite element stiffness matrix in the first place. This is accomplished by using strain gradient elements corrected for parasitic shear. In this way, we need not consider the elemental modeling errors from that point on.

In the next lesson, we see that elements exist that *must* be corrected using the strain gradient approach. If these elements are not corrected, qualitative errors can enter the solution that cannot be eliminated through mesh refinement. In this context, a qualitative error is an error in sign. For example, if the coupling term in an element should be positive, a qualitative error in this term would produce a negative quantity. In Lesson 12, the existence of qualitative errors due to parasitic shear terms is demonstrated in laminated composite Mindlin plate elements. These qualitative errors simply cannot be removed from isoparametric elements of this type. Thus, successful elements must be developed by other means.

As a final comment, this analysis indicates that if eight- or nine-node strain gradient elements are not available, the best solution is to use isoparametric elements evaluated

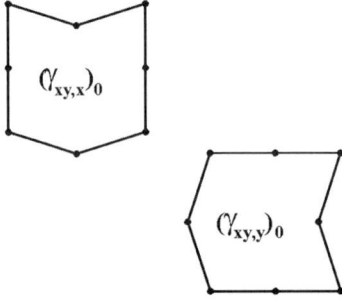

Figure 4. Spurious zero-energy mode shapes.

with either a two-by-two or a three-by-three Gaussian quadrature rule. The errors due to the parasitic shear terms in these elements can be removed by mesh refinement and no spurious zero-energy modes are created.

Nine-Node Element Strain Modeling Characteristics

When we substitute the displacement approximations for the nine-node element into the definitions of strain, we find the strain representations for this element to be:

Strain Representations in the Nine-Node Element

$$\varepsilon_x = (\varepsilon_x)_0 + (\varepsilon_{x,x})_0 x + (\varepsilon_{x,y})_0 y + (\varepsilon_{x,xy})_0 xy + 1/2(\varepsilon_{x,yy})_0 y^2 + 1/2(\varepsilon_{x,xyy})_0 xy^2$$

$$\varepsilon_y = (\varepsilon_y)_0 + (\varepsilon_{y,x})_0 x + (\varepsilon_{y,y})_0 y + 1/2(\varepsilon_{y,xx})_0 x^2 + (\varepsilon_{y,xy})_0 xy + 1/2(\varepsilon_{y,xxy})_0 x^2 y$$

$$\gamma_{xy} = (\gamma_{xy})_0 + (\gamma_{xy,x})_0 x + (\gamma_{xy,y})_0 y + (\varepsilon_{y,xx} + \varepsilon_{x,yy})_0 xy + 1/2\left[(\varepsilon_{x,xy})_0 x^2 + (\varepsilon_{y,xy})_0 y^2\right]$$

$$+ 1/2\left[(\varepsilon_{x,xyy})_0 x^2 y + \varepsilon_{y,xxy})_0 xy^2\right]$$

(11.7)

The strain representations of the nine-node element are identical to those for the eight-node element except for the addition of cubic terms to each of the strain representations. The additions to the normal strain terms are the correct Taylor series representations. The terms added to the shear strain expression are parasitic shear terms. They produce the same types of effects as the other parasitic shear terms. They can be completely removed from elements formulated using strain gradient notation without introducing spurious zero-energy modes. If they are not removed from the nine-node element, they add strain energy to the element and, hence, stiffen it. If attempts are made to remove the effects of the parasitic shear terms from the isoparametric form of the element with reduced-order Gaussian quadrature rules, the results are the same as for the eight-node element. If the two-by-two rule is used not all of the parasitic shear effects are removed. If the one-by-one rule is used, spurious zero energy modes are introduced into the element.

The strain representations for the nine-node element are given in Table 2. The parasitic shear terms are contained in the boxes. In the strain gradient formulation, the effects of parasitic shear terms are eliminated from the element by simply using the shear strain representation with the parasitic shear terms removed in the remainder of the element formulation. That is to say, the parasitic shear terms are identified and eliminated before the formulation continues. If, for example, there is a desire to compare the strain gradient element with parasitic shear effects to the isoparametric formulation, the parasitic shear terms can be included in the strain gradient formulation by not removing the parasitic shear terms. The strain models for the eight-node element are implicitly contained in Table 2. The eight-node strain representations do not include the last two rows of Table 2.

Strain Behavior under Rotation

The behavior of the strain models for the eight- and nine-node elements under rotation is similar to that of the four-node element. Specifically, the constant strain states and the linear strain states transform invariantly because they are complete polynomials. However, the higher order strain gradient terms do not transform invariantly because they are not represented by complete polynomials. If the reader chooses to see a demonstration of this lack of rotational invariance, the procedure used in Lesson 9 can be extended by taking higher order derivatives.

Table 2. Strain representations for the nine-node element.

i	$\{\varepsilon_i\}$	ε_x	ε_y	γ_{xy}
1	$(u_{rb})_0$	0	0	0
2	$(v_{rb})_0$	0	0	0
3	$(r_{rb})_0$	0	0	0
4	$(\varepsilon_x)_0$	1	0	0
5	$(\varepsilon_y)_0$	0	1	0
6	$(\gamma_{xy})_0$	0	0	1
7	$(\varepsilon_{x,x})_0$	x	0	0
8	$(\varepsilon_{y,x})_0$	0	x	0
9	$(\gamma_{xy,x})_0$	0	0	x
10	$(\varepsilon_{x,y})_0$	y	0	0
11	$(\varepsilon_{y,y})_0$	0	y	0
12	$(\gamma_{xy,y})_0$	0	0	y
13	$(\varepsilon_{x,xy})_0$	xy	0	$x^2/2$
14	$(\varepsilon_{x,yy})_0$	$y^2/2$	0	xy
15	$(\varepsilon_{y,xy})_0$	0	xy	$y^2/2$
16	$(\varepsilon_{y,xx})_0$	0	$x^2/2$	xy
17	$(\varepsilon_{x,xyy})_0$	$xy^2/2$	0	$x^2y/2$
18	$(\varepsilon_{y,xxy})_0$	0	$x^2y/2$	$xy^2/2$

Step 3 — Formulation of the Strain Energy Expression

A strain energy function expressed in terms of strain gradient quantities is now formed using the strain representations given in Eq. 11.6. The development transforms the strain energy expression for plane stress in the continuum to a function with a finite number of strain gradient variables. The matrix form of the strain energy expression given in Eq. 8.7 of Lesson 8 is reproduced here for convenience:

Matrix Form of the Continuous Strain Energy Expression for Plane Stress

$$U = \frac{1}{2} \int_\Omega \{\varepsilon\}^T [E] \{\varepsilon\} \, d\Omega$$

where $\{\varepsilon\}^T = [\, \varepsilon_x \quad \varepsilon_y \quad \gamma_{xy} \,]$

$$[E] = \frac{E}{(1-\nu^2)} \begin{bmatrix} 1 & \nu & 0 \\ \nu & 1 & 0 \\ 0 & 0 & \alpha \end{bmatrix}$$

(11.8)

$$\alpha = \frac{1-\nu}{2}$$

The strain energy expression is transformed to a function of a finite number of variables by replacing the strain terms in Eq. 11.8 with the polynomial representation of Eq. 11.4. We can write Eq. 11.4 in matrix form as

---

Strain Representation in Matrix Form

$$\{\varepsilon\} = [T]\{\varepsilon_{,}\}$$

where $\{\varepsilon\}^T = [\varepsilon_x \quad \varepsilon_y \quad \gamma_{xy}]$

$\{\varepsilon_{,}\}^T = [\text{See Table 2}]$

$[T] = [T_0 \quad T_1 \quad T_2 \quad T_3 \quad T_4 \quad T_5]$ (11.9)

$$T_0 = \begin{bmatrix} 0 & 0 & 0 \\ 0 & 0 & 0 \\ 0 & 0 & 0 \end{bmatrix} \quad T_1 = \begin{bmatrix} 1 & 0 & 0 \\ 0 & 1 & 0 \\ 0 & 0 & 1 \end{bmatrix} \quad T_2 = \begin{bmatrix} x & 0 & 0 \\ 0 & x & 0 \\ 0 & 0 & x \end{bmatrix}$$

$$T_3 = \begin{bmatrix} y & 0 & 0 \\ 0 & y & 0 \\ 0 & 0 & y \end{bmatrix} \quad T_4 = \begin{bmatrix} xy & y^2/2 & 0 & 0 \\ 0 & 0 & xy & x^2/2 \\ 0 & xy & 0 & xy \end{bmatrix} \quad T_5 = \begin{bmatrix} xy^2/2 & 0 \\ 0 & x^2y/2 \\ 0 & 0 \end{bmatrix}$$

The vector $\{\varepsilon_{,}\}$ contains the 18 strain gradient components in the same order as they are shown in Table 2. That is to say, the $\{\varepsilon_{,}\}$ vector in Eq. 11.8 is analogous to the $\{\varepsilon_{,}\}$ vector contained in Eq. 9.7 of Lesson 9. The transformation matrices of Eq. 11.9 do not contain the parasitic shear terms.

When we substitute the strain transformation given by Eq. 11.9 into the strain energy expression given by Eq. 11.8, the result is

---

Matrix Form of the Discrete Strain Energy Expression

$$U = \frac{1}{2} \int_\Omega \{\varepsilon_{,}\}^T [T]^T [E][T]\{\varepsilon_{,}\} \, d\Omega$$

$$U = \frac{1}{2} \{\varepsilon_{,}\}^T \left[\int_\Omega [T]^T [E][T] \, d\Omega \right] \{\varepsilon_{,}\}$$ (11.10)

$$U = \frac{1}{2} \{\varepsilon_{,}\}^T \overline{U} \{\varepsilon_{,}\}$$

This equation is a discrete approximation of the strain energy written with strain gradient quantities as the independent variables. The expansion of Eq. 11.10 produces simple integrals that are formed separately in the next step.

Step 4 — Integration of the Strain Energy Terms

The integrals that must be evaluated to form the stiffness matrix are contained in Eq. 11.10 in the following form:

Strain Energy Integrals

$$\overline{U} = \int_{\Omega} [T]^T [E][T] \, d\Omega \tag{11.11}$$

When we expand Eq. 11.11 using Eq. 11.9, the result is

Strain Energy Integrals

$$\overline{U} = \int_{\Omega} \begin{bmatrix} T_0^T E T_0 & T_0^T E T_1 & T_0^T E T_2 & T_0^T E T_3 & T_0^T E T_4 & T_0^T E T_5 \\ T_1^T E T_0 & T_1^T E T_1 & T_1^T E T_2 & T_1^T E T_3 & T_1^T E T_4 & T_1^T E T_5 \\ T_2^T E T_0 & T_2^T E T_1 & T_2^T E T_2 & T_2^T E T_3 & T_2^T E T_4 & T_2^T E T_5 \\ T_3^T E T_0 & T_3^T E T_1 & T_3^T E T_2 & T_3^T E T_3 & T_3^T E T_4 & T_3^T E T_5 \\ T_4^T E T_0 & T_4^T E T_1 & T_4^T E T_2 & T_4^T E T_3 & T_4^T E T_4 & T_4^T E T_5 \\ T_5^T E T_0 & T_5^T E T_1 & T_5^T E T_2 & T_5^T E T_3 & T_5^T E T_4 & T_5^T E T_5 \end{bmatrix} d\Omega \tag{11.12}$$

The components of the first row and the first column are zero in Eq. 11.12 because the matrix $[T_0]$ is zero. This result is as expected because $[T_0]$ represents the strain energy contained in the rigid body modes. When we evaluate the remaining components of Eq. 11.12 one at a time, we find that they contain the 28 integrals given in Table 3.

The integrals given in Table 3 have two significant features: (1) the relatively small number of integrals and (2) the relative simplicity of the integrands. In a nine-node element formed without strain gradient notation there are 171 separate integrals to be computed that have more complex integrands than do the integrals contained in Table 3. The form of the integrals is discussed in Appendix B of Lesson 8. The need to evaluate this large number of integrals motivates the use of the Gaussian quadrature integration

Table 3. Integrals contained in Eq. 11.12.

No.	Integral	No.	Integral
1	$I_1 = \int_{\Omega} d\Omega$	15	$I_{15} = \int_{\Omega} y^4 \, d\Omega$
2	$I_2 = \int_{\Omega} y \, d\Omega$	16	$I_{16} = \int_{\Omega} x^5 \, d\Omega$
3	$I_3 = \int_{\Omega} x \, d\Omega$	17	$I_{17} = \int_{\Omega} x^4 y \, d\Omega$
4	$I_4 = \int_{\Omega} y^2 \, d\Omega$	18	$I_{18} = \int_{\Omega} x^3 y^2 \, d\Omega$
5	$I_5 = \int_{\Omega} xy \, d\Omega$	19	$I_{19} = \int_{\Omega} x^2 y^3 \, d\Omega$
6	$I_6 = \int_{\Omega} x^2 \, d\Omega$	20	$I_{20} = \int_{\Omega} xy^4 \, d\Omega$
7	$I_7 = \int_{\Omega} x^3 \, d\Omega$	21	$I_{21} = \int_{\Omega} y^5 \, d\Omega$
8	$I_8 = \int_{\Omega} x^2 y \, d\Omega$	22	$I_{22} = \int_{\Omega} x^6 \, d\Omega$
9	$I_9 = \int_{\Omega} xy^2 \, d\Omega$	23	$I_{23} = \int_{\Omega} x^5 y \, d\Omega$
10	$I_{10} = \int_{\Omega} y^3 \, d\Omega$	24	$I_{24} = \int_{\Omega} x^4 y^2 \, d\Omega$
11	$I_{11} = \int_{\Omega} x^4 \, d\Omega$	25	$I_{25} = \int_{\Omega} x^3 y^3 \, d\Omega$
12	$I_{12} = \int_{\Omega} x^3 y \, d\Omega$	26	$I_{26} = \int_{\Omega} x^2 y^4 \, d\Omega$
13	$I_{13} = \int_{\Omega} x^2 y^2 \, d\Omega$	27	$I_{27} = \int_{\Omega} xy^5 \, d\Omega$
14	$I_{14} = \int_{\Omega} xy^3 \, d\Omega$	28	$I_{28} = \int_{\Omega} y^6 \, d\Omega$

approach in the isoparametric element formulation procedure. The effect on the accuracy of the element from using the mapping required to take arbitrary shapes onto the standard square required for the use of the Gaussian quadrature procedure is demonstrated later in this lesson.

Note that only 10 of the 28 integrals are nonzero when an element has a rectangular or parallelogram shape and it is aligned with the global axes with the centroid at the origin. In this case, the odd terms are all zero. Although this fact has not been exploited in the formulation of eight- or nine-node elements, it could be used in the formulation of a majority of the elements in many analyses.

The eight-node element formulation is contained implicitly in the nine-node formulation. The eight-node element does not contain the final row and column of Eq. 11.12.

Step 5 — Formulation of Coordinate Transformations

The strain energy is now transformed from strain gradient variables to nodal displacement coordinates to complete the formulation of the finite element stiffness matrix. This change of variables is accomplished by using the approximate displacement functions given in Eq. 11.2 and Table 1. The transformation is formed by evaluating the displacement approximations at the nodal points of the element. This transformation is then inverted to form the desired transformation from strain gradient to nodal coordinates.

We can write the transformation for all nine nodes as

Nodal Displacement to Strain Gradient Transformation

$$\{d\} = [\Phi]\{\varepsilon_,\}$$

where

$$\{d\}^T = \{u_1 \quad v_1 \quad u_2 \quad v_2 \quad \cdots \quad u_9 \quad v_9\}$$
$$[\Phi]^T = [[\Phi]_1^T \quad [\Phi]_2^T \quad \cdots \quad [\Phi]_9^T]$$
$$\{\varepsilon_,\}^T = [(u_{rb})_0 \quad (v_{rb})_0 \quad \cdots \quad (\varepsilon_{y,xxy})_0]$$

(11.13)

The idea of a transformation from nodal displacements to strain gradient quantities was originally presented in Lesson 8 as Eq. 8.17 for a three-node element. The alternate rows of this matrix are defined by the vectors u_i and v_i that are produced when the nodal coordinates x_i and y_i are substituted into Table 1. The columns of this matrix represent the deformed shape of the element for each of the individual strain states. The first three columns are the nodal displacements for the rigid body motions, u_{rb}, v_{rb}, and r_{rb}. The next three columns are the nodal displacements associated with the constant strain states, $(\varepsilon_x)_0$, $(\varepsilon_y)_0$, and $(\gamma_{xy})_0$. The remaining columns match the strain gradient quantities contained in Table 2. The deformations due to $(\gamma_{xy,x})_0$ and $(\gamma_{xy,y})_0$ were shown previously in Fig. 4.

The general form of the $[\Phi]$ matrix for the six-node element is presented as Eq. 10.14 in Lesson 10. The Φ matrix for a nine-node element with an even nodal spacing of 2.50 units and with the local origin at the central node, i.e., node 9, is given as

[Φ] for $h = 2.50$

	1	2	3	4	5	6	7	8	9	10	11	12	13	14	15	16	17	18
1	1.000	0.000	2.500	−2.500	0.000	−1.250	3.125	−3.125	0.000	6.250	0.000	3.125	0.000	−7.813	0.000	−7.813	0.000	9.766
2	0.000	1.000	−2.500	0.000	−2.500	−1.250	0.000	6.250	3.125	−3.125	3.125	0.000	7.813	0.000	7.813	0.000	9.766	0.000
3	1.000	0.000	2.500	2.500	0.000	−1.250	3.125	−3.125	0.000	−6.250	0.000	3.125	0.000	−7.813	0.000	7.813	0.000	9.766
4	0.000	1.000	2.500	0.000	−2.500	1.250	0.000	6.250	3.125	−3.125	3.125	0.000	−7.813	0.000	7.813	0.000	9.766	0.000
5	1.000	0.000	−2.500	2.500	0.000	1.250	3.125	−3.125	0.000	6.250	0.000	3.125	0.000	7.813	0.000	7.813	0.000	9.766
6	0.000	1.000	2.500	0.000	2.500	1.250	0.000	6.250	3.125	−3.125	3.125	0.000	7.813	0.000	7.813	0.000	9.766	0.000
7	1.000	0.000	−2.500	−2.500	0.000	1.250	3.125	−3.125	0.000	−6.250	0.000	3.125	0.000	7.813	0.000	−7.813	0.000	9.766
8	0.000	1.000	−2.500	0.000	2.500	−1.250	0.000	−6.250	3.125	−3.125	3.125	0.000	7.813	0.000	−7.813	0.000	9.766	0.000
9	1.000	0.000	2.500	0.000	0.000	−1.250	0.000	3.125	0.000	0.000	0.000	3.125	0.000	0.000	0.000	0.000	0.000	0.000
10	0.000	1.000	0.000	0.000	−2.500	0.000	0.000	0.000	0.000	0.000	3.125	0.000	0.000	0.000	0.000	0.000	0.000	0.000
11	1.000	0.000	0.000	2.500	0.000	0.000	3.125	0.000	0.000	0.000	0.000	0.000	0.000	0.000	0.000	0.000	0.000	0.000
12	0.000	1.000	2.500	0.000	0.000	1.250	0.000	0.000	3.125	−3.125	0.000	0.000	0.000	0.000	0.000	0.000	0.000	0.000
13	1.000	0.000	−2.500	0.000	0.000	1.250	0.000	−3.125	0.000	0.000	0.000	3.125	0.000	0.000	0.000	0.000	0.000	0.000
14	0.000	1.000	0.000	0.000	2.500	0.000	0.000	0.000	0.000	0.000	3.125	0.000	0.000	0.000	0.000	0.000	0.000	0.000
15	1.000	0.000	0.000	−2.500	0.000	0.000	3.125	0.000	0.000	0.000	0.000	0.000	0.000	0.000	0.000	0.000	0.000	0.000
16	0.000	1.000	−2.500	0.000	0.000	−1.250	0.000	0.000	3.125	−3.125	0.000	0.000	0.000	0.000	0.000	0.000	0.000	0.000
17	1.000	0.000	0.000	0.000	0.000	0.000	0.000	0.000	0.000	0.000	0.000	0.000	0.000	0.000	0.000	0.000	0.000	0.000
18	0.000	1.000	0.000	0.000	0.000	0.000	0.000	0.000	0.000	0.000	0.000	0.000	0.000	0.000	0.000	0.000	0.000	0.000

$$(11.14)$$

This transformation matrix is discussed in detail in the context of finite difference templates in Lesson 13.

The inverse of this transformation is required to complete the formulation of the finite element stiffness matrix. The inverse is written as

---

Strain Gradient to Nodal Displacement Transformation

$$\{\varepsilon_s\} = [\Phi]^{-1}\{d\} \qquad (11.15)$$

As discussed earlier, this inverse need not necessarily be computed numerically by direct inversion if that computation is considered a problem. We can form the inverse symbolically or through the use of the Gram-Schmidt orthogonalization process. These approaches have been used successfully to reduce the effort involved with the inversion.

The choice of the elemental coordinate system origin as the centroid simplifies the formulation of the inverse because the transformation given by Eq. 11.13 is made "more orthogonal." The minimum effect of choosing the centroid as the origin is to make the rigid body rotation vector orthogonal to additional strain gradient vectors. In some cases, other vectors also became orthogonal to each other. When a transformation contains a larger number of orthogonal vectors, a smaller effort is required to compute a fully orthogonal transformation which can then be inverted by transposition.

Step 6 — Formulation of Stiffness Matrix

With the availability of the strain energy expression in terms of strain gradient variables and the transformation from strain gradient variables to nodal displacements, we can now form the strain energy expression in terms of nodal displacements by substituting Eq. 11.15 into Eq. 11.10 to give

Transformed Strain Energy Expression

$$U = \frac{1}{2}\{d\}^T[\Phi]^{-T}\overline{U}[\Phi]^{-1}\{d\} \qquad (11.16)$$

The finite element stiffness matrix can be extracted from Eq. 11.16 according to the principle of minimum potential energy as (see Lesson 8, Eq. 8.25):

Nine-Node Finite Element Stiffness Matrix

$$[K] = [\Phi]^{-T}\overline{U}[\Phi]^{-1}$$

where $\qquad (11.17)$

$$\overline{U} = \int_\Omega [T]^T[E][T]\,d\Omega$$

This result is identical to the general definition of the strain-gradient-based finite element stiffness matrix given in Lesson 8, Eq. 8.21. As mentioned earlier, the eight-node element is implicitly contained in this expression. Equation 11.17 is modified to form the eight-node element stiffness matrix by removing the last row and column from Eq. 11.12. This eliminates the effects of the two strain gradient quantities added to account for the addition of the center node. The transformation matrix is also modified similarly.

Single-Element Modeling Comparisons

We now compare the modeling capabilities of eight-node isoparametric elements with those of geometrically identical strain gradient elements using the procedure developed in the previous lesson for evaluating single elements. These comparisons are possible because of the connection that exists between strain distributions and displacements in strain gradient notation. The relationship between the displacements and the strains is embodied in the [Φ] matrix, i.e., the columns of the [Φ] matrix contain the nodal displacement vectors that would impose a prescribed strain state in an accurate element.

With this capability to relate strains and displacements, we can compute the strain energy contained in an element and compare it to the strain energy it should contain, namely, the strain energy contained in the continuum. This comparison gives us an aggregate measure of the ability of the element to represent the continuum. We can then compare the strain distributions in the element to those in the continuum to give us a point-by-point measure of the modeling ability of the element. In brief, this evaluation process enables us to compare the strain energy content and the strain distributions contained in an isoparametric element to the conditions that should exist in the element. The strain energy content and strain distributions present in strain gradient elements exactly match those imposed on the continuum by the same strain states.

The results of evaluating the modeling capabilities of different configurations of eight-node isoparametric elements are presented next (see Note 1). We study four categories of initial deviation from a square, namely, (1) one or two edges curved, (2) straight edges with a parallelogram shape, (3) straight edges with a nonparallelogram shape, and (4) curved edges and a nonparallelogram shape. Samples of these shapes are shown in Figs. 5a–d. All of the strain states that an eight-node element can represent are then imposed on these configurations and analyzed one at a time.

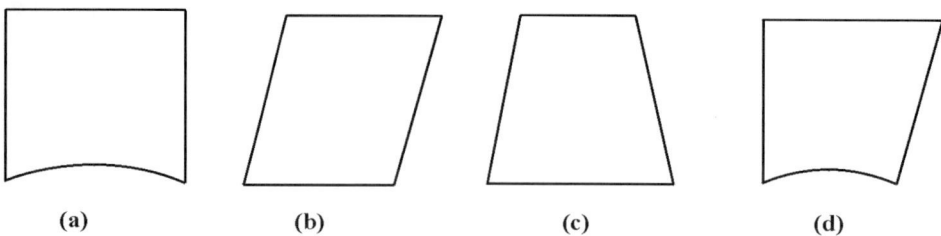

    (a)   (b)   (c)   (d)

Figure 5. Distorted eight-node elements.

We can summarize the conclusions resulting from the evaluation of 132 cases consisting of different element configurations attempting to represent different strain states as follows:

Summary of Conclusions

1. Eight-node isoparametric elements represent the constant strain states exactly regardless of the element configuration.

2. Eight-node isoparametric elements with either curved edges or nonparallelogram shapes (or both) do not represent the higher order strain states exactly.

These conclusions differ in one way from the results found for six-node elements. In six-node elements, all elements with straight edges represented all of the strain states correctly. This is not true for eight-node elements. If an eight-node element is not a parallelogram, it is not capable of accurately representing the higher order strain states. This modeling deficiency is in addition to the errors due to the parasitic shear terms. As noted in the discussion of Eq. 11.4, the parasitic shear terms in the eight-node element are $(\varepsilon_{x,xy})_0$ and $(\varepsilon_{y,xy})_0$. These erroneous shear strains contribute approximately 25 to 30% of the total strain energy in eight-node elements depending on the configurations when the element is modeling these strain states.

These conclusions are derived from the following type of analysis. The deformations associated with a specific strain state are imposed on an eight-node element. These deformation patterns are contained in the $[\Phi]$ matrix as columns. The strain energy content and the strain distributions produced in the element by these deformations are compared to those contained in a strain-gradient-based eight-node element. To better understand the behavior of Gaussian quadrature integration, the isoparametric element is formulated with three types of numerical integration: (1) the two-by-two Gaussian quadrature rule, (2) the three-by-three Gaussian quadrature rule, and (3) a fully converged numerical integration (typically using 400 subdivisions).

Strain Energy Results

A detailed analysis of the strain energy modeling capabilities of one sequence of initial configurations loaded with one strain state is now presented. The sequence of configurations studied is shown in Fig. 6. The initial shape is a square. The square is progressively distorted by changing the lengths of the top and bottom edges by 10 and 20%, respectively. The nodes are given a set of displacements associated with the strain distribution corresponding to $(\gamma_{xy,x})_0$. The results of this analysis are shown in Table 4.

The data contained in Table 4 are plotted in Fig. 7. The four curves are identified as $U_{2 \times 2}$, $U_{3 \times 3}$, U_A, and U_E. The first two designations identify the strain energy computed using the two-by-two and three-by-three Gaussian quadrature rules. The third designation identifies the strain energy actually contained in the isoparametric element when it is

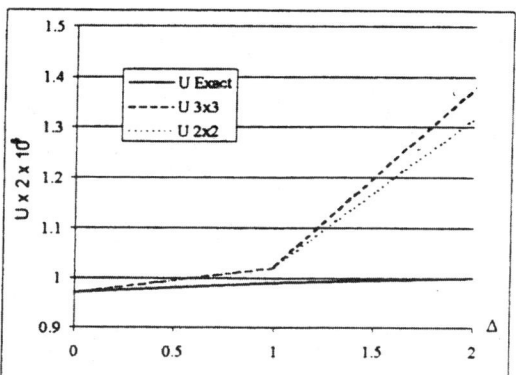

Figure 7. Strain energy vs distortion for a trapezoidal eight-node element representing $\gamma_{xy,z}$.

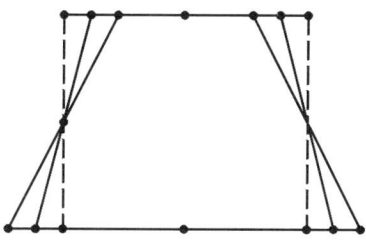

Figure 6. Trapezoidal eight-node element.

computed using the fully converged numerical integration. The curve labeled U_E identifies the strain energy contained in the continuum or in the strain gradient element. Any deviation from this final curve is an error in the strain energy content.

Some of the salient characteristics of isoparametric elements can be seen in these results by comparing the actual and the exact or strain gradient results. The strain energy content for the two cases is the same when the element is undistorted. This means that the eight-node isoparametric element represents the strain energy content correctly for this strain state when the element is not initially distorted. However, when the element is given an initial distortion, the strain energy contained in the element is greater than it should be. This means that the element is overly stiff. This modeling deficiency is due to the nonlinearity of the mapping from the actual shape of the element in global coordinates to the standard shape in the natural coordinates required for the Gaussian quadrature integration procedure.

The effect of this nonlinearity can also be seen in the behavior of the Gaussian quadrature integration procedure in evaluating the isoparametric strain energy expressions. The strain energy content found for the three methods of integration is the same for the undistorted case. That is to say, the two-by-two rule integrates the eight-node element exactly when the Jacobian is a constant. However, the two-by-two rule does not evaluate the strain energy correctly when the element is distorted. In this case, the three-by-three rule is close to integrating the strain energy expression accurately, but it is not exact either.

The case just discussed is typical of the modeling characteristics of a distorted eight-node element when it is representing many of the higher order strain states, (1) initially distorted isoparametric eight-node elements contain more strain energy than they should, i.e., more strain energy than in the continuum, and (2) the integration of the strain energy by the two-by-two Gaussian quadrature rule reduces the strain energy content and brings

Table 4. Strain energy vs distortion.

	Strain energy ($\times 10^{-5}$)			
Δ	2×2 rule	3×3 rule	Fully integrated	Strain gradient
0.0	9.600	9.600	9.600	9.600
0.1	10.54	10.55	10.52	9.974
0.2	13.65	13.75	13.72	11.12

it closer to the exact strain energy content. However, the two-by-two rule produces different results in other cases.

In some cases, integration by the two-by-two rule increases the strain energy in an isoparametric element so that it is greater than the strain energy computed when the strain energy is integrated correctly. This increases the modeling error. In the other cases, the strain energy is reduced to a level below that contained in the continuum when the two-by-two rule is used. In theory, this could cause an eight-node isoparametric element to violate the Rayleigh-Ritz criteria. That is to say, as the model is improved and the total strain energy content should get closer to the actual result, with this type of element the strain energy could increase as the answer improved, which is a violation of the Rayleigh-Ritz criteria. This is not a likely occurrence, but special circumstances could be envisioned that could produce this type of result. This condition exists in 14 of the 132 cases evaluated in the Ho thesis cited as Ref. 4.

Finally, a situation exists where the strain energy in the isoparametric element is below that of the continuum regardless of the integration procedure used to evaluate the strain energy. This condition exists in 8 of the 132 cases evaluated in Ref. 4. As discussed previously, an element displaying this characteristic could violate the Rayleigh-Ritz criteria. The strain energy versus distortion plots that depict these two cases are shown in Fig. 8.

Strain Distribution Results

The effect of the element distortion on the strain distributions in the eight-node element are discussed in this section. Figures 9 and 10 contain plots of the error distributions in the ε_x strain component in two elements when they are progressively distorted. In both cases, the elements are representing $\varepsilon_{x,y}$.

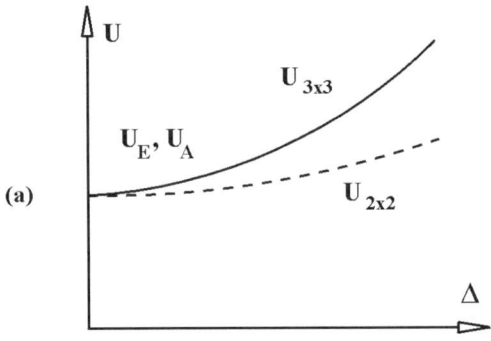

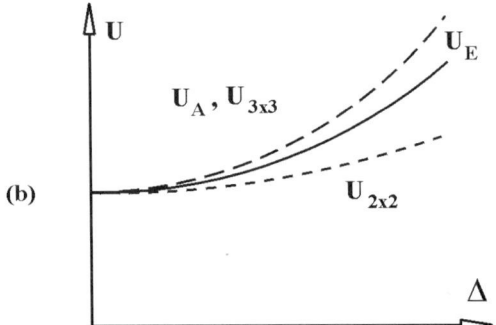

Figure 8. Strain energy vs distortion.

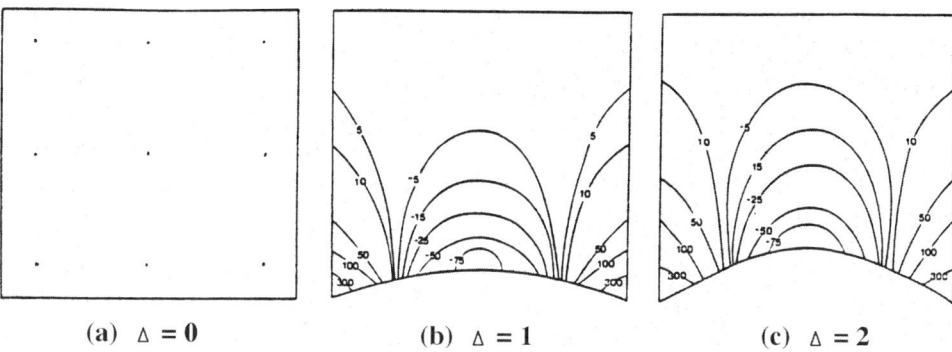

(a) Δ = 0   **(b)** Δ = 1   **(c)** Δ = 2

Figure 9. Strain distribution error vs distortion.

In Fig. 9a, we see that the undistorted element models the strains exactly throughout the element. The error distribution for the element with a distortion of level 1 is shown in Fig. 9b. The maximum error occurs in the corner and has a value of 487.5%. The maximum error occurring at the integration points of the three-by-three rule is 40.9%. The maximum strain error at the integration points for the two-by-two rule is −1.3%. The total strain energy error in this case is only 1.5%.

The error distribution for the element with a distortion of level 2 is shown in Fig. 9c. The maximum error occurs at the corner and has a value of 772.4%. The maximum error occurring at the integration points of the three-by-three rule is 69.1%. The maximum strain error at the integration points for the two-by-two rule is 0.3%. The strain energy error in this case is 3.8%.

The strain errors for this case are the highest that were found unless an element had more than one curved edge. This is interesting because this is likely to be one of the most common shapes used on a curved boundary. The next example is similar to this case except that the element is given a parallelogram shape in addition to the curved edge. We will see that the strain errors are substantially reduced.

In Fig. 10a, we see that the undistorted element models the strains exactly throughout the element. The error distribution for the element with a distortion of level 1 is shown in Fig. 10b. The maximum error occurs in the corner and has a value of −24.9%. The maximum error occurring at the integration points of the three-by-three rule is −13.3%. The maximum strain error at the integration points for the two-by-two rule is 4.4%. The total strain energy error in this case is only 0.8%.

The error distribution for the element with a distortion of level 2 is shown in Fig. 10c. The maximum error occurs at the corner and it has a value of −50.8%. The maximum error occurring at the integration points of the three-by-three rule is −29.4%. The maximum strain error at the integration points for the two-by-two rule is 8.9%. The strain energy error in this case is 3.5%.

The strain distribution errors for the remainder of the 132 cases are discussed in the Ref. 4. As mentioned earlier, the constant strain states are represented exactly in the eight-node element regardless of configuration. This means that as a model is refined and the majority of the strain energy is contained in the elements as constant strains, the errors in the strains are reduced.

Closure

This presentation has used strain gradient notation to provide insights into the modeling characteristics of the eight- and nine-node finite elements. We have seen that these

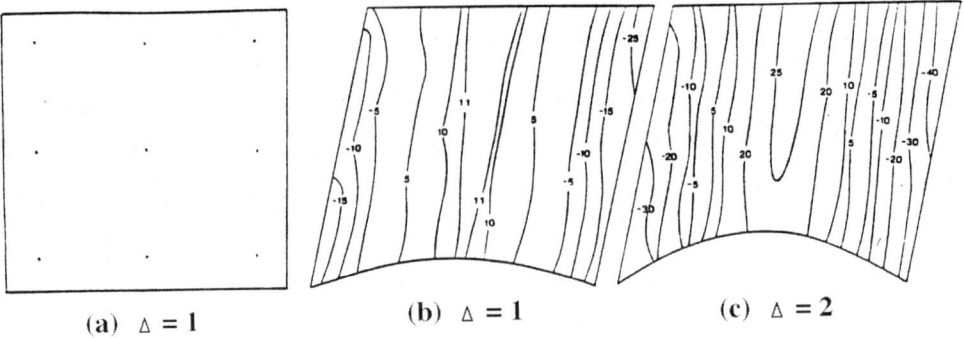

| (a) Δ = 1 | (b) Δ = 1 | (c) Δ = 2 |

Figure 10. Strain distribution error vs distortion.

elements contain parasitic shear as a result of being based on incomplete polynomials. In addition, the incomplete polynomials produce strain models that are not rotationally invariant.

In the process of analyzing the isoparametric form of these elements, we have seen that the parasitic shear terms cannot be removed from the isoparametric elements using reduced-order Gaussian quadrature integration. Furthermore, we saw that the isoparametric form of these elements does not accurately represent the total strain energy or the pointwise strain distributions unless the element is rectangular, i.e., elements with curved edges and nonrectangular shapes contain strain modeling errors because of the isoparametric mapping process. This deficiency only occurs in the higher order strain states so the elements converge to the correct results under mesh refinement.

When reduced-order Gaussian quadrature was used to integrate the shear strain energy in the eight- and nine-node isoparametric elements as an attempt to remove the parasitic shear terms, we found the following results. The use of a two-by-two rule did not remove all of the parasitic shear terms, so the elements still contained a modeling error. The use of a one-by-one rule removed all of the parasitic shear terms, but it also removed terms that belonged in the shear strain models. The removal of two of the legitimate terms from the shear strain energy expression added two zero-energy modes to the model. The effect of these spurious zero-energy modes on the finite element solution was shown with an example in Fig. 3. The errors in the displacements were directly related to the shapes of the strain states erroneously removed from the model.

The eight- and nine-node elements have basically the same characteristics. The analysis of the eight- and nine-node elements presented here leads to the following recommendation. Because of the errors inherent in the eight- and nine-node elements, it is recommended that if corrected strain gradient elements are not available, the six-node element be used instead of the eight- or nine-node elements.

As discussed in the development and application of error analysis procedures, it is recommended that nine-node elements or two six-node elements be used instead of eight-node elements. The reason for this is that the six- or nine-node elements produce a more uniform distribution of nodes, a situation that we will find to be important in the error analysis procedures developed in Part V. The absence of the central node in the eight-node element leaves a gap at regular intervals in the finite element nodal pattern.

Notes

1. The results presented here concerning the modeling capabilities of the eight-node element come from the earliest work of applying strain gradient notation to the finite

element method. In fact, in Ho's thesis (Ref.4), from which these results are drawn, the development of the strain gradient finite element formulation process follows the development of the finite element evaluation procedure. This is indicative of the fact that the discovery of the procedures for evaluating individual elements led to the development of an alternative way of formulating finite element stiffness matrices using strain gradient notation. In fact, the causes of parasitic shear and spurious zero-energy modes could not be identified until a strain gradient formulation process was available. These elemental errors were identified a year or so after the element formulation process was developed (see Ref. 1). Once the power of the new tool was understood, other modeling errors inherent in individual elements were identified. The next lesson discusses modeling errors in finite elements that have not been discussed in previous lessons, including some contained in bending elements.

References and Other Readings

1. Dow, J. O., and Byrd, D. E. "The Identification and Elimination of Artificial Stiffening Errors in Finite Elements," *International Journal for Numerical Methods in Engineering*, Vol. 26, 1988, pp. 743–762.

2. Dow, J. O., Ho, T. H., and Cabiness, H. D. "A Generalized Finite Element Evaluation Procedure," *ASCE Journal of Structural Engineering*, Vol. 111, No. 2, Feb. 1985, pp. 435–452.

3. Cabiness, H. D. "A Strain-Gradient Based Finite Element Evaluation Technique," M.S. Thesis, University of Colorado, 1985.

4. Ho, T. H. "A Generalized Finite Element Evaluation Procedure," Ph.D. Thesis, University of Colorado, 1984.

5. Byrd, D. E. "Identification and Elimination of Errors in Finite Element Analysis," Ph.D. Dissertation, University of Colorado, 1988.

6. Harwood, S. A. "Finite Element Error Analysis Using the Finite Difference Method," M.S. Thesis, University of Colorado, 1989.

Lesson 11   Problems

1. Form the shear strain expression for the nine-node element and identify the parasitic shear terms. *Hint*: See the expression for the strains in the eight-node element given by Eq. 11.4. Then add the effects of the two quadratic terms contained in u and v due to the x^2y^2 terms.

2. Form the shear strain energy expression. What order Gaussian quadrature rule does it take to integrate this expression for a rectangular nine-node element exactly? Would the same order Gaussian quadrature rule work for a quadrilateral? *Hint*: What is true about the Jacobian in both cases?

3. Will any spurious zero-energy modes be introduced if a two-by-two Gaussian quadrature rule is used? Will any spurious zero-energy modes be introduced if a one-by-one rule is used? *Hint*: The objective of this and the next problem is to reinforce the idea that the use of Gaussian quadrature integration introduces errors into most elements. The four-node quadrilateral, which serves as the model for the process, is the exception, not the rule.

4. Will any terms that introduce correct strain energy quantities into the strain energy expression be removed by a two-by-two Gaussian quadrature rule? By a one-by-one rule?

This set of problems is designed to introduce the common basis between the finite difference method and the finite element method.

5. Compare the $[\Phi]$ matrix for the nine-node finite element with the even nodal spacing of $h = 2.5$ given by Eq. 11.14 to the $[\Phi]$ matrix for the nine-node finite difference operator with a spacing of $h = 2.5$ given in Appendix A of Lesson 13. *Hint*: The objective of this comparison is to show that the finite difference method can represent any mesh or model that can be represented by a nine-node finite element.

6. The inverse of the $[\Phi]$ matrix given in Eq. 11.14 is contained in Appendix A of Lesson 13. Given a situation where the u and v displacements of node 1 in the finite element are 1.0 and 2.0, respectively, identify the strain gradient components represented by the finite element. Remember that the origin of the local coordinate system is at the center. *Hint*: This problem is designed to further reinforce the meaning of the strain gradient quantities. This problem shows that these basis functions are contained as components in the various possible nodal displacements.

7. Several times we have interpreted the meaning of the columns of the $[\Phi]$ matrix as the shape of the element given by a particular strain state. Give a meaning of a row of the inverse of the $[\Phi]$ matrix. *Hint*: Look first at rows 1 and 2 and then at row 3 before giving a general interpretation. This problem is designed to show that the idea central to the finite difference method (the use of difference operators to approximate the derivatives) is contained in the strain gradient formulation of the finite element method.

Lesson 12

Shear Locking, Aspect Ratio Stiffening, and Qualitative Errors

Purpose of Lesson   To demonstrate the source of the modeling deficiency known as shear locking, to show that aspect ratio stiffening is caused by an analogous mechanism, and to demonstrate the presence of a qualitative error in laminated plate elements.

Out-of-plane bending elements that include the effects of shear strains contain a modeling deficiency known as shear locking. Shear locking is characterized by sharp increases in element stiffness and condition number as the length to thickness ratio of an element becomes larger. Shear locking cannot be eliminated from the polynomial representation because the error is due to an inherent characteristic of the polynomial. However, shear locking can be controlled by placing geometric limits on the size of elements or by introducing a correction factor based on St. Venant's principle. This correction factor corresponds to the penalty functions developed by *ad hoc* means that are in current use. In out-of-plane applications, the clarification of the cause of shear locking assists in defining the appropriate uses for Kirchhoff and Mindlin plate bending elements. The polynomials introduce constraints that do not exist in the continuum. Plane elasticity elements with high length to width ratios exhibit similar behavior.

The effect of shear locking is exacerbated by the presence of parasitic shear, but the two errors are caused by two entirely different mechanisms. As previously demonstrated, the sources of parasitic shear can be eliminated from elements during the formulation stage using strain gradient notation. It will be shown that parasitic shear causes errors in laminated composite plate elements that *can only* be eliminated using strain gradient notation. These errors consist of coupling terms that cause deformations to move in the wrong direction. Cases have been identified that cannot be eliminated through mesh refinement. This is the first instance where the strain gradient formulation procedure must be used to produce correct models.

■   ■   ■

A well-known modeling error in plane elasticity problems is known as aspect ratio stiffening. Two elements with high aspect ratios and one with an aspect ratio of unity are shown in Fig. 1. Aspect ratio stiffening is characterized by an unnatural stiffness in elements with high aspect ratios. We will see that this excessive stiffness is due to an excess of strain energy contained in the shear mode. This source of error in finite element results is typically controlled by putting restrictions on the shape of allowable elements.

Aspect ratio stiffening cannot be definitively eliminated from elements because it is caused by an inherent characteristic in the polynomial displacement approximations that enables the polynomials to represent rigid body rotation. If the polynomials were corrected to enable the element to better represent shear strains and, hence, to eliminate aspect ratio stiffening, the element would not be capable of correctly representing rigid

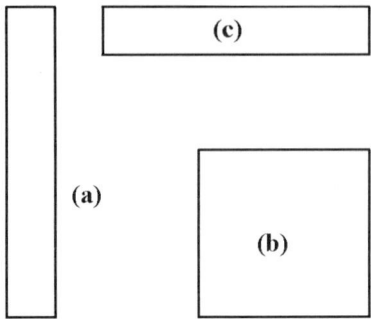

Figure 1. Planar finite elements with different aspect ratios.

body rotation. As discussed in the introduction to this section, an element must be capable of representing rigid body motions and constant strains to satisfy the convergence criteria.

Aspect ratio stiffening is caused by the same mechanism that produces an error known as shear locking in out-of-plane bending elements that are capable of representing through thickness shear. The physical examples for illustrating the phenomena of shear locking are straightforward and easy to demonstrate. Because of this and the fact that an understanding of shear locking is important when plate bending problems are being analyzed, the causes of shear locking are discussed in detail before an analysis of aspect ratio stiffening is undertaken.

Parasitic Shear and Shear Locking

Shear locking is characterized by a sharp increase in the element stiffness as the length to thickness ratio becomes larger. The increase in stiffness is counter to the expected physical behavior and is accompanied by an increase in the condition number for the element. The condition number is a function of the difference between the largest and smallest elements in a stiffness matrix and has consequences with respect to solution accuracy. Also, the increase in condition number can affect the stability of the computed results. These two distinctive characteristics of shear locking can be succinctly demonstrated by examining the behavior of the eigenvalues of a cantilevered Timoshenko beam element as the length changes. A cantilevered Timoshenko beam is shown in Fig. 2.

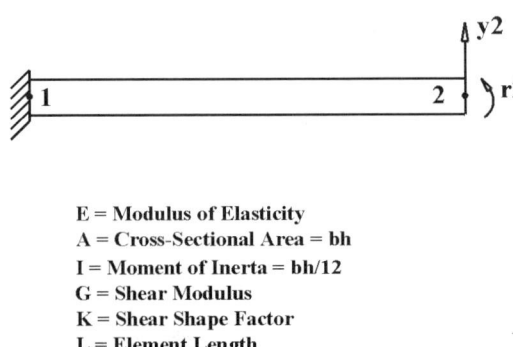

E = Modulus of Elasticity
A = Cross-Sectional Area = bh
I = Moment of Inerta = bh/12
G = Shear Modulus
K = Shear Shape Factor
L = Element Length

Figure 2. Cantilever beam.

The stiffness matrix for this restrained element is the following:

Restrained Timoshenko Beam Stiffness Matrix

$$\begin{bmatrix} \frac{KAG}{L} & \frac{KAG}{2} \\ \frac{KAG}{2} & \frac{KAGL}{4} + \frac{KAGL}{12} + \frac{EI}{L} \end{bmatrix} \begin{Bmatrix} v_2 \\ q_2 \end{Bmatrix} = \begin{Bmatrix} F_2 \\ M_2 \end{Bmatrix} \qquad (12.1)$$

The full stiffness matrix for a Timoshenko beam is developed in Appendix B where its modeling characteristics are investigated. The Euler-Bernoulli beam is developed in Appendix A so the modeling characteristics of the two beam elements can be contrasted. The parameters contained in the elements of the stiffness matrix are Young's modulus, E; the shear modulus, G; the area, A; the moment of inertia, I; the shear shape factor, K; and the element length, L. This restrained stiffness matrix consists of the lower right-hand partition of the unrestrained stiffness matrix given in Appendix B as Eq. B.19. It is found by forcing the displacement and rotation of the left end to be zero.

The stiffness element k_{22} of Eq. 12.1 is separated into three components. The first element, $KGAL/4$, is associated with the shear strain. The second element, $KAGL/12$, is associated with the erroneous parasitic shear term. The final element, EI/L, is associated with the flexure deformation.

The overall stiffness and condition number can be determined by evaluating the two eigenvalues, λ_{max} and λ_{min}, associated with this restrained stiffness matrix. The overall stiffness, given as the sum of the two eigenvalues or as $k_{11} + k_{22}$, varies with the length as shown in Fig. 3. The condition number, computed as $(\lambda_{max} - \lambda_{min})/\lambda_{min}$, varies with the length as shown in Fig. 4. These characteristics are computed for a beam with a unit cross section and a shape factor of 5/6.

The sharp rise in the overall stiffness with increasing length occurs as a result of the increase in k_{22} as L increases. The two terms in k_{22} with L in the numerator are directly proportional to L. The increase in condition number occurs because k_{22} increases and k_{11} decreases. The element k_{11} decreases because L is in the denominator.

The k_{22} term in the restrained stiffness matrix given by Eq. 12.1 contains the parasitic shear term $KAGL/12$. When this term is removed and the overall stiffness and condition number are plotted in Figs. 3 and 4, we can see the relation between the overall stiffness and condition number for the corrected and uncorrected stiffness matrices. The point at which the overall stiffness begins its ascent is slightly delayed. The rise in the condition

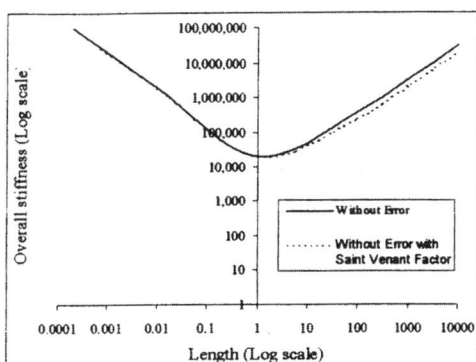

Figure 3. Overall stiffness vs length

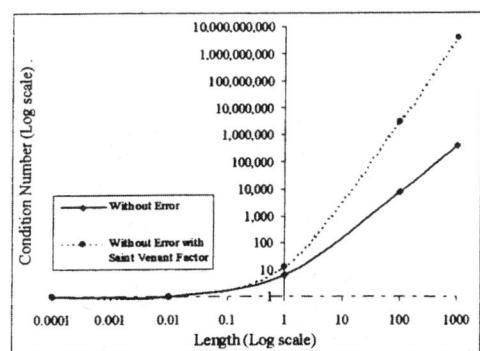

Figure 4. Condition number vs length.

number occurs sooner. This illustrates that the shear locking mechanism is independent of the parasitic shear phenomenon and that shear locking is caused by a second modeling deficiency.

The Cause of Shear Locking

The effects of shear locking have been demonstrated and the terms in the restrained stiffness matrix that are responsible for these effects have been identified. The modeling defect that causes these terms to improperly represent the physical systems is now identified. This identification is made possible as a result of the close relationship between the strain gradient coefficients and the physical system being modeled. This ability to directly compare the mathematical model to the physical system it represents brings to light inconsistencies in the model and suggests an approach for eliminating the modeling error.

We will see that the cause of shear locking results from constraints imposed on the shear strain by the polynomials used to represent the physical model. The identification of the precise cause of this discretization error enables this deficiency to be controlled by introducing modifications based on physical arguments.

The cantilevered Timoshenko beam element is used as the vehicle to identify the source of shear locking. The restrained stiffness matrix corrected for parasitic shear is

---

Restrained Timoshenko Beam Stiffness Matrix with the Parasitic Shear Effect Removed

$$[K] = \begin{bmatrix} \frac{KAG}{L} & \frac{KAG}{2} \\ \frac{KAG}{2} & \frac{KAGL}{4} + \frac{EI}{L} \end{bmatrix} \qquad (12.2)$$

In Appendix B, the unrestrained version of the stiffness matrix for a Timoshenko beam is developed in terms of the individual components that contribute to the overall stiffness matrix. The contributions of the flexure and shear modes of deformation to the restrained stiffness matrix given by Eq. 12.2 are

---

Finite Element Stiffness Matrix Due to Flexure

$$[K]_1 = \frac{EI}{L} \begin{bmatrix} 0 & 0 \\ 0 & 1 \end{bmatrix} \qquad (12.3)$$

Finite Element Stiffness Matrix Due to Shear

$$[K]_2 = KGA \begin{bmatrix} 1/L & 1/2 \\ 1/2 & L/4 \end{bmatrix} \qquad (12.4)$$

Equation 12.3 is the component of the stiffness matrix due to flexure strain. As we can see, L is in the denominator of the term contained in this matrix. The stiffness effect due to flexure decreases as L increases. This matches the expectations from physical experience. This means that the flexure model does not contribute to the stiffening characteristic of shear locking. Since this term, which decreases with L, is summed with the shear term $KAGL/4$ of Eq. 12.4 that increases with L, the flexure effects do not contribute to the increase in condition number with L. Thus, we can conclude that the flexure model does not contain the modeling deficiencies that produce shear locking.

Equation 12.4 is the contribution of the shear model to the stiffness matrix. As we can see, the element k_{11} decreases with L and the element k_{22} increases with L. The two off-diagonal terms are not dependent on L. The k_{22} term is directly responsible for the increase in the overall stiffness as L increases. The growth of k_{22} and the decrease in k_{11} as L increases is directly responsible for the increase in the condition number as L increases. Thus, the shear model is directly responsible for the modeling deficiency known as shear locking.

The Shear Model and the Physical System

We now correlate the shear model with the physical behavior of a beam to determine in what way the mathematical representation does not accurately represent the shear in the physical system. In Ref. 1, a meticulous backtracking through the derivation of the element stiffness matrix shows that the term $KAGL/4$ is due to the $\partial u/\partial z$ component of the shear strain and that the KAG/L term is due to $\partial w/\partial x$.

We can draw a similar conclusion from the following intuitive argument. Note that the $KAGL/4$ term is associated with the nodal rotation q_2. A rotated section is shown in Fig. 5a. As can be seen, the displacement u varies across the section with z. Thus, the term $KAGL/4$ is associated with the $\partial u/\partial z$.

Similarly, the term KAG/L is associated with the nodal displacement w_2. A displaced beam is shown in Fig. 5b. As we can see, the displacement produces a change in w along the length of the beam.

The following physical arguments account for the increase in the shear strain energy contained in the beam with an increase in length as indicated by the $KAGL/4$ term and the decrease in the shear strain energy in the beam as the length increases as specified by KAG/L. An undistorted beam with a high aspect ratio is shown in Fig. 6a.

Now consider giving each end of this physical system a rotation. This can be considered as applying a $\partial u/\partial z$ at each end. The displacement model from which the finite element stiffness matrix is derived enforces a uniform distortion along the full length of the beam. As shown in Fig. 6b, each of the "elemental volumes" that constitute the beam is distorted the same amount. Even if the beam were infinitely long, the distortion of the ends would be carried along the full length of the beam. Thus, the strain energy contained in the total beam as embodied in the term $KAGL/4$ increases

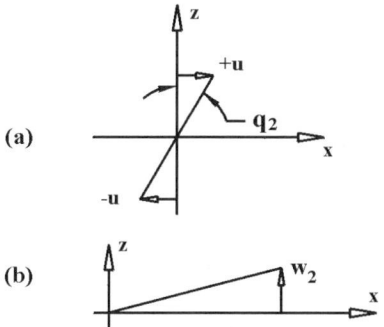

Figure 5. Nodal displacement kinematics.

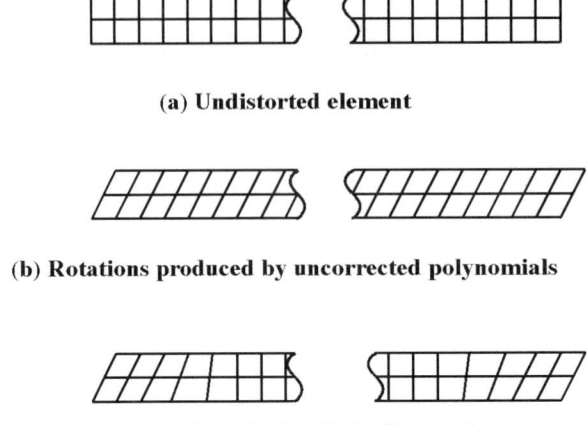

(a) Undistorted element

(b) Rotations produced by uncorrected polynomials

(c) Rotations produced using Saint Venant factors

Figure 6. A finite element and physical shear strain componets.

proportionally with the length as a result of the displacement model used in this development.

This unattenuated displacement does not fit our physical understanding of real physical behavior. If we rotated the end of a long beam, we would expect the rotation to lessen and die out depending on the length of the beam. The actual behavior would be more like that depicted in Fig. 6c. The end rotations would be damped out as they progress toward the center of the beam. As a result, the strain energy would be finite and bounded even if the beam were infinitely long. This localization of the strain concentration can be viewed as an example of St. Venant's principle.

Now consider giving the end of the undistorted beam a vertical displacement as shown in Fig. 7a. This can be considered as applying a $\partial w/\partial x$ deformation to the beam. The displacement model from which the finite element stiffness matrix is derived enforces such a uniform displacement along the full length of the beam as shown. However, in this case, the strain energy will not increase without bound as the length of the beam increases because the strain in each element decreases as the beam gets longer for a given vertical displacement. In fact, as the beam gets longer and longer, the total energy decreases because the strain is getting smaller. This phenomenon is captured in the term KAG/L, which decreases with length.

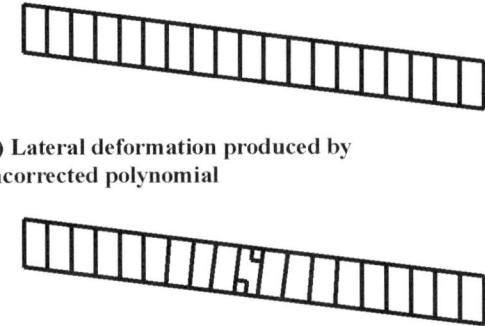

(a) Lateral deformation produced by uncorrected polynomial

(b) Lateral deformation produced by using Saint Venant factors

Figure 7. A polynomial induced and a physical shear strain component.

Although this strain energy model better fits our perception of nature, the actual deformation pattern is closer to that shown in Fig. 7b. The actual strain energy is less than that contained in the finite element model. This localized distortion can also be viewed as an example of St. Venant's principle.

Attenuation of Shear Locking

In the previous section, we saw that overconstraints due to the displacement polynomials cause the phenomena that characterize shear locking. However, the displacement polynomials cannot be directly modified because the same constraints accurately represent rigid body rotations. We can see this in the definitions of rotation and shear strain that are reproduced here for convenience:

Rigid Body Rotations and Shear Strains

$$q = \frac{1}{2}\left(\frac{\partial w}{\partial x} + \frac{\partial u}{\partial z}\right)$$

$$\gamma_{xy} = \left(\frac{\partial w}{\partial x} - \frac{\partial u}{\partial z}\right)$$

(12.5)

Thus, the modeling defects that characterize shear locking must be attenuated by applying a correction to the shear strain expression. Such a correction fits the idea of St. Venant's principle concerning the effects of local strain concentrations. Such a correction can be applied to the shear strain expression in the following manner:

Attenuated Shear Strain Expression

$$\gamma_{xy} = F_\gamma (\gamma_{xy})_0$$

$$= [0 \quad 0 \quad F_\gamma \quad 0]\{\varepsilon,\}$$

(12.6)

The multiplier F_γ is a shear strain attenuation factor that has a value less than unity. We call it a St. Venant factor so its physical nature is carried with it. In the next section, we discuss the computation of this attenuation factor. However, before doing this, the St. Venant factor is introduced into the stiffness matrix for the Timoshenko beam.

This is accomplished by referring to the derivation in Appendix B of this lesson. The correction factor enters the derivation in Eq. B.7. The second row of the transformation matrix is replaced by Eq. 12.6. Equation B.12 is modified by replacing the term u_{33} with *FKGAL*/2, where

---

Shear Strain Energy Attenuation Factor

$$F = \int F_\gamma^2 \, dL$$

(12.7)

The shear strain contribution to the overall stiffness given by Eq. B.18b is modified to become

Shear Contribution to the Finite Element Stiffness Matrix

$$[K]_2 = FKGA \begin{bmatrix} 1/L & -1/2 & -1/L & -1/2 \\ -1/2 & L/4 & 1/2 & L/4 \\ -1/L & 1/2 & 1/L & 1/2 \\ -1/2 & L/4 & 1/2 & L/4 \end{bmatrix} \qquad (12.8)$$

The only difference between this modified stiffness matrix and the original stiffness matrix is the inclusion of the shear strain attenuation factor as a multiplier of all of the terms. Summation of this modified stiffness matrix, which contains the shear strain contribution with the flexure component of the stiffness matrix given in Eq. B.18a, produces the following modified stiffness matrix:

Finite Element Stiffness Matrix without Parasitic Shear

$$[K] = \begin{bmatrix} \frac{FKGA}{L} & -\frac{FKGA}{2} & -\frac{FKGA}{L} & -\frac{FKGA}{2} \\ -\frac{FKGA}{2} & \frac{FKGA}{4}+\frac{EI}{L} & \frac{FKGA}{2} & \frac{FKGAL}{4}-\frac{EI}{L} \\ -\frac{FKGA}{L} & \frac{FKGA}{2} & \frac{FKGA}{L} & \frac{FKGA}{2} \\ -\frac{FKGA}{2} & \frac{FKGAL}{4}-\frac{EI}{L} & \frac{FKGA}{2} & \frac{FKGAL}{4}+\frac{EI}{L} \end{bmatrix} \qquad (12.9)$$

This result is corrected for both shear locking and parasitic shear. The attenuation factor F is designed to limit the growth of the strain energy as the length of the beam increases. In the next section, we demonstrate the computation of one version of this attenuation factor and show that it does control the negative characteristics of shear locking. As mentioned previously, this factor based on St. Venant's principle is identical to the penalty functions that have been developed by other considerations to control the modeling defects that constitute shear locking. The only difference is that the attenuation factor developed here is related to the exact cause of the modeling defect, the overconstraint of the displacement polynomials on the shear deformations. The knowledge of the sources of the modeling deficiencies that produce the characteristics of shear locking will enable analysts to make better informed choices between Euler-Bernoulli and Timoshenko beam models and by extension between Kirchhoff and Mindlin plate elements.

Computation of a St. Venant Factor

The remaining task is to evaluate the St. Venant factor in such a way that good results are produced and to demonstrate that the factor does, indeed, control the characteristics that identify shear locking. Any number of approaches are possible. An effective approach for computing the St. Venant factor is to force the displacements of a loaded element to match the results predicted by the exact theory. This is done for the case of a cantilevered element with an end shear load. This problem is chosen because it includes both flexure and shear deformations. This approach is similar to one previously used by Tessler and Hughes (see Ref. 1).

The single-element finite element representation of a cantilevered Timoshenko beam with an end load is given as

$$\begin{bmatrix} \frac{FKAG}{L} & \frac{FKAG}{2} \\ \frac{FKAG}{2} & \frac{FKAGL}{4} + \frac{EI}{L} \end{bmatrix} \begin{Bmatrix} w_2 \\ q_2 \end{Bmatrix} = \begin{Bmatrix} P \\ O \end{Bmatrix} \tag{12.10}$$

Solving for the tip deflection yields

$$w_2 = \frac{PL^3}{4EI} + \frac{PL}{FKGA} \tag{12.11}$$

From Timoshenko beam theory, the tip deflection for a cantilevered beam with an end shear load is

$$w_{\text{exact}} = \frac{PL^3}{4EI} + \frac{PL}{KGA} \tag{12.12}$$

When the finite element and the exact solutions are equated, the following expression for the St. Venant factor is found:

A St. Venant Factor

$$F = \frac{1}{1 + (KGAL^2/12EI)} \tag{12.13}$$

As we can see, this factor equals unity for a zero-length element and it approaches zero as the length approaches infinity. This factor is identical to the shear correction factor, developed by Tessler and Hughes (Ref. 1), which was developed as a correction to the shear shape factor.

The effect of the St. Venant factor on the characteristics of the stiffness matrix is illustrated by studying the overall stiffness and condition number as a function of length for the Timoshenko beam without parasitic shear. The results are shown in Figs. 8 and 9, respectively. We see that neither the overall stiffness nor condition number increase as the length increases. The St. Venant factor controls the unwanted characteristics of shear locking and produces a model that more nearly represents the physical problem.

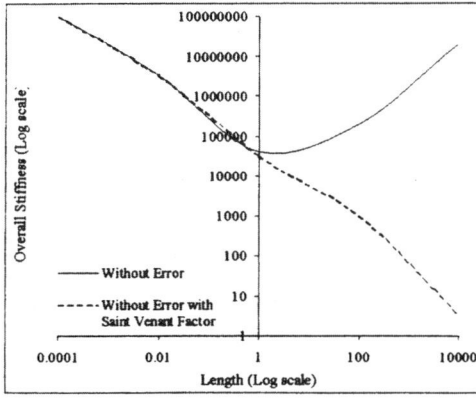

Figure 8. Overall stiffness vs length.

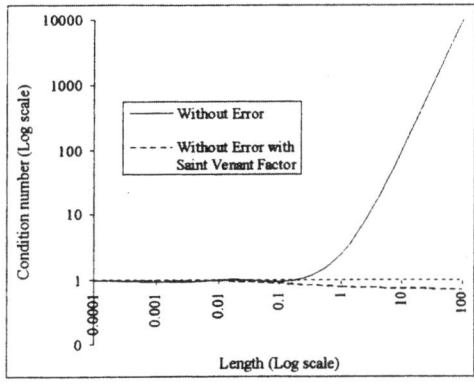

Figure 9. Condition number vs length.

Application of the St. Venant Factor

The ability of the St. Venant factor to eliminate the overstiffening and ill-conditioning associated with shear locking and produce accurate results is now demonstrated with two problems. The solution to the first problem shows the improvements to Timoshenko beam models that result from removing the effects of parasitic shear and adding a St. Venant correction factor to the shear strain model. The second problem demonstrates similar improvements to solutions of a more complex problem as a result of removing parasitic shear and correcting for shear locking. In addition, the problem provides an example of ill-conditioning produced by shear locking and the subsequent removal of this modeling deficiency through the application of the St. Venant factor.

Problem 1 — Accuracy Improvement in a Cantilevered Beam

The first problem is a cantilevered beam with an end load as shown in Fig. 10. The problem is solved for beams with a 1 × 1 cross section and two different total lengths, namely, 1 and 10 units. Each of these two problems is solved with three different meshes and three different elements. The first mesh contains a single element, the second contains 5 elements, and the third 10 elements. The first element type contains both parasitic shear and shear locking. In the second, the parasitic shear is removed but the element still contains the flaw due to shear locking. The final element type has the parasitic shear removed and a St. Venant correction term applied to the shear strain expression.

The results for the case where the beam has an overall length of unity are given in Table 1. In this case, the beam is short and stubby and a large portion of the deformation is due to shear effects. This is the situation where the Euler-Bernoulli beam fails to provide a good solution and for which the Timoshenko beam was developed.

The effect of the two corrections can be clearly seen in Table 1. The results for the uncorrected elements improve with the refinement of the finite element model, as expected. In the next column, the effect of removing parasitic shear is seen to improve the results. The correction for shear locking further improves the results shown in the final column. In this case, the two corrections produce improvements of approximately the same size. As we can see, the St. Venant factor increases toward unity as the length decreases.

The results for the case where the length of the beam is increased by a factor of 10 is given in Table 2. In this case, the parasitic shear term severely stiffens the element. As the length of the individual elements get smaller, the effect of the parasitic shear term is reduced as was the case in the previous example. The element model is improved more

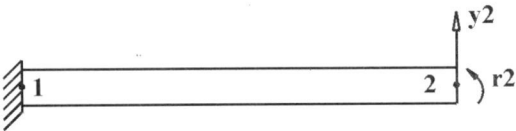

E = Modulus of Elasticity
A = Cross-Sectional Area = bh
I = Moment of Inerta = bh/12
G = Shear Modulus
K = Shear Shape Factor
L = Element Length

Figure 10. Cantilever beam problem.

Table 1. Results of cantilever beam analysis ($L = 1$).

Number of elements	St. Venant factor	Percentage of exact tip deflection		
		Element 1	**Element 2**	**Element 3**
1 @ 1.0	0.750E + 0	75.1	85.8	100.0
5 @ 0.2	0.987E + 0	98.7	99.6	100.0
10 @ 0.1	0.997E + 0	100.0	100.0	100.0

Note element 1 is uncorrected, element 2 has parasitic shear removed, and element 3 is fully corrected.

Table 2. Results of cantilever beam analysis ($L = 10$).

Number of elements	St. Venant factor	Percentage of exact tip deflection		
		Element 1	**Element 2**	**Element 3**
1 @ 10.0	0.291E − 1	2.9	75.4	100.0
5 @ 2.0	0.427E + 0	43.0	99.3	100.0
10 @ 1.0	0.750E + 0	75.4	100.0	100.0

Note element 1 is uncorrected, element 2 has parasitic shear removed, and element 3 is fully corrected.

by the elimination of parasitic shear than by the correction of the shear locking. The largest improvement caused by the addition of the St. Venant factor occurs for the longer beam, as expected. This correlates with the values of the St. Venant factors given.

The overall conclusions that we can draw from these results are the following: (1) the correction due to parasitic shear is mandatory and (2) the correction for shear locking is not mandatory but the finite element models constructed with corrected elements will be more efficient in terms of the level of accuracy achieved per degree of freedom than those without the correction. Furthermore, these results imply that if coarse representations of bending models are used, the model would best be represented with Euler-Bernoulli beams or Kirchhoff plate elements. This recommendation is based on the following argument. If the element is large, the aspect ratio is high and the shear effects are small. The degrees of freedom dedicated to the shear model would better be utilized in the flexure representation.

Problem 2 — Accuracy Improvement in a Simply Supported Beam

The second problem is a simply supported beam with equal and opposite shear loads at the one-third points as shown in Fig. 11. The problem is analyzed for five different lengths with either 3, 6, or 12 elements in the model. Three elements is the minimum number of elements possible in this model because of the loading condition. The deflections for the first three cases are given in Table 3. The quantities given in the table are the percentages of the actual values of the displacements at the points A and B. The displacements are antisymmetric because of the loading, which was chosen to produce a complex deformation in the center section of the beam. These three cases are shown separately because no ill-conditioning surfaces in these results. The next two cases are presented for the purpose of demonstrating the effects of ill-conditioning produced by shear locking.

The displacements produced by the Timoshenko beam including parasitic shear are unacceptable except for the 12-element discretization for the beam with the length of 3.0.

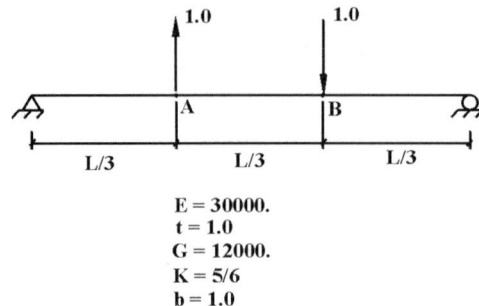

Figure 11. Simply supported beam problem.

For practical problems, this relative length is too short to allow for adequate modeling without an excessive number of elements. Thus, the conclusion reached in the analysis of the previous problem that Timoshenko beams with parasitic shear should never be used is reaffirmed.

The improvements due to the correction for shear locking are greater than in the previous example. Thus, the effect of the St. Venant correction factor is more important than indicated in the previous examples. This result implies that it is risky to use these elements without a correction for shear locking. Furthermore, this result reinforces the recommendation concerning the use of Euler-Bernoulli beams and Kirchhoff plate elements made in the previous example.

When this example is extended to longer beams with the same cross section, the effects of ill-conditioning are observed. The two problems that follow are included not because of their practical nature, but because they provide a compact demonstration of ill-conditioning and its control through the use of the St. Venant factor.

The deflection for the cases presented in Table 4 for element type 2, the element with the parasitic shear term removed but containing no correction for shear locking, were computed in both single and double precision. The results in Table 3 were all computed in single precision. As we can see, discrepancies exist between the supposedly equal magnitude deflections at points A and B for the single-precision solution. The difference between the two deflections indicates the presence of ill-conditioning.

Table 3. Results of simply supported beam analysis.

Length of beam	Number of elements	Percentage of exact deflection		
		Element 1	**Element 2**	**Element 3**
3.0	3	75.1	80.0	100.0
	6	92.4	95.1	100.0
	12	98.0	98.9	100.0
12.0	3	15.7	59.1	100.0
	6	42.7	88.4	100.0
	12	75.0	96.8	100.0
48.0	3	1.2	50.2	100.0
	6	4.5	87.5	100.0
	12	15.8	96.7	100.0

Note element 1 is uncorrected, element 2 has parasitic shear removed, and element 3 is fully corrected.

When the problem was solved using double precision, the effect of the ill-conditioning did not appear. This further indicates that the discrepancies in the single-precision solution were due to ill-conditioning.

The deflections for element type 3, which is corrected for shear locking by the incorporation of the St. Venant factor, were computed in single precision and no effects of ill-conditioning were seen in any of the analyses. These results demonstrate that the St. Venant factor eliminates the ill-conditioning due to shear locking, as it was designed to do. Although these two cases were designed to demonstrate the effects of ill-conditioning, the corrected elements produced results that matched the theoretical solutions in every case.

Aspect Ratio Stiffening

Plane stress elements with high aspect ratio geometries, as shown in Fig. 1, exhibit modeling deficiencies. High-aspect-ratio elements are sometimes overly stiff and they are ill-conditioned. The objective of this analysis is to identify the conditions that cause a high-aspect-ratio element to have these characteristics.

Plane stress elements based on incomplete polynomials were shown in earlier lessons to contain parasitic shear. When parasitic shear and the modeling deficiencies associated with high aspect ratio elements are considered together, it can be concluded that they are similar to the modeling deficiencies contained in a Timoshenko beam element. Because of these similarities, it is no surprise to discover that aspect ratio stiffening is caused by a mechanism similar to that which causes shear locking in Timoshenko beam elements (see Ref. 2). Thus, a high-aspect-ratio element is overly stiff only when it represents shear strains. This explains why the elements are sometimes overly stiff and sometimes not. If the element is not representing a high level of shear strain, it will produce a satisfactory representation. The development that identified the cause of aspect ratio stiffening is summarized here.

In the analysis presented in Ref. 1, the stiffness matrix for a rectangular four-node plane stress element is developed in symbolic or closed form. That is to say, the final form of the stiffness matrix is expressed in terms of the lengths of the sides of the element and the physical properties. In this symbolic development, the flexure and shear strain components of the stiffness matrix are formed separately, as was done for the constrained beam in Eqs. 12.3 and 12.4.

The eight-degree-of-freedom element is then constrained so that it behaves as a four-degree-of-freedom beam, as shown in Fig. 12.

Table 4. Results of simply supported beam analysis.

			Percentage of deflection			
			Element 2			
			Single precision		Double	
Length of beam	Number of elements	Element 1	A	B	A and B	Element 3
	3	0.0	50.7	49.4	50.1	100.0
567.0	6	0.0	85.9	89.8	87.8	100.0
	12	0.0	95.5	97.3	96.8	100.0
	3	0.0	56.8	42.9	49.8	100.0
2304.0	6	0.0	138.6	39.6	87.2	100.0
	12	0.0	103.9	70.3	96.5	100.0

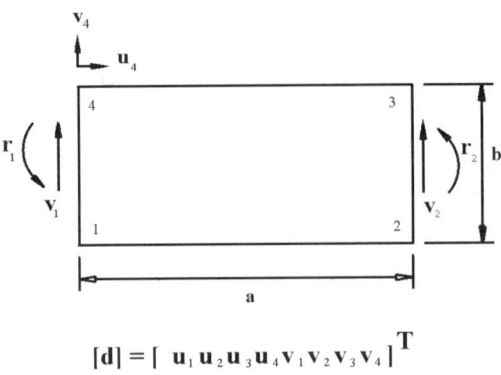

$$[d] = [\ u_1 u_2 u_3 u_4 v_1 v_2 v_3 v_4\ ]^T$$

Figure 12. Beamlike plane stress element.

When the shear strain component of the stiffness matrix is transformed to this beamlike coordinate system, that matrix has the same structure as Eq. 12.8. That is to say, some of the elements increase in size as the length of the beamlike structure, the parameter a, increases, some decrease, and some are independent of a. This means that the shear stiffness matrix due to the shear strain contribution will become overly stiff and ill-conditioned as the length increases. The stiffness matrix due to the flexure contribution decreases in size with an increase in the length, as expected. The form of the flexure component of the beamlike stiffness matrix is similar to that for the Timoshenko beam, so the analogy holds.

This means that the behavior of a high-aspect-ratio plane stress element is analogous to that of a Timoshenko beam that is not corrected for shear locking. If a high-aspect-ratio element is subjected to extension in either direction or flexure along either axis, it is not overly stiff. But, if it is subjected to a loading condition that induces shear, a high-aspect-ratio element is overly stiff.

It is definitely possible to correct this modeling deficiency in an individual element with a St. Venant factor, but the uncertainties connected to the constraints supplied by adjacent elements are difficult to gauge. As a result of this uncertainty and the ease with which this problem can be avoided, the use of the St. Venant factor to correct this error has not been studied in enough detail to recommend its use in production situations.

Note that the analysis of the displacements in high-aspect-ratio elements in finite element solutions can identify whether or not these elements are overly stiff. The participation of the different deformation modes can be identified through the use of strain gradient notation. If the amount of shear strain is small, a high-aspect-ratio element can be expected to perform well. If the amount of strain is high, the element will perform badly. The determination of the behavior of individual elements will be discussed in detail in later lessons on the *a posteriori* error analysis of finite element solutions.

This section has identified the cause of overstiffening in high-aspect-ratio elements as a deficiency in the representations of the shear strain by the displacement polynomials. The same polynomials representations provide correct models for the rigid body motions and the other strain states. This explains why high-aspect-ratio elements are sometimes overly stiff and sometimes provide accurate results.

Qualitative Errors in Laminated Composite Plate Elements

The presence of qualitative errors in laminated composite plate elements due to parasitic shear have been identified using strain gradient notation. A qualitative error is an error that misrepresents the sign or direction of a displacement in a problem. Such a situation is

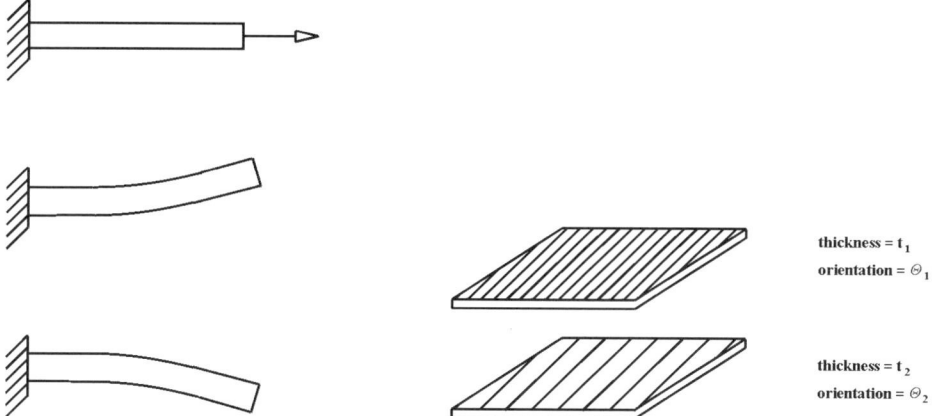

Figure 13. A qualitative error. **Figure 14.** Nonsymmetric two-layer laminate.

shown in Fig. 13 for a beamlike structure with coupling between the axial and the bending modes of deformation. Let us assume that the actual structure should deform, as shown in Fig. 13 (middle diagram) when it is given an axial load. The model contains a qualitative error if the finite element analysis produces a result, as shown in Fig. 13 (bottom diagram). Note that the bending is downward instead of upward.

Furthermore, some of the cases identified are such that the qualitative modeling error *is not* absorbed when the finite element model is refined. That is to say, the error is not attenuated when the mesh is refined. This is the first type of elemental error identified where the strain gradient approach must be used to eliminate the modeling error (see Refs. 3 and 4).

The concept of a qualitative error is demonstrated here with a two-layer nonsymmetric angle-ply laminate shown in Fig. 14 when it is subjected to a uniform twisting load. Specifically, the laminate consists of two layers of thicknesses of 0.10 in. and 0.12 in. and orientations of $+30$ degrees and -30 degrees with respect to the longitudinal edges of the plate. Due to the nonsymmetry with respect to its middle surface, this laminate possesses coupling between bending and extension. In addition, this laminate possesses coupling between bending and twisting because the fibers in the layers are not oriented in the direction of the edges of the plate.

The presence and nature of the qualitative errors introduced by parasitic shear into finite element stiffness matrices for laminated composite plates are now demonstrated. This demonstration is accomplished by comparing the motions of the finite element models with and without parasitic shear shown in Fig. 15. The displacements are compared for models consisting of 4, 16, 64, and 256 elements.

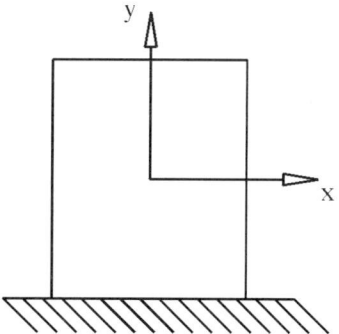

Figure 15. Problem configuration.

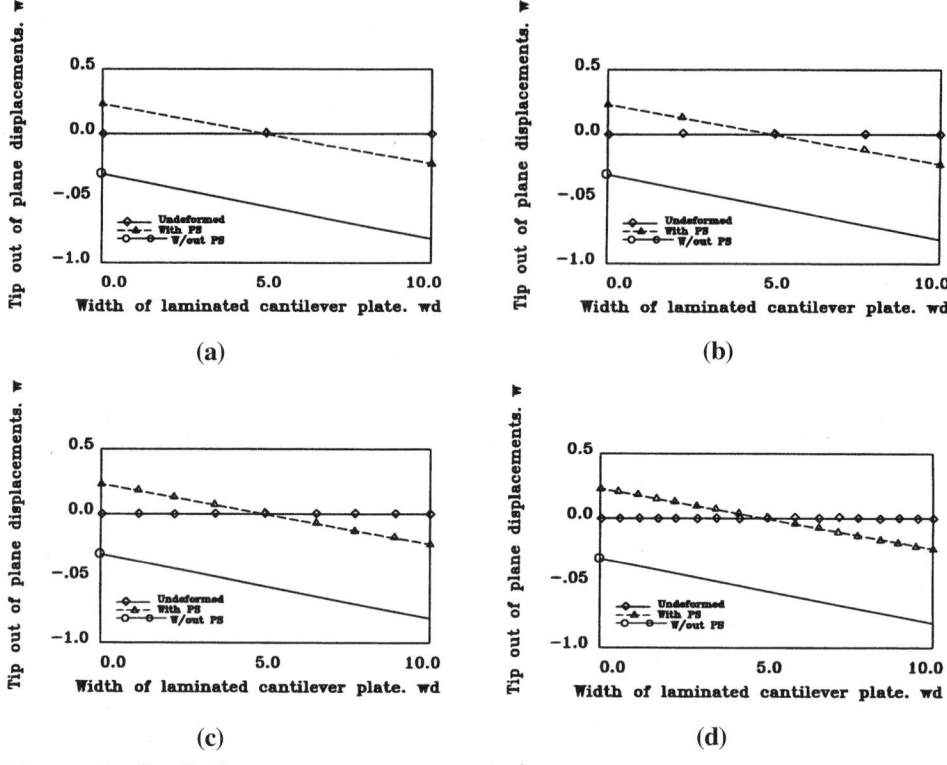

Figure 16. Qualitative error convergence study.

The qualitative error in this case is a suppression of bending in the element with parasitic shear, as shown in Fig. 16. Figure 16 shows the displacements of the end of the plate element. Twisting is observed through the rotation of the tip and bending is observed through its out-of-plane downward displacement. These displacements show that the element is twisting with very little bending. Therefore, the twist-bending coupling is not represented by the model containing parasitic shear. This qualitative error is removed when the element is corrected for parasitic shear. As we can see in Fig. 16, when the parasitic shear terms are removed, both twisting and bending are represented simultaneously and the size of the displacements are increased by an order of magnitude.

Figures 16a through 16d depict the variation of the out-of-plane displacement w of the tip of the plate across its width as the mesh is refined. These results show that the model containing parasitic shear is not capable of representing the twist-bending coupling regardless of the level of refinement. In other words, these results show that refinement does not have any attenuating effect on the qualitative error in this case. Thus, we can conclude that the qualitative error contained in this element cannot be removed by mesh refinement.

The parasitic shear terms must be removed from the laminated composite plate elements during element formulation if the resulting models are not going to misrepresent the behavior of the problem. This can be accomplished only in the general case with strain gradient formulation procedures or something equivalent. The parasitic shear terms cannot be removed from elements of this type using reduced-order Gaussian quadrature integration.

Closure

The procedure for formulating strain-gradient-based finite element stiffness matrices developed in Lesson 8 is applied to Euler-Bernoulli and Timoshenko beam elements in Appendices A and B. The formulation of these elements can be viewed as an introduction to the development of out-of-plane bending elements. The analysis of the resultant stiffness matrix for the Timoshenko beam identify the modeling deficiencies that occur when through-thickness shear strain is added to out-of-plane bending elements. Furthermore, Appendix B shows that if the displacement polynomial is one order higher than the polynomial that represents the rotation that the parasitic shear term was not present.

The stiffness matrix for the Timoshenko beam was then analyzed to identify the cause of shear locking. We find that the displacement polynomials added constraints to the shear deformations that increased the strain energy content of the element to levels that would not occur in nature.

We find that this modeling deficiency could not be directly corrected because the constraints on the displacement polynomials that caused the excess of shear strain energy are needed to produce the required rigid body rotations. However, a correction factor based on physical considerations is introduced into the shear strain expression and shown to correct the overstiffening and ill-conditioning that result from shear locking. This correction factor is called a St. Venant factor to identify the physical nature of the correction factor.

The analysis applied to the Timoshenko beam is extended to explain the "intermittent" nature of aspect ratio stiffening in plane stress elements. We find that the geometry of high-aspect-ratio elements makes it possible for the element to contain too much shear strain energy and, hence, be overly stiff. Since loading conditions exist that do not induce shear strains, the element would not be overly stiff when it was not representing shear strains. This explained the intermittent nature of aspect ratio stiffening.

In the last section, we saw that elements exist that must be corrected using strain gradient notation. Laminated composite Mindlin plate models were used to show that modeling errors exist that are not attenuated as the mesh is refined. This provides an imperative for the use of the strain gradient approach.

Appendix A
The Euler-Bernoulli Beam

The Euler-Bernoulli beam is the simplest out-of-plane bending element. Although this element is often considered as part of matrix structural analysis and not as part of finite element analysis, the development of the stiffness matrix for this line element provides a valuable background for the formulation of plate bending elements. As we see here, the Euler-Bernoulli beam element is amenable to development using strain gradient notation.

The constraints on the cross section of an Euler-Bernoulli beam enable the behavior of the element to be totally defined in terms of the transverse displacement. The three constraints require that (1) the planar cross sections normal to the neutral axis remain planar, (2) the cross sections remain normal to the neutral axis, and (3) the cross sections are inextensible through the thickness.

The requirement that the plane sections remain plane after deformation means that the in-plane displacements along the length of the beam are linear through the section. The requirement that the cross sections remain normal to the midplane means that the rotations of the cross sections are equal to the slope of the neutral axis. The constraint on the through thickness deformations eliminates the need for independent variables other than the midplane displacements.

A two-node Euler-Bernoulli beam element is shown in Fig. A.1. Each node has two degrees of freedom, the transverse displacement in the z direction and the rotation of the neutral axis around the y coordinate. The axial deformation of the neutral axis is not included because the flexural deformations are uncoupled from the longitudinal deformations. Because of the constraints on the cross sections, the displacements in the element can be defined in terms of the following two equations. The rotation of the neutral axis is given as

---

Definition of Rotation

$$q(x) = -\frac{dw(x)}{dx} \tag{A.1}$$

The first two constraints on the cross section enable us to define the displacements in the x direction to be

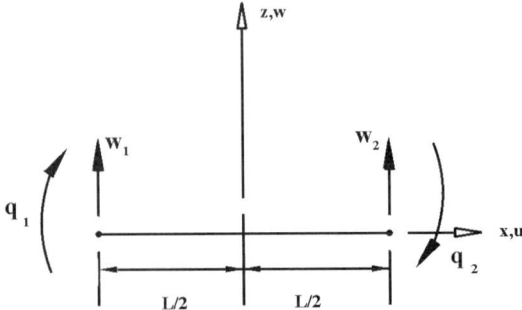

Figure A.1. Euler-Bernoulli beam coordinate definition.

Definition of u

$$u(x,z) = zq(x)$$

$$= -z\frac{dw(x)}{dx} \tag{A.2}$$

With these kinematic relationships, the transverse shear strain vanishes. This can be seen by applying the definition of shear strain to get

Shear Strain Model

$$\gamma_{xz} = \frac{\partial u}{\partial z} + \frac{\partial w}{\partial x}$$

$$= -\frac{\partial w}{\partial x} + \frac{\partial w}{\partial x} \tag{A.3}$$

$$= 0$$

The stiffness matrix for this element is now formulated. This is done by following the six-step process identified in the closure of Lesson 8.

Strain Gradient Displacement Polynomial Identification

Since the displacements and, hence, the strains are totally defined by the deformation of the neutral axis, the four-degree-of-freedom element is derived from a displacement polynomial of the following form:

$$w = c_1 + c_2 x + c_3 x^2 + c_4 x^3 \tag{A.4}$$

This polynomial is expressed in strain gradient notation as

Unreduced Strain Gradient Displacement Approximation

$$w(x) = (w_{rb})_0 + (\gamma_{xz}/2 - q_{rb})_0 x$$
$$+ [(\gamma_{xz,x} - \varepsilon_{x,z})/2]_0 x^2 + [(\gamma_{xz,xx} - \varepsilon_{x,zx})/6]_0 x^3 \tag{A.5}$$

Since the though thickness shear strain is zero, as shown in Eq. A.3, the transverse displacement expression reduces to

Strain Gradient Displacement Approximation

$$w(x) = (w_{rb})_0 - (q_{rb})_0 x - (\varepsilon_{x,z})_0/2x^2 - (\varepsilon_{x,zx})_0/6x^3 \tag{A.6}$$

This polynomial contains the four linearly independent strain states that this four-degree-of-freedom element is capable of representing. When these variables are extracted from Eq. A.6, they are

---

Linearly Independent Strain Gradient Quantities

$$(w_{rb})_0 \quad (q_{rb})_0$$
$$(\varepsilon_{x,z})_0 \quad (\varepsilon_{x,zx})_0$$

The first two terms represent the rigid body motions. The first deformation term represents the flexure or curvature produced by a constant moment. It plays the role of the constant strain term in this element. The second deformation term represents the rate of change of curvature along the length of the element produced by a linearly varying moment. Such a moment is caused by a shear load at the end of a cantilever beam.

Strain Model Formulation and Evaluation

The strain model for the two-node Euler-Bernoulli beam is formed by substituting Eqs. A.2 and A.6 into the definition of normal strain to give the following strain model:

Euler-Bernoulli Strain Model

$$\varepsilon_x = \frac{\partial u}{\partial x} = -z\frac{d^2 w(x)}{dx^2} \tag{A.7}$$

$$= (\varepsilon_{x,z})_0 z + (\varepsilon_{x,zx})_0 xz$$

This strain representation matches the model that is expected for an Euler-Bernoulli beam. The normal strains vary linearly across the section and the flexure is represented by a constant and a term that varies linearly in x. The through-thickness shears are absent because they were eliminated by the initial assumptions. There are no extra terms in the strain expressions. We can conclude that the two-node, four-degree-of-freedom Euler-Bernoulli beam represents the strains it was designed to model.

We can put the strain representation given by Eq. A.7 in matrix form as

---

Strain Representation in Matrix Form

$$\varepsilon_x = [T]\{\varepsilon_,\}$$

where

$$\{\varepsilon_,\}^T = [\,(w_{rb})_0 \quad (q_{rb})_0 \quad (\varepsilon_{x,z})_0 \quad (\varepsilon_{x,xz})_0\,] \tag{A.8}$$

$$[T] = [[T_0]\,|\,[T_\varepsilon]] = [[0 \quad 0]\,|\,[z \quad xz\,]]$$

Equation A.8 is the discrete model of the strains contained in the beam element. The model represents the strain in the continuum as discrete quantities that contain the rigid body motions and two flexure terms. The rigid body motions are introduced into this equation with the null partition in the transformation matrix. They are included in the strain expressions even though they make no contribution to the strain energy expression. This is done in preparation for a later step in which the strain energy is transformed to nodal displacements to form the finite element stiffness matrix. The rigid body motions must be included in the finite element stiffness matrix so the element satisfies the convergence criteria.

Strain Energy Expression in Generalized Coordinates

The strain energy function for the Euler-Bernoulli beam is dependent only on the normal strain in the x direction and can be written as

---

Strain Energy Expression

$$U = \frac{1}{2} \int_{\Omega} E \varepsilon_x^2 \, d\Omega \tag{A.9}$$

When we substitute the discrete representation of strain given in Eq. A.8 into Eq. A.9, the result is

---

Strain Energy Expression in Terms of Strain Gradient Quantities

$$U = \frac{1}{2} \{\varepsilon_,\}^T \left[E \int_{\Omega} [T]^T [T] \, d\Omega \right] \{\varepsilon_,\}$$

$$U = \frac{1}{2} \{\varepsilon_,\}^T [\overline{U}] \{\varepsilon_,\} \tag{A.10}$$

where

$$[\overline{U}] = E \int_{\Omega} [T]^T [T] \, d\Omega$$

Equation A.10 is a discrete approximation of the strain energy with strain gradient quantities as the independent variables. We can view the strain gradient components as the "generalized coordinates" for the Rayleigh-Ritz method.

Integration of Strain Energy Terms

When we expand the expression for $\overline{U}$ in Eq. A.10, the result is

---

Strain Energy Integrals

$$[\overline{U}] = E \int_{\Omega} \begin{bmatrix} [T_0]^T [T_0] & [T_0]^T [T_\varepsilon] \\ [T_\varepsilon]^T [T_0] & [T_\varepsilon]^T [T_\varepsilon] \end{bmatrix} d\Omega \tag{A.11}$$

Since the transformation $[T_0]$ is a null matrix, any term containing it is zero. Thus, the lower right partition $\overline{U}_{22}$ of Eq. A.11 contains the only nonzero elements. When we expand this partition, it has the following form:

Expanded Form of the Strain Energy Integrals

$$[\bar{U}_{22}] = EI \begin{bmatrix} I_1 & I_2 \\ I_2 & I_3 \end{bmatrix}$$

where

I is the second moment of the cross-section.

L is the length of the element.   (A.12)

$$I_1 = \int_L dx$$

$$I_2 = \int_L x \, dx$$

$$I_3 = \int_L x^2 \, dx$$

When we evaluate Eq. A.12 for the element shown in Fig. A.1, the result has the following form:

Strain Energy Integrals Evaluated

$$[\bar{U}] = EI \begin{bmatrix} 0 & 0 & 0 & 0 \\ 0 & 0 & 0 & 0 \\ 0 & 0 & L & 0 \\ 0 & 0 & 0 & \frac{L^3}{12} \end{bmatrix}$$ (A.13)

The fact that the first two rows and columns are zero results from the fact that the first two generalized coordinates are rigid body motions. The integral I_2 is zero because the origin of the local coordinate system is at the center of the element.

Transformation to Nodal Displacements

We are now in the position to transform the strain energy given by Eq. A.10 from a function of strain gradient quantities to nodal displacements. This transformation of the strain energy will produce the finite element stiffness matrix. The first step in this process is to form the transformation from nodal quantities to strain gradient variables using Eqs. A.1 and A.6. The general result is given as

Displacement and Rotation relations

$$\begin{Bmatrix} w_i \\ q_i \end{Bmatrix} = \begin{bmatrix} 1 & -x & -\frac{x^2}{2} & -\frac{x^3}{6} \\ 0 & 1 & x & \frac{x^2}{2} \end{bmatrix} \begin{Bmatrix} (w_{rb})_0 \\ (q_{rb})_0 \\ (\varepsilon_{x,z})_0 \\ (\varepsilon_{x,zx})_0 \end{Bmatrix}$$ (A.14)

When we substitute the nodal coordinates for the element shown in Fig. A.1 into Eq. A.14, the result is:

Nodal to Strain Gradient Quantity Transformation

$$\{d\} = [\Phi]\{\varepsilon_,\}$$

where

$$\{d\}^T = \{\, w_1 \quad q_1 \quad w_2 \quad q_2 \,\}$$

$$\{\varepsilon_,\}^T = \{\, (w_{rb})_0 \quad (q_{rb})_0 \quad (\varepsilon_{x,z})_0 \quad (\varepsilon_{x,zx})_0 \,\} \tag{A.15}$$

$$[\Phi] = \begin{bmatrix} 1.0 & L/2 & -L^2/8 & L^3/48 \\ 0.0 & 1.0 & -L/2 & L^2/8 \\ 1.0 & -L/2 & -L^2/8 & -L^3/48 \\ 0.0 & 1.0 & L/2 & L^2/8 \end{bmatrix}$$

This transformation must be inverted to give the relationship needed to change the strain energy expression given by Eq. A.10 to nodal displacements. When this is done, the result is

---

Strain Gradient to Nodal Quantity Transformation

$$\{\varepsilon_,\} = [\Phi]^{-1}\{d\}$$

where

$$[\Phi]^{-1} = \begin{bmatrix} \frac{1}{2} & -\frac{L}{8} & \frac{1}{2} & \frac{L}{8} \\ \frac{3}{2L} & -\frac{1}{4} & -\frac{3}{2L} & -\frac{1}{4} \\ 0.0 & -\frac{1}{L} & 0.0 & \frac{1}{L} \\ -\frac{12}{L^3} & \frac{6}{L^2} & \frac{12}{L^3} & \frac{6}{L^2} \end{bmatrix} \tag{A.16}$$

Equation A.16 is the relation needed to transform the strain energy expression from strain gradient quantities to nodal displacements.

Strain Energy Transformation and Stiffness Matrix Formulation

The strain energy can now be transformed to a function of nodal displacements to give the finite element stiffness matrix. We accomplish this by substituting Eq. A.16 into Eq. A.10 to give

Strain Energy Expression in Terms of Nodal Quantities

$$U = 1/2\{d\}^T[\Phi]^{-T}[\overline{U}][\Phi]^{-1}\{d\}$$

$$U = 1/2\{d\}^T[K]\{d\}$$

(A.17)

where $[K]$ is the finite element stiffness matrix

When we use Eqs. A.11 and A.16 to compute the finite element stiffness matrix defined in Eq. A.17, the result is

Finite Element Stiffness Matrix

$$[K] = \frac{EI}{L}\begin{bmatrix} 12 & -6L & -12 & -6L \\ -6L & 4L^2 & 6L & 2L^2 \\ -12 & 6L & 12 & 6L \\ -6L & 2L^2 & 6L & 4L^2 \end{bmatrix}$$

(A.18)

Closure

This appendix has developed the stiffness matrix for an Euler-Bernoulli beam using the strain-gradient-based procedure developed in Lesson 8. The resulting stiffness matrix is identical to the one generated by other formulation processes. This development is similar to that required to form the stiffness matrix for a Kirchhoff plate element. A Kirchhoff plate can be viewed as the two-dimensional extension of the Euler-Bernoulli beam.

Appendix B
The Timoshenko Beam

The Timoshenko beam is the simplest out-of-plane bending element that includes through-thickness shear. In the Euler-Bernoulli beam, the shear is eliminated as a consequence of representing the rotation of the neutral axis as the slope of the displacement polynomial. The through-thickness shear is added to the Timoshenko beam by representing the rotation of the neutral axis as an independent quantity.

The constraints on the cross section of the Timoshenko beam require that (1) the planar cross sections normal to the neutral axis remain planar and (2) the thickness of the beam does not change. In the case of the Timoshenko beam, the cross sections are free to rotate independently of the slope of the neutral axis as the result of the inclusion of an independent rotation polynomial, $q(x)$.

The requirement that the plane sections remain plane after deformation means that the in-plane displacements along the length of the beam are linear across the section. The constraint on the through-thickness deformations eliminates the need for an independent function to represent this deformation.

A two-node Timoshenko beam element is shown in Fig. B.1. Each node has two degrees of freedom, the transverse displacement in the z direction, w, and the rotation of the neutral axis around the y coordinate, q. The axial deformation of the neutral axis is not included in this model because the flexural deformations are uncoupled from the longitudinal deformations. Because of the constraints on the cross sections, the kinematic quantities in the element can be defined in terms of the displacement and rotation of the neutral axis.

The requirement that the cross section remains planar enables us to define the displacements in the x direction as

---

Definition of u

$$u(x, z) = zq(x) \tag{B.1}$$

With these kinematic relationships, the transverse shear strain does not vanish. We can see this by applying the definition of shear strain to get

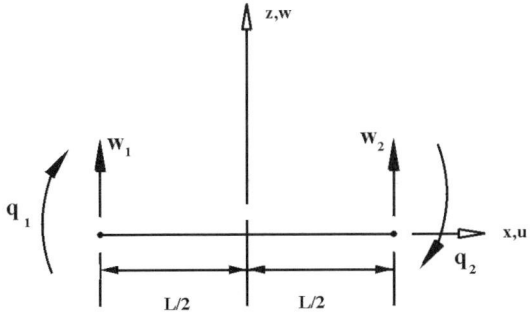

Figure B.1. Timoshenko beam coordinate definition.

Shear Strain Model

$$\gamma_{xz} = \frac{\partial u}{\partial z} + \frac{\partial w}{\partial x}$$

$$= q + \frac{\partial w}{\partial x}$$

(B.2)

As we can see in Eq. B.2, the through-thickness shear strain need not be zero since q and w are independent of each other. Now that the basic characteristics of the element have been defined, the stiffness matrix for the element will be formulated. This is accomplished by following the six-step process identified in the closure of Lesson 8.

Strain Gradient Displacement Polynomial Identification

As we can see in Fig. B.1, the element contains a total of four degrees of freedom. Since the displacements and rotations are independent of each other, the two quantities are represented by linear functions with a total of four independent variables as

Displacement and Rotation Polynomials with Arbitrary Coefficients

$$w = c_1 + c_2 x$$

$$q = d_1 + d_2 x$$

(B.3)

We find the displacement polynomials by comparing Eqs. B.1 and B.3 with the strain gradient expansions for displacements to give

Strain Gradient Displacement Polynomials

$$u(x) = (\gamma_{xz}/2 + q_{rb})_0 z + (\varepsilon_{x,z})_0 xz$$

$$w(x) = (w_{rb})_0 + (\gamma_{xz}/2 - q_{rb})_0 x$$

(B.4)

These polynomials contain the four linearly independent strain states that this four-degree-of-freedom element is capable of representing. When we extract these variables from Eq. B.4, they are

Linearly Independent Strain Gradient Quantities

$$(w_{rb})_0 \quad (q_{rb})_0$$

$$(\gamma_{xz})_0 \quad (\varepsilon_{x,z})_0$$

(B.5)

The first two terms represent the rigid body motions. The two deformation terms represent constant shear and constant flexure in the element. We can view both terms as constant strain terms. The shear strain in the Timoshenko beam replaces the linear variation of the flexure contained in the Euler-Bernoulli beam model. Thus, the complexity of the flexure representation in the two-node Euler-Bernoulli beam is higher than that contained in the two-node Timoshenko beam.

Strain Model Formulation and Evaluation

The strain models contained in the two-node Timoshenko beam are formed by substituting the displacement polynomials contained in Eq. B.4 into the definitions of normal and shear strain to give the following strain models:

Timoshenko Beam Strain Models

$$\varepsilon_x = \frac{\partial u}{\partial x} = (\varepsilon_{x,z})_0 z$$

$$\gamma_{xz} = \left(\frac{\partial w}{\partial x} + \frac{\partial u}{\partial z}\right) = (\gamma_{xz})_0 + (\varepsilon_{x,z})_0 x$$

(B.6)

The normal strain varies linearly across the section and the flexure is constant along the length of the beam. As mentioned earlier, this model is simpler than the flexural model for the Euler-Bernoulli beam. The through-thickness shear is represented by two terms. The first is a legitimate term. It represents constant shear strain in the element. The second term is a parasitic shear term. This normal strain term does not belong in the shear strain expression. It has the effect of stiffening the element. Later, we see that the parasitic shear term is caused by the fact that the displacement and rotation polynomials have the same order. If the displacement polynomial is one order higher than the rotation polynomial, parasitic shear will not exist. This is demonstrated in a later example.

We can put the strain representation given by Eq. B.7 in matrix form as:

Strain Representation in Matrix Form

$$\{\varepsilon\} = [T]\{\varepsilon_,\}$$

where

$$\{\varepsilon^T\} = \begin{bmatrix} \varepsilon_x & \gamma_{xy} \end{bmatrix}$$

(B.7)

$$\{\varepsilon_,\}^T = \begin{bmatrix} (w_{rb})_0 & (q_{rb})_0 & (\gamma_{xy})_0 & (\varepsilon_{x,z})_0 \end{bmatrix}$$

$$[T] = [[T_0] \,|\, [T_\varepsilon]] = \begin{bmatrix} \begin{bmatrix} 0 & 0 \\ 0 & 0 \end{bmatrix} \Big| \begin{bmatrix} 0 & z \\ 1 & x \end{bmatrix} \end{bmatrix}$$

Equation B.7 is the discrete model of the strains contained in the beam element. The model represents the strain in the continuum as discrete quantities that contain the rigid body motions, constant shear, and constant flexure. The rigid body motions are introduced into this equation with the null partition in the transformation matrix. They are included in the strain expressions even though they make no contribution to the strain energy expression. This is done in preparation for a later step in which the strain energy is transformed to nodal displacements to form the stiffness matrix. The rigid body motions must be included in the finite element stiffness matrix so the element satisfies the convergence criteria. The parasitic shear term is represented in Eq. B.7 by the x variable in the second row of the transformation matrix.

Strain Energy Expression in Generalized Coordinates

The strain energy function for the Timoshenko beam is dependent on the through-thickness shear and the normal strain in the x direction and can be written as

---

Strain Energy Expression

$$U = \frac{1}{2} \int_{\Omega} \{\varepsilon\}^T [E] \{\varepsilon\} \, d\Omega$$

where   (B.8)

$$[E] = \begin{bmatrix} E & 0 \\ 0 & G \end{bmatrix}$$

The elements of the constitutive matrix are Young's modulus, E, and the shear modulus, G. When we substitute the discrete representation of strain given in Eq. B.7 into Eq. B.8, the result is:

Strain Energy Expression in Terms of Strain Gradient Quantities

$$U = \frac{1}{2} \{\varepsilon_{,}\}^T \left[\int_{\Omega} [T]^T [E][T] \, d\Omega \right] \{\varepsilon_{,}\}$$

$$U = \frac{1}{2} \{\varepsilon_{,}\}^T [\overline{U}] \{\varepsilon_{,}\}$$ (B.9)

where

$$[\overline{U}] = \int_{\Omega} [T]^T [E][T] \, d\Omega$$

Equation B.9 is a discrete approximation of the strain energy with strain gradient quantities as the independent variables. We can view the strain gradient components as the "generalized coordinates" for the Rayleigh-Ritz method.

Integration of Strain Energy Terms

When we expand the expression for $\overline{U}$ in Eq. B.9 the result is

---

Strain Energy Integrals

$$[\overline{U}] = \int_{\Omega} \begin{bmatrix} [T_0]^T [E][T_0] & [T_0]^T [E][T_\varepsilon] \\ [T_\varepsilon]^T [E][T_0] & [T_\varepsilon]^T [E][T_\varepsilon] \end{bmatrix} d\Omega$$ (B.10)

Since the transformation $[T_0]$ is a null matrix, any term containing it is zero. Thus, the lower right partition $\overline{U}_{22}$ of Eq. B.10 contains the only nonzero elements. When we expand this partition, it has the following form:

Expanded Form of the Strain Energy Integrals

$$[\overline{U}_{22}] = \begin{bmatrix} GI_1 & GI_2 \\ GI_2 & EI_4 + GI_3 \end{bmatrix}$$

where

$$I_1 = KA \int_L dx$$

$$I_2 = KA \int_L x\,dx \qquad (B.11)$$

$$I_3 = KA \int_L x^2\,dx$$

$$I_4 = I \int_L dx$$

The elements of the energy matrix contain the length of the beam, L, and the moment of inertia, I, the area, A, and the shear shape factor, K, of the cross section. The term containing the integral I_3 is due to the parasitic shear. It can be eliminated by removing the x term from Eq. B.7.

When we evaluate Eq. B.11 for the element with origin at the center as shown in Fig. B.1, the result has the following form:

Strain Energy Integrals Evaluated

$$[\overline{U}] = \begin{bmatrix} 0 & 0 & 0 & 0 \\ 0 & 0 & 0 & 0 \\ 0 & 0 & u_{33} & 0 \\ 0 & 0 & 0 & u_{44} \end{bmatrix} \qquad (B.12)$$

where $u_{33} = KGAL$

$$u_{44} = EIL + KGAL^3 = u'_{44} + u''_{44}$$

The fact that the first two rows and columns are zero results from the presence of the rigid body modes as the first two generalized coordinates. The integral I_2 is zero because the origin of the local coordinate system is at the center of the element. The element u_{33} represents the strain energy for the shear. The u'_{44} component of the element u_{44} represents the beam flexure. The u''_{44} component is the parasitic shear term.

In later analyses of the stiffness characteristics of the Timoshenko beam, it is convenient to isolate the contributions of the different components of Eq. B.12. Thus, we can write the $\overline{U}$ matrix in component form as

---

Strain Energy Integrals in Component Form

$$[\overline{U}] = [\overline{U}]_1 + [\overline{U}]_2 + [\overline{U}]_3$$

where

$$[\overline{U}]_1 = \begin{bmatrix} 0 & 0 & 0 & 0 \\ 0 & 0 & 0 & 0 \\ 0 & 0 & u_{33} & 0 \\ 0 & 0 & 0 & 0 \end{bmatrix}$$

$$[\overline{U}]_2 = \begin{bmatrix} 0 & 0 & 0 & 0 \\ 0 & 0 & 0 & 0 \\ 0 & 0 & 0 & 0 \\ 0 & 0 & 0 & u'_{44} \end{bmatrix} \tag{B.13}$$

$$[\overline{U}]_3 = \begin{bmatrix} 0 & 0 & 0 & 0 \\ 0 & 0 & 0 & 0 \\ 0 & 0 & 0 & 0 \\ 0 & 0 & 0 & u''_{44} \end{bmatrix}$$

Transformation to Nodal Displacements

We are now in the position to transform the strain energy given by Eq. B.10 from a function of strain gradient quantities to nodal displacements. This transformation of the strain energy will produce the finite element stiffness matrix. The first step in this process is to form the transformation from nodal quantities to strain gradient variables using Eqs. B.1 and B.6. The general result is given as

---

Displacement and Rotation relations

$$\left\{ \begin{array}{c} w_i \\ q_i \end{array} \right\} = \begin{bmatrix} 1 & -x & \frac{x}{2} & 0 \\ 0 & 1 & \frac{1}{2} & x \end{bmatrix} \left\{ \begin{array}{c} (w_{rb})_0 \\ (q_{rb})_0 \\ (\gamma_{xy})_0 \\ (\varepsilon_{x,z})_0 \end{array} \right\} \tag{B.14}$$

When the nodal coordinates for the element shown in Fig. B.1 are substituted into Eq. B.14, the result is

Nodal to Strain Gradient Quantity Transformation

$${d} = [\Phi]{\varepsilon_,}$$

where

$${d}^T = { w_1 \quad q_1 \quad w_2 \quad q_2 }$$
$${\varepsilon_,}^T = { (w_{rb})_0 \quad (q_{rb})_0 \quad (\gamma_{xy})_0 \quad (\varepsilon_{x,z})_0 }$$

$$[\Phi] = \begin{bmatrix} 1.0 & L/2 & -L/4 & 0.0 \\ 0.0 & 1.0 & 1/2 & -L/2 \\ 1.0 & -L/2 & L/4 & 0.0 \\ 0.0 & 1.0 & 1/2 & L/2 \end{bmatrix}$$

(B.15)

This transformation must be inverted to give the relationship needed to change the strain energy expression given by Eq. B.9 to nodal displacements. When we do this, the result is

Strain Gradient to Nodal Quantity Transformation

$${\varepsilon_,} = [\Phi]^{-1}{d}$$

where

$$[\Phi]^{-1} = \begin{bmatrix} \frac{1}{2} & 0.0 & \frac{1}{2} & 0 \\ \frac{1}{2L} & \frac{1}{4} & -\frac{1}{2L} & \frac{1}{4} \\ -\frac{1}{L} & \frac{1}{2} & \frac{1}{L} & \frac{1}{2} \\ 0.0 & -\frac{1}{L} & 0.0 & \frac{1}{L} \end{bmatrix}$$

(B.16)

Equation B.16 is the relation needed to transform the strain energy expression from strain gradient quantities to nodal displacements.

Strain Energy Transformation and Stiffness Matrix Formulation

The strain energy can now be transformed to a function of nodal displacements to give the finite element stiffness matrix. We accomplish this by substituting Eq. B.16 into Eq. B.9 to give

Strain Energy Expression in Terms of Nodal Quantities

$$U = 1/2{d}^T[\Phi]^{-T}[\overline{U}][\Phi]^{-1}{d}$$

$$U = 1/2{d}^T[K]{d}$$

(B.17)

where $[K]$ is the finite element stiffness matrix

When Eqs. B.13 and B.16 are used to compute the finite element stiffness matrix defined in Eq. B.17, the result is given by the following three component matrices:

Finite Element Stiffness Matrix Due to $[U]_1$

$$[K]_1 = \frac{EI}{L} \begin{bmatrix} 0 & 0 & 0 & 0 \\ 0 & 1 & 0 & -1 \\ 0 & 0 & 0 & 0 \\ 0 & -1 & 0 & 1 \end{bmatrix} \qquad (B.18a)$$

Finite Element Stiffness Matrix Due to $[U]_2$

$$[K]_2 = KGA \begin{bmatrix} 1/L & -1/2 & -1/L & -1/2 \\ -1/2 & L/4 & 1/2 & L/4 \\ -1/L & 1/2 & 1/L & 1/2 \\ -1/2 & L/4 & 1/2 & L/4 \end{bmatrix} \qquad (B.18b)$$

Finite Element Stiffness Matrix Due to $[U]_3$

$$[K]_3 = \frac{KGAL}{12} \begin{bmatrix} 0 & 0 & 0 & 0 \\ 0 & 1 & 0 & -1 \\ 0 & 0 & 0 & 0 \\ 0 & -1 & 0 & 1 \end{bmatrix} \qquad (B.18c)$$

When the three component matrices are summed, the result is

Finite Element Stiffness Matrix with Parasitic Shear

$$[K] = \begin{bmatrix} \dfrac{KGA}{L} & -\dfrac{KGA}{2} & -\dfrac{KGA}{L} & -\dfrac{KGA}{2} \\ -\dfrac{KGA}{2} & \dfrac{KGA}{3}+\dfrac{EI}{L} & \dfrac{KGA}{2} & \dfrac{KGAL}{6}-\dfrac{EI}{L} \\ -\dfrac{KGA}{L} & \dfrac{KGA}{2} & \dfrac{KGA}{L} & \dfrac{KGA}{2} \\ -\dfrac{KGA}{2} & \dfrac{KGAL}{6}-\dfrac{EI}{L} & \dfrac{KGA}{2} & \dfrac{KGAL}{3}+\dfrac{EI}{L} \end{bmatrix} \qquad (B.19)$$

This stiffness matrix is identical to the one generated using the standard approach. When we eliminate the parasitic shear term, the stiffness matrix becomes

Finite Element Stiffness Matrix without Parasitic Shear

$$[K] = \begin{bmatrix} \frac{KGA}{L} & -\frac{KGA}{2} & -\frac{KGA}{L} & -\frac{KGA}{2} \\ -\frac{KGA}{2} & \frac{KGA}{4} + \frac{EI}{L} & \frac{KGA}{2} & \frac{KGAL}{4} - \frac{EI}{L} \\ -\frac{KGA}{L} & \frac{KGA}{2} & \frac{KGA}{L} & \frac{KGA}{2} \\ -\frac{KGA}{2} & \frac{KGAL}{4} - \frac{EI}{L} & \frac{KGA}{2} & \frac{KGAL}{4} + \frac{EI}{L} \end{bmatrix} \quad (B.20)$$

The only difference between the stiffness matrices given by Eqs. B.19 and B.20 is the contribution of the parasitic shear terms given by Eqs. B.18c to B.19. These terms add positive quantities to the diagonal terms of the stiffness matrix. These positive diagonal elements cause the overall stiffness of Eq. B.19 to be greater than the corrected matrix given by Eq. B.20. The effect of the additional stiffness is demonstrated in the main text. As we saw, the effect of parasitic shear can be eliminated at the formulation stage by removing the x term from Eq. B.7. In the standard finite element approach, parasitic shear can be eliminated by underintegrating the shear strain expression.

In the case of the plane stress elements developed earlier, it was found that parasitic shear was caused by the use of incomplete polynomials. We now see that parasitic shear is present in bending elements that represent through-thickness shear when the displacement and rotation polynomials have the same order. Parasitic shear does not occur when the displacement polynomial is one order higher than that of the rotation approximation. It can be shown that the same is true for Mindlin plate elements, the two-dimensional analog of the Timoshenko beam element.

The Five-Degree-of-Freedom Timoshenko Beam

This section develops the strain models for a five-degree-of-freedom Timoshenko beam to show that this element does not contain parasitic shear. This development identifies the source of parasitic shear in Timoshenko beam elements by inference since parasitic shear existed in the four node element. A five-degree-of-freedom Timoshenko beam is shown in Fig. B.2. As we can see, this element has three displacement degrees of freedom and two rotational degrees of freedom.

Since the displacements and rotations are independent of each other, the two quantities are represented by polynomial functions with a total of five independent variables as

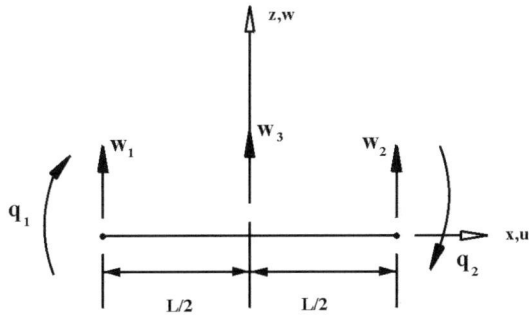

Figure B.2. A five-degree-of-freedom Timoshenko beam.

Displacement and Rotation Polynomials with Arbitrary Coefficients

$$q = d_1 + d_2 x$$

$$w = c_1 + c_2 x + c_3 x^2 \tag{B.21}$$

The strain gradient form of the displacement polynomials are found by comparing Eq. B.21 with the strain gradient expansions for displacements to give

Strain Gradient Displacement Polynomials

$$u(x) = (\gamma_{xz}/2 + q_{rb})_0 z + (\varepsilon_{x,z})_0 xz$$
$$w(x) = (w_{rb})_0 + (\gamma_{xz}/2 - q_{rb})_0 x \tag{B.22}$$
$$+ [(\gamma_{xz,x} - \varepsilon_{x,z})/2]_0 x^2$$

These polynomials contain the five linearly independent strain states that this element is capable of representing. When we extract these variables from Eq. B.22, they are

Linearly Independent Strain Gradient Quantities

$$(w_{rb})_0 \quad (q_{rb})_0$$

$$(\gamma_{xz})_0 \quad (\gamma_{xz,x})_0 \quad (\varepsilon_{x,z})_0 \tag{B.23}$$

The first two terms represent the rigid body motions. The three deformation terms represent constant shear, a linear variation in the shear, and constant flexure in the element.

The strain models contained in the two-node Timoshenko beam are formed by substituting the displacement polynomials contained in Eq. B.22 into the definitions of normal and shear strain to give the following strain models:

Timoshenko Beam Strain Models

$$\varepsilon_x = \frac{\partial u}{\partial x} = (\varepsilon_{x,z})_0 z$$

$$\gamma_{xy} = \left(\frac{\partial w}{\partial x} + \frac{\partial u}{\partial z}\right) = (\gamma_{xz})_0 + (\gamma_{xz,x})_0 x \tag{B.24}$$

The normal strain varies linearly across the section and the shear is represented by a constant and a linearly varying term. The normal strain model is identical to that for the four-node element. The shear model does not contain parasitic shear. The parasitic shear term was cancelled by the flexure term contained in the displacement polynomial. This example has demonstrated the general result that parasitic shear exists in the through-thickness shear representation when rotational and displacement polynomials have the same order and it does not exist when the displacement polynomial is one order higher than the rotation polynomial.

Closure

This appendix has developed stiffness matrices for Timoshenko beams using the strain-gradient-based procedure developed in Lesson 8. The resulting stiffness matrices are identical to those generated by other formulation processes. This appendix was developed so that the stiffness matrices for the Timoshenko beams are available in component form, as given in Eq. B.18 for use in the main body of this lesson. These results are used to analyze the stiffness characteristics of elements with through-thickness shear. We can extend the analysis presented here and in the main text to Mindlin plate elements.

References and Other Readings

1. Tessler, A., and Hughes, T. J. R. "An Improved Treatment of Transverse Shear in the Mindlin-Type Four Node Quadrilateral Element," *Computer Methods in Applied Mechanics and Engineering*, Vol. 39, 1983, pp. 311–335.

2. Dow, J. O., and Byrd, D. E. "The Identification and Elimination of Artificial Stiffening Errors in Finite Elements," *International Journal for Numerical Methods in Engineering*, Vol. 26, 1988, pp. 743–762.

3. Abdalla, J. E. "Qualitative and Discretization Error Analysis of Laminated Composite Plate Models," Ph.D. Thesis, University of Colorado, 1992.

4. Dow, J. O., and Abdalla, J. E. "Qualitative Errors in Laminated Composite Plate Models," *International Journal for Numerical Methods in Engineering*, Vol. 37, 1994, pp. 1215–1230.

5. Abdalla, J. E., and Dow, J. O. "An Error Analysis Approach for Laminated Plate Finite Element Models," *Computers and Structures*, Vol. 52, No. 4, Aug. 1994, pp. 611–616.

6. Shubert, M. K. "Formulation of Laminated Finite Elements using Strain Gradient Notation," Master's Thesis, University of Colorado, 1991.

Lesson 12   Problems

1. What is the polynomial required to define the lateral deformation $w(x, y)$ for a four-node Kirchhoff plate element? *Hint*: First define the degrees of freedom on the element, i.e., the nodal quantities.

2. Identify the strain gradient quantities defined by the polynomial defined in Problem 1. *Hint*: Remember that the through-thickness shear is not included as an unknown. It is assumed to be zero.

3. Form the displacement and rotation polynomials for a three-node, six-degree-of-freedom Timoshenko beam. Form the associated strain expressions. Discuss the strain modeling characteristics.

4. What are the polynomial expressions required to define the deformations and rotations of a four-node Mindlin plate element? *Hint*: Compare your result to the answer to Problem 1. Compare your result to the Timoshenko beam and to the Euler-Bernoulli beam.

5. Identify the strain gradient quantities contained in the representation found in Problem 4. Form the strain expressions and discuss the modeling characteristics of this element.

Part IV

The Strain Gradient Reformulation
of the
Finite Difference Method

Introduction

The theoretical basis and practical application of the error measures developed in Part V depend on the ability to form finite difference operators for the nodal patterns found in finite element meshes. The ability to form these derivative approximations emerges as a corollary to the strain gradient approach to the finite element method. We see in Part IV that the finite difference operators are contained in the transformations from nodal displacements to strain gradient quantities. The finite difference operators are produced when the [Φ] matrix is inverted. The ability to compute the finite difference operators in a routine manner ensures that the error measures developed in Part V are practical. Furthermore, this capability of computing derivative approximations for arbitrary nodal patterns may give the finite difference method a new life in solid mechanics applications.

Before the development of the finite element method, the finite difference method was used to find approximate solutions to a limited set of solid mechanics problems. Over the last 30 or 40 years, the finite element method has largely replaced the finite difference method in the solution of these problems because of the perception that the finite element method could better represent boundary conditions than the finite difference method. The developments in Part IV show that the finite difference method can represent any boundary condition that can be modeled by the finite element method.

The finite difference method is reformulated in this part of the book in terms of strain gradient notation. As a result, the reformulated finite difference method can represent any boundary condition that can be modeled by the finite element method. For example, the reformulated form of the finite difference method can routinely represent problems of the type shown in Fig. 1 with its complex boundary conditions and unevenly spaced mesh points. This reformulation of the finite difference method is presented so it can be used as the basis for developing error estimation procedures for the finite element method in Part V.

The finite difference method provides the basis for these error analysis procedures because it solves a different form of the governing differential equations than does the finite element method. The finite difference method solves the strong or differential form of the governing equations. The finite element method solves the weak or variational form of the governing equations.

Specifically, the finite difference form of the solution directly incorporates the natural boundary conditions. The finite element solution does not. The finite difference solution has smooth stress and strain fields at the nodal points. The finite element method does not. These differences enable aspects of the two types of solutions to be compared as a way of identifying errors in finite element solutions. Global, elemental, and pointwise error measures are developed in Part V based on these differences between the two solution techniques.

Common Basis for the Finite Element and Finite Difference Methods

In addition to the external similarity of being able to solve problems with the same mesh layout, the finite element and finite difference methods, *as developed here*, have the same internal basis. Both solution techniques are formulated here using truncated Taylor series expansions for the displacement approximation polynomials expressed in terms of strain gradient notation. As we saw in Part III, finite element stiffness matrices can be generated

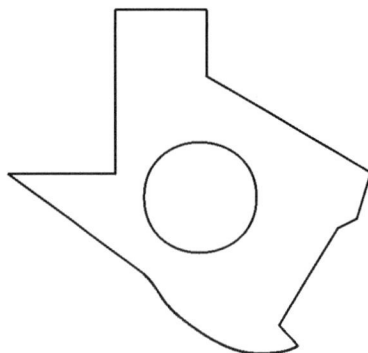

Figure 1. Finite difference model.

from [Φ] matrices formed using the strain gradient based approximation polynomials. In the next three lessons, the finite difference method is formulated from the [Φ] matrices used to form the finite element stiffness matrices. Thus, the two methods are put on a common basis.

Since the two methods have the same basis, the geometric flexibility of the two methods is identical. This means that finite element error estimators based on the finite difference method can be formed for any finite element model.

In addition to enabling the finite difference method to represent complex geometries with irregularly shaped finite difference templates, the reformulation of the finite difference method removes other limitations commonly attributed to the finite difference method. As is demonstrated, the version of the finite difference method presented here is capable of representing multimaterial interfaces and point loads.

An Overview of the Finite Difference Method

The finite difference method approximates the differential or strong form of the governing differential equations by approximating the solution of the differential equations at a finite number of mesh or nodal points on the interior and on the boundary of the problem. The differential equations are approximated by replacing the derivative quantities in the governing equations with finite difference approximations of the derivatives. That is to say, the derivatives at a mesh point are approximated as differences in the displacements of a set of points that constitute a finite difference template. The finite difference template consists of the point at which the derivatives are being approximated and a set of points in close proximity to this point.

Finite difference operators can be computed for points on the interior and on the boundary of complex regions such as the one shown in Fig. 2. The derivative approximations at the interior points are a function of the displacements of points on the domain of the problem. The derivative approximations at mesh points on the boundary are a function of the displacements of points on the domain and of points exterior to the region. The points outside of the boundary of the problem are called *fictitious points*.

The boundary conditions are introduced into the finite difference model by specifying the displacements or the applied stresses at points on the boundary. The representation of the boundary stresses in the finite difference models utilizes the fictitious points. The developments in the following three lessons explicate and extend the contents of this overview with a particular focus on the use of irregular templates and fictitious points. Part IV is designed to clarify the role of fictitious points in finite difference models.

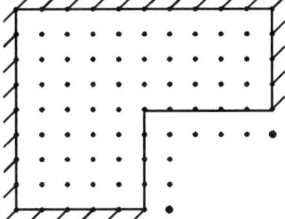

Figure 2. Fictitious mesh points.

The Strain Gradient Approach to the Finite Difference Method

The finite difference method, as developed here, differs in one major way from the finite difference method as it is generally presented. The strain gradient based finite difference method is *not* driven by the necessary use of nine-node central difference templates, although the nine-node template is used extensively. The approach presented here is based on the following three basic premises: (1) there must be at least one fictitious node per boundary node; (2) the finite difference templates must, as a minimum, represent rigid body motions, constant strains, and the six linear variations of the strain components; and (3) the location of the template nodes should, if possible, be ''topologically'' symmetric.

The first premise defines the minimum number and the approximate locations of the fictitious points. The second premise defines the minimum requirements for the overall finite difference templates needed to represent the plane stress problem. The six linear variations in the strains are required to represent the derivatives contained in the governing differential equations. The three constant strain terms are required to introduce the boundary conditions and the three rigid body terms are required to form a [Φ] matrix that will invert. Thus, the minimum number of degrees of freedom for a finite difference template is 12.

The first two premises remove the focus of the finite difference method from the exclusive use of the nine-node central difference template and generalize the treatment of boundary conditions. Then, in contrast to the first two premises, the third premise biases the method toward the use of nine-node templates. We will see that these three premises give the finite difference method a flexibility that makes it geometrically compatible with the finite element method while at the same time enabling the method to retain most of its usual characteristics.

The use of these three premises puts the formulation of the boundary condition models on a rational basis so that accurate and useable representations are routinely available. These premises give the finite difference method the modeling flexibility of the finite element method. The application of the three premises is now demonstrated with three small example mesh layouts. The first two meshes are even and uneven distributions of nodes on the interior of a region, respectively. These two examples exercise the second and third premises. The third example is a mesh layout on an outside or convex corner. This example brings the first premise concerning the minimum requirement for the number of fictitious nodes into play.

Templates for Interior Mesh Points

Let us consider the interior of a region with an evenly spaced mesh as shown in Fig. 3. The template required for the approximation of the governing differential equations at the

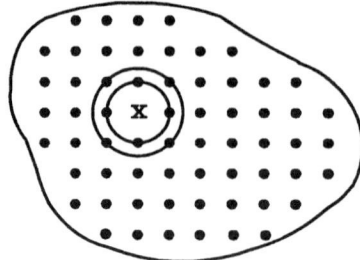

Figure 3. Interior region with an evenly spaced mesh.

point designated with the x is formed. There is no boundary adjacent to the point at which the governing equations are being approximated, so premise 1 is not applicable and no fictitious nodes enter the picture. Premise 2 requires that the template have a minimum of six mesh points so the 12 strain gradient terms can be represented. If we take the four points closest to the point under consideration, we have only five points defined for the template. We need at least one more point to complete the requirement of premise 2. Of the four points next closest to the central point, we have no reason to chose one point over the other three to complete a template with six mesh points. The third premise guides us in this situation. It tells us to take all of the topologically similar points. When we do this, we take all four of the next closest points. As we can see in Fig. 4, when we include these four points, we have a nine-node central difference template with even spacing.

In the case of the template just formed, the qualification given by the word "topological" was not needed. In this case, pure symmetry would have identified the nine-nodes we ended up with in our template. However, if the mesh had not been evenly spaced, pure geometric symmetry would not have defined a nine-node template. We can see this in Fig. 5. However, if we use the idea of "topological" symmetry or similar shapes, a distorted nine-node template is defined, as shown. Thus, the third premise is required to account for unevenly spaced meshes.

A Finite Difference Template for a Boundary Corner

An outside or convex corner is one of the most perplexing boundary conditions to represent in the finite difference method because of all the options available for modeling it. We form a *finite difference model* of the convex corner shown in Fig. 6 using the three premises to demonstrate the role of the fictitious points in boundary models. There is a minimum of one fictitious node for each boundary point, as specified by premise 1 and

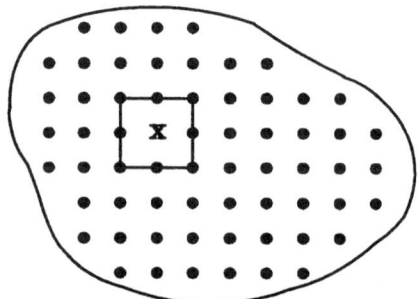

Figure 4. Nine-node finite difference template with even spacing.

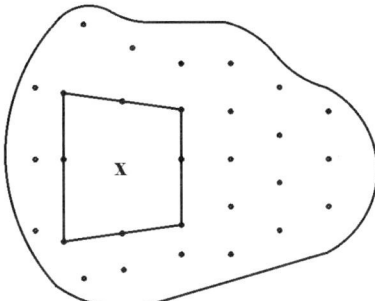

Figure 5. Nine-node finite difference templates with uneven spacing.

shown in Fig. 6. The fictitious nodes, the nodes that are outside of the boundary of the domain, are designated by the open circles. The displacement degrees of freedom associated with the fictitious nodes are introduced into the finite difference model by specifying normal and tangential stress boundary conditions at the boundary point associated with the fictitious node.

Let us now define templates for the corner node and for one of the nodes adjacent to it for the mesh shown in Fig. 6. The two mesh points closest to the corner node are enclosed in the smaller circle in Fig. 7. To form a template with at least six nodes, we must include the nodes next closest to the corner node. As shown in Fig. 7, the larger circle captures an additional four nodes to form a template with seven nodes. This template is not topologically symmetric. In the nodal configuration shown, there is no way to produce a topologically symmetric template.

We now form the template for a boundary point adjacent to the corner point. This template is shown in Fig. 8. The origin of the template (the point where the derivatives are being approximated) and the four points closest to the origin introduce five nodes into the template. If we include the points next closest to the point of interest, we include three more points in the template. The resulting eight-node template is not topologically symmetric. However, we can make this template topologically symmetric by adding the fictitious node associated with the corner node. This produces a nine-node finite difference template with unevenly spaced nodes.

It has been found that the templates for the outside corner shown in Figs. 7 and 8 produce displacement results that are acceptable. However, the strain components are not accurate and the equations that must be solved are ill-conditioned. This modeling deficiency can be corrected by imposing the topological symmetry premise on the template shown in Fig. 7.

The result of imposing this constraint is shown in Fig. 9. The two fictitious nodes shown by the x's are added. We now have a standard nine-node template. However, we

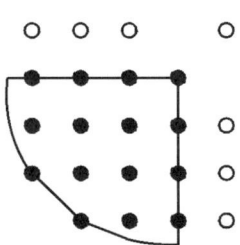

Figure 6. Outside or convex corner.

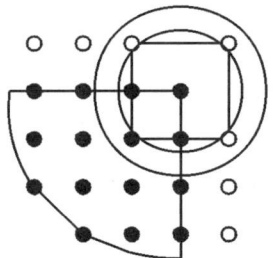

Figure 7. Finite difference template for the corner node.

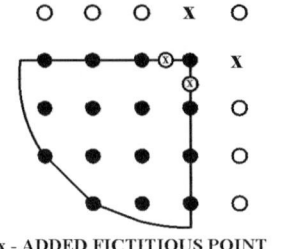

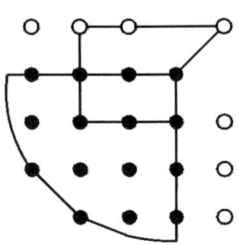

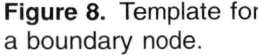

x - ADDED FICTITIOUS POINT
Ⓧ - ADDED BOUNDARY CONDITION

Figure 8. Template for a boundary node.

Figure 9. Convex corner with extra fictitious points.

have to find a way to introduce the four degrees of freedom associated with these additional fictitious nodes into the finite difference model. This has been successfully accomplished by specifying the boundary conditions at two additional points on the boundary. Note that these two points are not nodes in the finite difference models. In the examples used in Lesson 14, the points chosen are located between the two nodes of the finite difference model as shown in Fig. 9.

The addition of the two fictitious nodes also produces an evenly spaced template for the boundary node shown in Fig. 8. The new form of this template is shown in Fig. 10. Thus, the corner can be modeled with evenly spaced nine-node templates. This idea is extended to other corner models in Lesson 14. This short discussion was designed to introduce the idea that the boundary conditions in this reformulation of the finite difference method can be handled in a straightforward, rational manner. The evolution of the boundary conditions is discussed in detail in an appendix of Lesson 14.

Closure

These examples have introduced the procedure for identifying the finite difference templates required in the strain gradient approach to the finite difference method. The formulation of [Φ] matrices that will invert in every configuration of nodes is made possible by the use of strain gradient notation. As a result of the flexibility provided by formulating the finite element and finite difference methods from a common basis, a finite difference representation can be overlaid on most finite element models. This capability is used in Part V to improve current finite element error analysis procedures and to develop new error measures. In the three lessons contained in Part IV, however, the finite difference method is presented as a separate analysis technique.

Note that the ambiguities associated with representing corner nodes are not significant in a "real" problem. The corner node produces a stress concentration. If the location is critical to a given design, the sharp corner is rounded to lower the stresses. If the corner is not significant to the design, the representation used is not critical. Also note

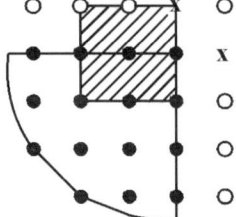

Figure 10. Finite difference template for a boundary point.

that regardless of the finite difference model used at critical points, the "special" nature of these points is identified in the error analysis process.

Contents of Part IV

This introduction was designed to show that the modeling capabilities of the finite difference method based on the use of strain gradient notation can represent any boundary condition that can be modeled by the finite element method. These capabilities are made possible as a result of the following: (1) the two methods have been put on a common basis via the use of the $[\Phi]$ matrix; (2) the focus of the finite difference method, as presented here, is on the role of the fictitious nodes and the needed derivative approximations; and (3) the place of the nine-node central difference template is clearly identified as secondary, not central.

Lesson Outlines

A brief outline of each of the three lessons presenting the strain gradient approach to the finite difference method is presented.

Lesson 13 introduces the finite difference approach for solving the *governing differential equations* that describe the plane stress problem. This lesson focuses on the finite difference representation of the governing differential equations and not on the boundary conditions. The lesson relies almost exclusively on fixed boundaries and the use of nine-node central difference templates. This approach is taken to clarify the formulation of the finite difference representation of the differential equations at individual nodes and the assembly of the overall model. The idea of fictitious nodes is introduced by finding the boundary stresses on a fixed boundary. The lesson contains two appendices. Appendix A details the formulation of finite difference operators and attempts to clarify their meaning. Appendix B demonstrates the assembly of a finite difference model in detail.

Lesson 14 focuses on finite difference *boundary condition models*. The general characteristics of boundary condition models and the role of fictitious nodes are discussed. The models of several boundary conditions are then developed. Specifically, models of straight boundaries, curved boundaries, outside or convex corners, and reentrant or concave corner models are formed. A general approach for introducing boundary loads into finite difference models is presented. Finally, the general structure of finite difference models is discussed. The lesson contains an appendix that outlines the evolution of the boundary condition models presented in this lesson.

Lesson 15 extends the modeling capabilities presented in the first two lessons by adding the following capabilities: (1) local mesh refinement, (2) point loads, (3) floating boundaries (boundaries not defined by a set of nodes) and (4) multimaterial boundaries.

Lesson **13**

Elements of the Finite Difference Method

Purpose of Lesson   To introduce the finite difference method for finding approximate solutions to plane stress problems.

In the finite difference method, the approximate solutions are found by solving a set of algebraic equations that are the discrete representation of the governing differential equations and the boundary conditions. The discrete representation is formed by replacing the derivatives in the governing equations and the boundary conditions with approximations expressed in terms of differences between nodal displacements. That is to say, "finite differences" replace the differentials or infinitesimals contained in the actual derivatives.

In this lesson, the finite difference method is developed using strain gradient notation. The use of this notation enables the boundary conditions to be easily validated and irregular meshes to be routinely used. These characteristics are used in the next part of the book to develop pointwise error measures and alternate strain extraction procedures for finite element results.

The finite difference method is developed for fixed boundary conditions in this lesson. In Lesson 14, the finite difference representation of the standard boundary conditions is formed. In Lesson 15, the method is extended to enable local mesh refinement and the solution of multimaterial problems.

■   ■   ■

The finite difference and finite element methods are related through the principle of minimum potential energy via the Euler-Lagrange equations, as shown in Fig. 1. The integrand of the potential energy functional is substituted into the Euler-Lagrange equations to produce the differential equations whose solution minimizes the potential energy of the system. These differential equations are the governing differential or equilibrium equations for the physical system. The results found by the finite element method directly minimize an approximation of the potential energy functional. These results indirectly provide an approximate solution to the governing differential equations. The finite difference method directly finds an approximate solution to the governing differential equations. The potential energy functional and the governing differential equations for the plane stress problem were developed in Lesson 3.

We can clearly see the relationship between the two methods by noting that the finite difference method models the governing differential equations with a discrete representation while the finite element method models the strain energy expression with a discrete representation. The error measures developed in Part V exploit the differences between the approximate solutions to the variational and differential forms of the plane stress problem, as outlined in the introduction to this part of the book.

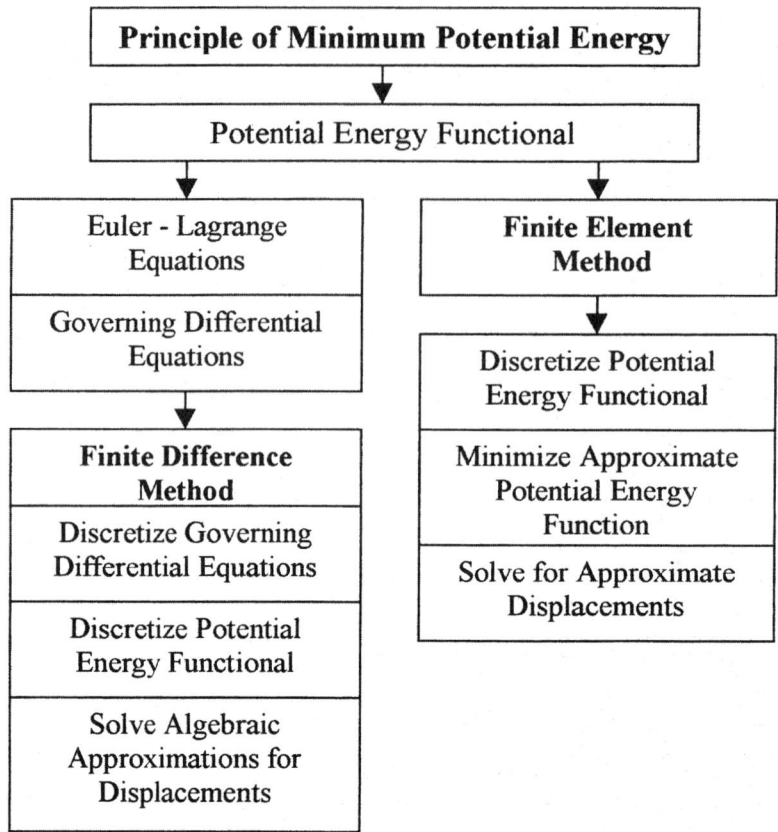

Figure 1. Relationship between the finite difference and the finite element methods.

Governing Differential Equations

The governing differential equations that describe the behavior of plane stress problems are the following:

The Governing Differential Equations

$$\left(\frac{E}{1-\nu^2}\right)\left[u_{,xx} + \nu v_{,xy} + \left(\frac{1-\nu}{2}\right)(u_{,yy} + v_{,xy})\right] = p_x$$

$$\left(\frac{E}{1-\nu^2}\right)\left[v_{,yy} + \nu u_{,xy} + \left(\frac{1-\nu}{2}\right)(v_{,xx} + u_{,xy})\right] = p_y$$

(13.1)

These second-order partial differential equations are the equilibrium equations for the plane stress problem in the x and y directions, respectively. Equation 13.1 is derived in Lesson 3 and is reproduced here for the convenience of the reader.

We can rewrite these equations in terms of strain gradient notation. For example, $u_{,xx}$ is equal to $\varepsilon_{x,x}$. When the governing differential equations given in Eq. 13.1 are expressed in terms of strain gradient notation and the leading coefficients are normalized, the result is

The Governing Differential Equations in Strain Gradient Notation

$$\varepsilon_{x,x} + \nu\varepsilon_{y,x} + \left(\frac{1-\nu}{2}\right)(\gamma_{xy,y}) = -\left(\frac{1-\nu^2}{E}\right)p_x$$

$$\varepsilon_{y,y} + \nu\varepsilon_{x,y} + \left(\frac{1-\nu}{2}\right)(\gamma_{xy,x}) = -\left(\frac{1-\nu^2}{E}\right)p_y$$

(13.2)

These equations contain the following six strain gradient quantities that must be approximated in the finite difference method:

Derivative Operators

$$\begin{array}{cc} \varepsilon_{x,x} & \varepsilon_{x,y} \\ \varepsilon_{y,x} & \varepsilon_{y,y} \\ \gamma_{xy,x} & \gamma_{xy,y} \end{array}$$

(13.3)

A Finite Difference Model

Before we develop the finite difference representations of the individual derivative terms given in Eq. 13.3 and later form the finite difference model of the equilibrium equations given in Eq. 13.2, we outline the process of developing a finite difference model with the example shown in Fig. 2. This example consists of a plane stress problem with fixed boundaries.

This problem is represented by a 10 unit by 8 unit rectangle. A mesh consisting of nine rows each with 11 nodes for a total of 99 nodes is superimposed on the region. The displacements in the two coordinate directions of the 36 nodes on the boundary are known to be zero. This leaves 63 interior nodes, for which 63 displacements in the x direction and 63 displacements in the y direction must be found.

The 126 equations required to find the 126 unknown displacements are supplied by the equilibrium equations given by Eq. 13.2. Each of the 63 interior nodes has two

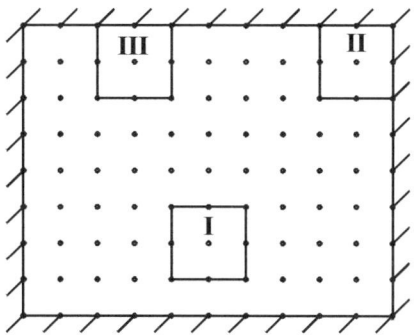

Figure 2. A simple plane stress problem.

equilibrium equations associated with it. Thus, our task is to find the discrete finite difference approximation of the equilibrium equations at each of the interior nodes for this problem.

The heart of the task is to approximate the six derivatives contained in Eq. 13.3 at the interior mesh points. This is accomplished through the use of nine-node central difference templates, as shown superimposed on three of the nodes in Fig. 2. Template I is totally located inside of the region. In this case, all 18 of the nodal displacements are unknowns. Template II is located in a corner. In this case, the displacements of the five nodes located on the boundary are known to be zero. Only the displacements of the four template nodes located on the interior of the region are unknown. Template III is located on a straight boundary. In this case, the displacements of the three boundary nodes are specified as zero. We must find the 12 displacements of the six template nodes on the interior of the region. We develop models for other types of boundary conditions in the next lesson. Note that the nodal pattern shown in Fig. 2 for this problem could equally well apply to a finite element representation for a nine-node element.

The Nine-Node Finite Difference Template

The three types of template shown in Fig. 2 are fundamentally the same. The only difference is that for two of the templates there are nodes located on the fixed boundary and the displacements of these nodes are known *a priori* to be zero. A representative template for an evenly spaced mesh is shown in Fig. 3. The polynomial terms that a template can represent are tentatively identified by superimposing the template configuration onto a Pascal's triangle. The results of this operation are shown in Fig. 3. This was originally done for a nine-node configuration in Lesson 6 and shown in Fig. 18 of that lesson (see Note 1). As a result of this comparison, we can see that the nine-node template can represent the following general polynomials:

General Polynomial Representation

$$u = a_1 + a_2x + a_3y + a_4x^2 + a_5xy + a_6y^2 + a_7x^2y + a_8xy^2 + a_9x^2y^2$$

$$v = b_1 + b_2x + b_3y + b_4x^2 + b_5xy + b_6y^2 + b_7x^2y + b_8xy^2 + b_9x^2y^2$$

$$(13.4)$$

Figure 3. A nine-node template.

When we replace the general coefficients by the strain gradient coefficients, the result is

---

Strain Gradient Polynomial Representation

$$u = (u_{rb})_0 + (\varepsilon_x)_0 x + \left(\frac{\gamma_{xy}}{2} - r_{rb}\right)_0 y + \frac{1}{2}(\varepsilon_{x,x})_0 x^2 + (\varepsilon_{x,y})_0 xy + \frac{1}{2}(\gamma_{xy,y} - \varepsilon_{y,x})_0 y^2$$

$$+ \frac{1}{2}(\varepsilon_{x,xy})_0 x^2 y + \frac{1}{2}(\varepsilon_{x,yy})_0 xy^2 + \frac{1}{4}(\varepsilon_{x,xxy})_0 x^2 y^2$$

$$v = (v_{rb})_0 + \left(\frac{\gamma_{xy}}{2} + r_{rb}\right)_0 x + (\varepsilon_y)_0 y + \frac{1}{2}(\gamma_{xy,x} - \varepsilon_{x,y})_0 x^2 + (\varepsilon_{9}y,x)_0 xy + \frac{1}{2}(\varepsilon_{y,y})_0 y^2$$

$$+ \frac{1}{2}(\varepsilon_{y,xx})_0 x^2 y + \frac{1}{2}(\varepsilon_{y,xy})_0 xy^2 + \frac{1}{4}(\varepsilon_{y,xxy})_0 x^2 y^2$$

(13.5)

This set of polynomials is capable of representing the following 18 strain gradient quantities:

Nine-Node Independent Strain States

$$(u_{rb})_0 \quad (v_{rb})_0 \quad (r_{rb})_0$$
$$(\varepsilon_x)_0 \quad (\varepsilon_y)_0 \quad (\gamma_{xy})_0$$
$$(\varepsilon_{x,x})_0 \quad (\varepsilon_{x,y})_0$$
$$(\varepsilon_{y,x})_0 \quad (\varepsilon_{y,y})_0$$
$$(\gamma_{xy,x})_0 \quad (\gamma_{xy,y})_0$$
$$(\varepsilon_{x,xy})_0 \quad (\varepsilon_{x,yy})_0$$
$$(\varepsilon_{y,xx})_0 \quad (\varepsilon_{y,xy})_0$$
$$(\varepsilon_{x,xxy})_0 \quad (\varepsilon_{y,xxy})_0$$

(13.6)

The displacement polynomials for the nine-node configuration contain the three rigid body motions, the three constant strain states, the six linear variations of the three strain components, four of the nine quadratic terms, and 2 of the 12 cubic terms.

As expected, the modeling characteristics of the nine-node central difference template are identical to that for a nine-node finite element. When the terms represented by a nine-node template are compared to the strain gradient quantities contained in the governing differential equations shown in Eq. 13.3, we can see that Eq. 13.6 contains all of the terms contained in Eq. 13.3. This means that this template is capable of representing the governing differential equations for the plane stress problem.

Finite Difference Operators

The finite difference approximation of the derivatives contained in Eq. 13.2 and given in Eq. 13.3 are developed in detail in Appendix A. In brief, the derivative approximations for these terms and those contained in Eq. 13.6 are developed by forming a coordinate transformation from nodal displacements to strain gradient quantities using Eq. 13.5 and inverting the resulting transformation.

The transformation, as derived in Appendix A, is written as

Nodal Displacement-Strain Gradient Quantity Transformation

$$\{d\} = [\Phi]\{\varepsilon_,\}$$

where

$$\{d\}^T = \{\, u_1 \quad v_1 \quad u_2 \quad v_2 \quad \cdots \quad u_9 \quad v_9 \,\} \tag{13.7}$$

$$[\Phi]^T = [\, [A]_1^T \quad [A]_2^T \quad \cdots \quad [A]_9^T \,]$$

$$\{\varepsilon_,\}^T = [\, (u_{rb})_0 \quad (v_{rb})_0 \quad \cdots \quad (\varepsilon_{y,xxy})_0 \,]$$

Equation 13.7 is the transformation from the 18 u and v displacements of the nodal points to the eighteen independent strain gradient quantities. The row vectors $[A_u]_i$ and $[A_v]_i$ and the column vector $\{\varepsilon_,\}$ of Eq. 13.7 are given in Table 1. The $[\Phi]$ matrix is the transformation matrix.

The finite difference operators are formed by inverting the transformation given in Eq. 13.7 to give

---

Finite Difference Operators

$$\{\varepsilon_,\} = [\Phi]^{-1}\{d\} \tag{13.8}$$

Table 1. The 18 coefficients for the nine-node finite difference template.

i	$\{\varepsilon_,\}$	$\{A_u\}_i^T$	$\{A_v\}_i^T$
1	$(u_{rb})_0$	1	0
2	$(v_{rb})_0$	0	1
3	$(r_{rb})_0$	$-y_i$	x_i
4	$(\varepsilon_x)_0$	x_i	0
5	$(\varepsilon_y)_0$	0	y_i
6	$(\gamma_{xy})_0$	$y_i/2$	$x_i/2$
7	$(\varepsilon_{x,x})_0$	$x_i^2/2$	0
8	$(\varepsilon_{y,x})_0$	$-y_i^2/2$	$x_i y_i/2$
9	$(\gamma_{xy,x})_0$	0	$x_i^2/2$
10	$(\varepsilon_{x,y})_0$	$x_i y_i$	$-x_i^2/2$
11	$(\varepsilon_{y,y})_0$	0	$y_i^2/2$
12	$(\gamma_{xy,y})_0$	$y_i^2/2$	0
13	$(\varepsilon_{x,xy})_0$	$x_i^2 y_i/2$	0
14	$(\varepsilon_{x,yy})_0$	$x_i y_i^2/2$	0
15	$(\varepsilon_{y,xy})_0$	0	$x_i y_i^2/2$
16	$(\varepsilon_{y,xx})_0$	0	$x_i^2 y_i/2$
17	$(\varepsilon_{x,xyy})_0$	$x_i^2 y_i^2/4$	0
18	$(\varepsilon_{y,xxy})_0$	0	$x_i^2 y_i^2/4$

The finite difference operators for the derivatives contained in the governing differential equations as shown in Eq. 13.3 are given for a template with an even spacing between the nodes as

---

General Finite Difference Operators for Evenly Spaced Nodes

$$\varepsilon_{x,x} = \frac{1}{h^2}\left(u_6 + u_8 - 2u_9\right)$$

$$\varepsilon_{y,y} = \frac{1}{h^2}\left(v_5 + v_7 - 2u_9\right)$$

$$\varepsilon_{y,x} = \frac{1}{4h^2}\left(v_1 - v_2 + v_3 - v_4\right)$$

$$\varepsilon_{x,y} = \frac{1}{4h^2}\left(u_1 - u_2 + u_3 - u_4\right)$$

$$\gamma_{xy,x} = \frac{1}{4h^2}\left(u_1 - u_2 + u_3 - u_4\right) + \frac{1}{h^2}\left(v_6 + v_8 - 2v_9\right)$$

$$\gamma_{xy,y} = \frac{1}{4h^2}\left(v_1 - v_2 + v_3 - v_4\right) + \frac{1}{h^2}\left(u_5 + u_7 - 2u_9\right)$$

(13.9)

A complete example of the process of forming the finite difference operators is presented in Appendix A.

The coordinate transformation approach for formulating the finite difference operators just presented differs from the standard approach for generating these derivative approximation (see Note 2). The approach just shown has two distinct advantages. One, it implicitly demonstrates the close relationship between the finite difference and the finite element methods. This is used to advantage in the next part of the book when the error analysis procedures previously discussed are improved and extended as a result of the recognition of this relationship. And two, the procedure used here is directly applicable to templates with unevenly spaced nodes, as demonstrated in Appendix A. The usefulness of this capability is demonstrated in subsequent lessons.

Modeling the Plane Stress Problem

We are now in a position to form the finite difference representation of the plane stress problem. We have seen that the nine-node template represents the derivative quantities contained in the governing differential equations and we have the finite difference approximations of these derivatives. The finite difference representation of the plane stress problem is formed by replacing the derivative operators contained in Eq. 13.2 with the finite difference operators given in Eq. 13.9.

This process is demonstrated in full algebraic detail for the case of the first equation contained in Eq. 13.2. Then the finite difference models are put in a matrix structure that is conducive to easy programming.

When we substitute the finite difference operators in place of the derivative functions in the first equation of Eq. 13.2, the result is

Formulation of an Equation Template

Differential Representation | Finite Difference Representation

$$\varepsilon_{x,x}$$

$$\frac{1}{h^2}(u_6 + u_8 - 2u_9)$$

$$+\nu\varepsilon_{y,x}$$

$$+\nu\frac{1}{4h^2}(v_1 - v_2 + v_3 - v_4)$$

$$+\left(\frac{1-\nu}{2}\right)\gamma_{xy,y}$$

$$+\left(\frac{1-\nu}{2}\right)\left[\frac{1}{4h^2}(v_1 - v_2 + v_3 - v_4) + \frac{1}{h^2}(u_5 + u_7 - 2u_9)\right]$$

$$= -\left(\frac{1-\nu^2}{E}\right)p_x$$

$$= -\left(\frac{1-\nu^2}{E}\right)p_x$$

$$(13.10)$$

The left-hand side of Eq. 13.10 is the differential equation being modeled. The right-hand side is the finite difference representation for a template with an even spacing between the nodes, which is designated as h. As we can see, the quantities contained in the finite difference approximation consist of the physical parameters of the problem, the spacing of the nodes in the finite difference template and the nodal displacements of the finite difference template. The finite difference method has replaced a differential equation with a set of algebraic equations.

One purpose for developing the full algebraic form of the finite difference representation of the differential equation as given in Eq. 13.10 is to clearly present the structure of a finite difference model. The finite difference representation of this single differential equation is a single algebraic equation in the form of a row vector. When we develop the finite difference model of the second equilibrium equation of Eq. 13.2, we will see that it also consists of a single row vector. This row-by-row structure of finite difference models makes the formulation of finite difference representations simpler than the formulation of finite element models, which means that the assembly of the global coefficient matrix for a finite difference model is accomplished one row at a time. This makes the formulation process easy to implement on parallel processing computers. This characteristic is seen in problems contained at the end of this lesson.

The row vector structure of the single governing differential equation is also seen when the equation is written in matrix (vector) form. The coefficients and the differential operators that define the differential equation are each written as vectors and multiplied to form the representation. This structure leads to a convenient and efficient computational form. When this is done, we can write the equilibrium equation in the x direction given in Eq. 13.2 and on the left-hand side of Eq. 13.10 as

Matrix Representation

$$[C']\{\varepsilon_,'\} = \{f_x\}$$

where

$$[C'] = [\,1 \quad \nu \quad (1-\nu)/2\,]$$

$$\{\varepsilon_,'\}^T = [\,(\varepsilon_{x,x})_0 \quad (\varepsilon_{x,y})_0 \quad (\gamma_{xy,y})_0\,]^T$$

$$\{f_x\} = (1-\nu^2/E)p_x$$

$$(13.11)$$

When we replace the derivative operators by the finite difference approximations, Eq. 13.11 represents the finite difference model of the equilibrium equation in the x direction of the plane stress problem in matrix (vector) form. The three coefficients of the differential equation are the only three elements of the $[C']$ matrix. The three finite difference operators are contained as the three elements of the $\{\varepsilon'\}$ matrix. These operators are defined in Eq. 13.9. When we multiply these two vectors, the right-hand side of Eq. 13.10 results.

Equation 13.10 defines the relationship between the differential equation and the finite difference model. The finite difference model is put in compact matrix form in Eq. 13.11. This equation adequately expresses the finite difference representation, but it is not as computationally convenient as it could be. A more efficient form of the finite difference representation is expressed in terms of the general form of the finite difference operators given in Eq. 13.8. When this is done, the result is

---

Strain Gradient Representation of Equilibrium Equation

$$[C_x]\{\varepsilon_,\} = \{f_x\}$$

where

$$[C_x] = [0] \tag{13.12}$$

except for

$$C_7 = 1 \qquad C_{10} = \nu \qquad C_{12} = (1 - \nu/2)$$

The analogous representation of the second governing differential equation, the equilibrium equation in the y direction, contained in Eq. 13.2 can be combined with Eq. 13.12 to give

---

Equilibrium Equations in Strain Gradient Notation

$$[C]\{\varepsilon_,\} = \{f\}$$

where

$$[C] = \begin{bmatrix} C_x \\ C_y \end{bmatrix} = [0] \tag{13.13}$$

except for

$$C_{1,7} = 1 \qquad C_{1,10} = \nu \qquad C_{1,12} = (1 - \nu/2)$$
$$C_{2,11} = 1 \qquad C_{2,10} = \nu \qquad C_{2,9} = (1 - \nu/2)$$

We can transform Eq. 13.13 to a function of nodal displacements by introducing the transformation given by Eq. 13.8 to give

Finite Difference Representation of the Plane Stress Problem

$$[\Omega]\{d\} = \{f\}$$

where

$$[\Omega] = [C][\Phi]^{-1}$$

(13.14)

Equation 13.14 represents the two governing differential equations for the plane stress problem at a single point in the finite difference model. We call the $[\Omega]$ matrix the equation template for the problem. This template is developed in detail in Appendix B.

By comparing the two equations that make up Eq. 13.14, it is seen that the difference between the two equations is totally contained in the elements of the coefficient matrices (vectors). The equation template can be changed by simply modifying the coefficient matrix (vector). This simple representation of a differential equation gives the finite difference representation a flexibility that does not exist in the finite element method. The differential equation being represented can be changed by modifying the input data that defines the coefficient matrix. No changes need to be made to the program.

Finite Difference Global Equation Assembly

Equation 13.14 contains the finite difference approximation of the two governing differential equations for the plane stress problem at the central node of a nine-node central difference template expressed in terms of local coordinates. To represent a problem, a number of these templates must be combined as functions of global coordinates. This process is best presented with a simple example.

Consider the plane stress problem with fixed boundaries represented with the two 9-node central difference templates centered at nodes 1 and 2 as shown in Fig. 4. This simple representation contains 12 nodes with 24 degrees of freedom. The two central difference templates are shown separately in Fig. 5.

The governing differential equations at nodes 1 and 2 are represented by Eq. 13.14. The $\{d\}$ matrices of the finite difference representation at both nodes are the same in *local coordinates*. That is to say, the $\{d\}$ matrix for the central difference template centered at node 1 is $\{u_1\ v_1 \ldots u_9\ v_9\}$ and the $\{d\}$ matrix for the central difference template centered at node 2 is $\{u_1\ v_1 \ldots u_9\ v_9\}$. However, when the local coordinates of the two templates, as shown in Fig. 3, are related to the global coordinates, as shown in Fig. 5, we can see the difference between the global identification on the nodes in the two templates. That is to say, the displacement at local node 1 of the central difference template with its origin (local node 9) located at the global node 1 correspond to the displacements at the node labeled as node 3 in the global system, etc. Similarly, node 1 of the second template corresponds to the global node 4. When we carry the process to

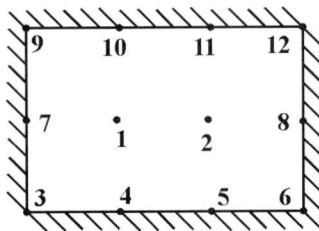

Figure 4. A simple plane stress problem.

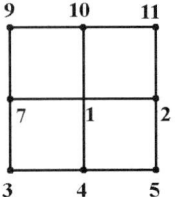 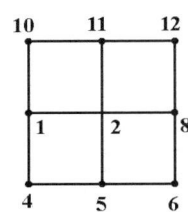

Figure 5. Template with global coordinates.

completion, the displacement vector for the central difference template centered at node 1 expressed in *global coordinates* is

---

Global Displacement Vector for Template 1

$$\{d\}_1^T = \begin{bmatrix} u_3 & v_3 & u_5 & v_5 & u_{11} & v_{11} & u_9 & v_9 & u_4 & v_4 & u_2 & v_2 & u_{10} & v_{10} & u_7 & v_7 & u_1 & v_1 \end{bmatrix}^T$$

(13.15)

The displacement vector for the central difference template centered at node 2 expressed in *global coordinates* is

---

Global Displacement Vector for Template 2

$$\{d\}_2^T = \begin{bmatrix} u_4 & v_4 & u_6 & v_6 & u_{12} & v_{12} & u_{10} & v_{10} & u_5 & v_5 & u_8 & v_8 & u_{11} & v_{11} & u_1 & v_1 & u_2 & v_2 \end{bmatrix}^T$$

(13.16)

When we compare Eqs. 13.15 and 13.16, we see that the displacement vectors are now different and that they relate to the global coordinates in Fig. 4.

For the sample problem shown, the boundaries are fixed. This means that the displacements of nodes 3–12 are zero in both the u and the v directions. Thus, the displacement vectors for the two templates reduce to functions of u_1, v_1, u_2, and v_2. Since two nine-node templates are used and each one contributes two equations, the finite difference approximation of this simple problem reduces to a four degree of freedom problem. The final result of this formulation is

---

Finite Difference Model (Fig. 4)

$$[C]\{d\} = \{f\}$$

where

$$[C] = \begin{bmatrix} C_x \\ C_y \end{bmatrix}$$

(13.17)

$$\{d\}^T = \begin{bmatrix} u_1 & v_1 & u_2 & v_2 \end{bmatrix}^T$$

$$\{f\}^T = \begin{bmatrix} f_{x_1} & f_{y_1} & f_{x_2} & f_{y_2} \end{bmatrix}^T$$

The procedure for expressing the finite difference representation of plane stress problems in terms of global coordinates has been outlined. The finite difference representation resulting from this process for the simple two-template problem is developed in detail in Appendix B. This example demonstrates the row-by-row formation of the finite difference representation of the plane stress problem. The procedure for expressing the finite difference representation of plane stress problems in terms of global coordinates has been outlined.

Boundary Stress Computation — Fictitious Nodes

In the previous section, we found the displacements of the interior nodes for a problem with fixed boundaries. With this information and the knowledge that the displacements of the boundary nodes are zero, we can compute the strains at the interior points using the finite difference representation of the strains given by Eq. A.9. However, we do not currently have a procedure for determining the strains on the fixed boundary, which are important in engineering applications.

The objective of this section is to outline the procedure for computing the boundary strains. This is accomplished by introducing a central difference template on the boundary, as shown in Fig. 6. As we can see, this template has nodes that are located outside of the domain of the problem. These exterior nodes are called *fictitious nodes*. To compute the strain at the boundary using this template, we must know the nodal displacements of all nine nodes that make up the template. Thus, we must find the displacements for the fictitious nodes.

The boundary nodes defined in terms of the local template node numbering system in Fig. 6 as nodes 1, 4, 5, 7, and 9 are known to be zero for this case. The displacement of the local node 8 (the global node 2) in Fig. 4 was found by solving the finite difference model. The example problem presented in the next section will demonstrate that the displacements of local nodes 2 and 3 must be taken as zero. This can be argued intuitively by noting that the strains along the horizontal boundaries must be zero. We now know the displacements of all of the nodes except for node 6. We must determine these displacements so we can compute the strains on the boundary.

The displacements at the local node 6 are found by recognizing that the governing differential equations must be satisfied on the boundary. This means that the finite difference representation of the governing differential equations given by Eq. 13.14 must be satisfied for the template that is centered at node 8 of the global coordinate system shown in Fig. 4. Thus we have two auxiliary equations available to us with which to find the two unknown displacements at node 6 in Fig. 6.

When these auxiliary equations are solved and the displacements of node 6 are known, we can compute the strains on the boundaries. With the availability of the strains, the stresses can be computed with the aid of the constitutive relations. The boundary strains and stresses for the simple problem addressed in this and the previous section are found in Appendix B.

For the case of larger problems, a set of simultaneous auxiliary equations are available for finding the displacements of the fictitious nodes. This enables us to find the strains on the boundary of larger problems. This process is discussed in more detail in the next section.

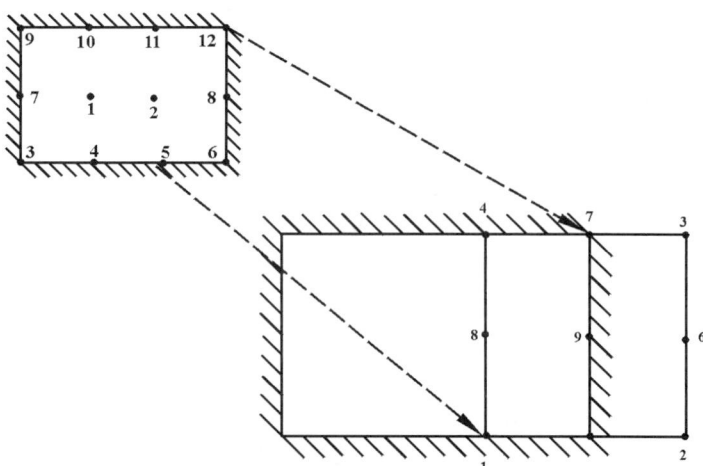

Figure 6. Boundary template superimposed on Fig. 4.

The concept of fictitious nodes was just introduced in the context of computing the boundary stresses for a problem with a fixed boundary. Fictitious nodes are present in all of the boundary models in the finite difference method. In one sense, they can be viewed as unnatural and may seem so to analysts familiar with the finite element method. However, in the development of the boundary conditions presented in the next lesson, we see that the fictitious nodes provide valuable assistance in defining the boundary conditions. The presence of fictitious nodes coupled with the guidance provided by strain gradient notation provides assistance in defining finite difference models.

Example

To demonstrate the generality of the procedure for computing the boundary strains just used to introduce the idea of fictitious nodes, the procedure for finding the boundary loads on the right-hand end of the problem shown in Fig. 7 is presented and validated. The arrows in the model represent the applied load. Note that the loads in a finite difference problem are the value of the distributed load applied at that point. There is no concept in the finite difference method that parallels the idea of an equivalent nodal load that is used in the finite element method.

In this case, the central difference templates must be applied to the right-hand boundary, as shown in Fig. 8. This case contains nine fictitious nodes. The displacements of the two fictitious nodes at the top and bottom in the figure are zero. This leaves the 14 displacements at the seven other fictitious nodes to be determined.

We find the displacements of the remaining seven fictitious nodes by forcing the governing differential to be zero at the boundary nodes. This gives the 14 equations required to solve for the 14 unknown displacements. This enables the boundary strains and boundary stresses to be computed as was our original intent.

Note at this point that the result of this procedure depends on an assumption. We have assumed that the displacements of the fictitious nodes at the top and bottom of the fixed boundary must be zero. We now see that this assumption has been validated.

This assumption is validated by solving the problem shown in Fig. 9. This is the same problem shown in Fig. 7 except that the boundary on the right-hand end is not fixed. It is free to move and has a load applied to it. The loads applied to this boundary are the loads computed for the case of the problem with the fixed boundary.

When the problem with the loaded boundary is solved, the results match those for the problem with the fixed boundary. That is to say, the displacements on the boundary are zero and the internal displacements are the same as for the problem with all of the boundaries fixed (Ref. 3). This serves as a validation of the procedure for computing the

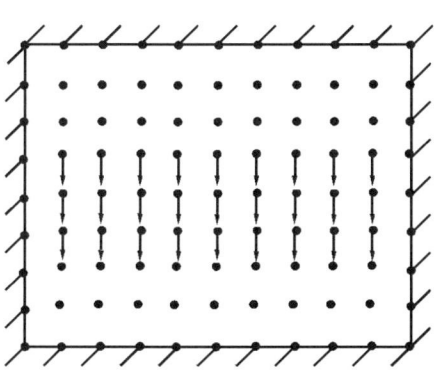

Figure 7. A finite difference problem.

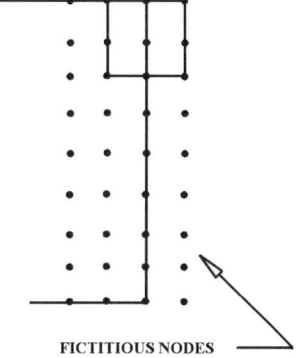

FICTITIOUS NODES

Figure 8. Central difference templates on the boundary.

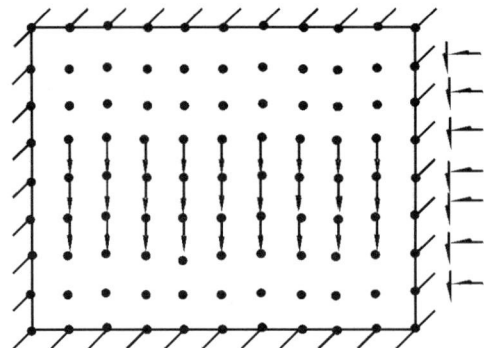

Figure 9. Finite difference model with one free boundary.

strains on the fixed boundary because the displacement results for the fixed case are correct. The finite difference model for the free boundary is developed and validated in the next lesson.

Closure

This lesson has developed the finite difference method in terms of strain gradient notation with the case of the plane stress problem with fixed boundaries. The finite difference method, as developed here, is based on the use of the [Φ] matrix. Since the finite element method presented earlier was developed in terms of the [Φ] matrix, the two approximate solution techniques are seen to have the same basis.

Both methods find displacement approximations at a finite number of points caused by applied loads. The finite difference method approximates the solution to the governing differential equations that provide the necessary conditions for minimizing the potential energy functional. The finite element method finds an approximate solution that directly tries to minimize the potential energy functional.

Although the two methods find the approximate displacements by different approaches, the heart of both methods can be seen to be related when they both are formulated with strain gradient notation. This is seen best when we compare the nine-node finite element to the nine-node central difference template. Both approaches require the formulation of the same displacement approximation in terms of strain gradient quantities. That is to say, the [Φ] matrix used in both approaches is the same. The inverse of this matrix plays a central role in both methods.

A further similarity is seen in the mesh or nodal point pattern that is used to define the structure of both problems. The nodal point patterns are identical. In the problems presented in this lesson, the meshes were evenly spaced. However, as shown in Appendix A, finite difference operators can be formed for templates with unevenly spaced nodes.

This similarity is exploited in the next part of the book. The error analysis procedures developed in the previous section are improved as a result of using the finite difference developments presented in this section. These improvements include pointwise error measures and alternate procedures for extracting "super-convergent" strains and stresses.

Notes

1. The procedure for identifying the polynomials represented by a finite difference template is identical to the process for identifying the constituents of the [Φ] matrices for

the formulation of a strain gradient based finite element. The finite difference operators are found by inverting the [Φ] matrix. The [Φ] matrix for a nine-node finite element is identical to that being formed here for the finite difference operator. At this level, the two methods are identical. The development is reproduced here for the benefit of the finite element practitioner who is not yet familiar with the finite difference method.

2. The standard procedure for forming finite difference operators is largely an *ad hoc* process. That is to say, there is no standard process for forming the desired result. Taylor series expansions are formed for displacement expressions at and "around" a local origin. These expressions are added and subtracted to get a derivative expression. The remainder terms are inspected to decide if the expression is of the correct order. If the remainder terms are of too low of an order, more terms are added and subtracted. The procedure presented here formalizes the procedure for forming finite difference operators.

Appendix A
Finite Difference Operator Development

This appendix presents the detailed development of finite difference operators for nine-node central difference templates using strain gradient notation. Finite difference operators for templates with and without evenly spaced nodes, as shown in Fig. A.1, are developed. The operators for the evenly spaced nodes are identical to those developed using the usual procedure for forming finite difference operators (see Note 2 in main text). One of the primary values of using strain gradient notation is the ability it affords for routinely forming the finite difference operators for templates with unevenly spaced nodes. This capability is demonstrated after the finite difference operators for the templates with evenly spaced nodes are formed.

Displacement Polynomials

For the reader who is not concerned with the full development of strain gradient notation and is interested only in the procedure for formulating the finite difference operators, the displacement polynomials used in the development are reproduced here for convenience. The terms contained in the polynomials are identified by comparing the finite difference template to Pascal's triangle in the main text of this lesson and in Fig. 18b of Lesson 6.

The displacement polynomials with arbitrary coefficients used to generate the finite difference operators for nine-node templates have the following form:

General Polynomial Representation

$$u = a_1 + a_2x + a_3y + a_4x^2 + a_5xy + a_6y^2 + a_7x^2y + a_8xy^2 + a_9x^2y^2$$

$$v = b_1 + b_2x + b_3y + b_4x^2 + b_5xy + b_6y^2 + b_7x^2y + b_8xy^2 + b_9x^2y^2$$

(A.1)

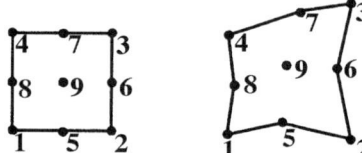

Figure A.1. Nine-node finite difference templates: (left) evenly spaced and (right) unevenly spaced.

When the polynomials are expressed in strain gradient notation, the result is

---

Strain Gradient Polynomial Representation

$$u = (u_{rb})_0 + (\varepsilon_x)_0 x + (\gamma_{xy}/2 - r_{rb})_0 y + 1/2(\varepsilon_{x,x})_0 x^2 + (\varepsilon_{x,y})_0 xy$$

$$+ 1/2(\gamma_{xy,y} - \varepsilon_{y,x})_0 y^2 + 1/2(\varepsilon_{x,xy})_0 x^2 y + 1/2(\varepsilon_{x,yy})_0 xy^2 + 1/4(\varepsilon_{x,xxy})_0 x^2 y^2$$

$$v = (v_{rb})_0 + (\gamma_{xy}/2 + r_{rb})_0 x + (\varepsilon_y)_0 y + 1/2(\gamma_{xy,x} - \varepsilon_{x,y})_0 x^2 + (\varepsilon_{y,x})_0 xy$$

$$+ 1/2(\varepsilon_{y,y})_0 y^2 + 1/2(\varepsilon_{y,xx})_0 x^2 y + 1/2(\varepsilon_{y,xy})_0 xy^2 + 1/4(\varepsilon_{y,xxy})_0 x^2 y^2$$

(A.2)

The polynomials contained in Eq. A.2 are expressed in an alternative form in Table A.1. The coefficients of the displacement approximation for u are given in the third column. These coefficients multiply the algebraic quantities given in column 2. The coefficients for v are given in the fourth column. These displacement approximations represent the movement of node i as linear combinations of strain gradient quantities.

Coordinate Transformation

The polynomials just presented provide the basis for forming the transformation from nodal coordinates to strain gradient quantities. This is best seen when we write Eq. A.2 and, consequently, Table A.1 with the strain gradient quantities as the independent variables to give

---

Nodal Displacement/Strain Gradient Relationship

$$\{d_i\} = [A]_i \{\varepsilon,\}$$

where

$$\{d_i\} = \begin{Bmatrix} u_i \\ v_i \end{Bmatrix} \qquad [A]_i = \begin{bmatrix} A_u \\ A_v \end{bmatrix}_i$$

(A.3)

Table A.1. The 18 coefficients for the nine-node finite difference template.

i	Term	a_i for $u(x,y)$	b_i for $v(x,y)$
1	1	$(u_{rb})_0$	$(v_{rb})_0$
2	x	$(\varepsilon_x)_0$	$(\gamma_{xy}/2 + r_{rb})_0$
3	y	$(\gamma_{xy}/2 - r_{rb})_0$	$(\varepsilon_y)_0$
4	x^2	$(\varepsilon_{x,x}/2)_0$	$[(\gamma_{xy,x} - \varepsilon_{x,y})/2]_0$
5	xy	$(\varepsilon_{x,y})_0$	$(\varepsilon_{y,x})_0$
6	y^2	$[(\gamma_{xy,y} - \varepsilon_{y,x})/2]_0$	$(\varepsilon_{y,y}/2)_0$
7	$x^2 y$	$(\varepsilon_{x,xy}/2)_0$	$(\varepsilon_{y,xx}/2)_0$
8	xy^2	$(\varepsilon_{x,yy}/2)_0$	$(\varepsilon_{y,xy}/2)_0$
9	$x^2 y^2$	$(\varepsilon_{x,xxy}/4)_0$	$(\varepsilon_{y,xxy}/4)_0$

Equation A.3 contains the polynomial representation of the u and v displacements at a given nodal point in terms of strain gradient quantities. The composition of the row vectors designated as $[A_u]_i$ and $[A_v]_i$ and the column vector given as $\{\varepsilon_{,}\}$ are given in Table A.2. Tables A.1 and A.2 both represent the displacement polynomials in the x and y directions. The only difference is that in Table A.2 the strain gradient quantities are written as the independent variables. In Table A.1, these same quantities are written as the coefficients of the terms in x and y.

We now use Eq. A.3 and Table A.2 to form the finite difference operators for the template with the evenly spaced nodes, as shown in Fig. A.2. This is accomplished by substituting the nodal locations of the nine nodes, one at a time, into Eq. A.3 to form a transformation from nodal coordinates to strain gradient coordinates. In the case of node 1, for example, the x and y coordinates are given as $x_1 = -2.50$ and $y_1 = -2.50$, respectively. When we combine the expressions for all nine nodes, the result is given as

---

Nodal Displacement/Strain Gradient Transformation

$$\{d\} = [\Phi]\{\varepsilon_{,}\}$$

where $\qquad\qquad\qquad\qquad\qquad\qquad\qquad\qquad\qquad\qquad\qquad\qquad\qquad$ (A.4)

$$\{d\}^T = \{\, u_1 \quad v_1 \quad u_2 \quad v_2 \quad \cdots \quad u_9 \quad v_9 \,\}$$
$$[\Phi]^T = [\, [A]_1^T \quad [A]_2^T \quad \cdots \quad [A]_9^T \,]$$

The $[\Phi]$ matrix for the evenly spaced template with a spacing of 2.50 units and with the local origin at the central node, i.e., node 9, is

---

Table A.2. The coefficients for the nine-node finite difference template.

i	$\{\varepsilon_{,}\}$	$\{A_u\}_i^T$	$\{A_v\}_i^T$
1	$(u_{rb})_0$	1	0
2	$(v_{rb})_0$	0	1
3	$(r_{rb})_0$	$-y_i$	x_i
4	$(\varepsilon_x)_0$	x_i	0
5	$(\varepsilon_y)_0$	0	y_i
6	$(\gamma_{xy})_0$	$y_i/2$	$x_i/2$
7	$(\varepsilon_{x,x})_0$	$x_i^2/2$	0
8	$(\varepsilon_{y,x})_0$	$-y_i^2/2$	$x_i y_i/2$
9	$(\gamma_{xy,x})_0$	0	$x_i^2/2$
10	$(\varepsilon_{x,y})_0$	$x_i y_i$	$-x_i^2/2$
11	$(\varepsilon_{y,y})_0$	0	$y_i^2/2$
12	$(\gamma_{xy,y})_0$	$y_i^2/2$	0
13	$(\varepsilon_{x,xy})_0$	$x_i^2 y_i/2$	0
14	$(\varepsilon_{x,yy})_0$	$x_i y_i^2/2$	0
15	$(\varepsilon_{y,xy})_0$	0	$x_i y_i^2/2$
16	$(\varepsilon_{y,xx})_0$	0	$x_i^2 y_i/2$
17	$(\varepsilon_{x,xyy})_0$	$x_i^2 y_i^2/4$	0
18	$(\varepsilon_{y,xxy})_0$	0	$x_i^2 y_i^2/4$

$[\Phi]$ for $h = 2.50$

	1	2	3	4	5	6	7	8	9	10	11	12	13	14	15	16	17	18
1	1.000	0.000	2.500	−2.500	0.000	−1.250	3.125	−3.125	0.000	6.250	0.000	3.125	0.000	−7.813	0.000	−7.813	0.000	9.766
2	0.000	1.000	−2.500	0.000	−2.500	−1.250	0.000	6.250	3.125	−3.125	3.125	0.000	7.813	0.000	7.813	0.000	9.766	0.000
3	1.000	0.000	2.500	2.500	0.000	−1.250	3.125	−3.125	0.000	−6.250	0.000	3.125	0.000	−7.813	0.000	7.813	0.000	9.766
4	0.000	1.000	2.500	0.000	−2.500	1.250	0.000	6.250	3.125	−3.125	3.125	0.000	−7.813	0.000	7.813	0.000	9.766	0.000
5	1.000	0.000	−2.500	2.500	0.000	1.250	3.125	−3.125	0.000	6.250	0.000	3.125	0.000	7.813	0.000	7.813	0.000	9.766
6	0.000	1.000	2.500	0.000	2.500	1.250	0.000	6.250	3.125	−3.125	3.125	0.000	7.813	0.000	7.813	0.000	9.766	0.000
7	1.000	0.000	−2.500	−2.500	0.000	1.250	3.125	−3.125	0.000	−6.250	0.000	3.125	0.000	7.813	0.000	−7.813	0.000	9.766
8	0.000	1.000	−2.500	0.000	2.500	−1.250	0.000	−6.250	3.125	−3.125	3.125	0.000	7.813	0.000	−7.813	0.000	9.766	0.000
9	1.000	0.000	2.500	0.000	0.000	−1.250	0.000	3.125	0.000	0.000	0.000	3.125	0.000	0.000	0.000	0.000	0.000	0.000
10	0.000	1.000	0.000	0.000	−2.500	0.000	0.000	0.000	0.000	0.000	3.125	0.000	0.000	0.000	0.000	0.000	0.000	0.000
11	1.000	0.000	0.000	2.500	0.000	0.000	3.125	0.000	0.000	0.000	0.000	0.000	0.000	0.000	0.000	0.000	0.000	0.000
12	0.000	1.000	2.500	0.000	0.000	1.250	0.000	0.000	3.125	−3.125	0.000	0.000	0.000	0.000	0.000	0.000	0.000	0.000
13	1.000	0.000	−2.500	0.000	0.000	1.250	0.000	−3.125	0.000	0.000	0.000	3.125	0.000	0.000	0.000	0.000	0.000	0.000
14	0.000	1.000	0.000	0.000	2.500	0.000	0.000	0.000	0.000	0.000	3.125	0.000	0.000	0.000	0.000	0.000	0.000	0.000
15	1.000	0.000	0.000	−2.500	0.000	0.000	3.125	0.000	0.000	0.000	0.000	0.000	0.000	0.000	0.000	0.000	0.000	0.000
16	0.000	1.000	−2.500	0.000	0.000	−1.250	0.000	0.000	3.125	−3.125	0.000	0.000	0.000	0.000	0.000	0.000	0.000	0.000
17	1.000	0.000	0.000	0.000	0.000	0.000	0.000	0.000	0.000	0.000	0.000	0.000	0.000	0.000	0.000	0.000	0.000	0.000
18	0.000	1.000	0.000	0.000	0.000	0.000	0.000	0.000	0.000	0.000	0.000	0.000	0.000	0.000	0.000	0.000	0.000	0.000

The alternate rows of this matrix are defined by the vectors $\{A_u\}_i$ and $\{A_v\}_i$, given in Table A.2. The columns of this matrix have the following interpretation. The first three columns are the nodal displacements for the rigid body motions, u_{rb}, v_{rb}, and r_{rb}. The next three columns are the nodal displacements associated with the constant strain states, ε_x, ε_y, and γ_{xy}. The remaining columns match the strain gradient quantities contained in the remainder of column 1 of Table A.2.

Finite Difference Operators

The finite difference operators for nine-node templates are found by inverting the $[\Phi]$ matrix defined in Eq. A.4 to give

Finite Difference Operators

$$\{\varepsilon,\} = [\Phi]^{-1}\{d\} \tag{A.5}$$

For the case of the template with the evenly spaced nodal points being evaluated, the finite difference operators are found by inverting the $[\Phi]$ matrix where $h = 2.50$, given earlier in numerical form, to give

$[\Phi]^{-1}$ for $h = 2.50$

	u_1	v_1	u_2	v_2	u_3	v_3	u_4	v_4	u_5	v_5	u_6	v_6	u_7	v_7	u_8	v_8	u_9	v_9
1	0.000	0.000	0.000	0.000	0.000	0.000	0.000	0.000	0.000	0.000	0.000	0.000	0.000	0.000	0.000	0.000	1.000	0.000
2	0.000	0.000	0.000	0.000	0.000	0.000	0.000	0.000	0.000	0.000	0.000	0.000	0.000	0.000	0.000	0.000	0.000	1.000
3	0.000	0.000	0.000	0.000	0.000	0.000	0.000	0.000	0.100	0.000	0.000	0.100	−0.100	0.000	0.000	−0.100	0.000	0.000
4	0.000	0.000	0.000	0.000	0.000	0.000	0.000	0.000	0.000	0.000	0.200	0.000	0.000	0.000	−0.200	0.000	0.000	0.000
5	0.000	0.000	0.000	0.000	0.000	0.000	0.000	0.000	0.000	−0.200	0.000	0.000	0.000	0.200	0.000	0.000	0.000	0.000
6	0.000	0.000	0.000	0.000	0.000	0.000	0.000	0.000	−0.200	0.000	0.000	0.200	0.200	0.000	0.000	−0.200	0.000	0.000
7	0.000	0.000	0.000	0.000	0.000	0.000	0.000	0.000	0.000	0.000	0.160	0.000	0.000	0.000	0.160	0.000	−0.320	0.000
8	0.000	0.040	0.000	−0.040	0.000	0.040	0.000	−0.040	0.000	0.000	0.000	0.000	0.000	0.000	0.000	0.000	0.000	0.000
9	0.040	0.000	−0.040	0.000	0.040	0.000	−0.040	0.000	0.000	0.000	0.000	0.160	0.000	0.000	0.000	0.160	0.000	−0.320
10	0.040	0.000	−0.040	0.000	0.040	0.000	−0.040	0.000	0.000	0.000	0.000	0.000	0.000	0.000	0.000	0.000	0.000	0.000
11	0.000	0.000	0.000	0.000	0.000	0.000	0.000	0.000	0.000	0.160	0.000	0.000	0.000	0.160	0.000	0.000	0.000	−0.320
12	0.000	0.040	0.000	−0.040	0.000	0.040	0.000	−0.040	0.160	0.000	0.000	0.000	0.160	0.000	0.000	0.000	−0.320	0.000
13	−0.032	0.000	−0.032	0.000	0.032	0.000	0.032	0.000	0.064	0.000	0.000	0.000	−0.064	0.000	0.000	0.000	0.000	0.000
14	−0.032	0.000	0.032	0.000	0.032	0.000	−0.032	0.000	0.000	0.000	−0.064	0.000	0.000	0.000	0.064	0.000	0.000	0.000
15	0.000	−0.032	0.000	0.032	0.000	0.032	0.000	−0.032	0.000	0.000	0.000	−0.064	0.000	0.000	0.000	0.064	0.000	0.000
16	0.000	−0.032	0.000	−0.032	0.000	0.032	0.000	0.032	0.000	0.064	0.000	0.000	0.000	−0.064	0.000	0.000	0.000	0.000
17	0.026	0.000	0.026	0.000	0.026	0.000	0.026	0.000	−0.051	0.000	−0.051	0.000	−0.051	0.000	−0.051	0.000	0.102	0.000
18	0.000	0.026	0.000	0.026	0.000	0.026	0.000	0.026	0.000	−0.051	0.000	−0.051	0.000	−0.051	0.000	−0.051	0.000	0.102

The finite difference operators are contained in the rows of the $[\Phi]^{-1}$ matrix. For example, the finite difference operators for $\varepsilon_{x,x}$ are contained in row 7 of the $[\Phi]^{-1}$ matrix as

---

Finite Difference Operator for $\varepsilon_{\mathbf{x,x}}$

$$\varepsilon_{x,x} = 0.160u_6 - 0.320u_9 + 0.160u_8$$

$$= \frac{1}{h^2}(u_6 - 2u_9 + u_8)$$

(A.6)

The first row of Eq. A.6 is the derivative operator for the template with a spacing of $h = 2.50$. The second expression is the general definition of this operator for an evenly spaced template.

The following six differential operators are contained in the governing differential equation for the plane stress problem: $\varepsilon_{x,x}$, $\varepsilon_{y,y}$, $\varepsilon_{y,x}$, $\varepsilon_{x,y}$, $\gamma_{xy,x}$, and $\gamma_{xy,y}$. The finite difference approximations of these derivatives are contained in rows 7, 11, 8, 10, 9, and 12 of the $[\Phi]^{-1}$ matrix, respectively. When we extract these operators for the case of the evenly spaced template, they are given as

---

Finite Difference Operators for $h = 2.50$

$$\varepsilon_{x,x} = 0.160u_6 + 0.160u_8 - 0.320u_9$$
$$\varepsilon_{y,y} = 0.160v_5 + 0.160v_7 - 0.320v_9$$
$$\varepsilon_{y,x} = 0.040v_1 - 0.040v_2 + 0.040v_3 - 0.040v_4$$
$$\varepsilon_{x,y} = 0.040u_1 - 0.040u_2 + 0.040u_3 - 0.040u_4$$
$$\gamma_{xy,x} = 0.040u_1 - 0.040u_2 + 0.040u_3 - 0.040u_4 + 0.160v_6 + 0.160v_8 - 0.320v_9$$
$$\gamma_{xy,y} = 0.040v_1 - 0.040v_2 + 0.040v_3 - 0.040v_4 + 0.160u_5 + 0.160u_7 - 0.320u_9$$

(A.7)

Note that all 18 of the displacements associated with the nine-node template are contained in the six derivative approximations shown in Eq. A.7. This means that all nine nodes are required to form the finite difference operators contained in this representation of the plane stress problem.

The finite difference operators for the general case of a template with evenly spaced nodes are given as

General Finite Difference Operators for Evenly Spaced Nodes

$$\varepsilon_{x,x} = \frac{1}{h^2}(u_6 + u_8 - 2u_9)$$

$$\varepsilon_{y,y} = \frac{1}{h^2}(v_5 + v_7 - 2u_9)$$

$$\varepsilon_{y,x} = \frac{1}{4h^2}(v_1 - v_2 + v_3 - v_4)$$

$$\varepsilon_{x,y} = \frac{1}{4h^2}(u_1 - u_2 + u_3 - u_4)$$

$$\gamma_{xy,x} = \frac{1}{4h^2}(u_1 - u_2 + u_3 - u_4) + \frac{1}{h^2}(v_6 + v_8 - 2v_9)$$

$$\gamma_{xy,y} = \frac{1}{4h^2}(v_1 - v_2 + v_3 - v_4) + \frac{1}{h^2}(u_5 + u_7 - 2u_9)$$

(A.8)

As shown in the main text of this lesson, these six finite difference operators are combined to represent the governing differential equations at each of the nodes in the finite difference model.

Strain Approximations

In elasticity problems, the displacements are the quantities sought in the solution process. However, these are not the unknowns of primary interest in most elasticity applications. The strains and, by extension, the stresses are usually of primary interest. The three strains are approximated by the following finite difference operators for the evenly spaced nine-node template as

Finite Difference Strain Operators

$$\varepsilon_x = \frac{1}{2h}(u_6 - u_8)$$

$$\varepsilon_y = \frac{1}{2h}(-v_5 + v_7)$$

$$\gamma_{xy} = \frac{1}{2h}(-v_5 + v_6 + v_7 - v_8)$$

(A.9)

These operators are used in the main text to evaluate the stresses on the fixed boundary of the problem solved as an example.

Finite Difference Operators for Unevenly Spaced Templates

To further demonstrate the efficacy of strain gradient notation, the finite difference operators for a template with unevenly spaced nodes are now presented. A template with unevenly spaced nodes is shown in Fig. A.2. The finite difference operators are found by inverting the [Φ] matrix for this template to give

$[\Phi]^{-1}$ for an Unevenly Spaced Mesh

	u_1	v_1	u_2	v_2	u_3	v_3	u_4	v_4	u_5	v_5	u_6	v_6	u_7	v_7	u_8	v_8	u_9	v_9
1	0.000	0.000	0.000	0.000	0.000	0.000	0.000	0.000	0.000	0.000	0.000	0.000	0.000	0.000	0.000	0.000	0.100	0.000
2	0.000	0.000	0.000	0.000	0.000	0.000	0.000	0.000	0.000	0.000	0.000	0.000	0.000	0.000	0.000	0.000	0.000	0.100
3	−0.020	−0.005	0.048	−0.036	0.002	0.028	0.000	0.000	0.051	0.012	−0.010	0.248	−0.147	−0.002	0.000	−0.028	0.076	−0.217
4	−0.011	0.000	−0.071	0.000	0.057	0.000	0.000	0.000	0.023	0.000	0.496	0.000	−0.004	0.000	−0.056	0.000	−0.434	0.000
5	0.000	0.039	0.000	−0.095	0.000	−0.004	0.000	0.001	0.000	−0.103	0.000	0.021	0.000	0.295	0.000	−0.001	0.000	−0.152
6	0.039	−0.011	−0.095	−0.071	−0.004	0.057	0.001	0.000	−0.103	0.023	0.021	0.496	0.295	−0.004	−0.001	−0.056	−0.152	−0.434
7	−0.005	0.000	−0.032	0.000	0.025	0.000	0.000	0.000	0.010	0.000	0.220	0.000	−0.002	0.000	0.074	0.000	−0.292	0.000
8	0.000	−0.020	0.000	−0.062	0.000	0.126	0.000	−0.038	0.000	0.071	0.000	−0.072	0.000	−0.118	0.000	0.018	0.000	0.094
9	−0.020	−0.005	−0.062	−0.032	0.126	0.025	−0.038	0.000	0.071	0.010	−0.072	0.220	−0.118	−0.002	0.018	0.074	0.094	−0.292
10	−0.020	0.000	−0.062	0.000	0.126	0.000	−0.038	0.000	0.071	0.000	−0.072	0.000	−0.118	0.000	0.018	0.000	0.094	0.000
11	0.000	−0.039	0.000	0.095	0.000	0.004	0.000	−0.001	0.000	0.103	0.000	−0.021	0.000	0.205	0.000	0.001	0.000	−0.348
12	−0.039	−0.020	0.095	−0.062	0.004	0.126	−0.001	−0.038	0.103	0.071	−0.021	−0.072	0.205	−0.118	0.001	0.018	−0.348	0.094
13	−0.036	0.000	−0.023	0.000	0.075	0.000	0.051	0.000	0.058	0.000	−0.023	0.000	−0.116	0.000	−0.025	0.000	0.039	0.000
14	0.023	0.000	0.099	0.000	0.078	0.000	−0.026	0.000	−0.083	0.000	−0.185	0.000	−0.071	0.000	0.025	0.000	0.140	0.000
15	0.000	0.023	0.000	0.099	0.000	0.078	0.000	−0.026	0.000	−0.083	0.000	−0.185	0.000	−0.071	0.000	0.025	0.000	0.140
16	0.000	−0.036	0.000	−0.023	0.000	0.075	0.000	0.051	0.000	0.058	0.000	−0.023	0.000	−0.116	0.000	0.025	0.000	0.039
17	0.043	0.000	0.037	0.000	0.047	0.000	0.034	0.000	−0.063	0.000	−0.075	0.000	−0.074	0.000	−0.033	0.000	0.084	0.000
18	0.000	0.043	0.000	0.037	0.000	0.047	0.000	0.034	0.000	−0.063	0.000	−0.075	0.000	−0.074	0.000	−0.033	0.000	0.084

The finite difference operators are contained in the rows of the $[\Phi]^{-1}$ matrix. For example, the finite difference operators for $\varepsilon_{x,x}$ is contained in row 7 of the $[\Phi]^{-1}$ matrix as

---

Finite Difference Operator for $\varepsilon_{x,x}$

$$\varepsilon_{x,x} = 0.220u_6 - 0.292u_9 + 0.074u_8$$

$$- 0.005u_1 - 0.032u_2 + 0.025u_3 \qquad \text{(A.10)}$$

$$+ 0.000u_4 + 0.100u_5 - 0.002u_7$$

When we compare Eq. A.10 to Eq. A.6, we can see that the operator for the unevenly spaced nodes has the same basic form as the operator for the evenly spaced case but additional terms are contained in the expression. That is to say, the u_6, u_9, and u_8 terms dominate the operator, but in the unevenly spaced templates other terms modify the approximation.

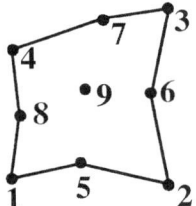

Figure A.2. An unevenly spaced template.

Closure

In one sense, we can view this appendix as redundant. The procedure for forming the $[\Phi]$ matrix is presented in other lessons. The procedure is originally defined in Lesson 6 for trusses and used several times in the formulation of finite element stiffness matrices. The procedure for forming the finite difference operators could have been covered with one sentence, namely, *The finite difference operators are contained in the inverse of the $[\Phi]$ matrix for the finite elements with the same nodal configurations.*

The reason for presenting this third context for the formulation of the $[\Phi]$ matrices is to provide slightly different interpretations of the meaning of the $[\Phi]$ matrix and of strain gradient notation. In this way, the connection between the physical systems being modeled and the finite difference and finite element representations is hopefully made clear. Simultaneously, these developments are designed to show the close connection between the finite difference and the finite element methods.

Appendix B

Finite Difference Model Formulation

The finite difference model of a simple plane stress problem with fixed boundaries is developed in this appendix. This presentation expands on the formulation procedure discussed in the main text. The purpose of this presentation is to demonstrate the row-by-row nature and the simplicity of the finite difference assembly process and the characteristics of the finite difference coefficient matrix.

The formulation of a finite difference model consists of the following four steps:

1. Generation of equation templates in local coordinates
2. Transformation of equation templates to global coordinates
3. Introduction of boundary conditions
4. Assembly of global coefficient matrix

After we complete the assembly process and solve the primary problem, we demonstrate procedures for extracting strains in the interior and on the boundary. The extraction of the interior strains consists of a simple substitution of known nodal displacements into the finite difference strain operators. The determination of the boundary strains requires the superposition of a nine-node template on the boundary, which introduces fictitious nodes into the problem. The auxiliary calculations required to compute the boundary strains defines the use and interpretation of the fictitious nodes.

Problem Definition

The plane stress problem to be solved is a rectangular region with fixed boundaries as shown in Fig. B.1. The finite difference model of this problem consists of two finite difference templates with even spacing. We should not expect that this simple representation will produce accurate results. This small model is designed to provide a simple introduction to the detailed formulation of a finite difference coefficient matrix.

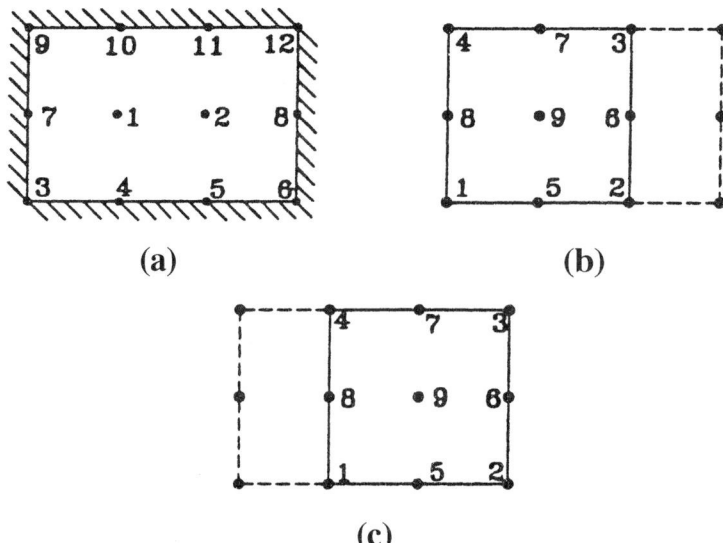

Figure B.1. A finite difference problem.

The nodal points of the finite difference model are superimposed on Fig. B.1a. The nodal numbering in Figs. B.1b and B.1c reflects the numbering of the local system. The numbering of the nodes reflects the *a priori* knowledge that only nodes 1 and 2 are unconstrained. The displacements at nodes 3–12 are known to be zero because they are located on the fixed boundary and will ultimately be eliminated from the analysis. The finite difference templates 1 and 2 are highlighted in Figs. 1b and 1c, respectively.

Equation Template Formulation

The finite difference equation templates that approximate the governing differential equations of the plane stress problem are formed by replacing the derivative operators in the differential equations with finite difference operators. This was done explicitly in the main text for the plane stress equilibrium equation in the x direction (Eq. 13.10) to give:

Formulation of an Equation Template

Differential Representation	Finite Difference Representation
$\varepsilon_{x,x}$	$\dfrac{1}{h^2}(u_6 + u_8 - 2u_9)$
$+\nu\varepsilon_{y,x}$	$+\nu\dfrac{1}{4h^2}(v_1 - v_2 + v_3 - v_4)$
$+\left(\dfrac{1-\nu}{2}\right)\gamma_{xy,y}$	$+\left(\dfrac{1-\nu}{2}\right)\left[\dfrac{1}{4h^2}(v_1 - v_2 + v_3 - v_4) + \dfrac{1}{h^2}(u_5 + u_7 - 2u_9)\right]$
$=-\left(\dfrac{1-\nu^2}{E}\right)P_x$	$=-\left(\dfrac{1-\nu^2}{E}\right)P_x$

$$(B.1)$$

The left-hand side of Eq. B.1 is the differential equation being modeled. The right-hand side is the finite difference representation, i.e., the derivatives have been replaced with finite difference operators. As we can see, the right-hand side consists of the physical parameters of the problem, the spacing of the nodes in the finite difference template and the nodal displacements of the finite difference template. The finite difference method replaces a differential equation with an algebraic equation.

For the problem solved here, Poisson's ratio, ν, is taken as 0.3 and the template spacing, h, and Young's Modulus, E, are kept as parameters. When the value of Poisson's ratio is introduced into Eq. B.1 and the equation is multiplied by $4h^2$ to remove the denominator, the equation template for this equilibrium equation becomes

---

Equation Template in the x Direction

$$1.4u_5 + 4.0u_6 + 1.4u_7 + 4.0u_8 - 10.8u_9$$

$$+ 0.65v_1 - 0.65v_2 + 0.65v_3 - 0.65v_4$$

$$= -3.64h^2/EP_x$$

$$(B.2)$$

This equation clearly shows the structure of a finite difference representation of a differential equation. The finite difference representation of this single differential equation is a single algebraic equation in the form of a row vector. The value of this structure is demonstrated in a later section of this appendix.

For completeness and clarity in the subsequent assembly of the finite difference coefficient matrix, the equation template for the plane stress equilibrium equation in the y direction is now formed. The general equation template for this equation is

---

Formulation of an Equation Template

Differential Representation	Finite Difference Representation
$\varepsilon_{y,y}$	$\dfrac{1}{h^2}(v_5 + v_7 - 2v_9)$
$+\nu\varepsilon_{x,y}$	$+\nu\dfrac{1}{4h^2}(u_1 - u_2 + u_3 - u_4)$
$+\left(\dfrac{1-\nu}{2}\right)\gamma_{xy,x}$	$+\left(\dfrac{1-\nu}{2}\right)\left[\dfrac{1}{4h^2}(u_1 - u_2 + u_3 - u_4) + \dfrac{1}{h^2}(v_6 + v_8 - 2v_9)\right]$
$= -\left(\dfrac{1-\nu^2}{E}\right)P_y$	$= -\left(\dfrac{1-\nu^2}{E}\right)P_y$

$$\text{(B.3)}$$

When we simplify the finite difference representation given in the right-hand side of Eq. B.3 by specifying ν as 0.3 and multiplying by $4h^2$, the result is

---

Equation Template in the y Direction

$$4.0v_5 + 1.4v_6 + 4.0v_7 + 1.4v_8 - 10.8v_9$$

$$+ 0.65u_1 - 0.65u_2 + 0.65u_3 - 0.65u_4 \qquad \text{(B.4)}$$

$$= -3.64h^2/EP_y$$

Equations B.2 and B.4 are represented by Eq. 13.13 in the main text of this lesson. When we combine Eqs. B.2 and B.4, the result is

---

Equation Template for a Node

$$[\Omega]\{d\} = \{f\} \qquad \text{(B.5)}$$

The transpose of the coefficient matrix $[\Omega]$ of Eq. B.5 is given in Table B.1 as are the components of the displacement vector $\{d\}$ expressed as displacements in the local coordinate system. The transpose of the applied force vector, $\{f\}$, is given as $[(-4h^2((1-\nu^2)/E)f_x) \ (-4h^2((1-\nu^2)/E)f_y)]$. This equation is the explicit representation of Eq. 13.14 from the main text for this problem.

Table B.1. Constituents of the general equation template in local coordinates.

	$\{d\}$	Column 1 of $[\Omega]^T$	Column 2 of $[\Omega]^T$
1	u_1	0.0	0.65
2	v_1	0.65	0.0
3	u_2	0.0	-0.65
4	v_2	-0.65	0.0
5	u_3	0.0	0.65
6	v_3	0.65	0.0
7	u_4	0.0	-0.65
8	v_4	-0.65	0.0
9	u_5	1.40	0.0
10	v_5	0.0	4.0
11	u_6	4.00	0.0
12	v_6	0.0	1.40
13	u_7	1.40	0.0
14	v_7	0.0	4.00
15	u_8	4.00	0.0
16	v_8	0.0	1.40
17	u_9	-10.80	0.0
18	v_9	0.0	-10.80

Coordinate Transformation

Equation B.5 and Table B.1 present the equation template for any evenly spaced finite difference representation of the plane stress problem with $\nu = 0.3$. It applies to both templates 1 and 2, shown in Fig. B.1, because it is defined in terms of local coordinates.

The next step in the formulation process is to relate this generally applicable equation template to a specific node in the problem. This is accomplished by replacing the local coordinates in Table B.1 with global coordinates. The relationship between the local and global nodes is shown in Fig. B.1. Local node 1 of template 1 is the same as global node 3 etc. Local node 1 of template 2 is the same as global node 4 etc. In this problem, the global displacement numbers are directly related to the global node numbers. That is to say, the displacement in the x direction at node i is given as u_i and the displacement in the y direction at node i is denoted as v_i. For example, the displacements at node 11 are u_{11} and v_{11}.

When these transformations are substituted into the general equation template given by Eq. B.5 and Table B.1, the resulting specific equation templates are given in Tables B.2 and B.3.

This transformation to global coordinates introduces the differences in the locations of the origins of the two finite difference templates that make up the finite difference model given in Fig. B.1. The template centered at node 1 differs from the template associated with node 2 because the global coordinates are different. Tables B.2 and B.3 contain the four equations that constitute the finite difference representation of this problem. These four equations are available to solve for the four unknown displacements at the two unrestrained nodes. However, these four equations contain 24 unknown displacements. Thus, we have a system of four equations with 24 unknowns. We reduce this system of equations to a solvable form by introducing the boundary conditions.

Table B.2. Equation template for global node 1.

	$\{d\}$	Column 1 of $[\Omega]^T$	Column 2 of $[\Omega]^T$
1	u_3	0.0	0.65
2	v_3	0.65	0.0
3	u_5	0.0	-0.65
4	v_5	-0.65	0.0
5	u_{11}	0.0	0.65
6	v_{11}	0.65	0.0
7	u_9	0.0	-0.65
8	v_9	-0.65	0.0
9	u_4	1.40	0.0
10	v_4	0.0	4.0
11	u_2	4.00	0.0
12	v_2	0.0	1.40
13	u_{10}	1.40	0.0
14	v_{10}	0.0	4.00
15	u_7	4.00	0.0
16	v_7	0.0	1.40
17	u_1	-10.80	0.0
18	v_1	0.0	-10.80

Table B.3. Equation template for global node 2.

	$\{d\}$	Column 1 of $[\Omega]^T$	Column 2 of $[\Omega]^T$
1	u_4	0.0	0.65
2	v_4	0.65	0.0
3	u_6	0.0	-0.65
4	v_6	-0.65	0.0
5	u_{12}	0.0	0.65
6	v_{12}	0.65	0.0
7	u_{10}	0.0	-0.65
8	v_{10}	-0.65	0.0
9	u_5	1.40	0.0
10	v_5	0.0	4.0
11	u_8	4.00	0.0
12	v_8	0.0	1.40
13	u_{11}	1.40	0.0
14	v_{11}	0.0	4.00
15	u_1	4.00	0.0
16	v_1	0.0	1.40
17	u_2	-10.80	0.0
18	v_2	0.0	-10.80

Boundary Conditions and the Global Coefficient Matrix

The inclusion of the boundary conditions for this problem with its fixed boundaries is a simple and straightforward process. The displacements at the 10 nodes on the boundary are restrained to be zero. That is to say, the 20 displacements at nodes 3–12 are zero.

When these constraints are introduced into the equilibrium equations given in Tables B.2 and B.3 and the order of the nodal displacements is changed, the results are presented in Tables B.4 and B.5. These two tables contain the finite difference representations of the equilibrium equations at the two unrestrained nodes, respectively.

These four equations constitute the finite difference model for the problem shown in Fig. B.1. Table B.4 contains the two equations for node 1 and Table B.5 contains the two equations for node 2. We can assemble these two sets of two rows to give the finite difference representation as

Finite Difference Model

$$\begin{bmatrix} -10.80 & 0.00 & 4.00 & 0.00 \\ 0.00 & -10.80 & 0.00 & 1.40 \\ 4.00 & 0.00 & -10.80 & 0.00 \\ 0.00 & 1.40 & 0.00 & -10.80 \end{bmatrix} \begin{Bmatrix} u_1 \\ v_1 \\ u_2 \\ v_2 \end{Bmatrix} = -4h^2 \frac{1-\nu^2}{E} \begin{Bmatrix} (f_x)_1 \\ (f_y)_1 \\ (f_x)_2 \\ (f_y)_2 \end{Bmatrix} \qquad \text{(B.6)}$$

Equation B.6 contains the finite difference representation of the equilibrium equations at the two unrestrained nodes. The first two equations apply to the unrestrained node 1. The second two equations represent the equilibrium equations at node 2. This system of four equations and four unknowns enables us to determine the displacements of the unrestrained nodes.

Note that the large diagonal elements of the coefficient matrix and the load vectors are both negative. Thus, the displacements produced by this set of equations will be positive for a positive load vector. In this case, the coefficient matrix is symmetric. However, the coefficient matrices for finite difference models need not be symmetric. For example, if we reorder the nodal displacements at node 2 so that the displacement v_2 precedes u_2, the result is

Table B.4. Equation template with boundary conditions for global node 1.

	$\{d\}$	Column 1 of $[\Omega]^T$	Column 2 of $[\Omega]^T$
1	u_1	-10.80	0.0
2	v_1	0.0	-10.80
3	u_2	4.00	0.0
4	v_2	0.0	1.40

Table B.5. Equation template with boundary conditions for global node 2.

	$\{d\}$	Column 1 of $[\Omega]^T$	Column 2 of $[\Omega]^T$
1	u_1	4.00	0.0
2	v_1	0.0	1.40
3	u_2	-10.80	0.0
4	v_2	0.0	-10.80

Finite Difference Model

$$
\begin{bmatrix}
-10.80 & 0.00 & 0.00 & 4.00 \\
0.00 & -10.80 & 1.40 & 0.00 \\
4.00 & 0.00 & 0.00 & -10.80 \\
0.00 & 1.40 & -10.80 & 0.00
\end{bmatrix}
\begin{Bmatrix}
u_1 \\ v_1 \\ v_2 \\ u_2
\end{Bmatrix}
= -4h^2 \frac{1-\nu^2}{E}
\begin{Bmatrix}
(f_x)_1 \\ (f_y)_1 \\ (f_x)_2 \\ (f_y)_2
\end{Bmatrix}
\qquad (B.7)
$$

The form of the finite difference model contained in Eq. B.7 is presented as a warning. Many of the algorithms used to solve systems of algebraic equations are designed for use with positive definite, symmetric coefficient matrices. Finite difference models need not meet these conditions.

In this section, we have found the finite difference representation of the small problem shown in Fig. B.1. We have seen that the assembly process of a coefficient matrix is a row-by-row operation. Since there is no coupling between the rows, the assembly process is a simple operation to implement. As a result, the finite difference method is easier to program than the finite element method.

Strain Extraction

In most engineering applications, the displacements are not the primary unknowns that are sought. Most often, the strains and stresses are the quantities of interest. The strains at the interior nodes are available at this point in the analysis by simple substitution.

This procedure for extracting strains at interior nodes is now demonstrated for this problem. The finite difference operators for the strains in an evenly spaced nine node central difference template expressed in local coordinates are

Finite Difference Strain Operators

$$
\varepsilon_x = \frac{1}{2h}(u_6 - u_8)
$$

$$
\varepsilon_y = \frac{1}{2h}(-v_5 + v_7) \qquad (B.8)
$$

$$
\gamma_{xy} = \frac{1}{2h}(-v_5 + v_6 + v_7 - v_8)
$$

These expressions are reproduced from Appendix A for convenience. In the case of template 1, shown in Fig. B.1, all of the displacements are zero except those at global nodes 1 and 2. These nodes are related to the local nodes in Eq. B.8 as $(u_1)_g = (u_9)_l$, $(v_1)_g = (v_9)_l$, $(u_2)_g = (u_6)_l$, and $(v_2)_g = (v_6)_l$. This produces the following strain expressions in terms of global coordinates:

Strains at Node 1 (Global Coordinates)

$$
\varepsilon_x = \frac{1}{2h} u_2
$$

$$
\varepsilon_y = 0 \qquad (B.9)
$$

$$
\gamma_{xy} = 0
$$

Since the normal strain in the y direction and the shear strain at node 1 are zero for any loading condition, this model with only two internal nodes is not an adequate representation of this problem. However, the procedure for extracting strains that has been demonstrated is identical to that used in larger problems.

In engineering problems, the boundary strains and stresses are also of interest. We now demonstrate the procedure for extracting the boundary strains. We accomplish this by locating the origin of a nine node central difference template on a boundary node as shown in Fig. B.2. The three nodes located outside of the boundary of the problem are called fictitious nodes. They are called this because any displacements attributed them do not have any real physical meaning. The displacements at these points provide a mechanism for including the boundary conditions in a finite difference model.

The displacements of the five nodes on the boundary, nodes 1, 4, 5, 7 and 9, are zero by definition. The displacements at the local node 8 (global node 2) were found when Eq. B.8 was solved. In the main text, we demonstrated that the node designated with the x in Fig. B.2 could be taken as zero. This leaves the displacements at local node 6 to be found.

These displacements are found by forcing the governing differential equations at the origin of the central difference template shown in Fig. B.2 to be satisfied. When we substitute the nodal values just specified into Eq. B.5, the result is

---

Equation Template for the Boundary Node

$$4.0u_6 + 4.0u_8 = 0$$

$$v_6 = 0$$

(B.10)

In this set of equations, u_6 and v_6 are the displacements at the fictitious node. The displacement of the local coordinate u_8 corresponds to the global displacement u_2. Thus, the displacement of the fictitious coordinate u_6 is the negative of the global displacement u_2. The displacement of the fictitious coordinate v_6 is equal to zero. We are now in a position to compute the boundary strains. When we substitute these displacements into Eq. B.8, we find the strains at the boundary node to be

---

Strains at the Boundary Node (Global Coordinates)

$$\varepsilon_x = \frac{1}{2h}u_2$$

$$\varepsilon_y = 0$$

$$\gamma_{xy} = 0$$

(B.11)

We have developed an approach for computing the boundary strains. This required the introduction of the concept of fictitious nodes. The use of fictitious nodes is extended in the next lesson when the finite difference models of other boundary conditions are developed.

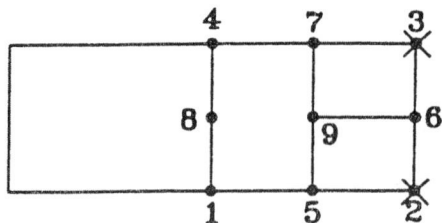

Figure B.2. A finite difference problem with a boundary template and its associated fictitious nodes.

Closure

This appendix has explicated many of the ideas presented and discussed in the main text of the lesson. In addition, some additional characteristics of the finite difference formulation were seen. The coefficient matrix of the finite difference representation need not be symmetric. When the matrix is not symmetric, we must be sure that the equation solver that we are using is designed for this case. Furthermore, the row-by-row formulation of the finite difference coefficient matrix was demonstrated. This is important because it enables us to easily write specialized programs for solving partial differential equations with the finite difference method.

The simplicity of the formulation of finite difference models partially explains why large finite difference programs that are maintained with great effort are not prevalent. The finite difference representations are simple enough that special application programs can be written by the user with little difficulty. In this book, the finite difference method is not used as a stand-alone solution technique. We use it to improve the error analysis and strain extraction procedures for the finite element method.

References and Other Readings

1. Dow, J. O., Jones, M. S., and Harwood, S. A. "A Generalized Finite Difference Method for Solid Mechanics," *International Journal of Numerical Methods for Partial Differential Equations*, Vol. 6, 1990, pp. 137–152.

2. Dow, J. O., Jones, M. S., and Harwood, S. A. "A New Approach to Boundary Modeling for Finite Difference Applications in Solid Mechanics," *International Journal for Numerical Methods in Engineering*, Vol. 30, 1990, pp. 99–113.

3. Dow, J. O., Harwood, S. A., Jones, M. S., and Stevenson, I. "Validation of a Finite Element Error Estimator," *AIAA Journal*, Vol. 29, No. 10, Oct. 1991, pp. 1736–1742.

4. Dow, J. O., and Stevenson, I. "An Adaptive Refinement Procedure for the Finite Difference Method," *International Journal of Numerical Methods for Partial Differential Equations*, Vol. 8, No. 6, Nov. 1992, pp. 537–550.

5. Dow, J. O., and Hardaway, J. L. "The Validation of Finite Difference Boundary Conditions Models for Solid Mechanics Applications," *AIAA Journal*, Vol. 30, No. 4, July 1992, pp. 1864–1869.

6. Harwood, S. A. "Finite Element Error Analysis Using the Finite Difference Method," Masters Degree Thesis, University of Colorado, 1989.

7. Stevenson, I. "A Generalized Adaptive Refinement Procedure for Finite Element and Finite Difference Analyses," Masters Degree Thesis, University of Colorado, 1990.

8. Hardaway, J. L. "Finite Difference Boundary Condition Modeling," Masters Degree Thesis, University of Colorado, 1991.

Lesson 13 Problems

1. Form a finite difference model of a uniformly loaded bar with both ends fixed. This problem is the one-dimensional analog of the two-dimensional problem that introduces Appendix B. The model should consist of five nodes on the bar itself. The displacements of the two end nodes will be zero because of the fixed constraints. This means that the problem will consist of three unknown displacements.

    a. Give the finite difference representation of the derivative contained in the governing differential equation of the bar.

    b. Form the equation template for the partial differential equation of the bar.

    c. Assemble the finite difference for the problem without boundary conditions.

    d. Introduce the boundary conditions.

    e. Find the nodal displacements.

    f. Find the strains at the nodes.

    g. Find the reactions at the two ends of the bar.

2. Comment on how the finite difference model would differ if the right-hand end where free. This boundary condition is discussed for the plane stress problem in detail in Lesson 14.

3. Find the boundary strains on the top edge of the problem in Appendix B.

Lesson 14
Finite Difference Boundary Condition Models

Purpose of Lesson   To present and validate the finite difference boundary condition models for straight edges, curved edges, and convex and concave corners that enable plane stress problems of any configuration to be represented.

In the previous lesson, an approach based on strain gradient notation for creating finite difference representations of plane stress models was introduced. The finite difference models used to illustrate the developments were rectangular regions with fixed boundaries based on nine-node central difference templates. In that lesson, the treatment of the fixed boundary was formalized and the procedure for extracting the stresses on fixed boundaries was introduced. These ideas are extended in this lesson to form the boundary condition models for unconstrained straight and curved edges. The specific cases considered include loaded and unloaded boundaries, reentrant (concave) and outside (convex) corners, and point loads.

As we saw in the previous lesson when the boundary stresses on a fixed boundary were computed, the use of a finite difference template on a boundary introduces fictitious nodes. The boundary models developed here focus first on the role of the fictitious nodes in these models. The fictitious nodes play the central role in these models. The place of the nine-node central difference template is important but secondary. In the development presented here, the fictitious nodes are integrated into the finite difference representation in a rational way so that they are a natural part of the finite difference model.

A point of view: As we see in the lesson, boundary conditions where singularities are possible, i.e., boundaries where there is a discontinuity in the slope, are awkward to represent using finite difference models. These models are discussed at length for the sake of completeness. If these boundary conditions were not represented, the reformulated finite difference method would be viewed as containing a deficiency. In practical reality, such "sharp corners" or singularities cannot exist. If such a boundary condition is found at a critical point in a model, the geometry is modified to a shape that does not produce a stress concentration, e.g., a sharp corner is replaced by a fillet. Such shapes are easily modeled. The error measures developed in Part V clearly identify such singularities.

■   ■   ■

In the previous lesson, finite difference models of rectangular plane stress regions with fixed boundaries were formed. The basic purpose of Lesson 13 was to develop the procedure for modeling the governing differential equations on the domain of the problem. The idea of fictitious nodes was introduced when the partial differential equations were approximated at nodes on the boundary as part of the process of computing boundary stresses. In this lesson, a generally applicable approach for incorporating the fictitious nodes into finite difference models is developed. Specifically, this means that loaded and unloaded models for straight edges, curved edges, outside or convex corners, and reentrant or concave corners are developed with the same representations. In this approach, every boundary condition is modeled in the same way. As a consequence of this generalization of boundary condition models, complex regions such as that shown in Fig. 1 can be routinely modeled. As we can see, this is a stylized outline of the state of Texas with a hole in it. This region has all types of boundary conditions: straight edges, curved edges, convex corners of various angles, concave corners of various angles, and an interior boundary.

Finite Difference Boundary Modeling Characteristics

As a prefatory comment, note that the finite difference boundary condition models presented here differ in one significant way from previous representations. Instead of having the locations of the fictitious nodes *exclusively* defined by the characteristics of the finite difference templates, the layouts of the finite difference templates are largely defined by the locations of the fictitious nodes. Then, if a nine-node template can be used in a straightforward manner, the template is modified to accommodate this configuration. This change in emphasis results in the uniformity in the treatment of the boundary conditions presented here. As a consequence of this change in emphasis, the central difference templates on the boundary are as likely as not to have unevenly spaced nodes and may have more than or less than nine nodes.

This uniformity of the treatment of boundary conditions is embodied in three premises that characterize the boundary condition models presented here. These three premises, which are discussed in the introduction of this part of the book, are the following: (1) there is at least one fictitious node per boundary node; (2) the finite difference templates must, as a minimum, represent rigid body motions, constant stresses, and the six linear variations of the strain components; and (3) the location of the template nodes should, if possible, be "topologically" symmetric.

The first premise defines the minimum number and approximate location of the fictitious nodes. The second premise defines the minimum number of nodes that must be contained in a template. Premise 3 biases the boundary models toward the use of nine-node central difference templates. These three criteria for defining the boundary templates provide a uniform approach for forming finite difference models that accurately represent boundary conditions in the problems being solved. Note that the nine-node templates need not consist of evenly spaced mesh points. Although the use of evenly spaced mesh points leads to computational efficiency, distorted meshes pose no difficulties, theoretical or computational.

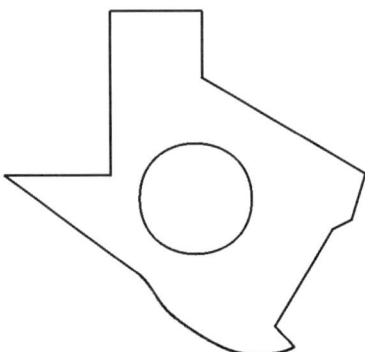

Figure 1. Boundary condition types.

As shown in previous lessons, the finite difference templates are formed by inverting the [Φ] matrices which represents the transformation from nodal displacements to strain gradient quantities. The only restriction on the mesh point layout for a boundary model is that the [Φ] matrix must invert. The fictitious nodes that are part of the templates on the boundary of the region are incorporated into the finite difference model by enforcing the normal and tangential stress boundary conditions. The details of this process are discussed later in this lesson.

Fictitious Nodes — Their Rationale and Location

Let us now demonstrate the ability of the three criteria just given to define the finite difference templates for a variety of finite difference boundary condition models. This is demonstrated here with five cases, namely, straight edges, curved edges, outside or convex corners, reentrant or concave corners, and "point loads."

Straight Boundary

The first example is a straight boundary with applied loads, as shown in Fig. 2. Figure 2a shows the boundary loaded with a normal and tangential load. The nodes on the interior and boundary of the region are shown. In Fig. 2b, the fictitious nodes specified by the first criterion are added to the model. One node is added for each node on the boundary. As we can see, an evenly spaced nine-node template applies to these situations. This template satisfies criteria 2 and 3 for the template definition. The template can represent the rigid body motions and the derivatives necessary to represent the governing differential equations and the boundary conditions.

The fictitious nodes are incorporated into the finite difference model by specifying the normal and tangential stresses on the boundary nodes, as shown in Fig. 2c. These applied forces must be balanced by the internal stresses on the boundary. The equations that define these boundary conditions are presented later. The purpose of this initial discussion is to define the template configurations for the various boundary condition models.

When the finite difference models of these two boundary equilibrium conditions at each boundary point are formed, they include the fictitious nodes. These finite difference representations of the boundary conditions are known as *auxiliary equations*. When the auxiliary equations are combined with the *primary equations*, the finite difference approximations of the governing differential equations, the complete finite difference model is formed. The relationship between the primary and auxiliary equations is discussed in detail in a later section of this lesson.

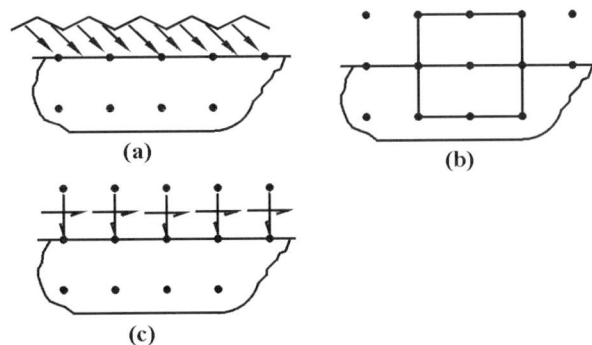

Figure 2. Straight boundary.

Curved Boundary

The second example consists of a curved boundary, as shown in Fig. 3. In principle, this boundary is treated exactly the same as the straight boundary. In detail, however, there are two differences. On a straight boundary, the orientation of the normal and tangential loads remains constant from point to point, as shown in Figs. 2a and 2b. On a curved boundary, the orientation of the normal and tangential loads change from point to point, as shown in Figs. 3a and 3b. This changing orientation is easily incorporated into the finite difference model with strain transformations as we see when the detailed development of the boundary condition models are formed later in this lesson and in Lesson 15. The second difference is in the structure of the nine-node central difference templates. On a straight boundary, the nine-node central difference template has an even nodal spacing. On a curved boundary, the mesh points are not evenly spaced, as shown in Fig. 3c. As demonstrated in Appendix A of Lesson 13, templates with uneven meshes pose no problem when they are formed using strain gradient notation. Examples of curved boundaries are solved in Lesson 15.

Convex Corner

The third example is that of an outside or convex corner. The first version of a finite difference model for this corner simply continues the progression started with Figs. 2 and 3. Let us continue to "bend" the straight edge of Fig. 2 until it forms a circle, as shown in Figs. 4a through 4c. As we can see, this model contains the same number of fictitious nodes as boundary nodes. Now let us further deform this circle into a square, as shown in Fig. 4d. Again, there are as many fictitious nodes as boundary nodes. Let us isolate one of the corners, as shown in Fig. 5. Fig. 5a shows a seven node finite difference template superimposed on this outside or convex corner. The nodes of this template are not evenly spaced. As mentioned earlier, the [Φ] matrix corresponding to this mesh configuration must invert. This was shown to occur for this configuration. The template centered on the

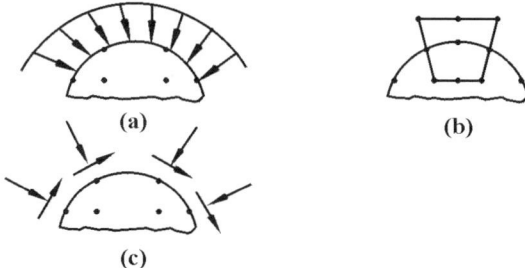

Figure 3. Typical curved boundary.

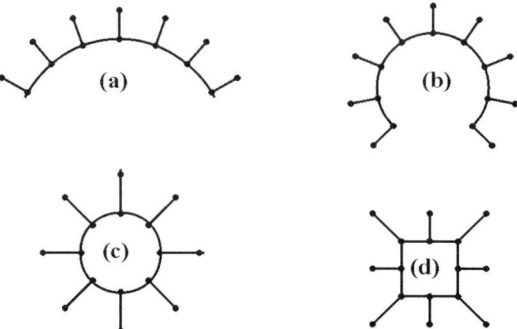

Figure 4. Curved boundary bent into a circle.

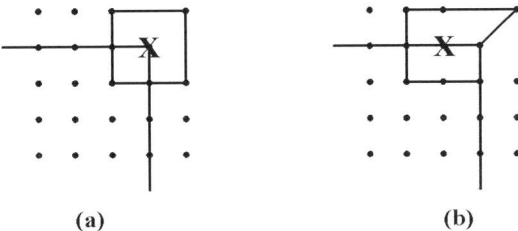

Figure 5. Template configuration for a convex corner.

boundary node adjacent to the corner node is shown in Fig. 5b. The [Φ] matrix for this template has also been found to invert.

The finite difference model shown in Fig. 5 has been found to be ill-conditioned. This deficiency is eliminated by augmenting the corner model shown in Fig. 5 with two additional fictitious nodes, which are shown as x's in Fig. 6. This produces a nine-node template. The additional four degrees of freedom associated with the fictitious nodes are incorporated into the finite difference model by specifying the boundary conditions at two additional points on the boundary of the model.

Note that the two points on the boundary at which the boundary conditions are specified are not nodes in the finite element model. In the cases used to validate these boundary condition models, the additional points are located midway between two nodes, as shown in Fig. 6. The addition of the two fictitious nodes shown in Fig. 6 produces an evenly spaced template for the boundary node, as shown in Fig. 7. Thus the convex or outside corner can be modeled with evenly spaced nine-node templates.

There are other successful models for representing the convex corner. The progression of ideas that led to this model are discussed in Appendix A. In the final analysis, the models presented here are used because they can be defined using a logical argument and they can be validated. The validation process is discussed in Appendix A. The stress models needed to apply the boundary conditions of boundary points that are not nodes in the finite difference templates are given in Lesson 15 by Eq. 15.2.

Concave Corner

The mesh layout for a reentrant or concave corner is now presented. This representation results from inverting the process used to arrive at the convex corner. In Figs. 8a through 8d, a straight boundary is bent into a circle with the fictitious nodes on the inside and then further deformed into a square. As we can see, it is possible for the dimensions to be such that the eight fictitious nodes are superimposed on each other. That is to say, several fictitious nodes can have the same coordinate locations.

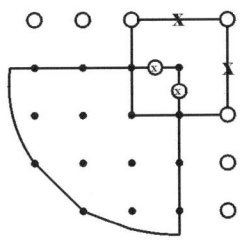

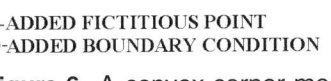

x-ADDED FICTITIOUS POINT
⊘-ADDED BOUNDARY CONDITION

Figure 6. A convex corner model.

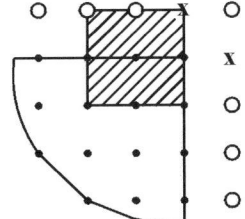

Figure 7. A typical boundary point.

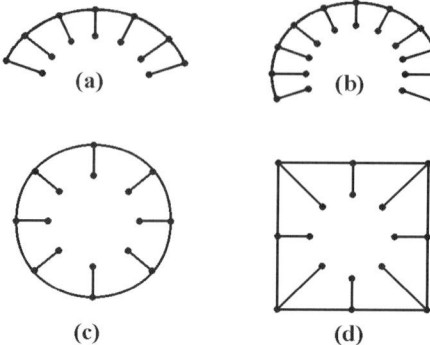

Figure 8. Curved boundary bent into a circle.

Let us retreat from this extreme case of considering four reentrant or concave corners at one time and consider only a single corner, as shown in Fig. 9a. In this figure, three fictitious nodes are seen to be attempting to occupy the same point. They are shown as a point and two x's. An unevenly spaced nine-node template is shown in Fig. 9a. In Fig. 9b, an unevenly spaced nine-node template centered at a node adjacent to the corner is shown. Similarly, an unevenly spaced nine-node template centered at the other node adjacent to the corner node is shown in Fig. 9c. If we understand that there are actually three fictitious nodes involved on the inside of this corner, we can superimpose them, as shown in Fig. 9d, and the three nine-node templates involved with this corner model will be evenly spaced. The three coincident fictitious nodes are incorporated into the finite difference model by imposing normal and tangential boundary stresses at the mesh point in the corner and at the two adjacent boundary points.

Convex Corner Revisited

We have presented the nodal layout for several boundary conditions. The mesh layout for the case of an acute convex or outside corner modeled with a nine-node template is shown in Fig. 10. There are three nodes *on* the region and six fictitious nodes. The template is sized so the nodes are evenly spaced. The fictitious nodes are incorporated into the finite difference model by enforcing the boundary conditions at the points indicated on the boundary. Some of these points are nodes of the finite difference mesh and some of them are not. This model has been validated using the procedures discussed in Appendix A.

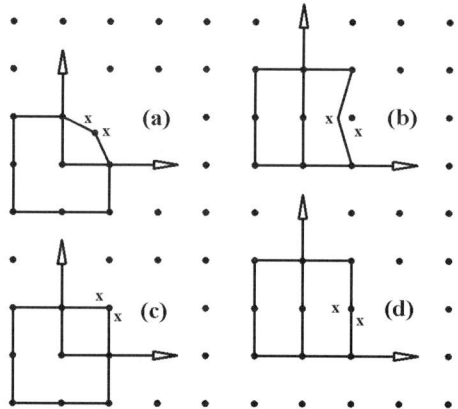

Figure 9. Reentrant or concave corner.

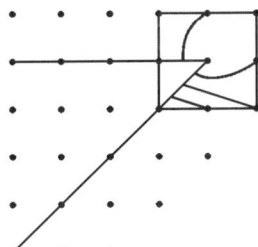

Figure 10. Nodal layout for an acute convex corner.

The Point Load

The procedure used to form the boundary conditions just discussed is now used to form a finite difference representation of a point load. The mesh layout is shown in Fig. 11. This is a nine-node template with four extra nodes. The load is shown as the triangular region in Fig. 11. The boundary loads at the two additional points on the boundary are taken as zero. The area of the load is equal to the total point load. We can view this model as a type of "Dirac delta function."

This model has been validated by comparing it to an equivalent finite element model. The displacements for the two solutions are nearly identical. The strain models have not been compared because the purpose of this model is to demonstrate the flexibility that the new approach gives to finite difference representations.

The template shown in Fig. 11 contains 13 nodes. The finite difference templates for the mesh points adjacent to the central node shown in Fig. 11 also have 13 nodes. The templates for the next boundary points have 11 nodes. The ability to routinely compute the finite difference approximations of the derivatives with templates formed from unevenly spaced nodes and a varying number of nodes is the key to the procedures presented here. The use of 11- and 13-node templates is further demonstrated in Lesson 15, where procedures for locally refining finite difference meshes are developed.

It is also possible to represent the point load with the finite difference mesh shown in Fig. 12. This template differs from the template shown in Fig. 11 in that the two nodes on the boundary are not fictitious nodes. The advantage of this representation is that the rest of the templates in the area of the point load have nine nodes. At this time, this model has not been validated. The key question to be answered concerns whether this model will provide a good representation of the strains in the area of the point load. This idea is presented to illustrate the flexibility that exists in finite difference modeling (see Lesson 20, Note 4).

The Boundary Load Equations

The equations used to model the boundary condition loads are introduced with the case of an evenly spaced mesh, as shown in Fig. 2. The first step is to form the finite difference

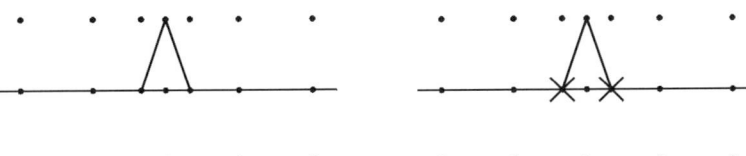

Figure 11. Point load.   **Figure 12.** Point load.

representations for the strains at a boundary point. These equations were shown in Appendix B of Lesson 13 for the case of a fixed boundary. The same equations hold for the boundary nodes discussed in this lesson. These strain equations are reproduced here for convenience as

---

Finite Difference Strain Operators

$$\varepsilon_x = \frac{1}{2h}(u_6 - u_8)$$

$$\varepsilon_y = \frac{1}{2h}(-v_5 + v_7)$$

$$\gamma_{xy} = \frac{1}{2h}(-v_5 + v_6 + v_7 - v_8)$$

(14.1)

We can generalize these equations so they apply to evenly or unevenly spaced templates by writing them in terms of the finite difference template. When Eq. 14.1 is written in operator form using the finite difference template, the result is

---

Finite Difference Strain Template

$$\{\varepsilon\} = [D][\Phi]^{-1}\{d\}$$

where

$$\{\varepsilon\} = \begin{Bmatrix} \varepsilon_x \\ \varepsilon_y \\ \gamma_{xy} \end{Bmatrix}$$

(14.2)

$$[D] = \begin{bmatrix} 0 & 0 & 0 & 1 & 0 & 0 & 0 & 0 & 0 & 0 & 0 & 0 & 0 & 0 & 0 & 0 & 0 & 0 \\ 0 & 0 & 0 & 0 & 1 & 0 & 0 & 0 & 0 & 0 & 0 & 0 & 0 & 0 & 0 & 0 & 0 & 0 \\ 0 & 0 & 0 & 0 & 0 & 1 & 0 & 0 & 0 & 0 & 0 & 0 & 0 & 0 & 0 & 0 & 0 & 0 \end{bmatrix}$$

The [D] matrix extracts the finite difference strain representation at the boundary node from the finite difference template. For example, the first row of the [D] matrix contains the value of unity as the fourth element. This element extracts the fourth row from the finite difference template, which is the finite difference representation of ε_x. The second and third rows of the [D] matrix extract the finite difference operators for ε_y and γ_{xy} from the finite difference template. The [D] matrix would have a different number of columns if the template being used had other than nine nodes.

This operator form of the strain representation is extended in Lesson 15 to compute the strains at points other than the central node. In Eq. 14.2, the product $[D][\Phi]^{-1}$ can be viewed as the strain template for plane elasticity problems. Equation 14.2 holds whether the finite difference template consists of evenly or unevenly spaced nodes. Thus, Eq. 14.2 is more general than Eq. 14.1 because it applies to models with unevenly spaced templates.

We can compute the stresses at the boundary points by introducing the strain approximations into the constitutive relation for the plane stress problem, which is reproduced here as

Plane Stress Constitutive Relation

$$\{\sigma\} = [E]\{\varepsilon\}$$

where

$$\{\sigma\} = \left\{ \begin{array}{c} \sigma_x \\ \sigma_y \\ \tau_{xy} \end{array} \right\} \qquad (14.3)$$

$$[E] = \frac{E}{(1 - \nu^2)} \begin{bmatrix} 1 & \nu & 0 \\ \nu & 1 & 0 \\ 0 & 0 & \frac{(1-\nu)}{2} \end{bmatrix}$$

We compute the stresses in terms of the finite difference operators by substituting Eq. 14.2 into Eq. 14.3 to give

Finite Difference Stress Template

$$\{\sigma\} = [E][D][\Phi]^{-1}\{d\} \qquad (14.4)$$

The expressions for extracting the three stress components just presented apply to all boundary nodes. As we have seen, the normal and tangential stresses are the quantities desired. When the boundary is rotated with respect to the coordinate axes, as shown in Fig. 13, the normal and tangential stresses can be extracted from Eq. 14.4. We can compute the normal and tangential stresses on any boundary with the following equation:

Normal and Tangential Boundary Stresses

$$\{\sigma_B\} = [T]\{\sigma\}$$

where

$$\{\sigma_B\} = \left\{ \begin{array}{c} \sigma_n \\ \tau_n \end{array} \right\} \qquad (14.5)$$

$$[T] = \begin{bmatrix} \cos^2 \theta & \sin^2 \theta & 2\sin\theta\cos\theta \\ -\sin\theta\cos\theta & \sin\theta\cos\theta & 1 - 2\sin^2\theta \end{bmatrix}$$

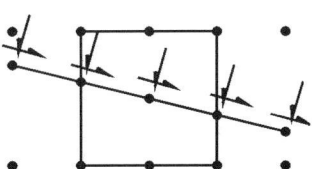

Figure 13. Normal and tangential stresses on a boundary.

We can put Eq. 14.5 in terms of the nodal displacements by substituting Eq. 14.4 into Eq. 14.5 to give

---

Normal and Tangential Boundary Stresses

$$\{\sigma_B\} = [\Omega_B]\{d\}$$

where

$$[\Omega_B] = [T][E][D][\Phi]^{-1}$$

(14.6)

Equation 14.6 represents the normal and tangential stresses at a boundary point. The coefficient matrix $[\Omega_B]$ is the boundary template for the normal and tangential stresses. This equation is the principle building block for the auxiliary equations in a finite difference model. There is one set of boundary equations for each fictitious node.

The boundary templates given by Eq. 14.6 are expressed in local coordinates. Because the boundary templates must be solved simultaneously, the auxiliary equations for each node must be expressed in global coordinates and assembled in the same manner as the equation templates were assembled in Lesson 13. When the individual boundary templates are expressed in global coordinates and assembled, the auxiliary equations for a free or loaded boundary result. We can write the assembled auxiliary equations as

---

Auxiliary Equations for a Loaded Boundary

$$[\Omega_B^R]\{d\} = \{\sigma_B^R\}$$

(14.7)

where $\{d\}$ represents global displacements.

This equation represents the boundary condition model for the total loaded boundary on the region. This is designated by the superscript R. The coefficient matrix $[\Omega_B^R]$ contains two rows for each fictitious node. We can assume that the displacement vector contains all of the unknown displacements of the problem, real and fictitious. The right-hand side of the equation contains the applied boundary stresses. For the case of an unloaded boundary, the right-hand side of Eq. 14.7 is zero.

Equation Structure for Finite Difference Models

In the development just completed, we formed the finite difference models for the boundaries of a region (the auxiliary equations) by assembling the boundary templates given by Eq. 14.6 for each fictitious node. In Lesson 13, we developed the finite difference representation of the governing differential equations on a region (the primary equations). The auxiliary equations and the primary equations are combined to form the full finite difference representation. This representation is solved to determine the displacement approximations on the region. The displacement results are then used to compute the finite difference approximations of the strains and stresses.

The solution process just outlined applies to both fixed and loaded boundaries. The finite difference models for these two types of boundary are, in principle, identical, but the solution process is slightly different for the two cases. The primary equations and auxiliary equations are uncoupled for the case of a fixed boundary. The structures of the finite difference models for the two types of boundary conditions are now presented one at a time.

A finite difference model consists of a primary and an auxiliary set of equations. The primary set of equations is made up of the finite difference approximation of the governing differential equations at each node on the region and on the boundary (see Note 1). The auxiliary set of equations introduces the boundary conditions into the model. The structure of the equations that constitute a finite difference model with fixed boundaries is shown in Fig. 14.

In Fig. 14, the primary and auxiliary equations are shown separately in the boxes enclosed with the double lines. The primary equations in this case consists of two components. The first component contains the representations of the governing differential equations *on the region*. Since this set of equations is not a function of the fictitious nodes and the displacements on the boundary are known, we can solve these equations independently for the displacements on the region. This is shown schematically in Fig. 14 by enclosing the finite difference model of the equilibrium equations on the region in a box with a single heavy line. We can also independently solve the equations in the other boxes in Fig. 14 outlined with a heavy line. The output of these independent equations is indicated with an asterisk. The displacements on the region are needed to solve the second component of the primary equations, the representation of the governing

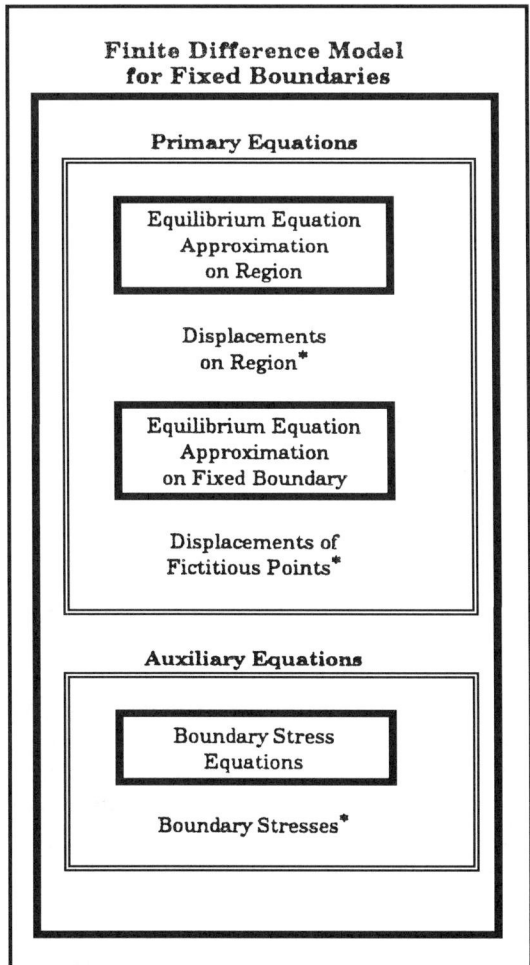

Figure 14. Equation structure of finite difference models with fixed boundaries (boundary displacements known, boundary stresses unknown). Note that * indicates output.

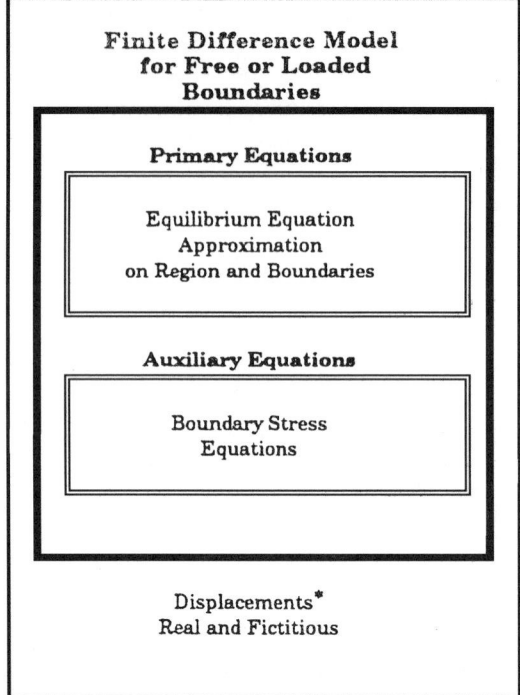

Figure 15. Equation structure of finite difference models with free or forced boundaries (boundary stresses known, boundary displacements unknown). Note that * indicates output.

differential equations *on the boundary*. We can then solve these equations independently for the displacements of the fictitious nodes.

The auxiliary equations shown in the second box enclosed with double lines in Fig. 14 represents the stresses on the boundaries. We can compute these quantities directly from the displacements found using the primary equations. The structure shown in Fig. 14 was exploited in Appendix B of Lesson 13 to introduce and demonstrate the procedure for forming the finite difference approximation of a plane stress problem. A problem with fixed boundaries was chosen so the finite difference method could be introduced without having to introduce the complication of explicitly including the boundary conditions in the representation.

The structure of the equations that constitute a finite difference model with free or loaded boundaries is shown in Fig. 15. The primary and auxiliary equations are shown separately in the boxes enclosed with double lines. The primary set of equations consists of a single component in this case. The primary set of equations represents the governing differential equations on both the region and the boundary so the displacements of the fictitious nodes are unknowns in the primary equations, which means that there are more unknowns than there are equations in the primary set of equations so we cannot solve these equations independently.

The auxiliary equations shown in the second box enclosed in double lines in Fig. 15 represent the stress boundary conditions. There are as many auxiliary equations as there are displacements associated with the fictitious nodes. Thus, we must combine the primary and auxiliary equations so we can solve for the displacements of the nodes on the region and on the boundaries (the real nodes) and of the fictitious nodes.

Figures 14 and 15 show that the auxiliary equations of all finite difference models are analogous to one another. They introduce the boundary stresses into the finite difference model. In the case of fixed boundaries, the boundary displacements are known and the boundary stresses are unknown. In the case of free or loaded boundaries, the boundary stresses are known and the displacements are unknown. However, the equations have the same structure. The primary difference between the finite difference models of problems with fixed boundaries and problems with other types of boundaries is the fact that we can solve the primary and auxiliary equations for the fixed boundary problem in separate steps.

A Modified Equation Structure for Loaded Boundary Models

As noted previously, the full finite difference model for a problem consists of the primary and auxiliary equations. In the case of a fixed boundary, we can compute the stresses on the boundary after the displacements are found, or we can solve the primary and auxiliary equations simultaneously.

Similarly, we can use the auxiliary equations for a loaded or free boundary, as given by Eq. 14.7, in two ways in the solution process. They can be combined as they stand with the primary equations to form the full finite difference model and solved for the displacements. These two sets of equations together contain $2N_R + 2N_F$ components, where N_R is the number of nodes on the interior and boundary of the region being analyzed and N_F is the number of fictitious nodes.

Alternatively, we can use the auxiliary equations to eliminate the fictitious nodes from the primary equations to reduce the size of the problem that must be solved. This is accomplished by partitioning Eq. 14.7 as follows:

Partitioned Auxiliary Equations

$$[[\Omega_{Bd}^R]|[\Omega_{Bf}^R]]\begin{Bmatrix} d_d \\ d_f \end{Bmatrix} = \{\sigma_B^R\}$$

where (14.8)

$\{d_d\}$ is the displacement on the domain.

$\{d_f\}$ is the displacement of the fictitious nodes.

We can rearrange this equation and solve for the fictitious nodes to give

---

Fictitious Node Equation

$$\{d_f\} = \{\sigma_B^R\} - [\Omega_{Bd}^R]^{-1}[\Omega_{Bf}^R]\{d_d\} \tag{14.9}$$

When we partition the primary equations in a manner similar to Eq. 14.8, we can use Eq. 14.9 to eliminate the displacements of the fictitious nodes in the primary equations. The displacements of the fictitious nodes are replaced with an expression containing the boundary stresses and the displacements on the region and boundary. The boundary stresses enter the problem as applied loads on the right-hand side of the finite difference model. The cost of eliminating the fictitious nodes from the finite difference model is the cost of inverting a matrix and of performing the additional manipulations of matrix equations required to eliminate the fictitious nodes. At this point, the structure of the resulting equations has not been studied in detail, so we have not determined whether or not the result can, in general, be made symmetric.

Closure

This lesson has developed a simple approach for forming finite difference boundary condition models that is applicable to all boundary condition types. In brief, the approach introduces one fictitious node for every mesh point on the boundary and then adds fictitious nodes to produce a template of the desired configuration.

The general boundary condition model is given in operator form by Eq 14.6. This equation has the same form as the differential equation template developed in Lesson 13 as Eq. 13.14. Both of these equations are based on the finite difference template $[\Phi]^{-1}$ Because of the vital role of the finite difference template in both of these equations and, hence, in the primary and auxiliary equations, we can say that the finite difference template is the primary building block for the finite difference method as formulated here.

As we have seen in Part III of this book, the finite element stiffness matrices also depend totally on the $[\Phi]$ matrix and its inverse. This common dependence leads to the contention made here that the finite element and finite difference methods have been put on a common basis.

This leads to a primary conclusion. Since the finite difference method has the same basis as the finite element method, the finite difference method can represent regions with the same complexity as the finite element method. In the next lesson, the flexibility of the strain gradient based finite difference method is further demonstrated.

The main purpose for developing the finite difference method in this book is its use in the next part of the book to improve error analysis techniques, to develop pointwise error measures, and to identify super convergent strain extraction procedures. To do this, the finite difference templates and boundary condition models must be able to overlay any finite element mesh. Because the $[\Phi]$ matrix serves as the common basis for the two methods, these improvements are possible.

Although the purpose of this finite development work was to complement the finite element method, this work has expanded the range of problems that we can solve by the finite difference method. There are two computational issues of interest that must be resolved before an absolute comparison of the two methods can be made: (1) Can the coefficient matrices of finite difference problems be made symmetric? and (2) Are the finite difference coefficient matrices more or less computationally robust than the corresponding finite element stiffness matrices?

As we have seen, the finite difference coefficient matrices are much simpler to form than finite element stiffness matrices and the differential equations that are being modeled can be changed more easily in a finite difference representation. Thus, when the geometric flexibility that the strain gradient approach provides is considered, the finite difference method can again provide competition to the finite element method.

The flexibility of the strain gradient formulation of the finite difference method is demonstrated in the next lesson when the tools necessary to improve the error analysis and strain extraction procedures are developed. Although it is not discussed here, the extension to three dimensions is straightforward.

Notes

1. The governing differential equations on a region must be satisfied at any point where the vertex of a triangle can be placed totally inside of the region. This means that the governing differential equations must be satisfied on corners as well as everywhere else on the domain of a ''reasonably'' shaped problem.

Appendix A
Finite Difference Boundary Model Evolution and Validation

The boundary condition models presented in this lesson provide a new perspective for forming finite difference models that gives the method a flexibility that was never before available. The approach that produced these new boundary condition models clarifies the role of the fictitious nodes, which, in turn, enables us to identify the general characteristics of finite difference templates. The new perspective follows from having the locations of the fictitious nodes define the layouts of the finite difference templates on the boundary instead of having the central difference templates define the locations of the fictitious nodes.

This appendix shows how this reversal of roles evolved and then validates the boundary models. By outlining the progression from having the boundary condition models defined by the template layout to having the template layouts defined by the boundary condition requirements, we can see the relationships between these two approaches for forming finite difference representations. An understanding of the connections between the two approaches provides insights into the characteristics of both models and provides the basis for using finite difference templates with other than nine nodes and unevenly spaced meshes.

The new perspective emerged from a study of boundary condition models based on nine-node central difference templates. For this reason, the discussion of the evolution of the boundary condition models starts with an outline of the procedures used to determine and validate the boundary condition models based on nine-node central difference templates.

Boundary Models Based on Nine-Node Central Difference Templates

In Refs. 1 and 2, boundary condition models for reentrant (concave) and outside (convex) corners based on nine-node central difference templates were developed and validated. Let us consider the rationale behind this work. As a starting point, the work in Refs. 1 and 2 assumed that the finite difference representation of a rectangular region with fixed edges produces accurate displacement and strain results on the domain of the problem because the finite difference model contains only fixed boundaries. That is to say, the models for fixed boundaries are well understood and no errors are introduced by the boundary models. Since this overall model produces good results, we use this problem as a control in Refs. 1 and 2 to validate the boundary conditions developed there.

Such a problem is shown in Fig. A.1. This control problem consists of a loaded rectangular region with fixed boundaries. Since there are no applied boundary loads to be introduced into the model, no fictitious nodes are present in the finite difference representation. The applied load is centered and varies as shown. The lines attached to the nodes are proportional to the magnitude of the load. We can solve this problem using evenly spaced nine-node templates with the mesh shown.

The finite difference model for this problem consists solely of the approximations of the governing differential equations at the nodes on the interior of the rectangular region because the displacements on the boundaries are known to be zero. In other words, the approximations of the governing differential equations provide as many equations as unknowns. This means that no loaded boundary condition models need be included to complete this finite difference model.

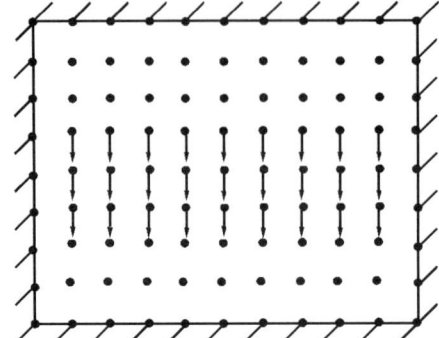

Figure A.1. Control problem.

However, this control problem implicitly contains subproblems with loaded reentrant and outside corners. This is shown in Fig. A.2 where the control problem (Fig. A.1) is separated into two regions with the boundary line shown. Region I contains the outside or convex corner and Region II contains the reentrant or concave corner.

When we separate these problems from the control problem, the subproblems are as shown in Fig. A.3. The meshes match the original nodal layout of Fig. 1. These subproblems are loaded with the original distributed load and boundary loads. The distributed load applied to the problem is shown in Fig. A.1. The boundary load consists of the internal loads found in the control problem. The boundary models developed in Refs. 1 and 2 are based on the layout of nine-node templates shown in Fig. A.3. As anticipated, when we solve these problems, the displacements and strains found in each subregion are identical to those found in the control problem. This means that

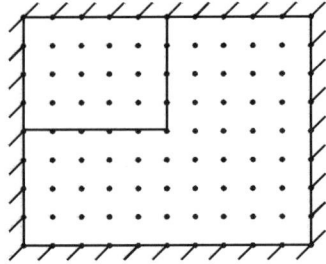

Figure A.2. Implicit subproblems.

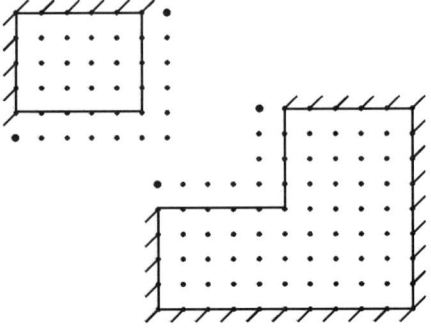

Figure A.3. Individual subproblems.

the boundary condition models applied to the subproblems accurately represent the embedded problems contained in the control problem. The boundary condition models used and the comparison of the results just outlined are presented in Refs. 1 and 2.

The corner models that successfully reproduced the results of the control problem are based on nine-node central difference templates. As discussed in the next section, the number of fictitious points contained in the concave corner model does not seem ideally suited to represent all of the possible applied load cases that are possible for this corner model. The questions concerning the correct number of fictitious nodes in the concave corner model led to the generalizations concerning the fictitious nodes and the finite difference templates presented in the main text of this lesson.

Nine-Node Central Difference Boundary Model Anomalies

The anomalies in the number of fictitious nodes contained in the two corner models are clearly seen when we closely examine the layout of the meshes shown in Fig. A.3. The mesh layouts for the two corners are shown in expanded form in Fig. A.4. The first anomaly we see in Fig. A.4a is for the concave or reentrant corner. There are three points on the boundary for which we should be able to specify normal and tangential loads. However, there is only one fictitious node associated with these three nodes. The presence of a single fictitious node means that we have provisions (one fictitious node) for introducing only one set of boundary loads at three possible sites.

This inability to introduce boundary loads at each of the available boundary nodes did not adversely affect the results produced by the subproblems. This underdetermined situation produced results that exactly matched the results from the control problem. In fact, the use of the load pairs existing in the control problem at any one of the three boundary nodes produced results in the subproblem that matched those of the control problem. Thus, we can conclude that the reentrant corner model with one fictitious node satisfactorily models the problem embedded in the control problem. The continuity requirements of the complete problem implicitly specify the other two load pairs.

However, we can envision a situation where we would like to be able to specify the boundary loads at each of the three boundary points. We cannot do this with the reentrant corner model containing only one fictitious node. Thus, the model with one fictitious node does not cover the general load case and, in this respect, the model is not satisfactory. The embedded problem does not represent all of the possible load cases for a problem with a reentrant corner because the loads at the three points are specified by one fictitious node.

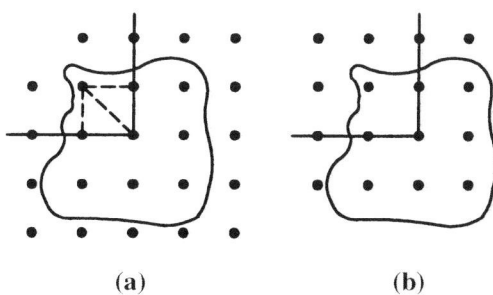

(a)   (b)

Figure A.4. Boundary model mesh layouts.

The second anomaly we see in Fig. A.4b is for the outside corner. In this case, the opposite problem prevails. There are three fictitious nodes associated with the one corner node. The problem in developing a model for an outside corner is to identify six boundary conditions to account for the six degrees of freedom associated with the three fictitious nodes. In Refs. 1 and 2, 15 different combinations of strain gradient quantities from the control problem produced results that matched those of the control problem.

The set of boundary conditions finally chosen in Refs. 1 and 2 to represent the boundary model are the following: σ_n, τ_t, $\sigma_{x,y}$, $\sigma_{y,x}$, $\tau_{xy,x}$, and $\tau_{xy,y}$. These boundary load terms were chosen because they can be extracted from the load distributions applied to the boundary of the problem. The higher order strain gradient quantities are simply the slopes of the applied loads. This approach to modeling the convex corner is not intuitively extendable to three dimensions.

Anomaly Resolution

The anomalies contained in the two corner models are connected to the difference in the number of fictitious nodes and the number of boundary nodes. The obvious correction is either to redefine the number of fictitious nodes associated with each of the models or to redefine the number of points at which boundary conditions are specified. In the work presented here, the decision was made to redefine the boundary models so nine-node templates could be used where possible. In the case of the reentrant corner, the decision to use nine-node templates if possible required that two extra fictitious nodes be added to the model so that boundary loads can be specified at the three boundary points. In the case of the outside corner, two nonnodal points are added to the boundary so the model contains only direct reference to boundary loads. The use of higher order strain gradient terms is eliminated and the model is easily extendable to three dimensions.

When these changes are made, the nodal layouts for the two corner models are as shown in Fig. A.5. The two x's in Fig. A.5a denote the presence of two additional fictitious nodes in the model. In Fig. A.5b, the two x's denote the two additional points (nonnodal) on the boundary at which the boundary stresses are specified.

The effect of the addition of the two fictitious nodes to the reentrant corner on the overall finite difference model can be seen in Fig. A.6. This figure shows the three finite difference templates associated with the three boundary nodes. In each case, a nine-node central difference template fits the configuration of the mesh points. The mesh layout for the new approach differs from the mesh layout in the embedded problem in that there is a different fictitious node associated with the template centered at each boundary point. The two added fictitious nodes enable us to specify the boundary loads at the three

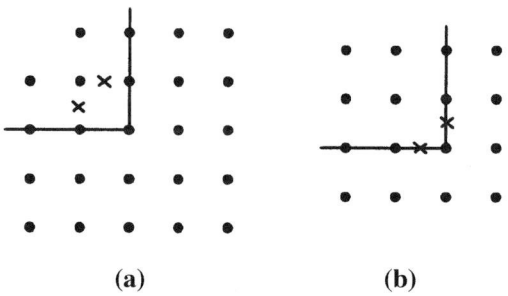

(a)   (b)

Figure A.5. Modified nodal layout for boundary models.

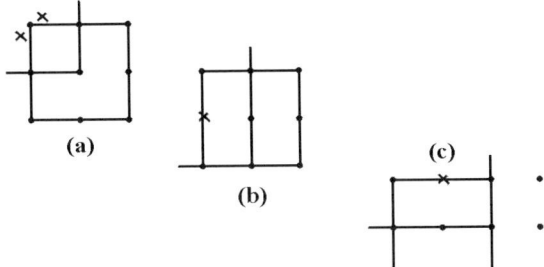

Figure A.6. Modified finite difference template configurations.

individual boundary points. If the corner is other than 90 degrees, the mesh will not be evenly spaced.

The effect of retaining the nine-node finite difference templates in the model for the outside corner is seen in Fig. A.7. In this figure, the two points at which the additional boundary stresses are enforced are shown. The three sets of boundary stresses can be successfully defined at the corner node, but the approach has an awkward feel to it. It is hard to visualize how one point can have three sets of stresses associated with it. This consideration led to the model defined here for the convex corner.

The new boundary models given in the main text of the lesson have been validated for the control problems shown in Figs. A.2 and A.3. Thus, these models can reproduce the results of the original models and the modeling deficiencies of the original models have been corrected.

The significance of these developments is not so much that they provide satisfactory results as that they eliminate the primacy of the template configuration in defining the nodal layout in the finite difference method and give the boundary modeling procedure an intuitively satisfactory nature. The central role in defining the boundary models is now invested in the fictitious nodes. The fictitious nodes are now responsible for defining the template configuration instead of the other way around.

Closure

The objective of this appendix has been met. The evolution from template-driven boundary condition models to fictitious-node-driven boundary condition models has been shown. The new point of view in which the fictitious nodes play the dominant role in defining the finite difference template layout on the boundaries provides general guidelines that enable boundary models to be routinely developed. These developments have shown that the nine-node template has an important but not necessary place in the formulation of finite difference models. The nine-node template has been chosen for use because of its effectiveness, not because its use is mandatory.

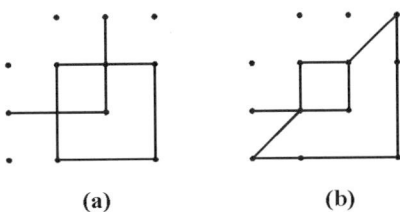

Figure A.7. Modified finite difference template configurations.

These developments have shown that alternative boundary models are available. One could envision a study to contrast the way in which different models represent the singularities at corners. Such a study might be necessary if the finite difference corner models were to be used in fracture mechanics work. However, the point of view taken here is that if the corners are the critical region in a problem, they should be represented by curved boundaries and locally refined meshes. In the next lesson, we see that localized mesh refinement can be routinely accomplished with nine-node finite difference templates with unevenly spaced nodes.

Note that the new approach makes the change to three dimensions a simple extension. One fictitious node with three degrees of freedom must be added at each boundary node. Other fictitious nodes can be added to produce a numerically stable or convenient configuration, such as a 27-node template. The three degrees of freedom at each fictitious node are incorporated into the finite difference model as the three components of the boundary load at either boundary nodes or at intermediate points that are not boundary nodes.

References and Other Readings

1. Dow, J. O., and Hardaway, J. L. "The Validation of Finite Difference Boundary Condition Models for Solid Mechanics Applications," *AIAA Journal*, Vol. 30, No. 4, July 1992, pp. 1864–1869.

2. Hardaway, J. L. "Finite Difference Boundary Condition Modeling," Masters Degree Thesis, University of Colorado, 1991.

3. Baishya, M. C. "A Generalized Finite Difference Method for the Plane Stress Problem," Ph.D. Dissertation, University of Colorado, 1993.

Lesson 14   Problems

Since this lesson primarily describes nodal layouts for various boundary conditions, there are no specific problems given here. As a way to reinforce the lesson, the reader can form mesh layouts for problems with various boundaries. For example, mesh layouts for specific portions of the outline of Texas shown in Fig. 1 could be formed.

Lesson **15**
Nonstandard Finite Difference Models

Purpose of Lesson   To present nonstandard modeling capabilities for the finite difference method that have been developed as a result of the use of strain gradient notation and the new boundary condition models.

The finite difference method has been largely supplanted by the finite element method in the analysis of most solid mechanics applications because of the perceived lack of flexibility with finite difference boundary models. The generalizations of the finite difference method presented in the previous two lessons enable us to develop the following extensions to the modeling capabilities of the finite difference method in this lesson: (1) local mesh refinements, (2) point loads, (3) moveable boundaries, and (4) multimaterial interfaces.

The purpose of these developments is to put the modeling capabilities of the finite difference method more in line with those of the finite element method. This similarity in modeling abilities is desired because some of the capabilities developed in this lesson are used in the next part of the book to improve finite element error analysis procedures. The reformulation of the finite difference and finite element methods in terms of strain gradient notation underscores the many common features that exist between the two methods.

■　　　■　　　■

In the two previous lessons, the finite difference method was reformulated using strain gradient notation. This reformulation resulted in two significant contributions. First is the ability to develop finite difference templates with other than nine nodes and unevenly spaced nodes became routine. Second is that the flexibility of the template formulation led to a new point of view regarding the role of fictitious nodes in boundary models. These changes make the development of boundary models a routine operation. These advances are exploited in this lesson to develop capabilities that were previously thought to be difficult or impossible to achieve in the finite difference method. Specifically, the following capabilities are developed:

1. Two approaches for locally refining finite difference meshes

2. An approach for introducing point loads on the boundary

3. An approach for forming boundary models that do not coincide with a line of nodes

4. The ability to model the interface between two different materials

Of the four extensions just outlined, only the capability of local mesh refinement is exploited in Part V to extend the finite difference method to the analysis of errors in the finite element method. However, the other extensions are presented to demonstrate the flexibility that the use of strain gradient notation and the new boundary condition models brings to the finite difference method.

New Capabilities

1. Local Mesh Refinement
2. Point Loads
3. Moveable Boundaries
4. Multi-Material Models

Local Mesh Refinement Procedures

The ability to locally refine finite difference meshes enables us to use a finer mesh in regions of high stress or strain concentrations. This means that we can use a coarse model in regions of slowly changing stresses or strains. The ability to locally refine meshes enables us to develop computationally efficient models. In the next part of the book, error analysis procedures applicable to finite difference results are introduced. These procedures identify regions of high error that must be represented with a finer mesh if more accurate results are desired. The developments presented in this section enable us to localize the required mesh refinement. Thus, the capability presented in this section constitutes one-half of an adaptive refinement procedure for the finite difference method.

The ability to locally refine finite difference meshes follows directly from the ability to develop finite difference templates with other than nine nodes and unevenly spaced nodes. Two approaches for refining two-dimensional finite difference meshes are presented. The first method uses specialized finite difference templates with more than nine nodes for use in the transition from coarse to fine meshes. The second refinement technique uses nine-node templates with highly irregular geometries or mesh layouts. The use of distorted templates in the transition zones eliminates the need for specialized transition elements. The two approaches are discussed in turn.

Mesh Refinement Using Specialized Transition Templates

The transition from a coarse to a fine finite difference mesh is shown in Fig. 1. The original coarse mesh with evenly spaced nodes is shown in Fig. 1a. In Fig. 1b, the bottom half of the region is refined by halving the nodal spacing. Standard nine-node templates with evenly spaced nodes are applicable within the transition zone in the region with the refined mesh, as shown in Fig. 1c. In the transition zone of the coarse mesh, special elements are required to accomplish the transition, as shown in Figs. 1d and 1e. The transition elements shown in Fig. 1d and 1e contain 11 and 13 unevenly spaced nodes, respectively. These templates are also used in the next section to introduce point loads on the boundary into finite difference models.

The use of strain gradient notation enables these transition templates to be formed in a straightforward manner. The required polynomial terms are shown for the two templates in Fig. 2 where Pascal's triangle is superimposed on the two meshes. These two templates represent all of the terms contained in the nine node template (see Lesson 13, Appendix A, Table A.1) plus those shown in Table 1 here. The x^3y and x^4y terms are not contained in the 11-node template.

Since the 13-node template contains 26 degrees of freedom, 26 independent strain gradient terms are required to form a transformation matrix between the strain gradient quantities and the nodal displacements. The first 18 independent strain gradient quantities are identical to those contained in the nine-node template (see Lesson 13, Eq. 13.6). The additional 8 strain gradient variables are the following:

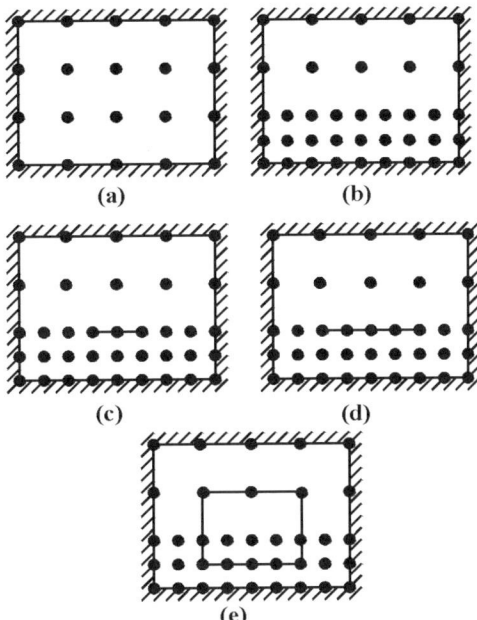

Figure 1. Localized finite difference mesh refinement.

Thirteen-Node Additional Strain States

$$
\begin{matrix}
(\varepsilon_{x,xx})_0 & (\gamma_{xy,xx})_0 \\
(\varepsilon_{x,xxx})_0 & (\gamma_{xy,xxx})_0 \\
(\varepsilon_{x,xxy})_0 & (\varepsilon_{y,xxx})_0 \\
(\varepsilon_{x,xxxy})_0 & (\varepsilon_{y,xxxx})_0
\end{matrix}
\qquad (15.1)
$$

The two displacement polynomials are evaluated at each of the nodes to form the transformation from nodal displacements to strain gradient quantities. This transformation, the $[\Phi]$ matrix, is inverted to form the finite difference template for the 13-node case.

If we consult the full Taylor series representation in Lesson 5, we see that the full strain gradient coefficient of the x^4 term of the v displacement expression contains both the $\gamma_{xy,xxx}$ and the $\varepsilon_{x,xxy}$ terms. Only the shear strain term is included in Table 1. We could have chosen to combine the two terms into a single generic independent variable. Any pair of terms multiplying any of the higher order terms (those above second order) could

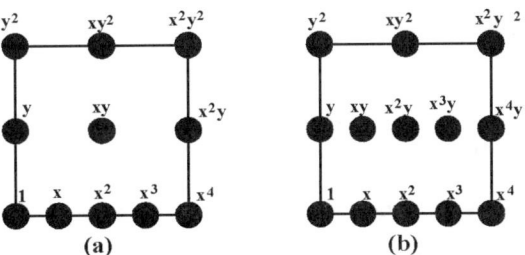

Figure 2. Pascal's triangle.

Table 1. The coefficients for the 11- and 13-node mesh refinement templates.

i	Term	a_i for $u(x, y)$	b_i for $v(x, y)$
10	x^3	$1/6(\varepsilon_{x,xx})_0$	$1/6(\gamma_{xy,xx} - \varepsilon_{x,xy})_0$
11	x^4	$1/24(\varepsilon_{x,xxx})_0$	$1/24(\gamma_{xy,xxx})_0$
12	x^3y	$1/6(\varepsilon_{x,xxy})_0$	$1/6(\varepsilon_{y,xxx})_0$
13	x^4y	$1/24(\varepsilon_{x,xxxy})_0$	$1/24(\varepsilon_{y,xxxx})_0$

have been combined because none of these derivatives are contained in the governing differential equations or the boundary conditions. The coefficients of the higher order terms are required only to complete the set of strain gradient coordinates so the $[\Phi]$ matrix will invert. A procedure utilizing only generic coefficients for the higher order terms is developed and used in Ref. 1.

In contrast, we can see that the coefficient of the x^3 term of the v displacement polynomial contains two terms. In this case, the shear strain term is a new term. It is one of the added degrees of freedom for both the 11- and the 13-node templates. The normal strain term is one of the independent degrees of freedom for the 9-node template, so it is not a new term.

Transition templates with other nodal configurations can be envisioned for unevenly spaced meshes. However, this development demonstrates that these additional templates can be formed if they are needed. For the example problems that are solved after the distorted 9-node transition templates are discussed, the templates presented here are sufficient.

Mesh Refinement Using Distorted Nine-Node Transition Templates

The process of mesh refinement is simplified if the need for specialized transition templates is eliminated. This can be accomplished through the use of distorted 9-node templates, i.e., templates with unevenly spaced nodes. The distorted 9-node equivalents of the 11- and 13-node transition templates are shown in Fig. 3. The formulation of the distorted 9-node templates is identical to that used to form an evenly spaced template, as was previously discussed.

Application of Mesh Refinement Templates

This section presents the results of applying the two different approaches for refining the mesh to two problems. These two problems are considered further in the next part of the book when the error analysis procedures are extended to finite difference models. This

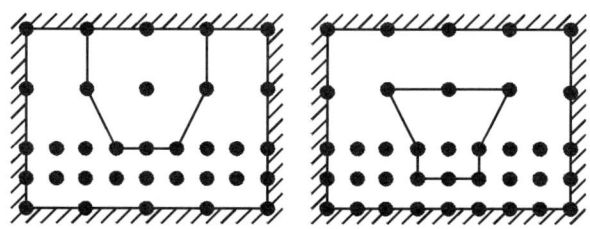

Figure 3. Distorted nine-node transition templates.

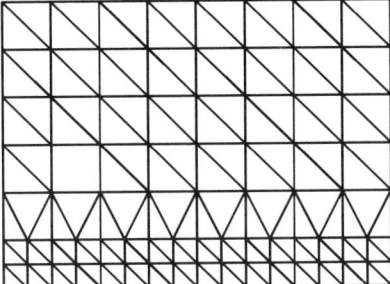

Figure 4. Rectangular region with coarse local refinement.

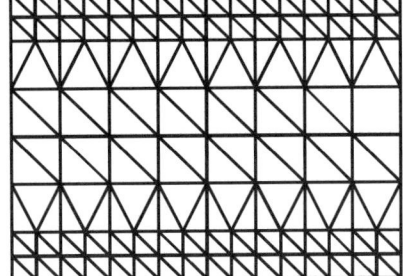

Figure 5. Rectangular region with fine local refinement.

means that the mesh refinement procedures presented in this section are one part of an adaptive refinement scheme.

The first problem solved is the rectangular region shown in Figs. 4 and 5. The loading consists of a uniform body force applied downward. The problem is solved for the two locally refined meshes shown in Figs. 4 and 5. Both refinement procedures are applied to the two locally refined meshes. The maximum displacements for the two mesh refinement procedures are shown in Table 2. The maximum displacement occurs at the center of the region. The different levels of shading refer to relative magnitudes of error estimates in the model. The error estimators are discussed at length in Part V.

As we see, the results of the two approaches are virtually identical. Thus, we can conclude that the use of specialized transition elements and the use of distorted nine-node templates give essentially the same results. The use of the distorted nine-node approach is the most logical approach to use since only one template type need be used and this will lead to the simplest programs.

To put the results of the locally refined meshes in perspective, the problem was also solved for the uniform coarse and fine meshes shown in Fig. 6. The spacing of the coarse mesh corresponds to the nodal spacing of the coarse regions in Figs. 4 and 5. Similarly, the fine mesh spacing corresponds to that of the locally refined regions in Figs. 4 and 5. In all of these analyses, the number of degrees of freedom listed includes those on the fixed boundary as well as those on the interior of the region.

The results of the uniformly refined mesh models are compared to the locally refined results in Table 3. As we see, the results of the locally refined meshes are bracketed by the results of the uniformly refined meshes. Thus we can conclude that the locally refined meshes produce accurate results.

Table 2. Results for two local refinement procedures — rectangular region.

Method of refinement	Degrees of freedom	Maximum displacement
Special	192	1.645
Distorted	192	1.646
Special	258	1.648
Distorted	258	1.650

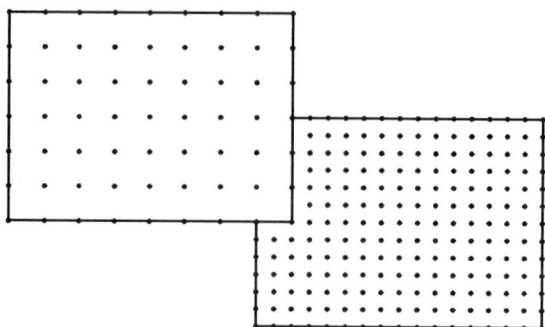

Figure 6. Uniformly refined meshes.

To demonstrate that the two refinement procedures apply to other than rectangular regions, the mesh refinement techniques are applied to the L-shaped problem shown in Fig. 7. The loading consists of a constant uniformly distributed body force acting at 45 degrees. The problem is solved for two locally refined meshes. Two levels of local refinement are shown in Figs. 7 and 8. Both the specialized and the distorted transition elements are used to solve each of the two problems. The maximum displacements for the four solutions are shown in Table 4. The maximum displacement occurs at the intersection of the dividing lines of the two rectangular regions (on the dividing line between the reentrant and outside corners).

We can reach the same conclusions for this case as those reached for the case of the rectangular region since both results are virtually identical. That is to say, since both approaches produce equivalent results, the use of the distorted nine-node approach is recommended because it is easier to implement.

Table 3. Results for uniformly and locally refined meshes — rectangular region.

Method of refinement	Degrees of freedom	Maximum displacement
Uniform	126	1.645
Distorted	192	1.646
Distorted	258	1.650
Uniform	442	1.650

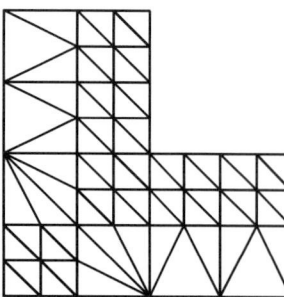

Figure 7. L-shaped region with coarse local refinement.

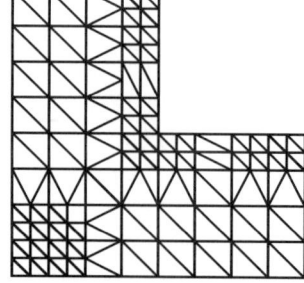

Figure 8. L-shaped region with fine refinement.

Table 4. Results for the two local refinement procedures — L-shaped region.

Method of refinement	Degrees of freedom	Maximum displacement
Special	94	0.0139
Distorted	94	0.0139
Special	224	0.0146
Distorted	224	0.0145

To put the accuracy of the locally refined models into perspective, the problem was also solved for three uniformly refined meshes that bracket the two locally refined meshes. That is to say, the problem was solved for the uniform meshes consisting of the three sizes of squares shown in Figs. 7 and 8. The results for the uniformly refined meshes and the two locally refined meshes using distorted nine node elements are shown in Table 5. As we can see, the results of the locally refined meshes are bracketed by the results of the uniformly refined meshes. Thus we can conclude that the locally refined meshes produce accurate results.

Table 5. Comparisons of results for uniformly and locally refined meshes.

Method of refinement	Degrees of freedom	Maximum displacement
Uniform	42	0.0129
Distorted	94	0.0139
Uniform	130	0.0141
Distorted	224	0.0143
Uniform	450	0.0145

Point Loads on the Boundary

This section adds a capability that is not usually associated with the finite difference method. A procedure for applying point loads to the boundary is shown. The finite difference model of a point load uses the specialized templates developed in the previous section for locally refining the mesh. This is seen in Fig. 9 where the finite difference representation of a point load is shown. Two nodes that bracket the point at which the point load is to be applied are added to the boundary. As a result, two additional fictitious points are required. This nodal configuration requires the use of the specialized 11- and 13-node templates developed in the previous section. The role of the point load as an idealization is discussed in Lesson 20, Note 4.

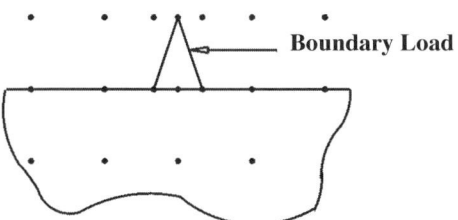

Figure 9. Finite difference point load model.

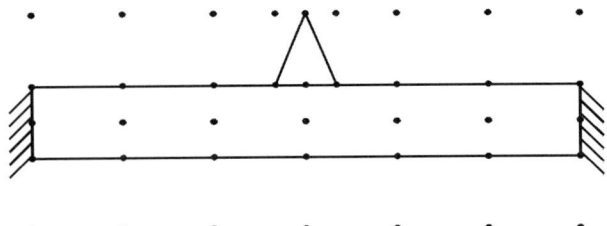

Figure 10. Beam problem with a point load.

The boundary stresses at the two bracketing points are zero. The boundary stresses at the center point of application are taken so the area under the triangular boundary loads equals the magnitude of the desired point loads. In actuality, this model is closer to a ''real'' point load than the model used in the finite element method. The finite difference representation consists of a high stress acting on a limited area. The specialized 11- and 13-node templates required to include the boundary loads in the finite difference model are identical to those used in the previous section.

This model of a point load has been validated for the case of a beam with fixed supports, as shown in Fig. 10 by solving an equivalent finite element model consisting of nine-node finite elements (see Ref. 3). The area under the triangular region is equal to the magnitude of the point load applied to the finite element model. The displacements produced by the finite difference and the finite element analyses of this problem are within 8% of each other. The finite element displacements are the largest. Similarly, the flexural stresses in the finite element model are approximately 8% greater than the stresses produced by the finite difference model.

This example problem has validated the finite difference representation of a point load presented in this lesson. This demonstration was meant to illustrate the flexibility of the modeling capabilities of the finite difference method based on strain gradient notation. That the finite difference model of a point load is somewhat awkward should not be taken as a deficiency in the method. The idea of a point load is an idealization that does not exist in nature. This example illustrates that this idealization can be included in a finite difference representation if it is so desired. However, if the point load is found to be applied at a critical point, the mesh must be refined and the point load will most likely be replaced with a more realistic representation.

Moveable Boundaries

The title of this section may at first seem somewhat misleading. The section could easily have been entitled, ''Boundaries not Coincident with Mesh Points.'' An example of a boundary that does not coincide with the mesh points is shown for a straight boundary in Fig. 11. However, as we can see, there is a fictitious node corresponding to each of the

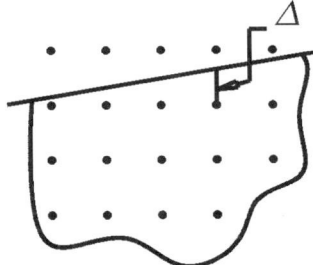

Figure 11. A moveable boundary.

nodes close to the boundary. Thus, we can apply a pair of point loads (normal and tangential) at an intermediate point between the fictitious node and the node closest to the boundary.

We incorporate the boundary loads into the finite difference model using the Taylor series nature of the displacement expansions to compute the strains. This is given as

---

Taylor Series Strain Expansion — Nine-Node Template

$$\{\varepsilon\} = [D]\{\varepsilon_{,}\}$$

where

$$[D] = \begin{bmatrix} 0 & 0 & 0 & 1 & 0 & 0 & \Delta x & 0 & 0 & \Delta y & 0 & 0 & \Delta x \Delta y & \Delta y^2/2 & 0 & 0 & 0 & 0 \\ 0 & 0 & 0 & 0 & 1 & 0 & 0 & \Delta x & 0 & 0 & \Delta y & 0 & 0 & 0 & \Delta x \Delta y & \Delta x^2/2 & 0 & 0 \\ 0 & 0 & 0 & 0 & 0 & 1 & 0 & 0 & \Delta x & 0 & 0 & \Delta y & 0 & 0 & 0 & 0 & 0 & 0 \end{bmatrix}$$

$$(15.2)$$

The Δ's provide the mechanism for locating the boundary point. Because of these Δ's we can use the idea of a moveable boundary. The location of the boundary depends on quantities other than the nodal locations. Thus, the boundary equations for incorporating the boundary stresses and the fictitious nodes into the finite difference model can be changed without disturbing the primary equations that satisfy the approximations of the governing differential equations. Equation 15.2 corresponds to Eq. 14.2 of Lesson 14. We validate the boundary model shown in Fig. 11 with the control problem shown in Fig. 12.

When the internal stresses along the boundary shown in Fig. 12 are applied to the free boundary shown in Fig. 13 as boundary loads, the displacements and strains everywhere in the problem shown in Fig. 13 match those of the control problem. Thus, we can conclude that Eq. 15.2 is a valid representation of a straight boundary. Equation 15.2 is also used to apply the boundary conditions to the boundary points that are not nodal points in the convex corner model developed in Lesson 14.

The boundary model given by Eq. 15.2 is capable of representing curved boundaries. This capability is demonstrated for two examples. The first example is shown in Fig. 14.

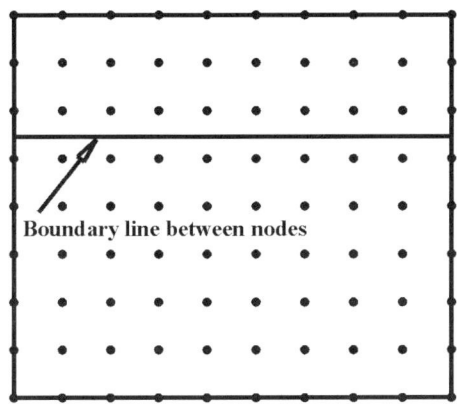

Figure 12. Control problem.

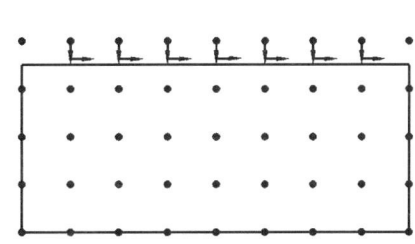

Figure 13. Moveable free boundary.

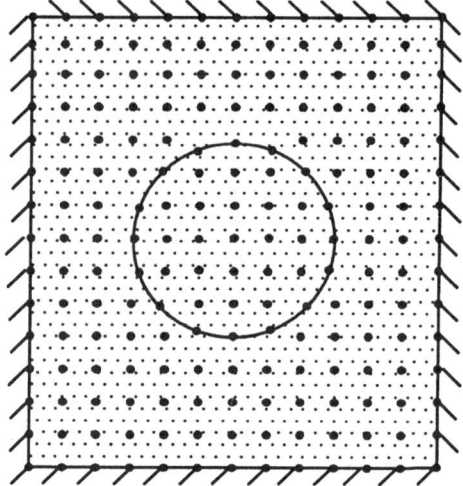

Figure 14. Control problem.

This figure is a control problem for a panel with a circular hole. In this case, note that the mesh points are arranged so that they lie on the circle. This means, of course, that the nine-node templates do not all contain evenly spaced nodes.

When the control problem is solved and the internal stresses are applied as boundary loads to the problem shown in Fig. 15, the displacements and strains everywhere in the panel are identical to those of the control problem. In this case, the Δ's in Eq. 15.2 are zero. That is to say, the modification to Eq. 14.2 given by Eq. 14.1 was not used. Note that there is one fictitious node for each boundary node on the circle. The mesh layout is such that the templates all have nine nodes but the nodes are not evenly spaced.

Now let us solve a similar problem with an evenly spaced mesh. The control problem is shown in Fig. 16. Note that all of the nodes do not lie on the boundary. When we apply

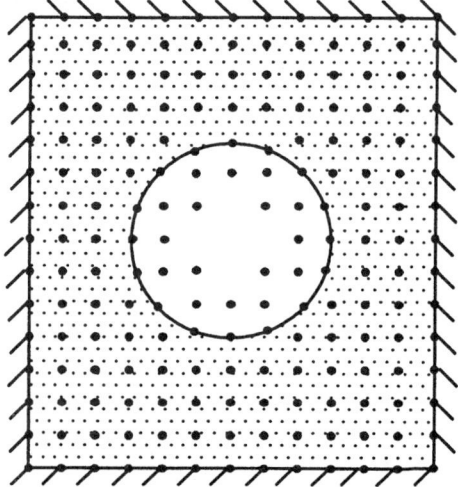

Figure 15. Problem with circular hole.

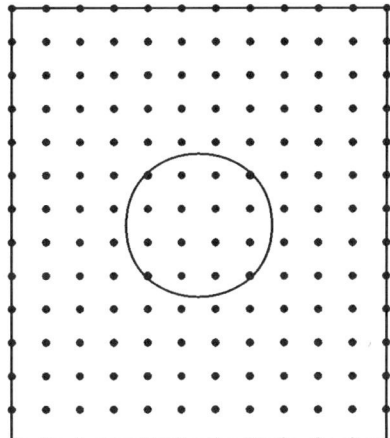

Figure 16. Control problem.

the internal loads on the circle to the problem shown in Fig. 17 using Eq. 15.2, the results match those of the control problem shown in Fig. 16.

In the case of a moveable boundary, the basic requirement for the number of fictitious nodes must be slightly modified. Since there may be no nodes on the boundary, the requirement that there is one fictitious node for each boundary node obviously does not hold. In this case, there is one fictitious node for each pair of boundary stresses included in the model.

This section has demonstrated the flexibility of the boundary models developed in Lesson 14. This form of boundary model enabled us to use the evenly spaced finite difference templates even when the boundary did not fall on mesh points. There is another possibility that has not been explored. Through the use of moveable boundaries, the shape of boundaries can be modified within limits by simply changing the Δ's in Eq. 15.1. This could possibly be used to optimize the configuration to fit some constraint, say, the minimization of a boundary stress or strain.

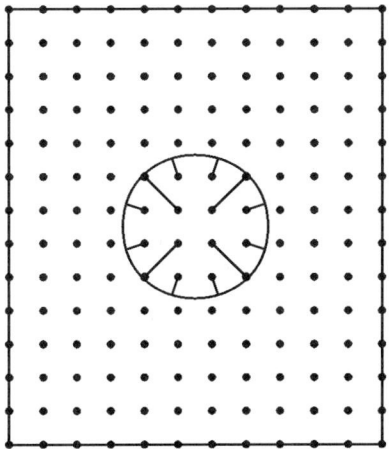

Figure 17. Problem with circular hole.

Multimaterial Models

This section adds another modeling capability that is not usually associated with the finite difference method. A procedure for representing the interface between two dissimilar materials is developed and validated. The boundary model is essentially the same as the models previously developed. This can be seen in Figs. 18 and 19. Figure 18 shows a region consisting of two materials connected at a surface that slopes with respect to the coordinate axes. A finite difference mesh is superimposed on the region so that nodes are located on the common boundary. The sloping boundary is used to graphically illustrate the use of distorted finite difference templates.

The two materials are shown separated in Fig. 19. The nodes associated with a typical interface point translate into six nodes when the two regions are separated. Each boundary point on the two regions has an interior point and a fictitious point associated with it. The finite difference approximation of the governing differential equations must be satisfied on the interior and boundary nodes for both materials. In addition to satisfying the approximations of the governing differential equations, the finite difference model must satisfy the displacement compatibility and boundary load equilibrium requirements at the boundary nodes. The displacement compatibility condition requires that the displacements be the same for the common boundary nodes. The boundary load equilibrium conditions require that the normal and shear stresses at the common boundary points be equal and opposite. We use the compatibility and equilibrium requirements to incorporate the fictitious nodes into the finite difference model.

The equilibrium equations that relate the forces between the two surfaces at the point of connection can be written in terms of the normal and tangential forces shown in Fig. 20. The equilibrium relations can be expressed analytically as:

$$(\sigma_n)_{\mathrm{I}} = (\sigma_n)_{\mathrm{II}}$$
$$(\tau_n)_{\mathrm{I}} = (\tau_n)_{\mathrm{II}} \tag{15.3}$$

These equations indicate that the normal and tangential stresses at each of the common points located along the boundary of the two regions must be equal and opposite.

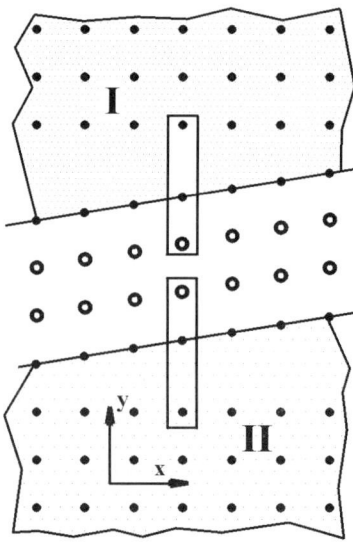

Figure 19. Interface region with fictitious nodes.

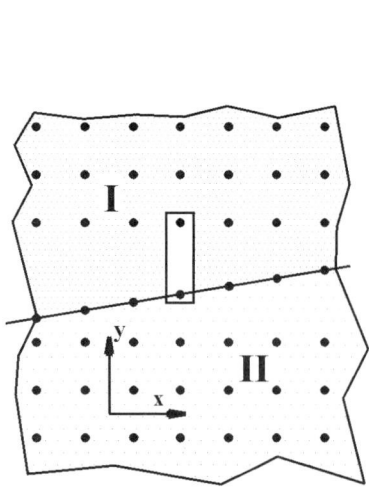

Figure 18. Interface region.

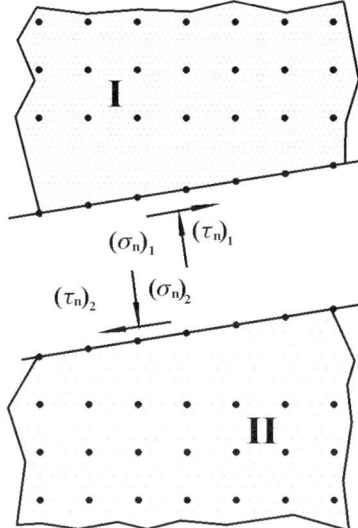

Figure 20. Interface with normal and tangential stresses.

The multimaterial boundaries just developed are now validated using a control problem consisting of a single material. The control problem shown in Fig. 21 was first solved for a single material. Then it was divided along the horizontal boundary line of nodes shown in Fig. 21. The two segments were then connected using the multimaterial boundary model. The displacement and strains were everywhere the same. The equality of the strains in the control problem and the multimaterial boundary model is illustrated in Fig. 22 for the ε_x strain component along the arbitrarily chosen strain evaluation line identified in Fig. 21. As we can see, the two strain results match exactly. These results mean that the multimaterial models are correct.

In the developments of the previous section, we saw that the boundary need not be defined by having a set of nodes located directly on the boundary. The same is true for the case of two connecting materials, as shown in Fig. 21. The boundary can be located between the mesh points. The displacement and equilibrium conditions on the boundary between the two materials can be satisfied using the Taylor series nature of the displacement and strain approximations. Again, the displacements and strains for the subdivided model matched the results of the control problem.

The final single material problem solved with the multimaterial boundary model for the corner model is shown in Fig. 21. The problem was divided along the boundary line

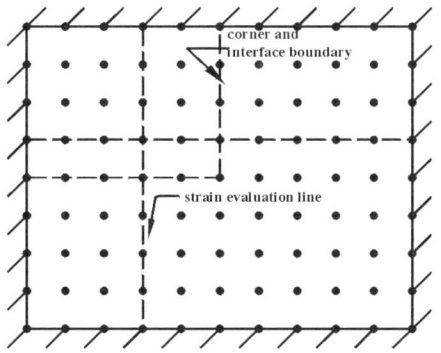

Figure 21. Control problem.

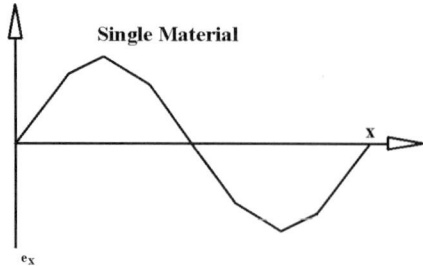

Figure 22. The ϵ_x strain component.

shown so corners exist in the two separate pieces. The two sections are connected using the multimaterial boundary model. The results for the displacements and the strains are everywhere the same as in the control problem. This also means that the strains along the arbitrary strain evaluation line match those for the other cases. Thus we can conclude that the multimaterial boundary model provides an accurate representation of corner models.

The multimaterial boundary models were then used to represent the problem shown in Fig. 23. This problem is geometrically identical to the control problem shown in Fig. 21. However, regions I and II consist of aluminum and steel, respectively. When the problem is solved, the ε_x strain component along the arbitrary strain evaluation line are shown in Fig. 24. As we can see, the strain components are larger than those of the single material problem, which was steel. Furthermore, there is a jump in the strain between the two materials at the interface, as expected.

A problem with a reentrant and an outside corner that formed a boundary between regions consisting of two different materials was also solved. The expected strain jumps were present. These results, plus the results from the single-material control problem, demonstrate that the multimaterial models accurately represent the physical system.

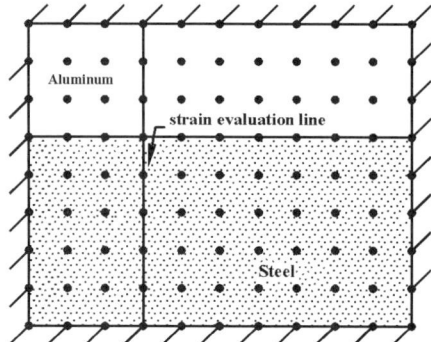

Figure 23. Multimaterial problem.

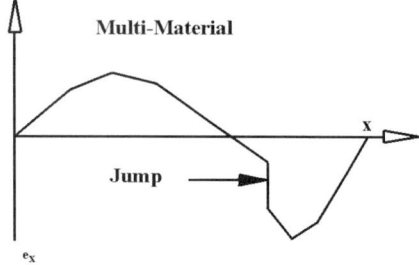

Figure 24. The ϵ_x strain component.

Closure

This lesson has presented approaches for representing new modeling capabilities for the finite difference method. Strain transformations for applying boundary loads to points that are not nodes in the mesh were given in Eq. 15.2. Templates for forming boundary models for point loads were given in Table 1 and Eq. 15.1. These same templates apply to mesh refinement procedures as several examples demonstrated. These capabilities demonstrate the flexibility that the new formulation procedure gives to the finite difference method. Finite difference meshes can be superimposed on practically any finite element mesh. This means that the error analysis techniques developed in the next part of the book are not limited to specialized finite element mesh layouts.

Furthermore, this lesson has extended the finite difference method to problems that are normally considered out of range for the method. We saw that multimaterial boundaries can be represented without difficulty. This lesson opened the door to several lines of interesting investigation.

For example, the possible use of the ''moveable boundaries'' in the determination of optimum boundary shape is intriguing. The direct connection between the strain representation and the nodal locations provides a direct connection for optimizing such a problem.

This part of the book has put the finite element and the finite difference methods on the same basis. Both methods have been reformulated in terms of the [Φ] matrices and both methods can represent the same boundary conditions. The equivalence of the two methods are discussed further in Part V.

References and Other Reading

1. Dow, J. O., Su, Z. W., Feng, C. C., and Bodley, C. S. ''An Equivalent Continuum Representation of Structures Composed of Repeated Elements,'' *AIAA Journal*, Vol. 23, No. 10, Oct. 1985, pp. 1564–1569.

2. Dow, J. O., and Hardaway, J. L. ''The Modeling of Multi-Material Interfaces in the Finite Difference Method,'' *International Journal for Partial Differential Equations*, Vol. 8, No. 5, Sept. 1992, pp. 493–503.

3. Baishya, M. C. ''Generalized Finite Difference Method for Plane Stress Problems,'' Ph.D. Dissertation, University of Colorado, 1993.

4. Hardaway, J. L. ''Finite Difference Boundary Modeling Procedures,'' M.S. Thesis, University of Colorado, 1991.

Lesson 15   Problems

1. Contrast the convex corner model for the embedded problem with the convex corner model for the free edge. *Hint*: This problem and the next problem are designed to provide further insights into the finite difference modeling of ''physically impossible''conditions. That is to say, a ''sharp''corner is a physical impossibility. See Lesson 20, Note 4, for further comments on the relationship between the finite difference method and the physical system it is modeling.

2. Contrast the concave corner model for the embedded problem with the concave corner model for the free edge.

Part V

A Posteriori Error Analysis Procedures:
Pointwise Error Measures
and
A New Approach for Strain Extraction

Introduction

The most effective method to obtain a finite element solution of ensured accuracy is to create the final model using a process of successive improvement. In this process, the accuracy of the initial finite element solution is evaluated with some type of *termination criteria*. If the desired level of accuracy is not present, the finite element mesh is improved by further subdividing the model in regions of unacceptable error under the guidance of a *refinement guide*. The new model is evaluated for accuracy and the process is continued until an acceptable result is attained. A flow chart for this iterative process, which is called *adaptive refinement*, is shown in Fig. 1. An example of a finite element model that has been adaptively refined is shown in Fig. 2.

The objective of Part V is to develop and compare refinement guides and termination criteria for use in the adaptive refinement process. The two most important approaches to compute these quantities to evaluate the accuracy of finite element results are the *smoothing* and the *residual* techniques. In brief, the smoothing approach evaluates how well the approximate solution satisfies continuity (and the natural boundary conditions), and the residual approach evaluates how well the approximate solution satisfies equilibrium. Pointwise smoothing and residual error measures are developed in the lessons that make up Part V. These new error estimators are compared to the aggregate error estimators that are in limited use today. The pointwise error estimators are more sensitive measures of errors because they have a higher resolution than the aggregate measures.

The idea of adaptive refinement is not new. Over 50 years ago, R. Courant clearly identified the need for refinement guides and termination criteria in a discussion of the interplay between the practical and theoretical points of view in mathematics with the following observation (see Ref. 1):

> Usually the solution of a difficult problem in analysis proceeds according to a general scheme: The given problem P with the solution S is replaced by a related problem P_n so simple that its solution S_n can be found with comparative ease. Then by improving the approximation P_n to P we may expect, or we may assume, or we may prove, that S_n tends to the desired solution S of P. The essential point in an individual case is to choose the sequence P_n in a suitable manner.

The process described by Courant is related to the finite element method in the following way. The initial finite element mesh plays the role of "*a related problem P_n so*

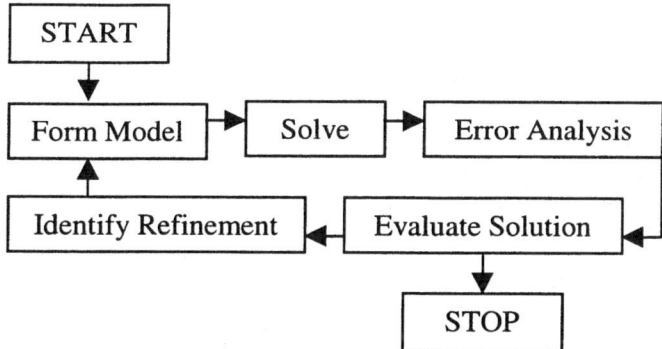

Figure 1. A schematic diagram of the adaptive refinement process.

simple that its solution S_n can be found with comparative ease.'' The refinement guides provide the capability of choosing ''*the sequence of P_n (the improved finite element models) in a suitable manner.*'' The termination criteria identify when to stop the adaptive refinement process, i.e., they enable us to expect, assume or prove ''*that S_n tends to the desired solution S of P. The essential point* . . . [of choosing] *the sequence P_n in a suitable manner*'' is the capability provided by the refinement guides and the termination criteria. Thus, we can conclude that the ability to estimate the errors contained in finite element results are crucial to the successful application of the finite element method.

Approaches for Computing Refinement Guides and Termination Criteria

Local or discretization errors exist in an approximate solution when the discrete model is incapable of representing the behavior of a continuous problem on a portion of the domain. As discussed next and shown in detail in the lessons that follow, the discretization errors in finite element models are identified by quantifying the amounts by which the finite element result fails to satisfy one or more of the known characteristics of the exact solution, namely, (1) the continuity of interelement stresses and/or strains, (2) the natural boundary conditions, and (3) the equilibrium conditions on a pointwise basis. In fact, we see in Part V that the continuity and equilibrium requirements are equivalent conditions. This means that evaluation measures based on continuity, i.e., the evaluation of the interelement jumps and the satisfaction of the boundary conditions in the finite element stress representations, also implicitly evaluate equilibrium at the nodes. Inversely, the evaluation of the equilibrium conditions at the nodes also implicitly evaluates the accuracy of the smoothed stress results.

The two approaches for computing refinement guides and termination criteria are presented, namely, the *smoothing* and *residual* techniques. The smoothing approach is described in a single stroke by saying that the error measures are formed by evaluating how well the stresses and/or strains in a finite element solution *satisfy continuity and boundary conditions*. The residual approach is compactly described by saying that the

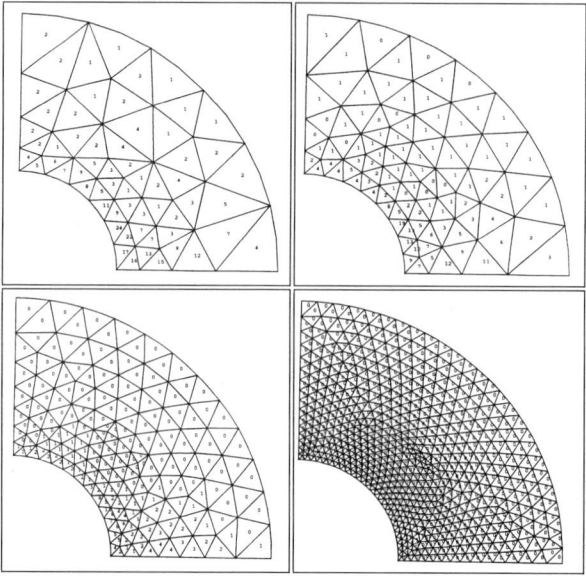

Figure 2. A sequence of models leading to acceptable stress results at a critical point.

error measures are formed by evaluating how well the finite element solution *satisfies pointwise equilibrium*. More complete descriptions of both techniques follow.

The Smoothing Approach

In the smoothing approach, the errors in the finite element result are estimated by comparing the discontinuous finite element stresses to a set of continuous stresses. This idea is illustrated in Fig. 3, where a smoothed solution is superimposed on a discontinuous finite element stress result for a problem with one unknown displacement. The estimated error is computed as a function of the "difference" between the two stress representations.

The smoothing approach provides an indirect way to assess the interelement jumps in the finite element representation. The intermediate step of forming the smoothed solution may seem superfluous if we are actually evaluating the interelement jumps. However, we see in the developments that follow that the smoothed solution is usually more accurate than the discontinuous stress representations. As a result, we can use it as the output of the finite element analysis. We identify the conditions for which the smoothed solution is of a higher order than the finite element stresses. That is to say, the smoothed solution is a better solution than the finite element result in certain situations. As we will see, this knowledge enables us to develop a solid theoretical foundation for evaluating measures based on the smoothing approach.

The *three primary* questions associated with the smoothing approach are: (1) How should the smoothed solution be formed? (2) How should the error in the finite element solution be quantified? and (3) What should be used for the termination criteria?

Many practical considerations influence the answers to these questions and raise other questions. For example, how shall the refinement guides and the mesh refinement process be related? Shall only the regions above some threshold level of error be refined or shall each element be refined in proportion to the magnitude of the error estimate existing in the element? We discuss these questions and others later in this introduction and in the lessons contained in this part of the book.

To put these questions in context, some of the developments that will be presented in the lessons are now being outlined. In the original development of the smoothing approach by Zienkiewicz and Zhu, the refinement guide consists of the estimates of the errors in the strain energy contained in individual finite elements and the termination criterion consists of an estimate of the error in the global strain energy (see Refs. 2 and 3). In Lessons 16 and 17, we see that the global error measure is not an ideal termination

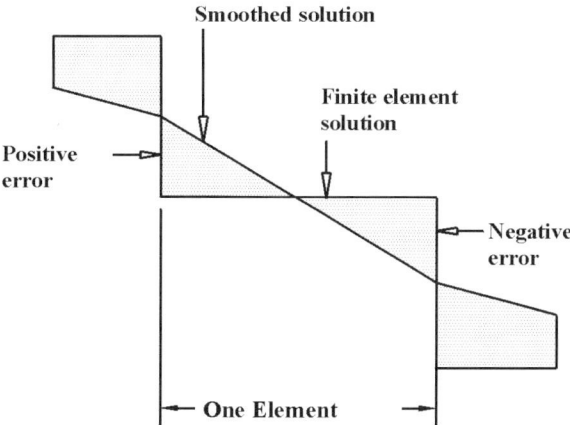

Figure 3. A generic smoothing approach.

criterion because the stresses at critical points are not closely related to the error in the overall strain energy. Other measures besides strain energy quantities are presented and demonstrated as possible refinement guides and termination criteria in Lesson 18.

Of particular interest are refinement guides based on the pointwise differences in stress quantities that are developed in Lesson 18. These pointwise measures have several advantages over the strain energy measures. Since the pointwise evaluation measures do not average the errors over the element, they exhibit a higher resolution than do the strain energy measures. In addition, pointwise measures can be tailored for particular applications. For example, we could specialize the refinement guides so they are related to some failure criteria for a given material, i.e., the von Mises criterion for steel or a shear stress criterion for ceramic materials.

The pointwise differences can also serve as termination criteria. As mentioned earlier, the smoothed solution is shown to be of a higher order than the discontinuous finite element stresses. As a result, the smoothed result can be interpreted as a "better" result than the finite element representation. With this knowledge, the differences between the two solutions can be interpreted as pointwise errors in the finite element stress representation. Because of the higher order of the smoothed stresses, we can also use the convergence of the smoothed stresses as termination criteria. The characteristics of the various refinement guides and termination criteria are compared in Lesson 19.

The Residual Approach

The residuals existing at the nodes of the finite element model can also serve as pointwise refinement guides and termination criteria. The residuals are the amounts by which the finite element solution fails to satisfy equilibrium at individual points. The *primary questions* in the residual approach are the following: (1) How should the residuals be computed? (2) What metrics should be used to most effectively quantify the errors in the finite element result? and (3) How should the error estimates be interpreted and used?

As we see in the lessons, residuals can be scaled so that they are expressed in the same units as the applied loads. As a result, the residuals can be interpreted as "fictitious applied loads" that cause the approximate solution to deviate from the actual or exact solution. This idea is shown in Fig. 4, where the residuals at the nodes in an out-of-plane plate bending problem are interpolated over the region to represent an applied distributed load. As the mesh is refined and the approximate solution approaches the exact result, the residuals, by definition, approach zero. Thus, the residuals are indicative of the level of errors contained in the approximate solution and can be used to form refinement guides

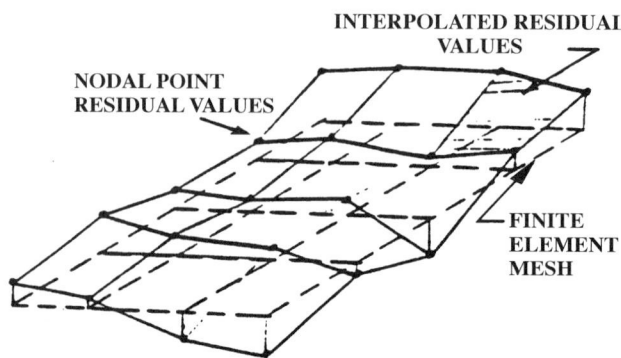

Figure 4. The residuals interpreted as "fictitious distributed loads."

and termination criteria. By interpreting the residuals as distributed loads, we can evaluate the relative error by relating them to parameters of the problem, i.e., the applied loads. The behavior of residuals at singularities is discussed in the lessons.

As described, the residuals can be used directly as error measures. However, the greatest appeal of the residual approach is the possibility of using the residuals to produce bounds for the pointwise errors in finite element results. When the residuals expressed as "fictitious" loads are applied to *an improved* model of the physical system being analyzed and this new "auxiliary" problem is solved, we can interpret the resulting displacements, stresses, and strains at individual points as pointwise bounds or as corrections to the original finite element solution. This implies that an alternative to the adaptive refinement approach is available for linear problems if we choose to use it. This approach was introduced by Kelly (see Refs. 4–7).

Kelly has also developed an approach using residuals that finds an exact correction for the finite element representation. In this procedure, the interelement jumps are interpreted as one of two components of the residuals. The other component is the amount by which the governing differential equations are not satisfied on the individual elements.

In this approach, the loads due to the stresses involved with the interelement jumps are distributed among the individual elements to produce equilibrium. Then, each individual element is loaded with the two residual components associated with the element and solved as a separate, uncoupled problem to find the exact solution. This procedure is demonstrated for a one-dimensional example in Lesson 19.

The approach just described has not found general usage. The most likely reason for this is that the individual problems that must be solved are basically identical to the original problem that could not be solved exactly, albeit, on a smaller scale. Furthermore, the process of distributing the interelement loads requires techniques that are not standard to the finite element method. Thus, it is not clear whether the approach developed by Kelly is as generally applicable as the adaptive refinement method. It might contain restrictions that do not allow it to be generally applicable. Nevertheless, this work is significant in that it interprets the interelement jumps as residual quantities and the residuals as fictitious applied loads.

In contrast, the adaptive refinement process described earlier uses standard finite element procedures to solve the improved representation of the problem. The adaptive refinement procedure sequentially improves the solution by repeatedly applying known techniques to models improved under the guidance of refinement guides until a result of acceptable accuracy is achieved as measured by some termination criterion.

In Part V, an approach for extracting residuals from finite element results that is straightforward and easy to apply is developed. This work shows that the interelement jumps in the stresses and the residuals are related. In Lesson 19, we see that the existence of continuous stresses, i.e., interelement jumps of zero, in the finite element solution is equivalent to the satisfaction of equilibrium. As a result, many of the questions that arise concerning the direct use of the residual approach in the evaluation and correction of finite element results can be easily approached.

A Unified Approach for Computing *A Posteriori* Error Estimators

The most important contribution to error analysis contained in Part V is the unification and improvement of the smoothing and residual approaches of evaluating finite element results. Specifically, a new technique for forming smoothed solutions which provides a new theoretical basis for error analysis is developed. This new point of view brings several theoretical and practical advantages to the evaluation of errors in finite element results:

1. It enables us to make improvements to the smoothing approach.

2. It enables us to make improvements to the residual approach.

3. It enables us to relate the smoothing and residual measures directly to each other, i.e., we can unify them.

4. It enables us to better understand existing evaluation procedures.

5. It enables us to indentify the conditions that ensure that the stresses extracted from the smoothed solution are of an order higher than the discontinuous stresses contained in the finite element model.

As we see in the lessons, these developments provide other far-reaching implications for computational mechanics. For example, the idea contained in the fifth item just presented implies that the nodal displacements are the only quantities produced by the finite element analysis that need to be directly used in the evaluation of the finite element results and to produce the stress outputs. This is the case because we can use the higher order smoothed stresses extracted from the smoothed solution as the stress representation from the finite element analysis. We never explicitly need to use the stresses from the underlying finite element model.

As a result, the specific finite elements that form a finite element model can be chosen purely for reasons of computational efficiency and robustness. Heuristic information concerning the ability of a specific element to provide good stress results for a given type of problem need not be considered. The smoothed stresses provide a better representation regardless of the element chosen.

A New Form of Smoothed Solution

The improvements already listed all stem from the new approach for forming the smoothed solution. In this approach, an augmented finite element result is constructed that satisfies the stress boundary conditions on free boundaries and the equilibrium conditions on fixed boundaries. These conditions are imposed on the finite element solution using techniques borrowed from the finite difference method. Finite difference templates are first superimposed on the boundary nodes of the finite element model. This step introduces a set of fictitious nodes into the finite element mesh. That is to say, the finite element mesh is augmented with a set of fictitious nodes. The displacements of these nodes are determined by enforcing the stress and equilibrium conditions on the boundary nodes of the finite element solution.

As a result, the augmented finite element solution exhibits the form of a finite difference result and it is interpreted as an approximate finite difference result. This augmented solution is not an actual finite difference solution because the finite difference approximations of the governing differential equations are not satisfied at all of the nodes.

This interpretation of the augmented solution provides the basis for the improvements to the evaluation of finite element results developed in this work. By interpreting the augmented solution as an approximate finite difference solution, we acquire the basis for forming new kinds of pointwise evaluation measures. This opportunity exists because we now have two approximate stress representations available to us that are related to each other at the most basic level. As we know from the first part of the book, the approximated finite difference solution (the augmented finite element solution) is a solution of the differential form of the problem being solved and the finite element solution is a solution of the variational form of the problem.

Since both of these approximate solutions must converge to the same result as the discrete model is improved, the differences between the two solutions must be due to the inability of the discrete model to accurately represent the continuum. Thus, we can view the differences between the two results as quantifying the discretization errors.

Note that this identification of the discretization errors does not require us to consider one result to be more accurate than the other. As a result, we can say that the evaluation of the discretization errors is separate from the evaluation of the accuracy of the stresses. This does not mean that we cannot evaluate the accuracy of the stresses. It simply means that the availability of the refinement guides emerge from the derivation of evaluation measures at an earlier stage than do accuracy measures. As a result, we have two approaches for quantifying the discretization errors. One is based on stress differences, and the other is based on the residuals contained in the augmented finite element solution, the approximate finite difference result. We discuss these two approaches for forming evaluation measures in turn.

Stress-Based Evaluation Measures

Interpreting the augmented finite element result as an approximate finite difference result enables us to develop refinement guides and termination criteria based on the differences in stress quantities that perform better than the strain energy measures. This improved performance is possible for two reasons: (1) we can compute the stresses on a pointwise basis and (2) we can identify the conditions that ensure that the smoothed strains are of an order higher than the finite element representations. Both characteristics result from the Taylor series nature of the finite difference and the finite element methods. The Taylor series nature becomes obvious when we recall that both methods have a common basis in strain gradient notation.

Pointwise refinement guides can be formed by comparing the finite element stresses to the stresses extracted from the smoothed solution on a pointwise basis. The smoothed stresses are extracted by superimposing finite difference templates on the nodal points and computing the Taylor series coefficients. The most logical points to choose for this comparison are the nodes. However, comparisons at the Gaussian points would also produce reliable refinement guides. The refinement guides computed at either of these points would identify discretization errors because of the argument presented earlier, i.e., the two forms of the solution must converge to the same result, so differences in the solutions must be due to deficiencies in the discrete model.

The interpretation of the augmented finite element result as an approximate finite difference solution also allows us to evaluate the accuracy of the stresses in the finite element representation. In Lesson 19, we identify the conditions that must exist for the smoothed stresses, i.e., the stresses extracted from the augmented finite element solution, to be of an order higher than the discontinuous finite element stress representations. As a result, we can claim that the smoothed stresses are more accurate than the stresses in the finite element representation. In this situation, the differences between the two results can be interpreted as an estimate of the errors in the finite element stresses. We can, however, reach a more significant conclusion from this result.

Since the smoothed stresses are of an order higher than the discontinuous finite element stresses, we can use these stresses as the output from the finite element analysis. Therefore, we are interested in the accuracy of the smoothed stresses. One way we can judge the accuracy of the smoothed stresses is to evaluate the convergence of these stresses as the mesh is refined in the region where the accuracy is being studied.

The observation that the smoothed stresses are the quantities of interest suggests that the differences in the stresses in the finite element result and the smoothed stresses be used as refinement guides and that the convergence of the smoothed stresses be used as the termination criteria. The estimated errors in the finite element result could be used as termination criteria but this could be overly conservative since the finite element stresses are of an order lower than the smoothed representations. This conservatism come from the fact that the smoothed stresses may converge long before the finite element stresses reach a constant value. This is demonstrated with an example in Lesson 19.

Residual Evaluation Measures

The interpretation of the augmented finite element result as an approximate finite difference result provides a new theoretical foundation and new techniques for forming residual-based evaluation measures. The residuals are extracted from the nodes of the augmented solution with the following procedure.

A finite difference template is first superimposed on each node of the finite element mesh. Then the displacements from the augmented solution are substituted into the finite difference operators for the templates. These operators are, in turn, substituted into the approximations of the governing differential equations. The amounts by which these approximations fail to satisfy the equilibrium equations are the residuals. This process is similar to the way in which the smoothed stresses are extracted from the augmented finite element solution.

This procedure for computing the residuals has two primary advantages over the Kelly approach, one of which is theoretical and the other is computational. The theoretical advantage is that there is no question about the meaning of the residuals. These quantities measure how far the approximate finite difference solution differs from the actual finite difference result. Since the approximate result is derived by augmenting the finite element result with boundary conditions, we can also interpret the residuals as identifying how far the finite element result differs from the actual finite difference result. Since the finite element and the finite difference solutions must converge to the same result, the residuals are a direct measure of the discretization error. Thus, we can use the residuals directly as refinement guides.

In addition, we can use the residuals as termination criteria. This is the case because the residuals measure how far the augmented solution deviates from an actual finite difference result. When we use the residuals as termination criteria, we are evaluating the smoothed stresses, not the discontinuous finite element representations. This means that we must use the smoothed stresses as the output of the finite element analysis if the result is evaluated using residuals.

The computational advantage of the new approach derives from the fact that the residuals are pointwise in nature and of only one type. In contrast, two different types of residuals must be computed in the Kelly approach and they are not pointwise quantities: (1) residuals similar to the ones contained in the augmented result must be computed over the domain of each finite element and (2) the jumps in the stresses on the boundaries between the elements must be computed. Another advantage of the new approach is that there is no question of whether the interelement jumps should be considered as residuals. That question simply disappears because we are evaluating the smoothed stresses at the nodes. There are no interelement jumps in these pointwise representations.

In Lesson 19, two example problems are used to demonstrate that the residuals possess the characteristics we would like to see in an evaluation measure. In regions of singularities, the residuals increase as the mesh is refined. This can be interpreted as identifying regions where the geometry should be modified, i.e., sharp corners should be rounded etc. In more typical regions, the residuals decrease under mesh refinement.

The size of the residuals can be put in perspective by relating them to parameters of the problem in two simple ways. The residuals can be interpreted as fictitious applied loads, *à la* Kelly, as shown in Fig. 4. In this interpretation, we can compare the residuals in magnitude to the applied loads to provide a relative size of the errors in the stresses. If the residuals are relatively small when compared to the applied loads, the problem is sufficiently accurate. Or, with the same interpretation of the residuals, we can apply them to the existing linear finite element model to get an *approximation* of the errors in the stresses.

The stresses found by applying the fictitious loads directly to the existing model *cannot* be used to correct the stresses found in the original analysis because of orthogonality considerations (see Refs. 4–7). This means that the finite element model would represent the stresses that produced the fictitious loading if it could. Since the original model cannot capture these stress distributions, the model must be refined to represent these more complex stresses. After the model is refined, we can apply the fictitious loads to the model to get a correction for linear problems. Whether the computation of a correction is a more effective way of improving the finite element result than simply applying the actual loads to the model is not studied here, but it is a question worth pursuing. Note that the residuals can be used to guide the refinement of the finite element model since the residuals quantify the inability of the model to represent these stress distributions.

As we can see, the new approach puts the use of residuals on a firm theoretical foundation and it provides an effective way for computing these quantities. This new capability opens the door to several interesting research topics. For example, the question of whether to use residual-based or stress-based evaluation measures can be approached in a straightforward manner using the idea of the augmented solution as an approximate finite difference result. The questions of what refinement guide and of what termination criteria to use in practice are discussed in general terms in Lesson 19.

Relationships between the Finite Element and the Finite Difference Methods

In this section, the various relationships that exist between the finite element and the finite difference methods are presented. We see that these relationships provide the basis for new approaches for evaluating finite element results and a new technique for solving finite difference problems. In fact, a careful comparison of the two methods requires that we reevaluate whether a problem should be solved with the finite element method or with the finite difference method.

Same Goal, Different Approach

In Part I of this book, we saw that the finite element and the finite difference methods share a common goal of trying to minimize the strain energy in the structure being analyzed. However, it is important to emphasize that the two methods use different approaches to reach this common goal. The finite element method attempts to achieve this goal directly by minimizing an approximate representation of the strain energy in the problem. On the other hand, the finite difference method tries to minimize the strain energy indirectly by approximating the solution of the governing differential equations for the problem. As demonstrated in Part I, the governing differential equations are conditions that must be satisfied if the strain energy is to be a minimum.

The existence of two different approaches for achieving the same goal provides the theoretical basis for two new methods for quantifying discretization errors in finite element models, namely, a new smoothing approach and a new residual approach. The new smoothing approach computes the pointwise differences in the two stress representations to form refinement guides. The new residual approach uses the residuals in the augmented finite element result as refinement guides and termination criteria.

Same Expansion, Different Coefficients

Besides trying to reach the same objective in different ways, the two methods have a common polynomial basis. Both methods are derived from the same Taylor series representation of the displacements. However, the coefficient matrices for the two methods are formed from different terms contained in the Taylor series expansions. The finite element method uses the first-order coefficients or the strains in its representation of the strain energy. The finite difference method uses the second-order coefficients or the derivatives of the strains in its representation of the governing differential equations.

Specifically, we saw in Parts III and IV that both methods could be formed by relating the nodal displacements and the nodal locations in terms of the Taylor series expansions in the form of [Φ] matrices. For example, we saw that the nine-node finite elements and nine-node finite difference templates are developed from the same [Φ] matrix. This means that the finite difference templates can represent the same geometries as the nine-node finite element. The existence of a common basis for the two methods means that both methods can represent the same set of problems.

Same Displacements, Different Stress Representations

In the outline of the new approach for evaluating finite element results, we saw that the displacement outputs of the two methods can be related. By augmenting the finite element result with fictitious nodes as a way of introducing additional boundary conditions into the representation, we saw that a finite element solution can be converted into an approximate finite difference result. In fact, Lesson 19 demonstrates that innate characteristics of the two solutions are related. In Lesson 19, we prove that when the interelement jumps in the finite element stresses are zero, the residuals in the approximate finite difference solution are zero. However, this does not imply that the converse is true. This is shown with a one-dimensional example in which the residuals are zero and the interelement jumps are not zero in the finite element representation of the stresses. In addition to clarifying the relationship between the interelement jumps and the residuals, this example illustrates the fact that the finite difference stresses can be of an order higher than the finite element representations.

Related Solution Techniques

Not only do both methods share the same theoretical goal of minimizing the strain energy and the same Taylor series basis for the formulation of their coefficient matrices, the solution of a finite element version of the problem can also be directly related to the solution of the associated finite difference problem. In Lesson 19, a procedure for iteratively solving finite difference problems with adaptively refined finite element models is presented. In this procedure, a finite element model is improved until the residuals in the augmented solution reach an acceptable level. This result then becomes the solution to the finite difference problem.

The ability of the adaptive refinement process to solve finite difference problems is brought to the attention of the reader for several reasons. On the one hand, it provides another tool with which to solve finite difference problems. On the other hand, it shows that the two methods are equivalent with respect to the problems that the two methods can solve. This means, for example, that if we chose one method over the other because we believed that the chosen method could represent a wider range of boundary conditions, that reason is no longer valid. As a result, the use of either method in any field because it is ''common practice'' can be legitimately questioned. The decision of which method to use must now be made for more basic reasons, such as computational efficiency or computational robustness.

A Universal Evaluation Procedure

The ability to solve finite difference (FD) problems using the adaptive refinement of a finite element (FE) model suggests the possibility of using stress-based refinement guides to evaluate finite difference results. Stress-based refinement guides for the finite difference results can be formed by subdividing the finite difference template into three- or four-node segments. These segments are then treated as finite elements. That is to say, a set of discontinuous stresses is extracted from these subdivisions. Then, these discontinuous stresses are compared to the smoothed stresses extracted from the finite difference result to form refinement guides. This procedure is demonstrated in Lesson 20.

Application to FE Results

The process just described for evaluating finite difference results by subdividing the finite difference templates can be extended to apply to the evaluation of finite element results. In fact, the extension is direct in that there are no changes in the procedure. The discontinuous stresses used to form the evaluation measures are extracted from the subdivisions of the finite difference templates used to form the smoothed stresses. As a result, the stresses in the underlying finite element model never need to be computed.

The only output from the finite element analysis used explicitly in the evaluations are the displacements. As a result of extracting the discontinuous stresses from the overlaid finite difference template, the refinement guides we compute by this process are independent of the underlying finite element model. This procedure is not demonstrated explicitly, however, the success of applying the approach to finite difference models in Lesson 20 demonstrates that the procedure is, indeed, independent of the underlying finite element model. That is to say, when we apply the procedure to a finite difference model, its independence from the underlying finite element model is ensured because there is no underlying finite element model.

The importance of being able to develop a universal evaluation procedure is twofold. On the one hand, it lets us choose the finite elements used in the finite element model for computational reasons instead of for a perceived ability to handle a certain type of problem well. This implies that the output stresses from the analysis are the smoothed stresses. On the other hand, it enables us to create specialized postprocessors for evaluating finite element results that can be appended to a wide variety of codes. The only quantities needed from the finite element analysis are the nodal displacements, the nodal locations, the material properties, and the boundary conditions.

Closure

The developments presented in the previous four parts of the book are integrated and extended here. In Part I, we saw that both the finite element and finite difference methods attempt to find an approximate solution that minimizes the strain energy in the problem being solved with different approaches. The developments presented in Part V depend on the common goal of the two approximate solution techniques and on the similarities of and the differences between the two methods. These similarities and differences are presented and used to advantage in the development of the error measures.

Augmented Solution

The similarities and differences between the two methods are exploited by augmenting finite element displacement results so that additional boundary conditions are satisfied. This augmented result is then interpreted as an approximate finite difference solution. The ability to form an approximate finite difference result from an associated finite element result provides the basis for new theoretical insights and new operational procedures.

Stress-Based Refinement Guides

We use the fact that both methods seek the same goal to claim that the two results must converge to the same result. We then compare the two stress representations and assume that any differences between them are due to the inability of the discrete model to accurately represent the actual continuous solution. The differences between the two stress representations at individual points are used as refinement guides. These pointwise measures enjoy certain advantages over strain energy measures: (1) they can be tailored for different purposes and (2) they have a higher resolution than the strain energy measures. The flexibility of the pointwise measures is due to the fact that they are not required to represent a fixed quantity like strain energy. They have a higher resolution because these pointwise measures are not diluted by the averaging that is implicit in the integration process.

A Universal Evaluation Procedure

We can develop a procedure for forming stress-based refinement guides that are independent of the underlying finite element model. This procedure is similar to the approach just described. The only difference is the source of the discontinuous stress representations used in the formation of the refinement guides. In the universal approach, the discontinuous stresses are extracted on a node-by-node basis from the finite difference template superimposed on the mesh to compute the smoothed stresses. The finite difference templates are subdivided and the discontinuous stresses are extracted by treating the subdivisions as finite elements. This procedure is demonstrated in Lesson 20.

Extraction of Residuals

The ability to form an approximate finite difference result from a finite element result provides a new way to evaluate finite element results with residuals. The residuals can be extracted by superimposing a finite difference template on the nodes of the augmented finite element mesh and determining if this solution satisfies the equilibrium equations at the nodes. Since the two solutions must converge to the same result, the amount by which the augmented solution fails to satisfy the governing differential equations identifies the discretization errors in the finite element model. The residuals can be used as both refinement guides and as termination criteria.

An Iterative Solution Procedure for FD Problems

The ability to extract the residuals from the augmented finite element result provides an opportunity to develop a new technique for solving finite difference problems. A finite element model can be adaptively refined until the residuals in the augmented result are adequately small. Then the augmented result can be taken as an actual finite difference solution. The result is a close representation of the actual solution because it satisfies both the underlying finite element model and the associated finite difference representation.

Same Polynomial Basis — Higher Order Stress Representations

In Parts III and IV, we formulated both methods from the same Taylor series expansions. This common basis enables us to compare the details of the stress representations in the two methods. This comparison lets us identify the conditions when the nodal stresses of the finite difference representation are of an order higher than the finite element stresses.

We exploit the different orders of the polynomial representations of the two stress models in two ways. The higher order nature of the finite difference representations enables us to use the smoothed stresses as high accuracy output for the finite element analysis. The condition also enables us to develop termination criteria based on the convergence of the smoothed solution. In addition, we could use the differences between the two results to estimate the accuracy of the finite element stresses. However, this criteria can be overly conservative if the output of the finite element analysis is presented as smoothed stresses.

Different Order Coefficients

Although the two methods are based on the same Taylor series approximations of the displacements, the coefficient matrices of the two solution techniques are based on terms of different orders from the Taylor series expansions. It seems possible that by symbolically forming the coefficient matrices for the two methods in terms of strain gradient notation that a close examination of the structures of the two matrices could identify the relative computational characteristics of the two methods. Such an analysis might clearly identify the circumstances when it is better to solve a problem with one method rather than with the other. This idea is not pursued here but the possibility surfaces because of the developments presented here.

Relationships between Evaluation Techniques

Because the stress representations for the two solution techniques emerge from the same polynomial basis, we can and do develop a relationship between the interelement jumps in the finite element stresses and the residuals in the augmented solution. In Lesson 19, we see that when the interelement jumps are zero, the residuals in the augmented solution are also zero. That is to say, both solutions have converged. However, a counterexample shows that the converse need not be true. A one-dimensional example shows that while a solution can have residuals of zero, the interelement jumps in the finite element solution need not be zero. This demonstrates that the smoothed solution can be of an order higher than the stresses in the underlying finite element model.

Conclusion

In conclusion, the developments presented in this part of the book provide new insights and capabilities for computational mechanics. We no longer need to choose between the finite element and the finite difference methods simply because one of the methods has been traditionally used to solve a given class of problems. Now we can choose for reasons of computational efficiency and robustness because we have a solid theoretical basis for understanding that both methods can solve problems with the same geometry. Furthermore, we can extract stress results of the same order from both solution techniques.

Furthermore, the developments presented here improve our understanding of evaluation procedures and our capabilities to compute evaluation measures. These new insights and capabilities mean that we can select the evaluation procedure that fits the needs of the problem being solved. And more importantly, we can use these evaluation procedures with confidence because these developments ensure that our results are of guaranteed accuracy.

Finally, note that the augmented finite element result provides the widest range of evaluation capabilities. When this approach to error analysis is chosen, either stress-based or residual-based evaluation measures can be applied. However, the stress-based measures can be used with other smoothing approaches. The results based on the augmented finite element result provide the theoretical basis for the less precise, but effective approaches. As a final note, if the residual approach is to be used, the residuals should be extracted from the augmented finite element result.

Contents of Part V

A brief outline of each of the five lessons which discuss *a posteriori* error measures is now presented.

Lesson 16 introduces the Zeinkiewicz and Zhu smoothing approach to error analysis, which is the first simple and practical approach for evaluating the errors in finite element results. The idea of using the differences between the discontinuous stresses in a finite element solution and a smoothed result to estimate the error in the strain energy content of individual finite elements is developed. Examples show that this quantity satisfactorily identifies the regions in a model that must be refined to improve the results. However, we see that the estimate of strain energy error is not an ideal termination criterion because the errors at critical points can be disguised in the integration process. In the Zeinkiewicz and Zhu procedure, the smoothing is accomplished by nodal averaging. We see that this form of smoothing underestimates the errors on the boundaries. The needs for ''sharper'' termination criteria and better evaluation of errors on boundaries provide the motivation for the developments presented in the remainder of the lessons. Note that the developments presented in this part of the book are extensions of the brilliant insight of Zienkiewicz and Zhu for forming strain-energy-based refinement guides.

Lesson 17 develops a new way to form the smoothed solution that improves the error estimates on the boundary. In this approach, the finite element solution is augmented with a set of fictitious nodes. The displacements of the fictitious nodes are found using techniques from the finite difference method that introduce additional conditions on the boundaries of the finite element solution. Specifically, the natural boundary conditions are satisfied on the nodes of the free boundaries and equilibrium is satisfied on the nodes of the fixed boundary. This new approach to forming a smoothed solution improves the error estimates for the elements on the boundary. When this augmented finite element solution is interpreted as an approximate finite difference result in Lesson 19, a new theoretical basis for evaluating finite element results is provided.

Lesson 18 demonstrates that pointwise error measures can be developed using the augmented finite element result presented in Lesson 17. This leads to the identification of several pointwise error estimators that are more effective and more efficient than the energy-based error estimators. They are more effective than the energy measures because they evaluate the stresses and/or strains at critical points. That is to say, these error measures focus on the quantities of interest to the analyst. The pointwise error estimators are more efficient than the energy measures because integrations do not have to be performed to compute them. Since the pointwise measures are not diluted by the integration process, they have a higher resolution than the strain energy measures.

Lesson 19 integrates and extends the developments contained in the previous portions of the book. This is accomplished by interpreting the augmented finite element solution formulated in Lesson 17 and applied in Lesson 18 as an approximate finite difference solution. The identification of this new relationship between the finite element and the finite difference results provides a theoretical basis for evaluating finite element results. The following results derive from the interpretation of the augmented finite element solution as an approximate finite difference result:

1. A new approach for extracting residuals from finite element results
2. A new iterative approach for solving finite difference problems
3. A new approach for using residuals in the evaluation of finite element results

4. The validation of the pointwise refinement guides formed in Lesson 18

5. The identification of the conditions for the smoothed strains to be of an order higher than the strain components in the finite element result

6. The identification of a relationship between the residuals in the finite element solution and those in the augmented result

7. The identification of a relationship between the interelement jumps in strains and the residuals in the augmented solution

8. An outline of a universal evaluation postprocessor for finite element results

The list of results just presented have significant implications for computational mechanics. They imply that (1) the choice of whether to use the finite element or the finite difference method for specific classes of problems must be revisited, (2) the inventory of finite elements can be evaluated for efficiency and reduced, (3) evaluation software can be created that is independent of the code to which it is applied, and (4) the analysis step in computer-aided design is closer to becoming automated.

Lesson 20 applies the universal evaluation procedure suggested in the previous lesson. This is done by identifying discretization errors in finite difference models. In addition to demonstrating a stress-based evaluation procedure for finite difference results, this application shows that stress-based refinement guides are available for finite element models that are independent of the underlying finite elements used to form the representation. This is significant because it validates the idea that a generally applicable strain extraction and error analysis postprocessor can be created that can be appended to a wide variety of finite element codes. Of equal importance, this development shows that adaptive refinement procedures can be applied to finite difference models because of the availability of distorted finite difference templates. Note that the residual approach to evaluating finite element results is, by its very nature, independent of the underlying finite element model.

References

1. Courant, R., "Variational Methods for the Solutions of Problems of Equilibrium and Vibrations," *Bulletin of the American Mathematical Society*, Vol. 49, 1943, pp. 1–23.

2. Zienkiewicz, O. C., and Zhu, J. Z., "A Simple Error Estimator and Adaptive Procedure for Practical Engineering Analysis," *International Journal for Numerical Methods in Engineering*, Vol. 24, 1988, pp. 337–357.

3. Zienkiewicz, O. C., and Zhu, J. Z., "The Superconvergent Patch Recovery and *A-Posteriori* Error Estimates, Parts I and II," *International Journal for Numerical Methods in Engineering*, Vol. 33, 1992, pp. 1331–1382.

4. Kelly, D. W., "The Self-Equilibration of Residuals and Complementary *A Posteriori* Error Estimates in the Finite Element Method," *International Journal for Numerical Methods in Engineering*, Vol. 20, 1984, pp. 1491–1506.

5. Kelly, D. W., and Isles, J. D., "Procedures for Residual Equilibration and Local Error Estimation in the Finite Element Method," *Communications in Applied Numerical Methods*, Vol. 5, 1989, pp. 497–505.

6. Kelly, D. W., and Isles, J. D., "A Procedure for *A Posteriori* Error Analysis for the Finite Element Method which contains a Bounding Measure," *Computers and Structures*, Vol. 31, No. 1, 1989, pp. 63–71.

7. Yang, J. D., Kelly, D. W., and Isles, J. D., "*A Posteriori* Pointwise Upper Bound Error Estimates in the Finite Element Method," *International Journal for Numerical Methods in Engineering*, Vol. 36, 1993, pp. 1279–1298.

8. Babuska, I., Strouboulis, T., Upadhyay, S. K., and Copps, K., "Validation of A Posteriori Error

Estimators by Numerical Approach,'' *International Journal for Numerical Methods in Engineering*, Vol. 36, 1994, pp. 1073–1123.

9. Dow, J. O., and Byrd, D. E., ''The Identification and Elimination of Artificial Stiffening Errors in Finite Elements,'' *International Journal for Numerical Methods in Engineering*, Vol. 26, Mar. 1988, pp. 743-762.

10. Dow, J. O., Harwood, S. A., Jones, M. S., and Stevenson, I., ''Validation of a Finite Element Error Estimator,'' *AIAA Journal*, Vol. 29, No. 10, Oct. 1991, pp. 1736–1742.

11. Dow, J. O., and Hamernik, J. D., ''An Improved Smoothing Approach for Estimating Errors on the Boundary,'' *International Journal for Numerical Methods in Engineering*, under review.

12. Hamernik, J. D., ''A Unified Approach to Error Analysis in the Finite Element Method,'' Ph.D. Dissertation, University of Colorado, 1993.

13. Dow, J. O., Hamernik, J. D., and Sandor, M. J., ''A New Approach for Computing Residuals and Super-Convergent Stresses in Finite Element Results,'' *International Journal for Numerical Methods in Engineering*, under review.

14. Dow, J. O., and Stevenson, I., ''An Adaptive Refinement Procedure for the Finite Difference Method,'' *International Journal of Numerical Methods for Partial Differential Equations*, Vol. 8, No. 6, Nov. 1992, pp. 537–550.

Lesson **16**

The Zienkiewicz/Zhu Error Estimation Procedure

Purpose of Lesson   To introduce smoothing approaches for estimating errors in individual finite elements and to demonstrate the use of error estimators to improve finite element models.

In 1987, Zienkiewicz and his graduate student Zhu developed a procedure for easily computing the relative error in individual finite elements that is brilliant in its simplicity (see Ref. 1). The errors in the individual elements are estimated by computing the strain energy contained in the difference between the discontinuous finite element solution and a smoothed solution formed from the finite element result. Finite elements found to contain high errors are subdivided to form a model that more accurately represents the actual problem. The analysis is terminated when an overall or global error measure reaches a predefined level.

The Z/Z approach, however, contains two significant deficiencies: (1) the smoothed solution used in their approach does not explicitly contain the actual boundary conditions of the problem being solved and (2) the method does not have the capability of estimating the accuracy of stresses or strains at individual points. As a result of the first deficiency, the method does not correctly estimate the errors in elements on the boundary, particularly in early refinements. The second deficiency means that high errors may exist in pointwise quantities even if the strain energy criteria for terminating the analysis is satisfied. These deficiencies slow the convergence process and can give an unwarranted confidence in the results of an analysis.

Both of these flaws are corrected in later lessons. Even with these deficiencies, the initial insight that led to the Z/Z error estimators, i.e., the comparison of the finite element result to a smoothed solution, must be considered as the starting point for the development of efficient error analysis procedures.

■   ■   ■

The analysis of the errors in the finite difference model shown in Fig. 1 illustrates the salient features of the Zienkiewicz/Zhu (Z/Z) error estimation procedure. In this model, the bar elements represent constant strains when they are deformed. The discontinuous stresses produced by the finite element model of the bar are shown by the horizontal lines in Fig. 2. In this figure, a discontinuous finite element stress solution for a simple problem is compared to a smoothed stress representation. The dot-filled regions in Fig. 2 represent the discretization errors contained in the individual elements. These errors are quantified by computing the strain energy contained in the differences between the two stress representations. The individual elemental errors are summed to give a global or overall error estimate.

The smoothed stress representation is formed by averaging the discontinuous stresses

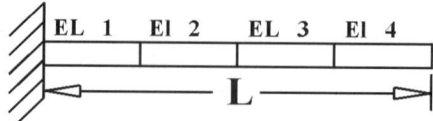

Figure 1. Bar represented with constant strain elements.

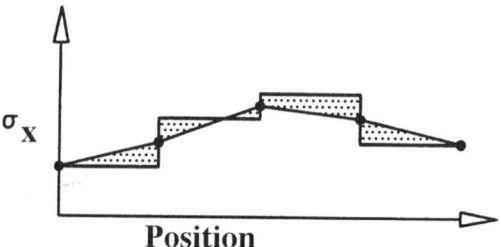

Figure 2. Finite element and smoothed stress representation.

at the nodes and interpolating the stresses between the nodes with the finite element *displacement* interpolation function. The displacement interpolation function is used to produce a continuous stress representation that is of one order higher than the finite element stress models. A discussion of alternative procedures for forming the smoothed solution is presented in a later section of this lesson. In Fig. 1, the stresses are interpolated with a linear function.

Applications of the Z/Z Error Estimators

The error estimates provided by the Z/Z approach have two primary functions: (1) the elemental error estimates are used to guide mesh refinement in adaptive refinement procedures and (2) the global error estimates are used to terminate an analysis. In the mesh refinement process, more elements are put in regions with high errors. We can subdivide existing elements into two or more elements to produce a child mesh or we can regenerate the mesh with more elements in regions of high error. If the existing elements are subdivided, the total strain energy will stay the same or decrease according to the Rayleigh-Ritz criteria. If the mesh is refined with a totally different mesh pattern, energy convergence is not guaranteed, but the error measures ensure the quality of the results.

The adaptive refinement process is terminated when a predefined level of global error is reached. Note that there is no direct relationship between the pointwise stresses or strains and the global error. The lack of connection between these two quantities drives the development of the pointwise error measures presented in Lesson 18.

We can see the reduction in the overall level of error that occurs when a mesh is subdivided by comparing Fig. 3 to Fig. 2. Figure 3 depicts the errors estimated by the Z/Z

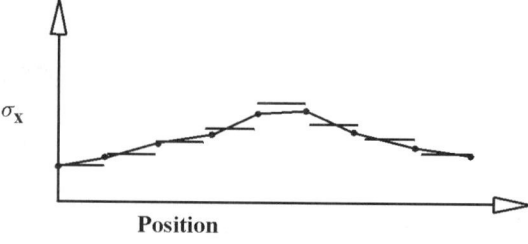

Figure 3. Finite element and smoothed stress representations for a uniformly refined mesh.

approach for the bar shown in Fig. 1 when the model is uniformly refined by halving the length of each element. As we can see, the total error in the problem is reduced as the finite element model is refined and better represents the actual problem. Uniform refinement is not commonly used in practice because it increases the concentration of elements throughout the mesh, not just in critical areas. This strategy is inefficient because it puts too many elements in regions of little interest.

This inefficiency in uniform refinement occurs because most structures have a few relatively small regions containing stress concentrations surrounded by large regions of relatively uniform stress. Thus, uniform refinement puts a majority of the new elements in regions of low error. This highlights the need for accurate error estimates to guide adaptive refinement schemes.

In adaptive refinement, the mesh is refined in regions of high error, while regions of low or acceptable error are ignored or eliminated from the analysis with submodeling techniques. Thus, elements added to the model are concentrated in locations where they are most needed to improve the accuracy of the solution. The Z/Z approach successfully guides adaptive refinement even though it underestimates the errors in elements on the boundary. This flaw often slows the convergence process because boundaries are the locations of stress concentrations in many cases. As a result, critical elements with high errors may not be given first priority in the subdivision process. The underestimation of errors in the elements on the boundary also causes us to underestimate the overall error. This is particularly true in the initial coarse mesh refinements found in the early stages of model refinement.

Underestimation of Element Errors in the Original Z/Z Approach

The reason that the original Z/Z approach underestimates the errors in elements on the boundary is clearly shown in Fig. 2. As we can see, the difference between the finite element solution and the smoothed solution is zero at the right-hand end. This occurs in the one-dimensional case because the smoothed solution equals the average of the finite element solution at the nodes and there is only one element connected to the end node. The error at the end point is obviously not zero because the stress given by the finite element model does not match the actual boundary condition, which is zero in this case. This means that the error in this element is underestimated.

This situation contrasts to the smoothed solution at the interior node of the end element where the results from two internal elements are responsible for the smoothed result. On the interior of the domain, we see that the inability of the finite element to represent the actual solution is a discontinuity in the finite element result. On the boundary, we see that the inability of the finite element model to represent the actual solution is an inaccuracy in the boundary stresses.

Although the one-dimensional case just discussed is extreme in that there is only one element connected to the boundary node, it contains the essence of the reason why the errors are underestimated on the boundary in early refinements by the Z/Z approach for a wide variety of problems. Finite element solutions typically produce incorrect approximations for stresses on the boundary in early mesh refinements. Correct results are not found until the problem is nearly converged. This characteristic of finite element solutions as they converge is discussed further in a later section of this lesson (see Fig. 14). Thus, this deficiency in the original version of the Z/Z approach is a result of the convergence process and the way in which the smoothed solution is formed on the boundary. The underestimation of errors in elements on the boundary is an important

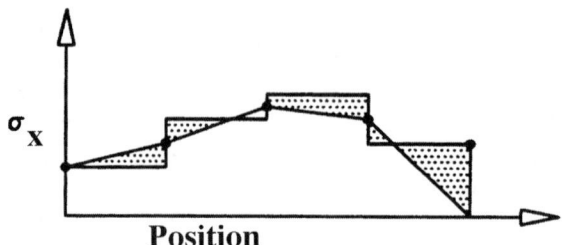

Figure 4. Finite element and a modified smoothed stress representations.

deficiency because many, if not most, of the stress concentrations in structural problems occur on or near a boundary.

Procedures for correcting this deficiency are discussed in more detail later in this lesson in the section entitled, A Discussion of Smoothing Techniques. As a prelude to this discussion, let us see how the inclusion of the actual boundary conditions improves the estimates of the errors in elements on the free boundary. When the actual boundary stress is included in the smoothed solution, the relationship between the finite element and the smoothed solution is as shown in Fig. 4. The averaged stress values are used to form the smoothed solution on the interior.

As we can see in Fig. 4, the difference between the finite element result and the smoothed solution at the right-end is not zero as it was in Fig. 1. Thus, the error estimate in the end element is not automatically underestimated as was the case when the boundary stresses were found from the finite element result. The error estimate is closer to the actual error when the actual boundary conditions are included. Corrections to the smoothing procedure at the fixed boundaries are developed in Lesson 17.

This discussion of the underestimation of the errors on the boundary highlights another problem with the Z/Z approach, namely, the need for a procedure to estimate the errors in stresses and/or strains at individual points. This discussion of error estimates on the boundary has shown that it is possible to have high errors in individual stress or strain components at a point while having low elemental error estimates. This simple observation identifies the reason why a pointwise error estimator is needed. Without an effective pointwise error estimator, an analyst can have a false sense of confidence in the accuracy of the structural analysis.

This situation is exacerbated if a global error estimate is used to terminate an analysis. Low errors in a majority of the elements surrounding a stress concentration can hide the presence of even a group of elements with high elemental errors because of an averaging effect. Similarly, the error in the stresses and strains can be high at a critical point if the global error estimate is low. This situation is demonstrated at length in a later portion of this lesson when the error measures are applied to adaptive refinement problems.

Even with these deficiencies, the original Z/Z approach is sufficient to guide mesh refinement. We can see this in Fig. 5, which shows the estimate of the total error in a cantilever beam problem for a sequence of uniformly refined meshes using the original Z/Z approach. The initial estimate of the total error is low. Then it increases and finally decreases, as it should when the mesh is refined. This pattern of behavior for the global or overall error occurs because of the inaccuracies in the estimation of the elemental errors on the boundary. Other examples of this type of behavior are shown in the next lesson. Procedures for improving the error estimates on the boundary are discussed after the procedures for computing the error estimates are presented.

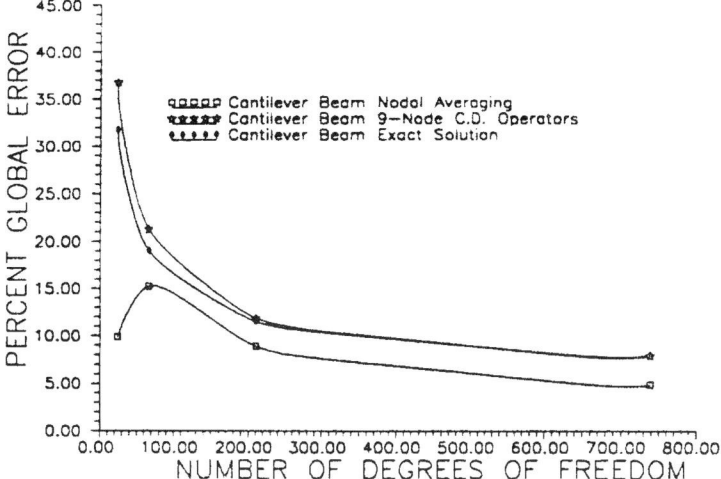

Figure 5. Global error vs mesh refinement.

Quantification of the Error Estimates

The elemental error measure in the Z/Z approach is quantified by computing the strain energy contained in the difference between the discontinuous finite element solution and the smoothed solution. This quantity is formally called the energy norm. The error measure or energy norm is given as

---

Elemental Strain Energy Error Measure

$$| e_i |^2 = \int_{\Omega_i} \{(\sigma^* - \sigma')^T [D]^{-1} (\sigma^* - \sigma')\} \, d\Omega$$

where

$|e|_i$ is the error energy norm in element i.   (16.1)

σ^* is the vector of smooth stresses.

σ' is the vector of finite element stresses.

$[D]$ is the constitutive matrix.

Ω_i is the volume of element i.

This quantity is applicable to one-, two-, and three-dimensional problems because of its vector form. The appropriate stress vector and constitutive relation must be used.

The reason that this error measure is used can be seen by inspecting Fig. 1 or Fig. 4. If the error measure is quantified as the area between the two stress curves, the error in elements with equal positive and negative areas would be reported as zero, which is obviously not the case. The error measure given by Eq. 16.1 essentially squares the difference between the two curves so the error is always positive. The energy measure is used so the error can be related to the total energy in the problem. This gives the error estimate a better physical feel. If the constitutive matrix $[D]$ is removed from Eq. 16.1, the resulting error estimate is called an L^2-norm. This error measure does not have the direct connection to the strain energy as does the energy error norm. The L^2-norm is used in some error estimates and is discussed in Lesson 18.

We can compute a global error estimator by summing the error estimates for the individual elements to give

Global Strain Energy Error Measure

$$| e |^2 = \sum_{i=1}^{n} | e_i |^2 \tag{16.2}$$

where n is the total number of elements.

We can normalize the global error measure in terms of the total energy in the system as

Normalized Global Error Measure

$$\eta = \frac{| e |}{(| U |^2 + | e |^2)^{1/2}} \times 100\% \tag{16.3}$$

where U is the strain energy contained in the finite element solution.

The normalized global error can be interpreted as the percentage of error in the strain energy in the problem. This quantity is often used to terminate an adaptive refinement process using the Z/Z approach. The process is terminated when a predetermined level of error is reached.

We can normalize the elemental error in a similar manner as

Normalized Elemental Error Measure

$$\eta_i = \frac{| e_i |}{(| U_i |^2 + | e_i |^2)^{1/2}} \times 100\% \tag{16.4}$$

The normalized error measure gives the percentage error in an element in terms of the error in that element. It gives a uniformity to the elemental error measures. It enables us to compare the errors contained in individual elements.

The use of the energy error norm is not limited to the Z/Z approach to error estimation. This quantity is closely related to error measures used in other approaches. For example, in Lesson 18 on pointwise error measures, the integrand of Eq. 16.1 is used as one measure of pointwise error. This pointwise quantity can be interpreted as the strain energy density.

A Discussion of Smoothing Procedures

In the discussion of elemental error analysis up to this point, we saw that the formulation of the smoothed solution was central to the computation of the Z/Z type error measure. In this section, we outline some of the procedures that have been used to form the smoothed solution and discuss the limitations of these approaches.

Smoothing in the Original Z/Z Error Estimator

In the first version of the Z/Z error estimator, a smoothed solution is formed that contains the same amount of strain energy as the finite element approximation and that minimizes the difference between the discontinuous finite element solution and the smoothed solution being sought. The nodal values for this smoothed solution are found using a least-squares technique. The stresses over the element are formed by using the *displacement* interpolation function used to form the original finite element. This approach is computationally expensive and it underestimates the errors in elements on the interior of the domain as well as on the boundary.

A Nodal Averaging Approach

In 1987, a modification to the Z/Z approach was implemented that simplifies the formulation of the smoothed solution (see Ref. 1). A simple averaging of the nodal stresses in the finite elements intersecting at a node is used to form the smoothed nodal stresses. The continuous stresses over the elements are formed using the displacement interpolation functions used to form the element as is the case in the Z/Z approach. This modification produced results that are essentially the same as those produced by the least-squares approach. Although the averaging approach is computationally more efficient than the minimization technique, it still contains the same deficiencies as the least-squares approach.

A Finite Difference Smoothing Technique

Both of the smoothing techniques just discussed were criticized for being excessively circular. That is to say, the smoothed stresses against which the finite element stresses are compared are formed from the discontinuous finite element stresses being evaluated. To counter this objection, Dow et al. showed that a smoothed solution formed using the finite element displacements in finite difference operators produced results nearly identical to those formed directly from the finite element stresses (see Ref. 3).

In this approach, a finite difference template is overlaid on the finite element model, as shown in Fig. 6 and the strain values are computed for the central node. This enables us to compute the stress values for use in the smoothed solution. The finite element displacements from nodes surrounding the central node are used in this computation. The procedure for smoothing the boundary stresses using the finite difference approach is outlined later in this lesson and is developed in Lesson 17.

The use of the nodal displacements removes the circularity of using the finite element stresses to evaluate these selfsame stresses contained in the Z/Z and the nodal averaging procedures. The nodal displacements are the most accurate quantities available in the finite element result. The displacements are one order more accurate than the strains. This is, of course, the case because the strains are computed as derivatives of the finite element displacements.

In addition to validating the use of the finite element stress values in forming the smoothed solution, the development of smoothing techniques based on the finite difference method has led to improvements to the smoothing type of error estimators and to the development of pointwise error measures. The use of finite difference techniques

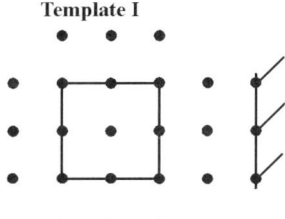

Figure 6. Finite difference templates overlaid on a finite element model.

enables us to introduce the actual boundary conditions of the problem directly into the smoothed solution. Furthermore, the use of finite difference techniques enables us to give the smoothed solution an alternate interpretation that leads to the development of pointwise error measures. The use of finite difference operations in error estimates are discussed in Lessons 17 and 18.

An Alternative Interpretation of the Smoothed Solution

In earlier interpretations, the smoothed solution was viewed as an ''improved solution.'' The intuitive argument for this interpretation is that the actual solution does not contain discontinuities, so the smoothed solution is an improvement over the discontinuous finite element solution. We see the weakness of this argument in the limited way in which the smoothed solutions are used in practice. The smoothed solution is used by most analysts only in the error estimation procedures. It is not used to extract stress values or pointwise error measures. This is the case because there is no theoretical justification for viewing the smoothed solution as an improvement in the pointwise values. In fact, the discontinuities being smoothed over indicate inadequacies in the finite element model. Note that nodal averaging is often used in the graphical presentation of finite element results.

When we form the smoothed solution using techniques from the finite difference method, however, the door is opened for interpreting the smoothed solution as an actual finite difference solution or as an approximate solution to the differential form of the problem being solved. This interpretation is discussed at length in Lesson 18.

When the finite difference approach for forming smoothed solutions was originally developed, the finite difference boundary models available for use were not as flexible as those available for the finite element method. This meant that the use of finite difference smoothing was not universally applicable. This limitation led to the development of the extension of the finite difference boundary modeling capabilities presented in Part IV of this book. The availability of these extended boundary condition models means that the finite-difference-based error analysis procedures are applicable to any problem that can be modeled with the finite element method.

The Revised Z/Z Smoothing Approach

In 1992, Zeinkiewicz and Zhu changed their approach for forming the smoothed solution so it was more efficient and did a better job of estimating the errors in elements on the boundary. This was accomplished with a local smoothing scheme. The smoothed solution at a node is formed by interpolating the stresses found at the Gaussian points surrounding the node, as shown in Fig. 7. The smoothed stresses over the element are found as before

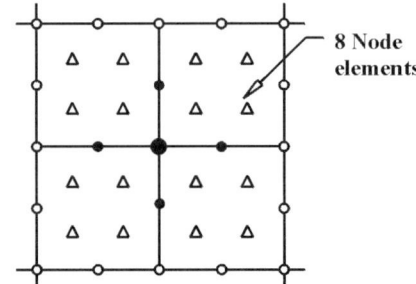

Figure 7. Modified Z/Z smoothing on the interior.

by interpolating the stresses with the displacement shape functions. The stresses at the Gaussian points are extrapolated to the boundary nodes as shown in Fig. 8. As we can see, the smoothed stress and the boundary stresses in the finite element model differ so the errors estimated for the elements on the boundary are increased. This treatment of the elements on the boundary is an improvement over the original approach, but it does not definitively contain the actual boundary conditions, as does the finite difference smoothing approach. The results of this improvement are not presented here (see Ref. 4 for examples of this approach).

In one sense, the modified Z/Z approach is similar to finite difference smoothing. In both cases, the smoothing uses quantities (stresses or displacements) from points surrounding the nodal point at which the smoothed solution is being computed. In the Z/Z approach such a strategy is not possible for smoothing the solution on the boundary. In the finite difference smoothing method, the inclusion of the actual boundary conditions enables a similar process of surrounding the boundary nodal points with a full set of nodal points.

A finite difference template is overlaid on a boundary node, as shown in Fig. 9. The nodal displacements of the six nodes on the domain of the problem are known from the finite element solution. The displacements of the fictitious nodes are found by enforcing the actual boundary conditions. This enables all three stress components to be smoothed on the boundary. The smoothing of the three stress components is shown to be important in Lesson 17 when the finite difference smoothing method is demonstrated. A somewhat similar approach is used to smooth the stresses on fixed boundaries using the finite difference approach.

● **Data to boundary nodes**

△ **Extrapolated Gauss point**

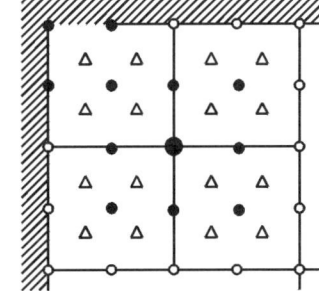

Figure 8. Modified Z/Z smoothing on the boundary.

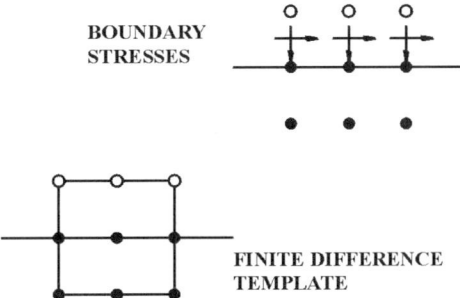

Figure 9. Finite difference template overlaid on a boundary node.

Use of Z/Z Error Estimates in Adaptive Refinement

In the previous discussion, we noted that (1) the use of global error estimates is not very effective in terminating analyses, (2) the stresses on boundaries are often critical to the design, and (3) the stresses on boundaries approach the actual boundary conditions in the convergence process. The third characteristic of the finite element analysis can be stated inversely as follows: the stress boundary conditions are likely to be poorly represented in the early refinements of a finite element model. This section presents an example problem that demonstrates this characteristic. Furthermore, this example demonstrates two additional adaptive refinement strategies.

Problem Definition

This demonstration utilizes the problem of a square shear panel of unit thickness with an interior circular hole. The geometry of the problem and the loading of 1000 lb/in. are shown in Fig. 10. The material properties are taken as $E = 30.0 \times 10^6$ and $\nu = 0.30$. The symmetry of the problem is such that it can be solved with a one-quarter symmetric representation. The problem is solved with a commercially available finite element code into which a Z/Z error estimator has been added as a postprocessor. The smoothed solution is formed using nodal averaging. The problem is modeled with six-node isoparametric linear strain elements. These results are taken from Refs. 5 and 6.

The problem is solved with three adaptive refinement schemes. In the first case, the full model is refined in regions of high error and then solved again until an acceptable level of error is reached. In this example, elements with 5% or greater error are arbitrarily defined as high error elements and the model is refined until all of the elements are at or below 5%. We see that this criteria is very conservative for this problem. The uncertainty of the level of strain energy error to choose to guarantee a given level of accuracy in the stresses and/or strains is a motivation for the development of a pointwise error measure.

We see that the elements with high error are concentrated near the region of high stress and that there are a large number of elements with little or no error. This suggests the removal of regions of high accuracy from repeated analysis. Two alternative strategies that eliminate regions of high accuracy from the analysis and increase the efficiency of the analysis are presented. We also see the problem that arises from using the global error measure as a criterion for terminating an analysis in these examples.

Full Model Refinement

The original quarter symmetry model for the 1:1 ratio ellipse, i.e., the circular hole, containing 430 degrees of freedom (DOF) is shown in Fig. 11 with the percentage

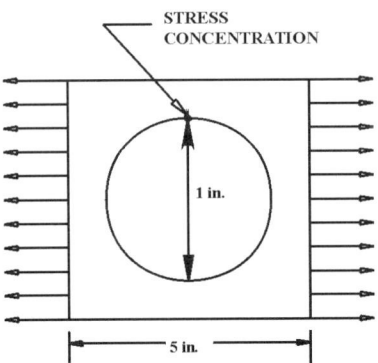

STRESS
CONCENTRATION

1 in.

5 in.

Figure 10. Problem definition.

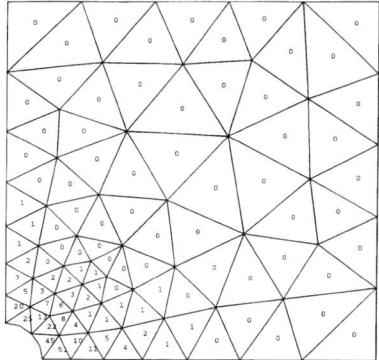

Figure 11. Initial finite element mesh with elemental error norms.

elemental error norms. The maximum stress of 2688 psi is 11.7% less than the theoretical maximum stress of 3047 psi. The analytic solution for the problem of a panel of finite dimensions with elliptical holes is taken from Ref. 7.

The global error norm for the model is 3.9%. If the quality of the result were judged in terms of the global error norm and in the absence of a known solution, we might conclude that the stress results produced by this analysis are dependable. However, the elemental error norms presented in Fig. 11 indicate that large errors exist in the elements adjacent to the hole. The maximum elemental error is 51.1%.

The 1:1 model shown in Fig. 11 is globally refined until the maximum elemental error is 5%. The peak value of σ_x is 3020 psi, which is 0.9% less than the analytic solution. The final finite element model contains 11,454 degrees of freedom and is shown in Fig. 12. This final result was the last step in a sequence of refinements. The aggregate results of these refinements are given in Table 1.

It is interesting to note that the peak stress found in the first refinement agrees with theory to within 2.1%. This means that the subsequent refinements are probably not needed. Note that the analytic solution is not available for most problems, so the accuracy of the approximate solution must be confirmed in other ways. This can be accomplished by evaluating the change in the peak stress from refinement to refinement or by specifying an elemental error criterion. In either case, the third and, perhaps, the fourth

Table 1. Results for a globally refined model.

DOF	Maximum elemental error (%)	$(\sigma_x)_{max}$ (psi)	Stress error (%)	Global error (%)
430	51.1	2,690	11.72	3.9
1,362	21.6	2,985	2.03	0.6
2,150	13.2	3,000	1.54	0.4
5,606	7.2	3,015	1.05	0.3
9,958	5.1	3,020	0.89	0.2
11,454	4.0	3,020	0.89	0.2

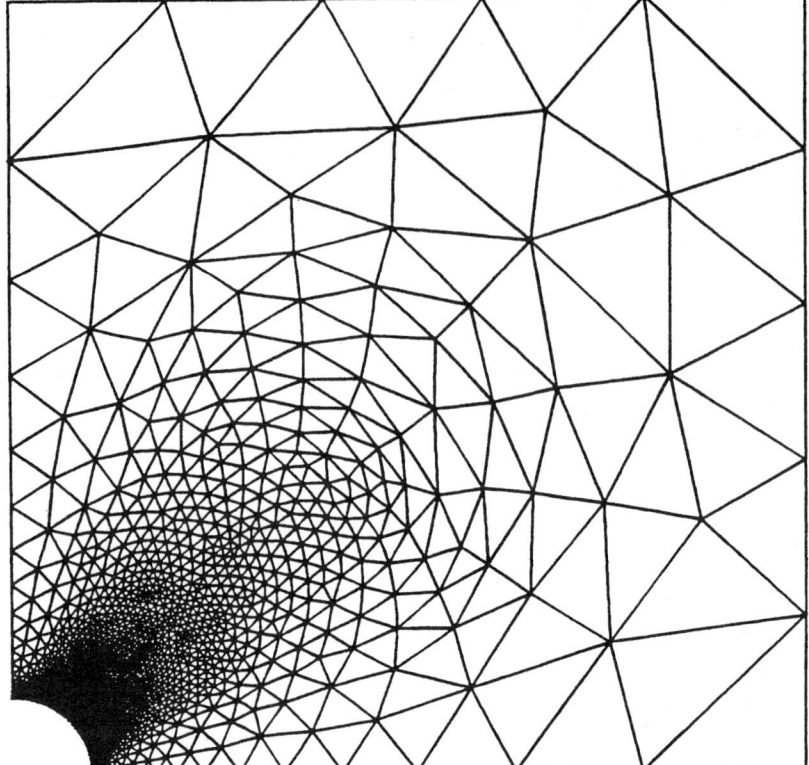

Figure 12. Final global mesh refinement.

iterations would have to be performed before convergence could be assumed. The third iteration has 5 times as many DOF as the original mesh and the fourth refinement has 13 times as many.

Another indicator of convergence is the agreement between the actual boundary stresses and those predicted by the finite element model. In the case being solved, there are no normal or tangential stresses applied to the boundary of the hole. When these stresses are computed for the finite element model, they should be identically equal to zero. The average nodal normal stresses for the first five models are plotted versus their angular position in Fig. 13 to show their convergence characteristics. As we can see, the approximate boundary condition approaches the actual result. The fact that the boundary stresses are misrepresented in the original coarse mesh shows the value of incorporating the actual stress boundary conditions in the error estimator, as mentioned in the discussion of other smoothing techniques presented earlier. The stress distribution for the final refinement is nearly identical to the previous two refinements shown.

A count of the low error elements shows that 81.4% have elemental errors of 5% or less. Since solution times are related to the square of the problem size, we can see that a penalty is paid for retaining unnecessary portions of the problem.

As we can see in the final column of Table 1, the global errors are not closely correlated to the errors in the maximum stresses or the maximum elemental error. As we can deduce, the global error is actually a better measure of the converged area of the overall problem. That is to say, if the boundary size were increased, the global error would be reduced, and the error in the maximum stress would remain the same. A similar result is seen in the next example problem solved.

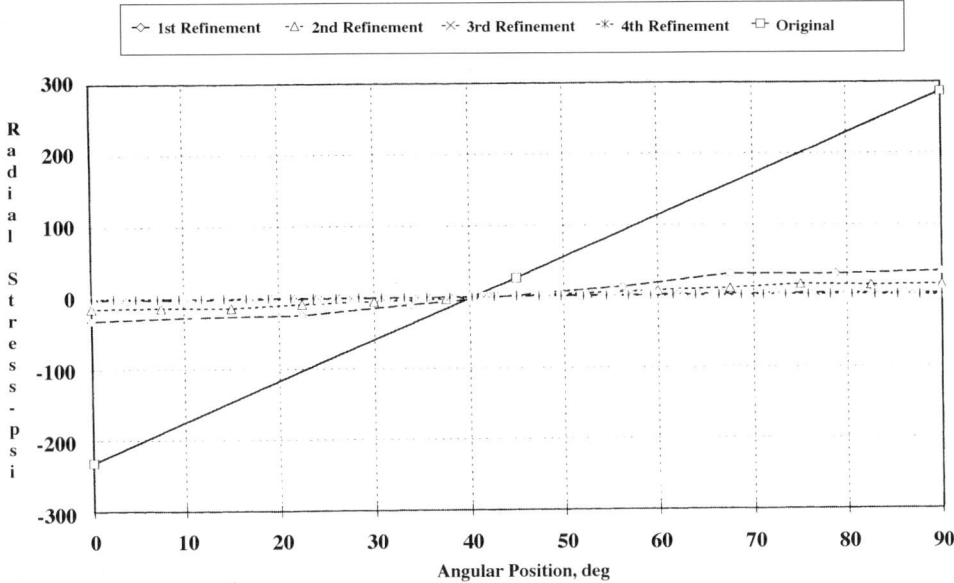

Figure 13. Finite element normal stress vs position.

Fixed Internal Boundary

As shown in the previous subsection, the reanalysis of the total problem contained a large region that did not need to be reanalyzed because the errors in this region are low. The available computational resources could be better expended on the critical part of the problem where the stresses and errors are higher if the region of little interest is removed from the subsequent analyses.

A large portion of the region of low error is separated from the original problem with an interior boundary that consists of the arc of radius of 1.65 in. shown in Fig. 14. The origin of the arc coincides with the center of the interior hole and the radius encompasses the element farthest from the center with an error measure of 5%. This provides an internal boundary that accurately represents the effect of the remainder of the problem on the submodel. The nodes on the boundary are given the displacements found in the original analysis.

The results of successively refining this submodel are given in Table 2. As we can see, the critical stress for even the initial refinement correlates well with the exact result. The accuracy of the results shown in Table 2 indicate that the interpolation of the displacements on the internal boundary from the initial solution of the whole problem accurately imposes the effects of the remainder of the problem on the subproblem. Again, we see that there is little correlation between the global error measure and the maximum elemental error or the accuracy of the stress at the critical point. In fact, global error loses its meaning in this context.

The distribution of the elemental errors in the first four refinements of the subproblem are shown in Fig. 15. As the refinement progresses, more and more of the region is populated by elements with little or no error. This suggests that the internal boundary need not be fixed and that the computational efficiency could be enhanced by introducing a strategy with a moving internal boundary.

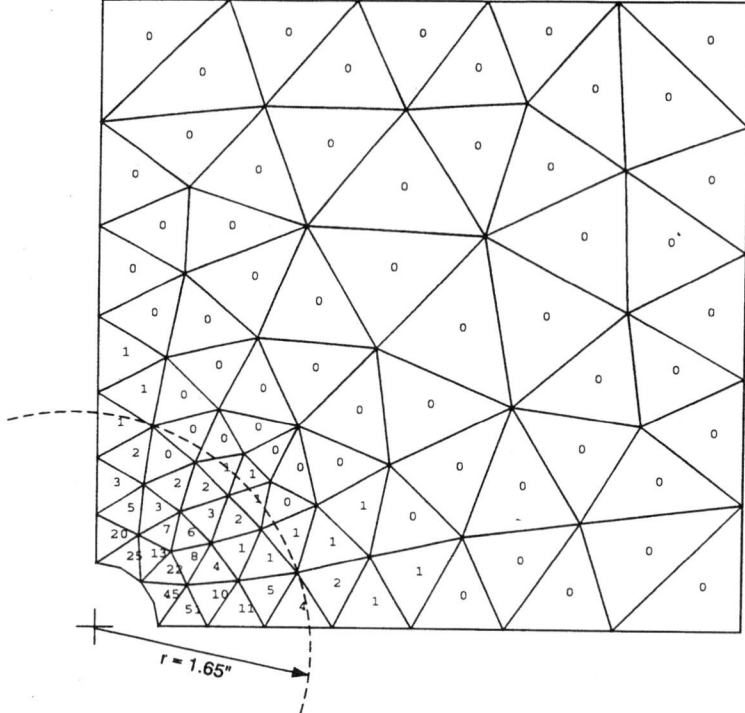

Figure 14. Initial boundary for the fixed radius submodel.

Floating Internal Boundary

As we can see in Fig. 15, large regions of low error exist as the problem is refined. This suggests the use of a floating internal boundary. That is to say, we can use the idea used in the previous example to reduce the size of the problem dynamically to continually shrink the size of the problem.

Two different floating boundary schemes were evaluated. Originally, the origin of the radial boundary coincided with the center of the hole. However, preliminary studies showed that a wider range of problems could be satisfactorily analyzed by the moving boundary strategy if the center of the arc was located at the point of maximum stress (see Refs. 5 and 6). In strategies involving moving boundaries, the boundary conditions are

Table 2. Results for a locally refined model.

DOF	Maximum elemental error (%)	$(\sigma_x)_{max}$ (psi)	Stress error (%)	Global error (%)
282	25.0	2,984	2.07	4.1
442	20.0	2,977	2.29	2.4
942	12.9	3,005	1.38	1.0
4,466	7.2	3,017	0.98	0.2
9,174	5.1	3,018	0.95	0.1
11,394	4.0	3,018	0.95	0.1

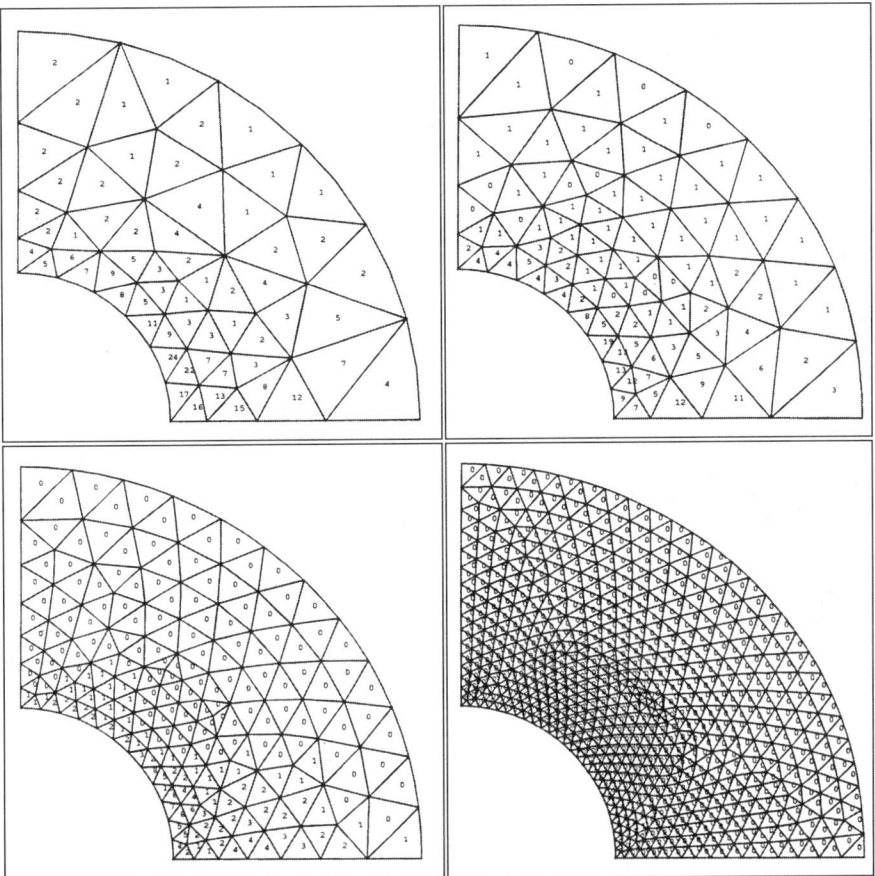

Figure 15. Refined meshes with elemental error estimates.

interpolated from the previous, locally refined mesh instead of from the original global model. This approach was shown to be necessary if accurate results were to be achieved. The results of this strategy for the 1:1 geometry are shown in Table 3.

When we compare these results to the previous two solutions of this problem, we can make two observations concerning accuracy. First, we see that approximately the same number of DOF are required to achieve the same level of accuracy in the maximum stress. Stress errors of 1.25% or less are found by the three approaches with models

Table 3. Results for a locally refined model with a moving internal boundary.

DOF	Maximum elemental error (%)	$(\sigma_I)_{\max}$ (psi)	Stress error (%)
1,400	26.6	2,926	3.97
1,356	19.5	2,975	2.36
1,282	15.2	2,989	1.90
1,218	13.1	2,989	1.90
2,946	9.1	3,002	1.48
5,402	5.3	3,009	1.25
5,160	4.6	3,009	1.25

consisting of 5606, 4466, and 5160 DOF, respectively. Second, we see that substantially fewer DOF are required to achieve the specified level of 5% for the maximum elemental error with the floating boundary approach than for the other two approaches. Only 5160 DOF are required versus 11,454 and 11,394 for the global model and the fixed internal boundary model, respectively. Thus, we can conclude that the approach of adaptively refining a local model with a moving boundary centered at the point of maximum stress produces efficient and accurate results.

Applications to More Severe Stress Concentrations

The approach just presented of using a floating interior boundary centered at the point of maximum stress has been applied to more severe geometries. The results of applying the procedure to stress concentrations in elliptical holes with aspect ratios of 2:1, 5:1, and 10:1 are now presented.

The coarse global model of the 2:1 stress concentration with the error distribution in the mesh is shown in Fig. 16. The mesh contains 672 DOF and yields a peak x component of stress of 2956 psi. The analytic result is 5017 psi so the stress error is 41%. The model has a maximum elemental error of 60%.

When the model is globally refined with the 5% elemental error as the target, the results are as shown in Table 4. As we can see, sufficiently accurate results were obtained before the elemental error criterion was reached. With a final refinement of 5232 DOF, a peak stress of 4952 psi is predicted. This is 1.3% less than the analytic solution.

When the floating radius internal boundary approach is applied to the 2:1 problem, the results are as shown in Table 5. After seven refinements, the peak x component stress

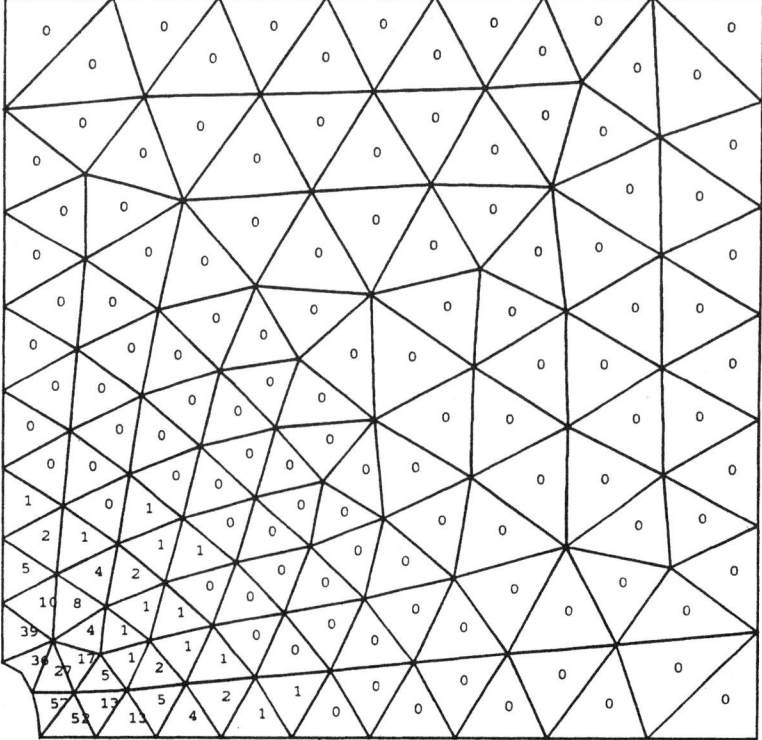

Figure 16. Initial mesh refinement for the 2:1 ellipse with error norms.

Table 4. Results for a globally refined 2:1 model.

DOF	Maximum elemental error (%)	$(\sigma_x)_{max}$(psi)	Stress error (%)
672	59.7	2,956	41.1
1,430	59.7	3,246	35.3
2,104	56.4	3,743	25.4
3,740	18.4	4,730	5.7
5,060	11.8	4,900	2.3
5,432	9.2	4,952	1.3

is 4954 psi, which is 1.26% less than the exact solution. This result was achieved with a model containing 2802 DOF. This is approximately one-half of the size of the model needed to achieve a comparable accuracy by refining the global model.

The coarse global model of the 5:1 stress concentration with the error distribution in the mesh and the location of the initial interior boundary is shown in Fig. 17. The mesh contains 246 DOF and a peak x component stress of 3224 psi. The analytic result is 10,948 psi so the approximate stress value is more than three times smaller than the actual value. This corresponds to a stress error of 70.6%, a maximum elemental error of 74%, and a global error of 11.3%. As is typical, the maximum error is near the hole and the errors drop off rapidly away from the hole.

The results obtained by refining the problem with the floating interior boundary centered at the point of maximum stress are summarized in Table 6. After seven refinements, a peak x component of stress of 10,507 psi is predicted. This is 4.03% less than the theoretical value. This result was achieved with a mesh of 6,736 DOF and a maximum elemental error of 3.6%.

Since this problem was not solved using global refinement, there is no basis for directly establishing the efficiency of the moving boundary technique for this problem. Based on the average element size in the final refinement, however, we can approximate the problem size for a uniformly refined mesh as follows. Inspection of the final mesh refinement for the 5:1 ellipse shown in Fig. 18 gives the size of the elements in the critical

Table 5. Results for a locally refined 2:1 model with a moving internal boundary.

DOF	Maximum elemental error (%)	$(\sigma_x)_{max}$ (psi)	Stress error (%)
246	74.0	3,224	70.55
592	65.7	3,797	24.32
754	27.2	4,511	10.09
720	18.7	4,774	4.84
644	11.8	4,853	3.27
666	10.0	4,889	2.55
1,326	7.3	4,938	1.58
2,802	3.8	4,954	1.26

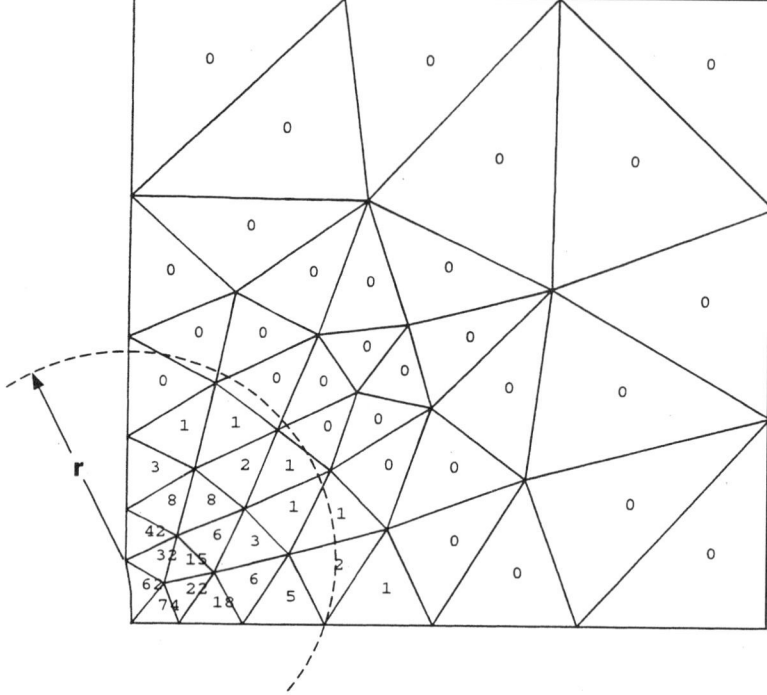

Figure 17. Initial finite element mesh with elemental error norms and internal boundary for 5:1 ellipse.

region. The length of the side shown is 0.00458 in. A uniformly refined 5 in. × 5 in. mesh would contain approximately 1.125×10^7 DOF. If we made an adjustment of, say, 100 to allow for a transition to larger elements toward the boundary (for the 1:1 geometry, an adjustment factor of 67 was found), the full model would contain approximately 112,500 DOF. Thus, we conclude that for the 5:1 stress concentration, the submodeling based adaptive refinement procedure provides a substantial reduction in the number of DOF required to accurately represent this problem.

The results of applying the procedure to the stress concentration consisting of a 10:1 ellipse produces the results shown in Table 7. After eight refinements, a peak stress

Table 6. Results for a locally refined 5:1 model with a moving internal boundary.

DOF	Maximum elemental error (%)	$(\sigma_x)_{max}$ (psi)	Stress error (%)
246	74.0	3,224	70.55
1,264	80.9	5,651	48.38
1,630	34.7	8,102	26.00
1,718	19.4	9,405	14.09
1,586	15.1	9,838	10.14
1,468	13.0	9,988	8.77
3,172	8.2	10,294	5.97
6,736	3.6	10,507	4.03

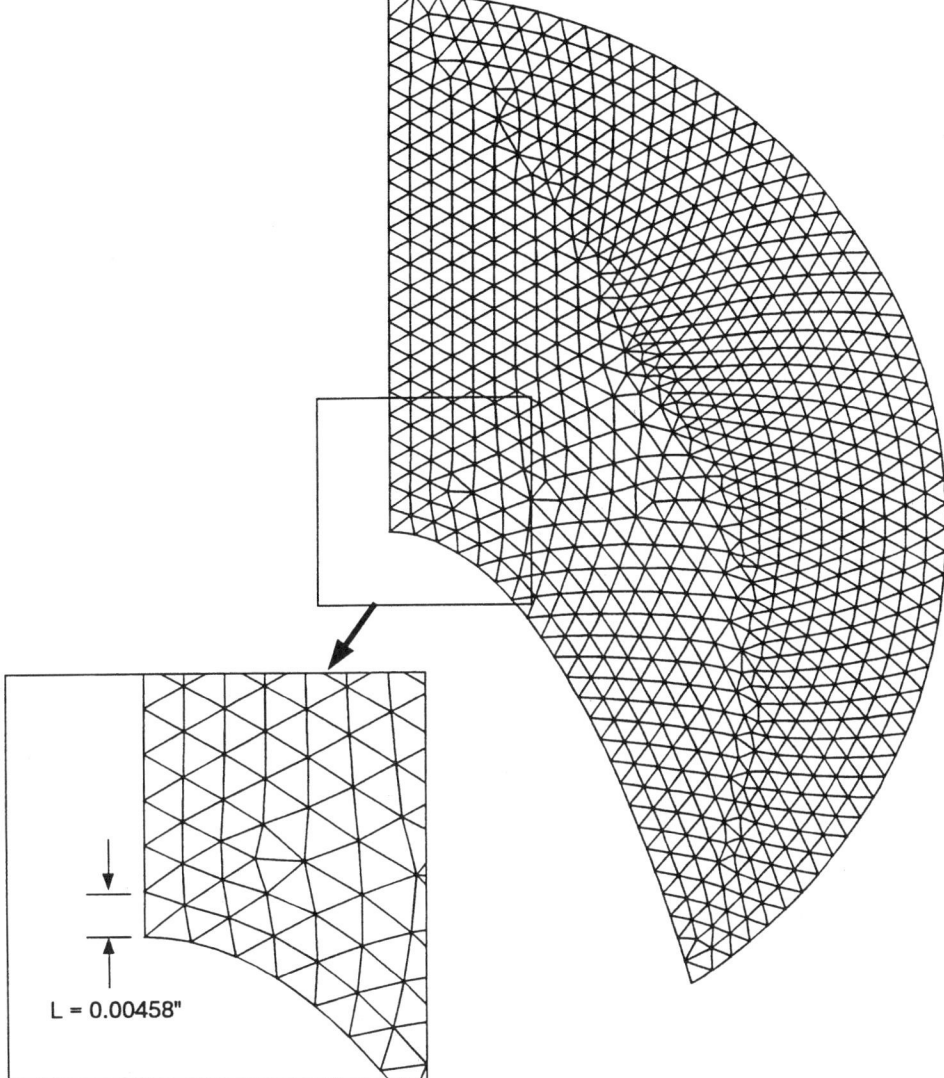

L = 0.00458"

Figure 18. Final mesh refinement for the 5:1 stress concentration.

component in the x direction of 20,039 psi is predicted. This is 3.92% less than the analytic result. This result was achieved with a mesh of 5430 DOF and a maximum elemental error of 0.8%.

Closure

This lesson has presented and applied the Z/Z approach to error estimation. The two major limitations of the method have been brought forward, namely, (1) the underestimation of the errors in the elements on or near the boundary and (2) the ineffectiveness of the global error measure as a termination criteria because of its lack of correlation to the accuracy of the stresses or strains.

The use of the elemental error measure as a termination criterion was introduced and shown to be more effective than the global error measure. The errors in the strain energy

Table 7. Results for a locally refined 10:1 model with a moving internal boundary.

DOF	Maximum elemental error (%)	$(\sigma_x)_{max}$ (psi)	Stress error (%)
1,328	65.7	3,055	85.35
1,668	64.2	4,546	78.20
1,876	71.3	8,393	59.76
1,784	53.5	12,000	42.46
1,624	33.6	14,771	29.18
1,570	25.8	16,043	23.08
3,450	18.8	17,720	15.04
7,450	11.2	19,110	8.37
5,430	0.6	20,039	3.92

in individual elements are, by definition, more closely related to the errors in the individual stress or strain components at critical points than the global estimate, but the correlation is still not ideal. This deficiency identifies the need to estimate the accuracy of the stress and strain components at individual points.

The solution to the problem of underestimating the elemental errors in elements on the boundary of free surfaces was outlined in this lesson. The idea of including the actual boundary conditions in the smoothed solution as a way of improving the error estimates on the boundary was introduced. In Lesson 17, we see the problem of underestimating the error in elements on the boundary to also include the estimation of errors on fixed boundaries.

In Lesson 17, the smoothing approach to error analysis, of which the Z/Z approach is the first practical example, is extended to improve the estimate of errors on the boundaries, both fixed and free. Concepts from the finite difference method developed in Part IV are used to impose conditions on both types of boundaries to improve the error estimates in elements on the boundaries. The ideas used to improve the boundary error estimates presented in Lesson 17 lead to the development of pointwise error estimates in Lesson 18. Also in Lesson 18, the pointwise error estimators are suggested as total replacements for the elemental error estimates, since the pointwise error estimates can be used both to guide refinement and to terminate the analysis. The use of pointwise error measures for these tasks eliminates the need to integrate over the individual elements to form the error estimates.

As discussed in the introduction to this section, it is imperative that error analysis be made universally available to the users of the finite element method. The Z/Z approach provided the first practical and effective way of satisfying this requirement. The prediction of what technical developments will be embraced in practice is rarely successful, so it is nearly impossible to know what error estimation procedure will finally be the practical standard. However, it is safe to say that the Z/Z approach will be seen as the starting point for the approach finally chosen for practical, production use. The developments presented here are offered as possibilities for the next candidate in this evolution and to widen the discussion of error analysis.

References and Other Readings

1. Zienkiewicz, O. C., and Zhu, J. Z., "A Simple Error Estimator and Adaptive Procedure for Practical Engineering Analysis," *International Journal for Numerical Methods in Engineering,* Vol. 24, 1987, pp. 337–357.

2. Byrd, D. E., "Identification and Elimination of Errors in Finite Element Models," Ph.D. Thesis, University of Colorado, 1988.

3. Dow, J. O., Harwood, S. A., Jones, M. S., and Stevenson, I., "Validation of a Finite Element Error Estimator," *AIAA Journal,* Vol. 29, No. 10, Oct. 1991, pp. 1736–1742.

4. Zienkiewicz, O. C., and Zhu, J. Z., "Superconvergent Recovery Techniques and A-Posterori Error Estimation in the Finite Element Method, Parts I and II," *International Journal for Numerical Methods in Engineering,* Vol. 33, 1992, pp. 1331–82.

5. Sandor, M. J., "Sub-Model Boundary Identification for Use in a Global/Local Adaptive Refinement Technique," Masters Degree Thesis, University of Colorado, August 1993.

6. Dow, J. O., and Sandor, M. J., "A Sub-Modeling Approach to Adaptive Mesh Refinement," *AIAA Journal,* Vol. 33, No. 8, pp. 1550–1553.

7. Young, W. C., *Roark's Formulas for Stress and Strain,* 6th ed. McGraw-Hill, Inc., New York, 1989, p. 725.

8. Dow, J. O., and Byrd, D. E., "The Identification and Elimination of Artificial Stiffening Errors in Finite Elements," *International Journal for Numerical Methods in Engineering,* Vol. 26, March 1988, pp. 743–762.

9. Harwood, S., "Finite Element Error Analysis Using the Finite Difference Method," M.S. Thesis, University of Colorado, 1989.

Lesson 16   Problems

The first four problems are designed to provide a simple venue for acquiring an active understanding of Z/Z error measures.

1. Compute the error in a one-element bar model fixed at one end and subjected to a concentrated load at the free end. Use nodal averaging.

2. Compute the error in a one-element bar model fixed at one end that is hanging under its own weight. Use nodal averaging.

3. Redo Problem 2 and use the actual stress boundary condition at the free end.

4. Redo Problems 2 and 3 for the case where the bar is divided into two equal length elements.

5. Compute the error in the shear strain in a square four-node element if the stress in the finite element model is 500 psi and the averaged nodal stresses are 180, 320, 650, and 450, respectively. Note: The objective of this problem is to clarify the averaging process that is inherent in an error measure based on strain energy quantities even when only one component is evaluated.

6. If you calculated the strain energy content in Problem 5 by directly integrating over the area, solve the problem again using Gaussian quadrature. If you used Gaussian quadrature, solve Problem 5 again using direct integration. Note: This problem is meant as a reminder that isoparametric elements do not generally provide stress and strain results at the nodes.

Lesson 17

Error Estimation Based on Finite Difference Smoothing

Purpose of Lesson   To improve the accuracy of Z/Z type error estimators by forming the smoothed solution using techniques from the finite difference method and to demonstrate the resulting improvements in the elemental and global error estimates.

The finite-difference-based smoothing is accomplished by overlaying finite difference templates on the finite element mesh and using the finite difference operators: first, to enforce the stress boundary conditions on the free boundaries and to satisfy the finite difference approximation of the governing differential equations on the fixed boundaries and, second, to extract the strains and, subsequently, find the stresses at the nodal points. The enforcement of these conditions provides the auxiliary equations that enable us to compute the displacements of the fictitious nodes. The smoothed stresses over the individual elements are then formed by interpolating the finite difference nodal stresses. The finite difference smoothing produces better results on the boundary than the Z/Z approach because more information concerning the problem is incorporated into the smoothed solution. The improvements in the error estimators produced by extracting the strains from the displacements of the finite element results and the displacements of the fictitious nodes using finite difference templates are demonstrated with three example problems.

The finite difference smoothing approach has a significance other than its ability to improve the error estimates for finite element models. The smoothed solution formed using the techniques from the finite difference method can be interpreted as an approximation to a finite difference solution. This interpretation is used in Lesson 18 to develop pointwise error estimators. The results from the example problems suggest that pointwise error measures might provide better termination criteria than global or elemental error measures.

■   ■   ■

The essence of finite difference smoothing is shown in Fig. 1, where nine-node central difference templates are overlaid on a finite element mesh. Some of the templates are completely on the domain and some of the templates extend beyond the boundary and contain fictitious nodes. The displacements of the fictitious nodes are found by enforcing conditions on the boundary that produce as many auxiliary equations as there are unknown fictitious displacements. The displacements on the domain are known from the finite element analysis. With all of the needed displacements available, the strain components at the finite element mesh points are extracted using finite difference operators. These strain approximations are used to compute the stresses at the finite element mesh points which serve as the basis for the smoothed solution. The smoothed solution formed *on the nodes inside of the domain* (not those on the boundary) using this approach produces results that are essentially the same as those produced by nodal averaging or by the modified Z/Z approach if the mesh is nearly uniform. However, the

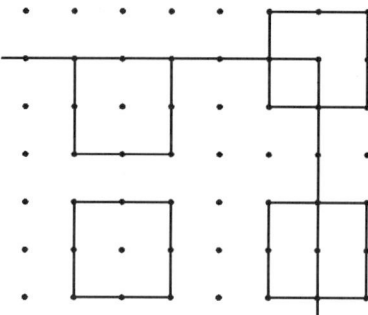

Figure 1. Finite difference templates overlaid on a finite element mesh.

smoothed stress quantities computed for the mesh points on the boundary are significantly different from those produced by nodal averaging or by the modified Z/Z approach, particularly in the early refinements.

Free and Loaded Boundary Smoothing

The smoothed solution formed using the finite difference approach produces significantly better error estimates for elements on free boundaries than do the Z/Z or nodal averaging approaches. We can see the reason for this in Fig. 1. The nodes of the finite difference template surround even the boundary nodes. This means that more information goes into the smoothing of the strain quantities on the boundary than is the case in other smoothing approaches. The displacements of the fictitious nodes associated with the free boundaries are evaluated by enforcing the stress boundary conditions at the boundary nodes. The process for computing the displacements of the fictitious nodes is presented in a later section of this lesson.

We can see the effect of the improvement on the smoothed solution for the case of a one-dimensional bar. In Fig. 2a, the boundary stress produced by the finite element model on the free boundary has a value other than zero in the early stages of refinement. When the solution is smoothed by nodal averaging, the smoothed solution matches the finite

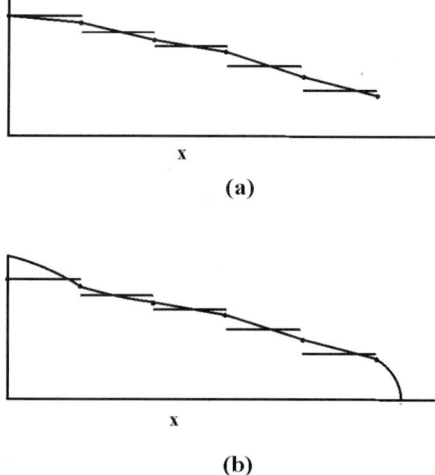

Figure 2. Error estimates by smoothing on a one dimensional bar: (a) nodal averaging and (b) finite difference smoothing.

element result and not the actual boundary condition. As a result, the error in this element is underestimated.

In the case of the finite difference smoothing, shown in Fig. 2b, the smoothed stress on the end of the bar is equal to the actual stress in the problem, i.e., the boundary stress is zero. When we compare Figs. 2a and 2b, we see that the improved stress representation in Fig. 2b increases the area between the finite element and the smoothed solutions. That is to say, the finite difference smoothing produces a larger and a more conservative error estimate.

The error estimates in elements on the boundary of higher dimension problems are improved for similar reasons as a result of using the finite difference techniques. The smoothed solution more closely resembles the actual solution as a result of the modification. The inclusion of the actual boundary conditions directly modifies two stress components and indirectly modifies the third stress component. Specifically, the finite difference smoothing approach is particularly effective in identifying stress concentrations and singularity points as regions of high error. These characteristics are demonstrated in the example problems that follow.

Note that the nodal averaging approach does not produce zero-error estimates on boundary nodes for higher dimension problems as a matter of course as is the case for one-dimension problems. This is the case because more than one element intersects at most boundary nodes so the average stress value can be different from the value in a specific element.

Fixed Boundary Smoothing

The basic philosophy used to form the finite-difference-based smoothed solution on a fixed boundary is identical to the philosophy used on free boundaries. A central difference template is overlaid on the boundary node as shown in Fig. 1. The displacements of the fictitious nodes are then evaluated by solving a set of auxiliary equations containing the fictitious displacements formed from conditions enforced on the fixed boundary.

In the case of a boundary node, three options recommend themselves as a way of introducing more information into the smoothed solution: (1) the enforcement of the nodal displacements, (2) the enforcement of the displacements at nonnodal points adjacent to the boundary nodes, and (3) the enforcement of the governing differential equations at the boundary node. These are precisely the conditions used in the formulation of finite difference models presented in Part IV. The stress boundary conditions cannot be used because the stresses are not known on constrained boundaries.

The first option is not useable in this application because the enforcement of this condition does not produce equations that include the fictitious nodes. The Taylor series expansion does not contain any derivative terms when it is evaluated at the central node. The second option has not been tried because the third option is successful. The use of displacements on the boundary but not at a node should improve the smoothed solution because the Taylor series expansion for displacements away from the central node contains derivative terms. The use of displacements at nonnodal points on the boundary is an extension of the procedure for forming finite difference models on fixed corners or cantilever boundaries. This idea is discussed further in Lesson 18 when higher order pointwise error estimators are developed.

The third option is used in the examples presented here because the formulation of these equations is analogous to the process used to smooth the solution on loaded boundaries. That is to say, the finite difference approximations of the governing differential equations are enforced at the boundary nodes. As shown in Fig. 2 at the left-hand end, the stress found with the finite difference smoothing approach differs significantly from the stress found with nodal averaging. This is demonstrated in the examples that follow where we see that this approach provides better estimates for elements next to the fixed boundaries and at singular points than do the other smoothing techniques.

This approach also applies if the displacements on a boundary have a specified value other than zero. Examples of displacements other than zero are settlement problems and boundary movements produced by kinematic conditions found in contact problems.

An Expanded View of Finite Difference Smoothing

The finite difference approach has a far-reaching significance beyond the improvements it brings to the error estimates in Z/Z or smoothing-type error estimators. We can interpret the smoothed solution formed using the finite difference approach as an approximation of the finite difference solution to the problem being solved, i.e., the solution to the differential form of the governing partial differential equations. This interpretation opens the door to developing pointwise error estimators.

Since the finite element and finite difference results should be the same in a converged result, any difference between the two approximate solutions indicates that one or both of the solutions has not converged and that an error exists in one or both of the approximations at that location. We can quantify this difference to give us a pointwise error estimate in the finite element result at that point. We can do this because (1) we can estimate how nearly the smoothed solution approximates a finite difference solution and (2) we can form a smoothed solution with the finite difference approach that is of an order higher than the finite element solution. This idea is exploited in Lesson 18.

Furthermore, the interpretation of the smoothed solution produced by the finite difference approach as an approximation to the finite difference solution enables us to define the conditions when the smoothed solution is, indeed, an actual improvement on the finite element solution. That is to say, we will be able to accurately define a superconvergent solution. This, too, is discussed in Lesson 18.

A Less Comprehensive Improvement to the Smoothed Solution

One might be tempted to modify the Z/Z or nodal averaging approaches by directly incorporating the stress boundary conditions into the smoothed solution. In the one-dimensional case, the process is identical to the finite difference smoothing approach on a free surface. However, in higher dimensions the two approaches differ.

In finite difference smoothing on a free surface in two dimensions, the two known boundary stresses are used to evaluate the displacements of the fictitious nodes. When the finite difference operators are evaluated, the third stress component differs from the finite element value if errors exist. Furthermore, the direct incorporation of the boundary stresses cannot be applied to fixed boundaries. There are no auxiliary conditions to apply unless the finite difference approach is used. The error estimates produced by the finite difference smoothing and the nodal averaging approach on the fixed boundary are compared in the example problems presented later in this lesson.

The improvements provided by the direct incorporation of the two boundary stress components are shown at their best in Fig. 3. In Fig. 3, we see that the boundary smoothing produces results that are as good as, if not better than the finite difference smoothing in early refinements. However, in later refinements, the results approach those of the Z/Z smoothing because (1) the stresses on the surface are converging, (2) there is no correction on the fixed boundary, and (3) the stresses on the fixed boundary are slow to

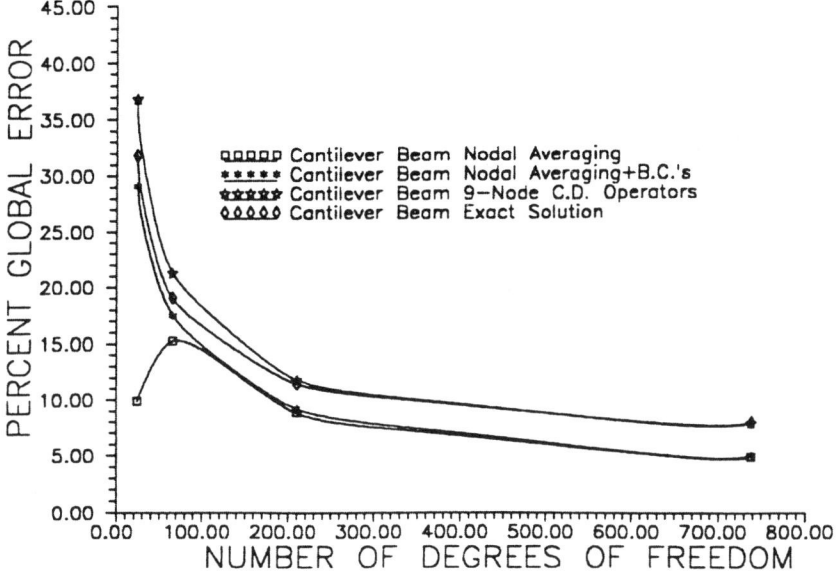

Figure 3. Global error estimates with finite difference smoothing compared to other approaches.

converge. The stresses on the fixed boundary are more resistant to convergence than the stresses on a free boundary because they are usually more complex. This complexity exists because the fixed boundary is solidly constrained. It cannot displace to accommodate the stresses. Instead, the stresses become more complex to accommodate the stringent displacement constraint. The fact that the error estimates approach those produced by the Z/Z method demonstrates the need to improve error estimates on fixed boundaries. This figure is discussed further in the first example.

The Implementation of Finite Difference Smoothing

In Part IV, we developed procedures for forming finite difference models. We formed approximations to the governing differential equations at each node on the domain, including the boundary nodes. This process produced as many equations as there are displacements at the mesh points on the domain. The templates overlaid on the mesh points on the boundary introduced fictitious nodes into the finite difference mesh. It was then necessary to form auxiliary equations equal in number to the displacements associated with the fictitious nodes. These auxiliary equations are formed by introducing the boundary conditions into the finite difference model. The auxiliary equations and the approximations to the governing differential equations result in a set of equations with as many equations as unknown fictitious and real displacements. These equations are solved for the displacements, which are, in turn, used to compute the stresses and strains that constitute the finite difference solution.

As an aside, the need to find the displacements of the fictitious nodes on the free boundaries means that more conditions must be satisfied by the finite difference solution than by the finite element solution. Whether this means that the finite difference solutions are more accurate than the finite element solutions for a given mesh is now open for meaningful debate. The question is open because the developments of Part IV enable the finite difference method to solve practically any solid mechanics problem that can be solved by the finite element method.

A Summary of Finite Difference Smoothing

The finite difference smoothing process overlays finite difference templates on the finite difference mesh as shown in Fig. 1. This operation introduces fictitious nodes into the calculation of the smoothed solution. This set of fictitious nodes is identical to the set that would be present if we chose to solve the problem with the finite difference method. The displacements of the fictitious nodes on loaded boundaries are evaluated by introducing the stress boundary conditions. The fictitious nodes on fixed boundaries are evaluated by satisfying the governing differential equations on boundary nodes. That is to say, we use procedures from the finite difference method to form the smoothed solution.

Once the fictitious displacements are found, these displacements and those from the finite element solution are substituted into the individual finite difference templates to compute the strains at the nodes on both the boundary and on the interior of the model. Then the stresses are computed at each node and used to form the smoothed solution. *The same concept is used to develop the error measures for finite difference smoothing* as was used for the other smoothing approaches. There is one difference in the elemental error measure as it is used in this lesson. The elemental error measure is normalized with respect to the total strain energy in the problem. This corresponds to the approach originally developed in the Z/Z approach. This contrasts to the elemental error measure in the previous lesson given by Eq. 16.4, which is normalized with respect to the strain energy in the individual element.

In this lesson, the elemental error estimates are computed using the following expression:

Elemental Error Measures

$$\eta_i = \frac{|e_i|}{(|U'|^2 + |e|^2)^{1/2}} \times 100\%$$

(17.1)

where e_i is the error in the ith element.

U' is the total strain energy in the problem.

e is the total strain energy error in the total problem.

The process of computing the error estimates based on finite difference smoothing is outlined in Fig. 4.

1. Solve the Finite Element problem.
2. Overlay templates on boundary nodes.
3. Enforce stress boundary conditions on loaded surfaces.
4. Enforce Gov. Diff. Eq. on fixed boundaries.
5. Solve for fictitious displacements.
6. Substitute FE and FD displacements into the Finite Difference templates.
7. Solve for nodal stresses.
8. Form smoothed solution.
9. Compute error measures.

Figure 4. An outline of the finite difference smoothing process.

Finite Difference Smoothing on Partial Regions

In some applications, the analyst may not desire to compute error estimates on the whole domain of the problem. This can be done with a simple modification to the process previously described for computing error estimates on the whole region. The modification is shown in Fig. 5. In this figure, the fictitious points are shown on the region where the error estimates are desired. The fictitious nodes are evaluated as described in the previous section with one difference. The stress or displacement boundary conditions are also enforced at the points on the boundary shown in the figure by the x's. These points are not nodes in the mesh. This is analogous to the process used to form the needed number of auxiliary equations for convex corners in the finite difference method.

This approach reduces the number of simultaneous equations that must be solved to find the displacements of the fictitious nodes. After the fictitious nodes are found, the process then proceeds as outlined in Fig. 4.

It is possible to further reduce the number of simultaneous equations that must be solved at one time to find the displacements of the fictitious nodes. For example, we could overlay a six-node template, as shown in Fig. 6, on each boundary node. The two unknown displacements at the fictitious nodes could then be computed by enforcing the stress boundary conditions. We could employ a similar process on fixed boundaries by enforcing the governing differential equations. We could then use this template to compute the three components of stress for the smoothed solution at the boundary node.

These two strategies for reducing the computation required for finite difference smoothing are presented to show that this approach does not necessarily require the solution of large sets of equations. Note that this approach for smoothing on partial regions has yet to be tested and other options are available. For instance, we could use nine-node templates one at a time and enforce boundary conditions at points other than at the central nodes, as already discussed.

Introduction to the Example Problems

The error estimation procedure based on finite difference smoothing is applied to three example problems. The results are compared to the nodal averaging and the augmented nodal averaging approaches. The first example is a cantilever beam with an applied shear stress at the free end. The second problem is a square shear panel with a circular hole loaded in tension. The final example is a square shear panel with a square hole loaded in tension.

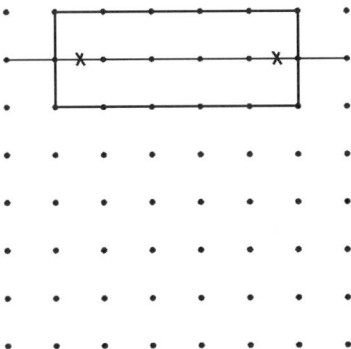

Figure 5. A finite difference mesh overlaid on a selected region of a finite element model.

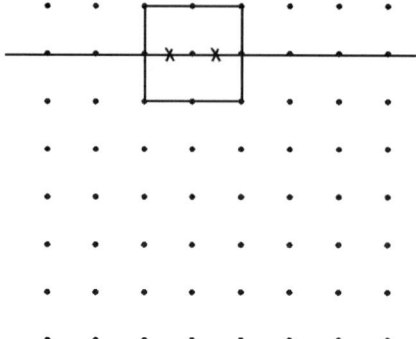

Figure 6. A single finite difference template overlaid at a boundary point.

All three problems are modeled with planar strain gradient four-node elements that have been corrected for parasitic shear. Four-node elements were chosen because they are *not* based on complete quadratic polynomials. This means that the constant strains are the highest complete representation in the elements. This is a consideration because this application of finite difference smoothing uses nine-node central difference templates. These templates contain a complete quadratic representation of displacements and, hence, a complete linear representation of strains. This is not of significance in this lesson, but it is significant in the development of pointwise error estimators in the next lesson. In the pointwise error estimator, the need for a smoothed solution with a higher order Taylor series representation is considered significant.

Cantilever Beam

The cantilever beam problem is chosen as an example for several reasons. In Lesson 12, we analyzed the strain modeling behavior of the individual elements in detail and we studied this problem closely. Thus, we know that in a model consisting of a single row of elements, the cantilever beam model representation cannot be accurate. By solving this problem, we can directly compare the error analysis results presented in the next section to our previous experience. Furthermore, this model contains several geometries commonly found in practice.

The problem contains an intersection of a fixed boundary with a free surface to produce a singularity point. It also contains a cross section with well-known stress conditions. The shear stress is zero at the boundaries and distributed across the section as a parabola. The normal stress is zero on the surface and the distribution of flexural stress is well understood. Furthermore, this example demonstrates the ability to model fixed and loaded boundary conditions for both straight edges and corners.

We see that the nodal averaging approach produces low estimates for the error at all levels of refinement. We see that the underestimation is especially severe when the problem is dominated by a free or loaded surface and when high errors are present. Furthermore, we see that as the stresses converge to the correct value on the free surfaces that the errors are dominated by the errors on the slow-to-converge fixed boundary. This shows the need for modifying the smoothed solution on fixed boundaries. Finally, we see that the finite difference approach is significantly better at identifying errors at singular points than the other approaches.

Circular Hole in a Square Panel

The circular hole in a square panel is chosen as an example problem because it was studied in detail in the previous lesson and because it contains a curved boundary and a stress concentration. It also demonstrates the ability to model "roller" boundary conditions and a further example of a loaded boundary.

In this example, we see that both the nodal averaging and the augmented nodal averaging procedures underestimate the error at all levels of refinement as compared to the finite difference approach. This shows the need to improve the smoothing of the third stress component on free boundaries.

Square Hole in a Square Panel

The problem of a square hole in a square shear panel is chosen because it contains a singular point at the intersection of two free straight surfaces. This type of problem enables the four-node element to perform at its best. We see that the results of the three methods are as close to each other at all levels of refinement as in any example studied to date. However, we see that the finite difference approach clearly identifies the presence of high errors at the singular point better than the other methods.

Example 1   The Cantilever Beam Problem

The cantilever beam problem solved as an example is shown in Fig. 7. The original mesh consists of 5 elements with 24 degrees of freedom. It is uniformly refined three times in both coordinate directions to produce models with 20, 80, and 320 elements and 66, 210, and 738 degrees of freedom, respectively. The global error estimates are presented first. Then we discuss the elemental error estimates.

The discussion of global errors should not be interpreted as a recommendation for their use in terminating solutions. The presentation in the previous lesson outlined the deficiencies of global error measures. In this discussion, the global error measures are used largely to compare the aggregate behavior of the different types of error measure.

Global Error Estimates

This section compares the global error estimates produced by finite difference smoothing to those produced by three other approaches, namely, (1) nodal averaging, (2) nodal averaging with boundary stresses, and (3) a Timoshenko beam solution (which compares favorably with a fully converged finite element solution).

The total strain energy for the Timoshenko beam solution of a beam with a rectangular cross section is given as:

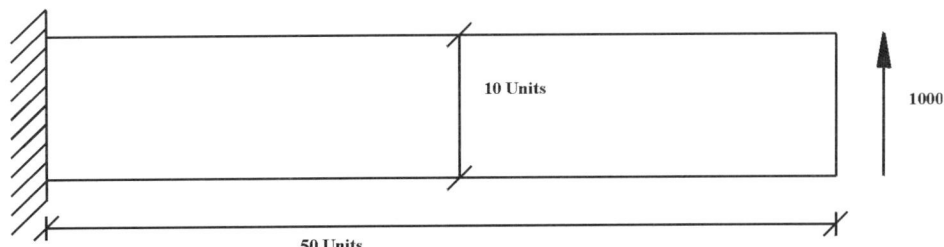

Figure 7. A cantilever beam with a distributed shear load at the free end.

Strain Energy in a Timoshenko Beam

$$U = U_{\text{flexure}} + U_{\text{shear}}$$
$$= \frac{Pl^3}{6EI} + \frac{3P^2l}{5GA}$$

(17.2)

where P is the applied shear load.

l is the length of the beam.

E is the modulus of elasticity.

I is the moment of inertia.

G is the shear modulus.

A is the cross-sectional area.

The global error estimates for the finite difference method are compared to the nodal averaging approach and the Timoshenko solution in Fig. 8.

Figure 8 shows that the finite difference approach definitively corrects the deficiency in the global error estimate predicted by the Z/Z smoothing type of error estimator. The nodal averaging approach shown by the lower curve underestimates the global error. This underestimation results from errors in estimating errors on the boundaries.

We see in the discussion of the pointwise errors in the next section that the underestimation in the Z/Z approach in the early refinements is dominated by errors in estimates on the free boundaries. We saw the reason for the underestimation in the initial mesh by the Z/Z approach in Lesson 12. The limited modeling capability of the four-node element does not let one element represent the expected parabolic distribution with other than a constant value. Thus, the errors in the shear stresses of the individual elements are significantly underestimated. The inability of the four-node element to represent the Poisson effect for flexure produces further errors in the normal stresses. The underestimations in later mesh refinements are dominated by errors in the element estimations on the fixed boundaries.

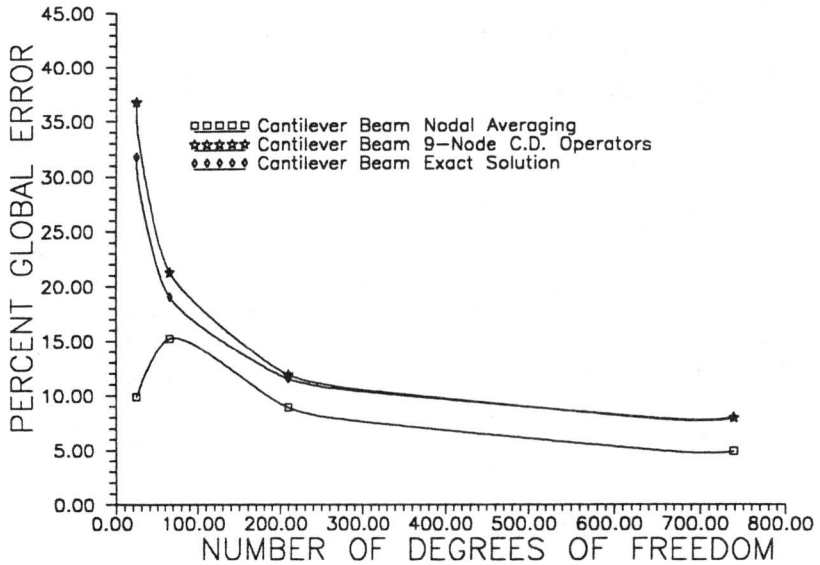

Figure 8. Global error estimates vs model refinement.

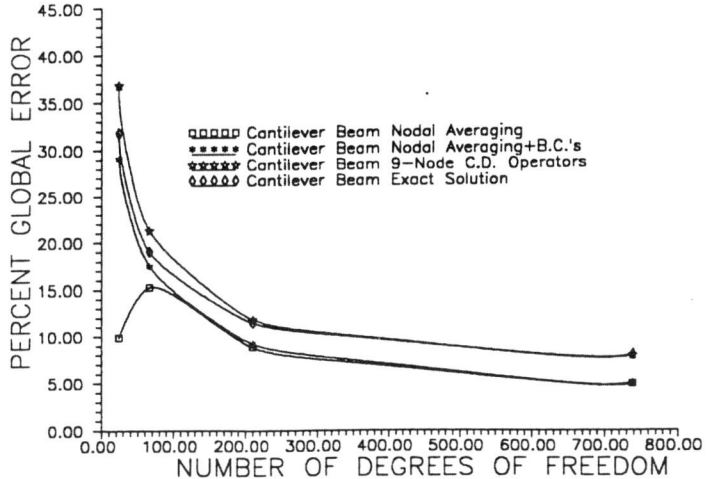

Figure 9. Global error estimates vs model refinement.

In Fig. 9, the error estimates produced by the nodal averaging approach when it is augmented with the actual stresses on the free boundaries is added to Fig. 8. As we can see, this modification significantly improves the estimates produced by the nodal averaging approach during the early refinements. This validates the idea that the underestimation by the Z/Z approach in the early meshes is largely due to underestimations on the free boundaries. However, the estimates produced by augmented nodal averaging match those of the nodal averaging approach in the later refinements. This implicitly validates the idea that the underestimations in the more refined meshes are due to errors in the error estimates on the fixed boundary.

Elemental Error Estimates

This section compares the elemental error estimates produced by finite difference smoothing to those produced by the nodal averaging approach and the nodal averaging approach augmented with stresses on the free boundaries. The elemental errors are shown on the individual elements in the figures presented in the following order:

Nodal averaging

Augmented nodal averaging

Finite difference smoothing

A blank entry represents an error estimate of less than 0.01%. That is to say, the strain energy contained in the elemental error measure is less than 0.01% of the total error in the energy norm in the overall problem. This differs from the elemental errors reported in the previous lesson, which were measured as a percentage of the strain energy in the individual elements.

Elemental Errors in the Initial Mesh.   In Fig. 10, the elemental errors are shown for the original mesh of the cantilever beam. As we can see, the nodal averaging approach produces nonconservative results in every element. The nodal averaging approach is nearest to the other approaches at the loaded end. It is farthest from the other approaches near the fixed boundary. At the loaded end, the stress on the boundary nodes are not zero, so the smoothed value produced by nodal averaging is closer to the actual boundary condition than it is on the unloaded free boundaries. In the next two elements closest to the loaded end, the nodal averaging approach is far below the other two estimators. These

0.20	0.20	0.20	0.20	0.20
1.63	3.63	2.23	1.30	0.46
8.05	4.01	1.95	1.10	0.57

Figure 10. Three types of elemental error estimates, initial mesh.

estimates are low because the nodal averaging approaches are using the finite element stresses in the smoothing on the boundary and not the actual boundary conditions. This and the fact that most nodes are on free boundaries are the causes of the original low value in the global error estimate in Figs. 8 and 9.

The augmented nodal averaging approach produces elemental error estimates that are slightly higher than those of the finite difference smoothing in the second and third elements from the loaded end. The estimates produced by these two methods are not identical because the third stress component in the finite difference smoothing differs from that used in the augmented nodal averaging approach.

The most significant difference between the three estimators occurs at the fixed boundary. This is as would be expected. The nodal averaging approach uses the finite element results at each node as does the augmented approach. That is to say, the augmented approach is no different from nodal averaging on the fixed boundary. This is the reason that the global estimate provided by the augmented approach coincides with the nodal averaging approach in the higher refinements in Fig. 9.

The finite difference smoothing produces a significantly higher error estimate on the boundary element than do either of the two other methods. This occurs because it is the only estimator that brings additional information about the problem into the smoothed stress representation on the fixed boundary. This explains the well-known error on fixed boundaries associated with the Z/Z type of error estimates. Specifically, these types of error estimators indicate that the highest errors occur one row away from a fixed boundary instead of on the boundary itself. This is the case because the Z/Z approach uses only the finite element stresses to form the smoothed solution. It does not introduce additional information concerning the fixed boundary as does the finite difference approach. As a result, the errors on the boundary are underestimated. The Z/Z approach successfully guides mesh refinement even with this deficiency because elements near locations of high error are also refined in the adaptive refinement process.

As a final note, the question could arise as to why the initial error estimates produced by nodal averaging are not zero in some of the elements. In the one-dimensional smoothing discussed earlier, nodal averaging produced the exact same stress on the boundary node as the finite element result. In the case of the cantilever beam, there are two elements attached to each boundary node. The stresses differ in the elements along the beam, so the average stresses are not identically equal to the stresses in the individual elements. This produces a nonzero error estimate, albeit, a low estimate.

Elemental Errors in the First Uniformly Refined Mesh. In Fig. 11, the elemental error estimates for the first uniformly refined mesh are shown. The error estimates for this mesh have an attenuated pattern that is similar to that found in the initial mesh layout. The estimates produced by nodal averaging are closer to the other two approaches than was the case in the initial mesh. This is as expected because the finite element result is beginning to converge.

0.15 0.13 1.09	0.33 0.35 0.34	0.20 0.28 0.24	0.17 0.22 0.19	0.12 0.18 0.16	0.09 0.14 0.12	0.06 0.11 0.09	0.04 0.08 0.07	0.02 0.07 0.06	0.02 0.03 0.04
0.15 0.13 1.09	0.33 0.35 0.34	0.20 0.28 0.24	0.17 0.22 0.19	0.12 0.18 0.16	0.09 0.14 0.12	0.06 0.11 0.09	0.04 0.08 0.07	0.02 0.07 0.06	0.02 0.03 0.04

Figure 11. Three types of elemental error estimates, first uniform refinement.

The biggest difference between the finite difference error estimates and the estimates provided by the other approaches is in the elements next to the fixed boundary. Again, this is expected because of the differences in the treatment on the fixed boundary. It is worth noting that the elements with the highest estimated errors in the nodal averaging approach and the augmented nodal averaging approach are in the elements one row away from the fixed boundary. The singularity point at the corner is not identified as the location of highest error by these error estimators. This can lead to inefficient intermediate refinements in nodal averaging procedures.

Elemental Errors in the Second Uniformly Refined Mesh. The elemental error estimates for the second uniform refinement are shown in Fig. 12. As we can see, the model is converging nicely. The error estimates are nearly identical everywhere except at the fixed boundary. The highest error estimates for the two methods that use nodal averaging are still one row away from the fixed boundary. Of particular interest are the elements at the two corners. The finite difference error estimates are approximately 16 times larger than those of the other two error estimators. Thus, we can conclude that the finite difference approach can identify singularity points more accurately than the other two approaches.

Elemental Errors in the Third Uniformly Refined Mesh. The elemental errors for the third and final uniform refinement are shown in Fig. 13. As we can see, the error estimates indicate that the model is fully converged in every element except in those near the singularity point at the corner. There, the finite difference error estimate is five times that of the other methods. This result again shows that the finite difference error estimator is better at identifying singular points that the other approaches.

.02 .03 .32	.02 .02 .02	.02 .02 .02	.02 .02 .02	.02 .01 .02	.01 .02 .02	.01 .02 .01	.01 .01 .01	.01 .01 .01	.01 .01 .01	.01 .01 .01	.01 .01 .01	.01 .01 .01	.01 .01 .01	.01 .01 .01	.01	.01	.01		
.01 .01 .03	.02 .02 .02	.02 .02 .02	.02 .02 .02	.02 .02 .01	.01 .01 .01	.01 .01 .01	.01 .01 .01	.01 .01 .01	.01 .01 .01	.01 .01 .01	.01 .01 .01	.01							
.01 .01 .03	.02 .02 .03	.02 .02 .02	.02 .02 .02	.02 .02 .01	.01 .01 .01	.01 .01 .01	.01 .01 .01	.01 .01 .01	.01 .01 .01	.01 .01 .01	.01 .01 .01	.01							
.02 .03 .32	.02 .02 .02	.02 .02 .02	.02 .02 .02	.02 .01 .02	.01 .02 .02	.01 .02 .01	.01 .01 .01	.01 .01 .01	.01 .01 .01	.01 .01 .01	.01 .01 .01	.01 .01 .01	.01 .01 .01	.01 .01 .01	.01	.01	.01		

Figure 12. Three types of elemental error estimates, second uniform refinement.

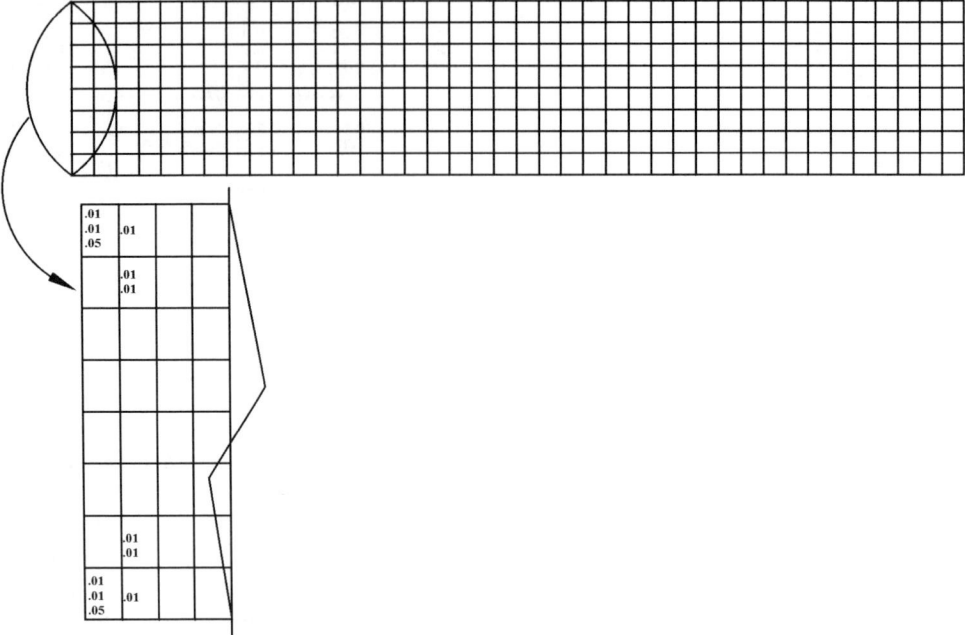

Figure 13. Three types of elemental error estimates, third uniform refinement.

Example 2   Square Plate with a Circular Hole

The square panel with the circular hole is shown in Fig. 14. Because of symmetry, it is represented using a quarter section in the finite element model with roller boundary conditions on the two edges. The problem is modeled with an original mesh of 8 elements and two uniform refinements of 32 and 128 elements, respectively. The three models contain 30, 90, and 360 degrees of freedom. This problem was solved with extreme refinement in the previous lesson. Because of this, the results presented here are designed to compare the characteristics of the three different error analysis methods, not to pursue the solution to convergence. The global error and the elemental errors are discussed in turn.

Global Error Estimates

This section compares the global error estimates produced by finite difference smoothing to those produced by (1) the nodal averaging procedure and (2) the nodal averaging approach augmented with boundary stresses. The three types of global error estimates for the three mesh refinements are shown in Fig. 15.

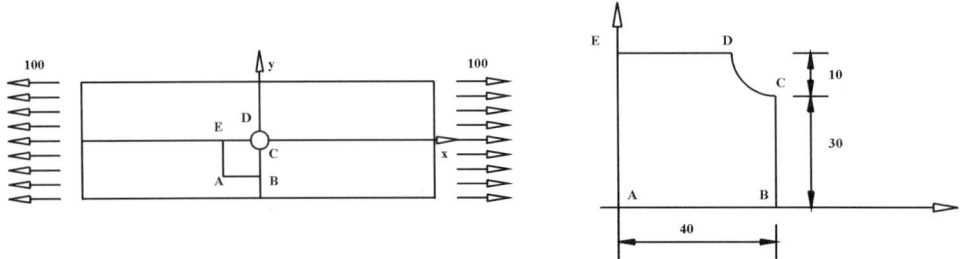

Figure 14. Model for a circular hole in a square panel.

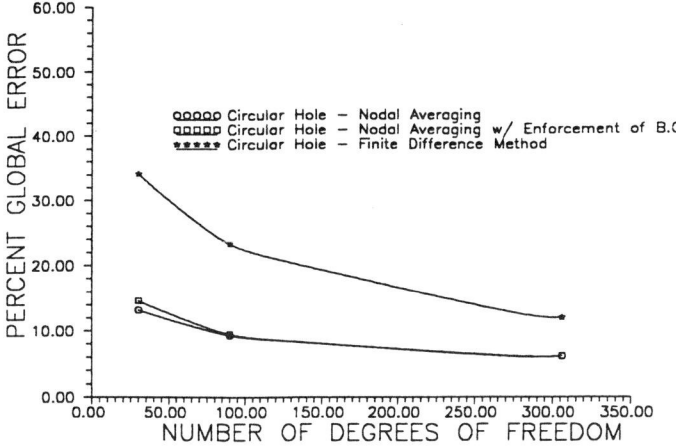

Figure 15. Global error estimates vs model refinement.

In this case, the finite difference error estimation procedure produces higher estimates than the other two methods that are consistent for the different levels of mesh refinement. As we can see, there is very little difference between the global error estimates produced by the nodal averaging approach with and without the augmentation of the stress boundary condition. This means that the inclusion of the surface tractions in the smoothed solution provides little improvement in this example problem. The reason for this is that the major errors are in the stress component with the highest values around the hole. This component is not affected by the enforcement of the boundary corrections. This is evaluated further in the next lesson when we discuss the errors in the individual stress components at the nodes.

Elemental Error Estimates

This section compares the elemental error estimates produced by finite difference smoothing to those produced by the nodal averaging approach and the nodal averaging approach augmented with stresses on the free boundaries. The elemental errors are shown on the individual elements in the figures presented in the following order:

Nodal averaging

Augmented nodal averaging

Finite difference smoothing

A blank entry represents an error estimate of less than 0.01%.

Elemental Errors. In Fig. 16, the elemental errors are shown for the original mesh of the circular hole problem. As we can see, the finite difference approach produces significantly higher estimates than the other two methods, particularly in the elements near the stress concentration. The two nodal averaging approaches do not identify the high error elements in the right-hand corner. This is the case because neither approach adds additional information on the roller boundary.

The elemental error estimates for the first refinement are shown in Fig. 17. Again, the different error estimation methods identify similar regions as having the highest errors. However, the error in the finite difference approach is significantly higher than in the other approaches. As we can see, the highest error is identified in the same region as in the highly refined model in the previous lesson. The errors on the boundary have been reduced as a result of this mesh refinement. The error reported at the stress concentration

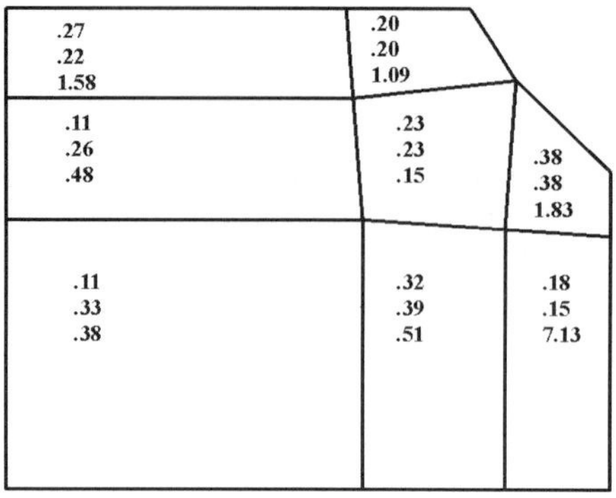

.27 .22 1.58	.20 .20 1.09		
.11 .26 .48	.23 .23 .15	.38 .38 1.83	
.11 .33 .38	.32 .39 .51	.18 .15 7.13	

Figure 16. Local error estimates for the initial mesh.

by the finite difference method is approximately 6 times that for the other two methods (0.56 vs 0.09). The finite difference error estimate in the region of highest error is approximately 25 times that estimated by the other two elements (3.04 vs 0.12).

The elemental error estimates for the second and final refinement are shown in Fig. 18. Again, the three error estimation methods identify similar regions as having the highest errors. However, the finite difference approach produces an error estimate that is significantly higher than the other approaches at the stress concentration. The finite difference error estimate is approximately 4 times the estimate produced by the other two methods (0.04 vs 0.01). In the element with the highest error estimate, the difference between the finite difference approach and the other methods is a factor of 15 times (0.89 vs 0.06).

As was the case in Lesson 16 with the highly refined models for this problem, the same area contained the highest error. This region, however, does not contain the point of highest stress. This has significance when planning a mesh refinement strategy. We must

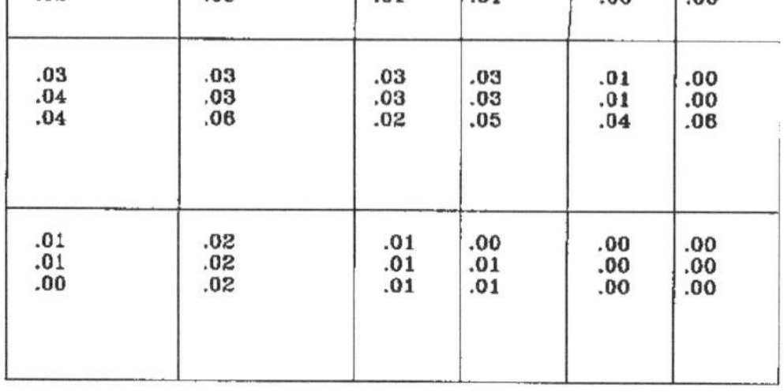

.02 .02 .01	.03 .03 .06	.01 .01 .16	.03 .03 .16		
.01 .01 .01	.02 .02 .05	.02 .02 .16	.06 .06 .95		
.01 .01 .01	.02 .02 .04	.01 .01 .01	.05 .05 .31	.12 .12 3.04	.09 .09 .56
.01 .02 .02	.03 .03 .03	.02 .02 .01	.02 .02 .01	.03 .03 .06	.05 .05 .06
.03 .04 .04	.03 .03 .06	.03 .03 .02	.03 .03 .05	.01 .01 .04	.00 .00 .06
.01 .01 .00	.02 .02 .02	.01 .01 .01	.00 .01 .01	.00 .00 .00	.00 .00 .00

Figure 17. Local error estimates for the first mesh refinement.

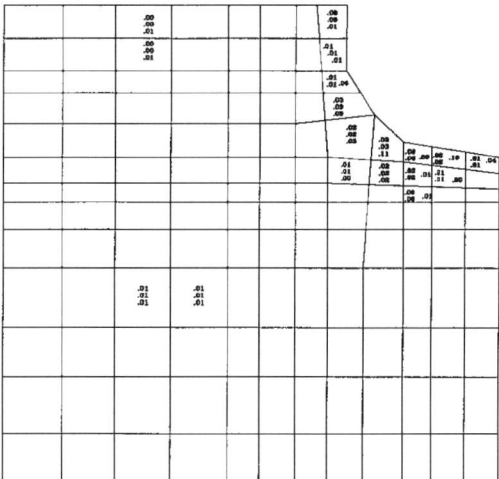

Figure 18. Local error estimates for the second mesh refinement.

decide whether the area of highest error influences the region that is critical to the design. In this case, the region of high error is so close to the critical point that it must be considered influential on the stress concentration. In other situations, this might not be the case and the refinement could be stopped or a substructuring approach such as that developed in the last lesson could be applied.

Example 3   Square Plate with a Square Hole

The square shear panel with the square hole is shown in Fig. 19. Because of symmetry, it is represented using a quarter section finite element model with roller boundary conditions. The problem is modeled with an original mesh and four uniform refinements.

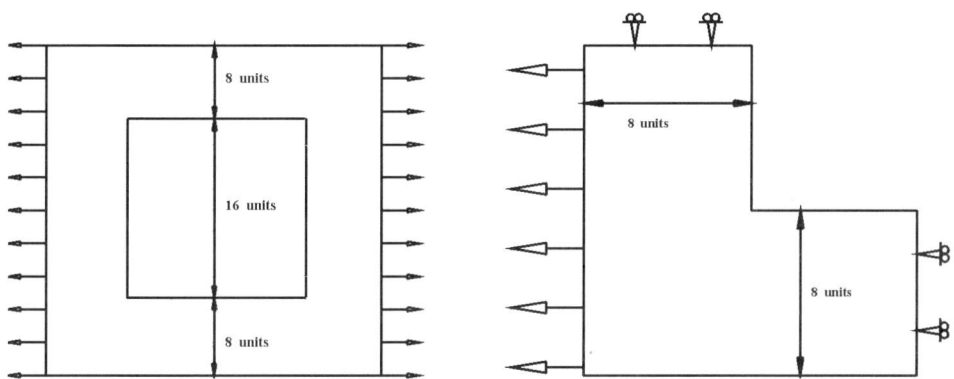

Figure 19. Model for a square hole in a square panel.

The models contain 3, 12, 48, 192, and 768 elements and 16, 42, 130, 225, and 833 degrees of freedom, respectively. The global and elemental errors are discussed in turn.

This problem evaluates the effectiveness of these procedures in estimating the errors in a re-entrant corner with its singular point. The global errors are presented in Fig. 20. As we can see, the three methods are closer to each other than in either of the previous cases. As usual, the two nodal averaging approaches are similar as the problem converges and the finite difference approach estimates a higher error.

The elemental errors in the five models are shown in Figs. 21 through 25. As we can see, the three error estimators converge to essentially the same results except in the area of the stress concentration. With each mesh refinement, the errors estimated by the finite difference approach grow larger with respect to the two averaging approaches. In the final refinement, the finite difference estimates are 7 to 20 times as large as the nodal averaging and the augmented nodal averaging results (0.69 vs 0.03, 0.07 vs 0.01, and 0.38 vs 0.04). Thus, we can conclude that the finite difference approach is more capable of identifying singular points than the other approaches.

Closure

This lesson has improved the error estimating capabilities of the smoothing approach by using techniques from the finite difference method. The finite difference techniques improved the estimates on free boundaries by introducing the actual boundary conditions into the smoothed solution. The estimates on fixed boundaries are improved by forcing the governing differential equations to be satisfied at the nodes on these boundaries. The finite difference approach also provides better error estimates on the domain for nonuniform meshes than does the nodal averaging approach. This results from the effects of the displacements surrounding the central node being weighted according to location.

We have seen that the results found using finite difference smoothing are more conservative and more accurate than those found using other smoothing approaches. That is to say, the error estimates produced by the finite difference approach are higher than those produced by the other approach and closer to the exact or converged results where

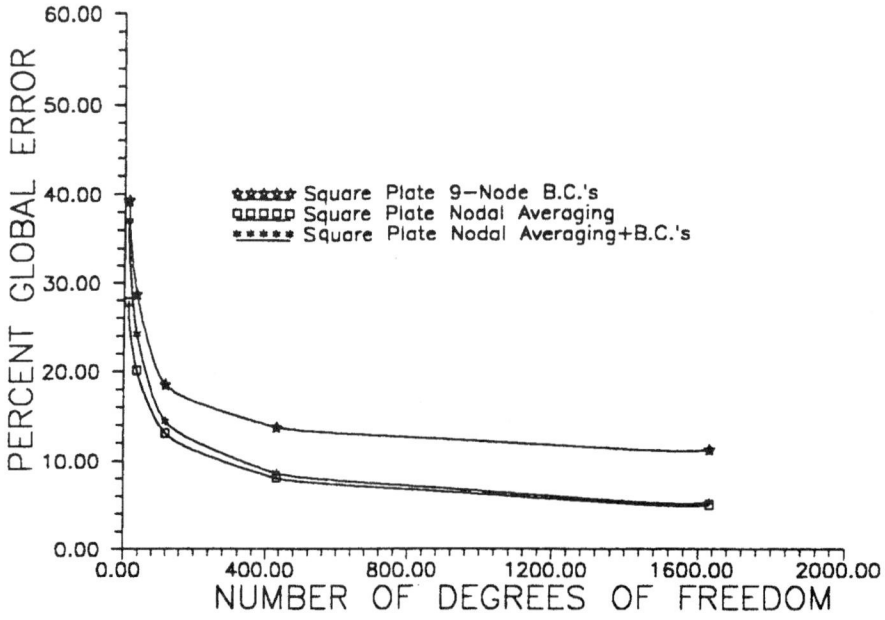

Figure 20. Global error estimates vs model refinement.

3.89 **4.83** **4.43**	
1.57 **2.23** **1.98**	**2.97** **8.86** **11.90**

Figure 21. Local error estimates for the initial mesh.

.22 .34 .31	.36 .25 1.74
.29 1.05 .83	1.18 2.06 3.23

.36 .31 .35	.25 .25 .42	.98 1.03 1.35	.10 .13 .28
.06 .07 .04	.17 .24 .17	.13 .18 .15	.12 .12 .10

Figure 22. Local error estimates for the first mesh refinement.

.02 .01 .02	.01 .01 .02	.01 .01 .02	.01 .01 .02				
.02 .04 .04	.03 .03 .02	.03 .03 .02	.03 .03 .04				
.04 .06 .07	.05 .05 .04	.07 .07 .05	.14 .11 .12				
.04 .06 .07	.05 .05 .04	.11 .11 .09	.21 .46 1.54				
.04 .05 .05	.05 .05 .04	.06 .06 .06	.06 .06 .18	.29 .33 .68	.02 .03 .03	.01 .01 .01	.01 .01 .01
.02 .02 .02	.02 .02 .02	.03 .03 .03	.07 .07 .05	.03 .03 .02	.02 .02 .02	.01 .01 .01	.01 .01 .01
		.01 .01 .01	.01 .01 .01	.01 .01 .01	.01 .01 .01	.01 .01 .01	.01 .01 .01
		.01 .01	.01 .01	.01 .01	.01 .01	.01 .01	.01 .01 .01

Figure 23. Local error estimates for the second mesh refinement.

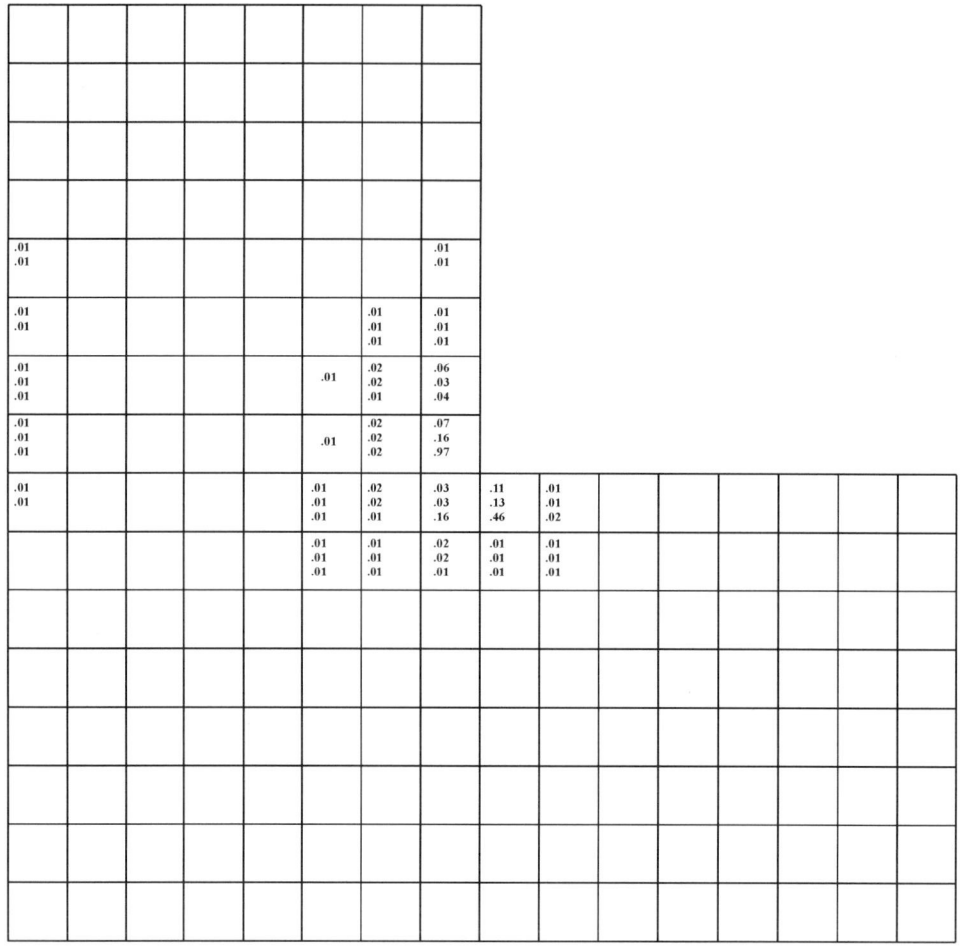

Figure 24. Local error estimates for the third mesh refinement.

these comparisons have been available. This is true for both global and individual elemental errors. The improvement in elemental error estimation is most clearly seen in elements at singular points.

This ability to better identify singular points was demonstrated for both 90 degree intersections between fixed and free boundaries at reentrant corners and between two free boundaries at convex corners. This provides solid evidence that the finite difference modifications to both free and fixed boundaries provide significant improvements to the error estimation process.

Furthermore, the capabilities developed and applied in this lesson show that the results of Lesson 16 can be extrapolated. In Lesson 16 we saw that the computation of global errors is heavily dependent on the size of the overall problem. This led to the recommendation that we use the elemental error measure for a termination criteria instead of the global error because the resolution of the error estimator is improved. In addition, we saw that the arbitrarily chosen elemental error of 5% as a termination criteria was overly conservative for the problems solved as examples. This criterion led to levels of refinement that were higher than actually required. This conclusion was reached because the stress values at the stress concentrations had ceased to converge.

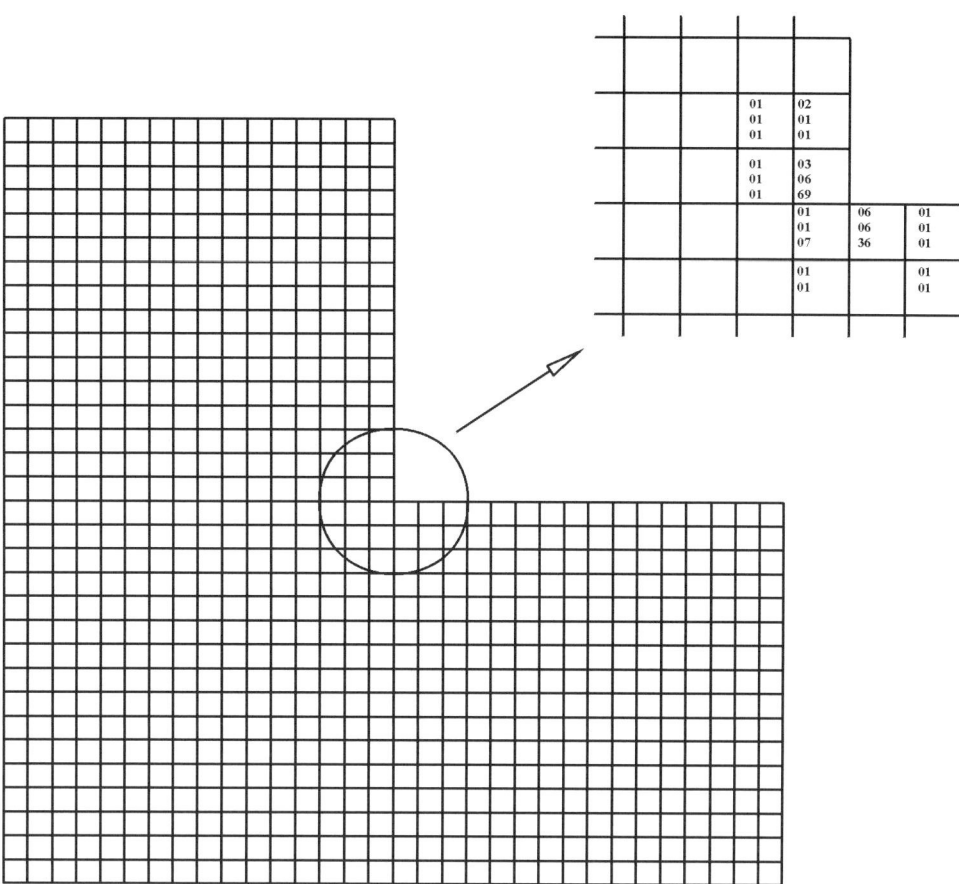

Figure 25. Local error estimates for the fourth mesh refinement.

This result implies that the best error measures to use as termination criteria should be based on pointwise error estimates in stress or strain values. The finite difference modification to the Z/Z smoothing approach can be extended to make this possible. This is the case because the smoothed result provided by the finite difference approach can be of a higher order than the finite element solution, and we can estimate the degree to which it satisfies the differential solution. Pointwise error estimates are developed and applied in the next lesson.

In Lesson 18, we see that we can replace the elemental error measures as a guide to mesh refinement with pointwise error estimates. This eliminates the need to integrate over each of the individual finite elements. This reduces the computational effort required and it also means that the pointwise error estimates are completely independent of the finite element formulation being evaluated for errors. The shape functions used to formulate the elements need not be involved in the error estimation procedure as is currently done during the integration process.

Reference

1. Hamernik, Jubal D., "A Unified Approach to Error Analysis in the Finite Element Method," Ph.D. Dissertation, University of Colorado, 1993.

Lesson 17   Problems

1. Compute the error in a one-element bar fixed at one end and hanging under its own weight using the finite difference approach. See Part IV for the governing differential equation and the finite difference operator.

2. Compare the results of Problem 1 to Problems 2 and 3 of Lesson 16.

3. Compute the error in Problem 5 of Lesson 16 if the actual boundary stresses is 100 psi so the nodal stresses for the smoothed solution are 100, 100, 650, and 450, respectively.

Lesson 18

Nonintegrated Pointwise Refinement Guides

Purpose of Lesson   To develop and demonstrate high-resolution pointwise refinement guides that are functions of the differences between the nodal values of the finite element stresses and the smoothed stresses extracted from the augmented finite difference result using nine-node central difference templates.

The use of pointwise values of stress and/or strain quantities enables us to develop a variety of metrics as refinement guides and termination criteria. In fact, we can use one metric as the refinement guide and another as the termination criteria. The availability of a variety of measures means that we can tailor metrics to the problem being solved. Termination criteria are discussed in Lesson 19.

The development of pointwise refinement guides is motivated by the limitations of strain energy metrics. We can form pointwise refinement guides that have the following advantages over strain-energy-based measures: (1) they are more flexible, e.g., measures can be tailored for different applications; (2) they have a higher resolution, i.e., the metrics exhibit a wider range of values and larger magnitudes in critical regions; and (3) they are more efficient to compute because no integration is necessary in their formulation. The increase in resolution is important because it makes proportional refinement easier to implement, i.e., the level of refinement specified can be a simple function of the magnitude of the refinement guide.

The flexibility of the pointwise approach is demonstrated in this lesson by developing and applying two different refinement guides and postulating others. One of the pointwise refinement guides demonstrated is an absolute measure of discretization error. Its value depends solely on the level of discretization error contained in each element regardless of the level of stress. The other refinement guide demonstrated is a relative measure of discretization error. In this measure, the absolute level of the discretization error is normalized with respect to the maximum stresses contained in the overall problem. This refinement guide focuses on regions of high stresses containing high discretization errors, i.e., the regions that must be refined in practical problems to ensure the accuracy of stresses critical to a successful design.

We compare the absolute and relative pointwise refinement guides to a strain energy measure in two problems that are each refined twice. We see that the resolutions of the pointwise refinement guides are substantially higher than the resolution of the strain-energy-based refinement guide. This difference occurs because the error estimates at points with high error are not diluted by being averaged with the error estimates at points with lower errors. Pointwise refinement guides are put on a solid theoretical foundation in the next lesson, where pointwise termination criteria are developed.

■   ■   ■

The strain-energy-based termination criteria demonstrated in Lesson 16 are not closely correlated with the estimated errors in the stresses at critical points. This is shown in Tables 1–7 of Lesson 16, where we see that the smoothed stresses are converged and the elemental strain energy measures are continuing to decrease as the meshes are refined. The situation is not improved even when the accuracy of the strain-energy-based error estimations are improved in Lesson 17 using the augmented finite element result to form the smoothed solution.

This lack of a direct relationship between the errors at critical points and the strain energy measures is not surprising because of the averaging process inherent in the computation of the strain energy measures. The estimate of the error in the strain energy is formed by integrating the pointwise differences between the smoothed stresses and the finite element stresses existing in an element. The integration process averages the pointwise differences, thus submerging large pointwise values in an averaged value. This dilution of the high values of estimated error at individual points is illustrated in Fig. 1.

In this figure, an assumed error distribution with a high gradient typical of the errors in an element located at a critical point is contrasted to the average error in the element. As we can see, the high estimate at the critical point has been submerged in the average that represents the estimated error in the strain energy. The recognition of how the integration process inherent in strain energy measures absorbs the high estimates at individual points suggests the obvious. If we want error estimates that are closely correlated to errors at critical points, we should develop error measures that are based on pointwise quantities.

This is the course of action that we follow in this lesson. We develop pointwise refinement guides that are based on the pointwise differences between the smoothed stresses and the finite element stresses at the nodal points. Pointwise termination criteria are developed in the next lesson based on the convergence of nodal stresses. In addition, another type of pointwise refinement guide and termination criterion are developed in Lesson 19. These error measures are based on the residuals contained in the augmented finite element result. The use of the residual measures to evaluate finite element results is demonstrated in the next lesson.

Two Nonintegrated Refinement Guides

Two nonintegrated, i.e., pointwise, refinement guides are developed in this section. One is an absolute measure of discretization error and the other is a relative measure of this

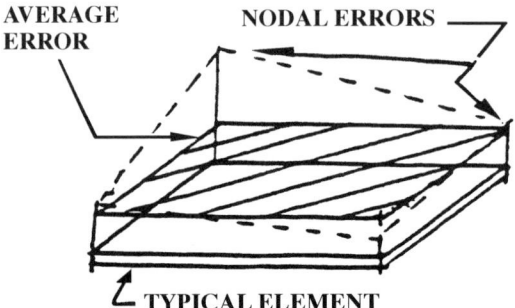

Figure 1. A high gradient point wise error distribution contrasted to the average value.

modeling deficiency. These two distinctive measures of discretization error are developed to demonstrate the flexibility provided by pointwise refinement guides. Refinement guides other than these two are postulated in the discussions that follow.

The two pointwise refinement guides both use the same metric for estimating the discretization error. They differ in the weighting factor used to normalize the measure of the discretization error. The discretization errors in the finite element model are quantified as the differences between the finite element and the smoothed stresses at the nodes. The discretization errors could be quantified equally well and with less computational effort by directly using the interelement jumps in the finite element stresses. However, we see in the next lesson that the smoothed stresses can be used to advantage in extracting stresses and for forming termination criteria. Because of these uses, the added computational expense involved in forming the smoothed stresses is acceptable in the formulation of pointwise refinement guides. The metric for quantifying the discretization errors that serves as the basis of the two refinement guides is formed in the next subsection.

The first refinement guide developed is designed to focus attention on regions with high discretization errors and high stresses. This refinement guide is *not* a pure measure of discretization error. This metric is a relative measure of discretization error because the level of error in each element is scaled with respect to the same outside standard, the maximum stresses in the problem. Its objective is to identify regions that must be improved to provide accurate results at critical points in practical structural problems. The discretization error measure is normalized by dividing it by the sum of the highest stress components contained in the overall problem. As a result, the refinement guide is presented as a percentage of the maximum stresses in the problem. This refinement guide is designed to identify regions of most interest to analysts.

The second refinement guide is designed to quantify the discretization error in the individual elements regardless of the level of stress in the element. This *absolute* measure is normalized with respect to the maximum stress values contained in each individual element. As a result, the metric is expressed as a percentage of the elemental stresses so the discretization error can be compared on an element-to-element basis.

The resolution for the absolute discretization measure is found to be higher than that for the relative measure. This characteristic occurs because the denominator for most elements will typically be smaller than the denominator formed from the maximum global stresses for the relative measure. We do not expect the refinement guide based on the absolute measure of discretization error to be used in practical structural applications. However, it can be used as a research tool to ensure that a solution is fully converged.

An Unscaled Discretization Error Measure

This section forms the unscaled measure of discretization error used in the relative and absolute refinement guides formed in this lesson. The discretization errors are measured as a function of the differences between the finite element result and the smoothed stresses at the nodes. The nodal stresses for the smoothed solution are immediately available as a consequence of superimposing the finite difference (FD) templates on the nodes of the finite element (FE) model. In finite element models using strain gradient based finite elements, the nodal stresses are readily available as a consequence of the formulation. In finite element models using isoparametric finite elements, we can find the nodal stresses directly from the stress representations or by extrapolating the stresses from the Gaussian points. The pointwise differences in the strains at a typical node are given at the nodes as

Nodal Stress Differences for a Typical Node

$$\left\{ \begin{array}{c} \Delta\sigma_x \\ \Delta\sigma_y \\ \Delta\tau_{xy} \end{array} \right\}_i = \left\{ \begin{array}{c} \sigma_x \\ \sigma_y \\ \tau_{xy} \end{array} \right\}_i^{FE} - \left\{ \begin{array}{c} \sigma_x \\ \sigma_y \\ \tau_{xy} \end{array} \right\}_i^{FD} \qquad (18.1)$$

The discretization errors in an element are quantified in the two refinement guides developed here by summing the absolute value of the largest differences for each stress component within an individual element. For example, consider the case of a four-node element with the following stress conditions: the largest difference in the x component of the normal stress is at node 1, the largest difference in the y component of the normal stress is at node 4, and the largest difference in the shear stress is at node 2. The discretization error measure is taken as the sum of the absolute values of these maximum quantities. This measure is used as the numerator in the refinement guide and is given as

Numerator of the Refinement Guide

$$\left| \Sigma\Delta\sigma_{max} \right|_{element} = \left\{ \left| \Delta\sigma_x \right|_{max} + \left| \Delta\sigma_y \right|_{max} + \left| \Delta\tau_{xy} \right|_{max} \right\}_{element} \qquad (18.2)$$

We chose this quantity to quantify the discretization error in the individual elements for two reasons. First, the quantity is not the ''kernel'' for the strain energy, i.e., it is not the strain energy density. That is to say, it cannot be integrated over the element to form a strain energy measure. This enables us to see that quantities other than strain energy can be used to evaluate discretization errors in finite element results. Second, this numerator is designed to produce a refinement guide with a resolution that is higher than the strain energy density measure. In the case of the strain energy density, the pointwise refinement guide depends on a sum containing the absolute values of the stresses at one of the nodes of an element.

The strain energy density, in general, has a lower resolution than the quantity given by Eq. 18.2 because the extreme differences in each stress component would not necessarily exist at a single node. Thus, the metric contained in Eq. 18.2 produces a discretization error measure that demonstrates three points: (1) the flexibility of forming refinement guides from pointwise quantities, (2) the fact that strain energy need not be used as the error estimation metric, and (3) the fact that we can form a useful measure that has a higher resolution than the strain energy density.

There is nothing wrong with using the strain energy density as a refinement guide. This measure was avoided here to reinforce the idea that refinement guides need not be based on strain energy quantities and to demonstrate the flexibility of the pointwise approach to error estimation. Pointwise discretization error measures not based on Eq. 18.2 or strain energy quantities are also possible. We could form a refinement guide based on a single stress component. For example, in ceramic materials, it might be advantageous to focus on the representation of the shear stress as the refinement guide.

The metric formed in Eq. 18.2 for quantifying discretization errors can be further specialized for specific tasks. In the next two subsections, globally normalized and locally normalized refinement guides are formed using Eq. 18.2 as the numerator. The two refinement guides differ only in the choice of the scaling factor that is used for the denominator. The scaling factors enable us to express the refinement guides as percentages. The globally normalized refinement guide has units of percentage of maximum overall stress and the locally normalized measure has units of percentage of maximum elemental stress.

A Globally Normalized Refinement Guide — A Relative Measure of Discretization Error

The refinement guide formed in this subsection is designed for use in practical structural applications. Its purpose is to ensure that regions of high stress receive priority in the refinement process. This goal is accomplished by dividing the discretization error measure of Eq. 18.2 for each element with the same scaling factor, i.e., the denominator for each elemental refinement guide is the same number.

The denominator chosen for this globally normalized refinement guide consists of the sum of the absolute values of the maximum finite element stresses contained anywhere in the domain of the problem being solved. That is to say, the maximum value of the x component of the normal stress can come from one element and the maximum y component can come from another element etc. This denominator is given as

Denominator for the Globally Normalized Refinement Guide

$$\left| \Sigma \sigma_{max} \right|_{global} = \left\{ \left| \sigma_x \right|_{max} + \left| \sigma_y \right|_{max} + \left| \tau_{xy} \right|_{max} \right\}_{global} \qquad (18.3)$$

This normalization is designed to produce large refinement guides at critical points. Since large differences in the finite element and the smoothed stresses are more likely to occur in regions of high stress than in regions of low stress, this relative measure of discretization error will accomplish its goal of identifying regions critical to structural analysis problems that must be refined.

When the numerator given by Eq. 18.2 and the denominator given by Eq. 18.3 are combined, the globally normalized refinement guide for the ith element becomes

A Globally Normalized Refinement Guide

$$[\eta_{rel}]_i = \frac{\left| \Sigma \Delta \sigma_{max} \right|_{element}}{\left| \Sigma \sigma_{max} \right|_{global}} \times 100 \qquad (18.4)$$

This refinement guide is referred to as a relative measure of discretization error because it is scaled with a global quantity, namely, the largest stresses in the domain. Thus, the level of discretization error is reported as a percentage of maximum stress existing in the overall problem. The behavior of this refinement guide is demonstrated in the next section, where we see that it accomplishes its goal of identifying regions of critical interest in structural problems.

Locally Normalized Refinement Guide — An Absolute Measure of Discretization Error

The refinement guide formed in this subsection is designed as an absolute measure of discretization error. Its purpose is to identify the levels of discretization error in individual elements as a percentage of stress in the element. A metric for accomplishing this is formed by dividing the discretization error measure given by Eq. 18.2 with a measure that quantifies the level of stress in each individual element. In this way, the discretization error in the individual elements can be compared to each other on an element-by-element basis in terms of the percentage of the maximum stresses in the elements.

The denominator chosen for this locally normalized refinement guide has a *form* that is analogous to the numerator as given by Eq. 18.2. The denominator consists of the sum of the maximum stress values existing at any nodes in the element being evaluated. That is to say, the maximum value of the x component of the normal stress can be at one node and the other maximum stress components can be at other nodes in the elements. This measure of elemental stresses is expressed as

Denominator for the Locally Normalized Refinement Guide

$$\left| \Sigma \sigma_{\max} \right|_{\text{element}} = \left[\left| \sigma_x \right|_{\max} + \left| \sigma_y \right|_{\max} + \left| \tau_{xy} \right|_{\max} \right]_{\text{element}} \qquad (18.5)$$

When the numerator given by Eq. 18.2 and the denominator given by Eq. 18.5 are combined, the absolute refinement guide for the ith element becomes

A Locally Normalized Refinement Guide

$$[\eta_{\text{abs}}]_i = \frac{\left| \Sigma \Delta \sigma_{\max} \right|_{\text{element}}}{\left| \Sigma \sigma_{\max} \right|_{\text{element}}} \times 100 \qquad (18.6)$$

This refinement guide is referred to as an absolute measure of discretization error because the discretization measure in each element is scaled with respect to the stresses existing in that element. This normalization has the effect of producing large refinement guides whenever the discretization error is large with respect to the level of stress existing in the element. The behavior of this refinement guide is demonstrated in the next section.

We see that this absolute measure accomplishes its goal. It quantifies the discretization error in a way that does not depend on the relative level of stress existing in the individual elements. The elements identified by the absolute measure as having large discretization errors are usually different from those identified by the relative refinement guide.

The two refinement guides just formulated are demonstrated in the next section. We see that we have achieved our goals. We can tailor refinement guides to identify regions with the characteristics we specify and we see that the pointwise refinement guides have a higher resolution than the strain energy measures. This demonstrates the flexibility that pointwise quantities give us in forming error measures. If a refinement guide that focuses on different characteristics is desired, it can be formed.

Application of the Pointwise Refinement Guides

The characteristics and effectiveness of the two pointwise refinement guides just formed is demonstrated with two example problems. The first example is a cantilever beam with an applied shear stress at the free end. This problem is chosen as an example because it contains two types of singularity. The first type occurs at the reentrant corners where the fixed boundary and the beam model intersect. The second type of singularity occurs at the convex corners at the free end where the load is applied. A successful refinement guide should be expected to identify both of these singularities.

The second example problem is a square shear panel with an internal circular hole. The edge of the panel is loaded in tension on the two opposite edges. This problem is chosen as an example because it contains a stress concentration at a known location that is not a singularity. A successful refinement guide should be able to identify high discretization errors at such a critical point.

These examples demonstrate the behavior of the pointwise refinement guides in two types of stress concentration that are of interest to analysts. The two pointwise refinement guides are compared to each other and to the strain energy based refinement guide developed in Lesson 16. The strain energy refinement guide consists of the estimated error in the strain energy in an element divided by the total strain energy in the problem. This is the refinement guide developed by Zienkiewicz and Zhu. The smoothed solution used for each of the refinement guides is formed using finite difference techniques.

Both example problems are modeled using four-node finite elements that have been corrected for parasitic shear. The example problems are of unit thickness and the material properties are given as $E = 30{,}000$ units and $\nu = 0.3$.

Example 1   The Cantilever Beam Problem

The cantilever beam problem is shown in Fig. 2. The original mesh consists of 5 elements and 24 degrees of freedom. The problem is uniformly refined twice in both coordinate directions to produce models consisting of 20 and 80 elements and 66 and 210 degrees of freedom, respectively.

Mesh 1

The three different elemental refinement guides for the initial mesh are presented in Fig. 3. In all of the figures that follow, the refinement guides are shown in the following order: top, relative pointwise; middle, strain energy; and bottom, absolute pointwise.

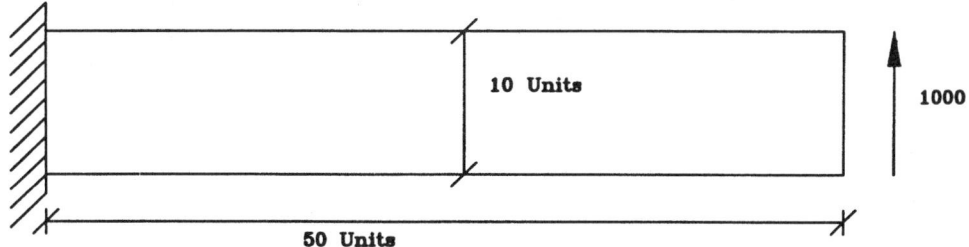

Figure 2. A cantilever beam with a distributed shear load at the free end.

54.83	33.02	26.54	20.06	13.57
11.38	3.34	2.07	1.22	0.42
54.83	42.12	46.73	57.01	100.00

Figure 3. Three types of refinement guides for the initial mesh.

As we can see, the pattern of errors estimated by the relative pointwise refinement guide matches the pattern produced by the energy based refinement guide. This similarity exists because both refinement guides are relative measures, i.e., they have been normalized by global quantities.

The two relative refinement guides both identify the element next to the fixed boundary as having the largest refinement guide, i.e., it is the element in the most need of refinement. This is as expected because this element is located in the region of maximum moment due to the applied load and the element contains two singularities. As a result of these two conditions, the stresses in this element are high and have a more complex distribution than a single element can represent.

The two relative refinement guides then decrease monotonically as they approach the free end. This corresponds to the fact that the stress levels decrease toward the free end of the beam because of the loading conditions. Thus, we can conclude that the relative pointwise refinement guide identifies the elements of interest to analysts, i.e., elements with high stresses and high discretization errors.

The percentage values produced by the two relative refinement guides differ markedly. In Fig. 3, the pointwise measure reports a maximum value of 54.8% and a minimum value of 13.6%. Similarly, the maximum and minimum values for the strain energy measure are 11.4 and 0.4%, respectively. As we can see, the resolution of the pointwise measure is higher than that for the strain energy measure. Thus, it is performing the job for which it was designed.

The data contained in Fig. 3 show that the results for the absolute pointwise refinement guide differ from the results of the relative measures. The absolute measure identifies the element at the free end as having the highest level of discretization error. This is not surprising because this element contains two singularities and a stress imposed on the free surface. The stresses on the boundary are not free to seek their own levels as is the case on a fixed boundary where they are not specified.

The absolute refinement guide decreases toward the center as the effect of the loaded boundary is reduced, as predicted by St. Venant's principle, and increases again toward the fixed boundary. Note that the value of the relative and the absolute pointwise refinement guides are the same at the fixed boundary. This means that the denominators for the two measures are the same in this element because both refinement guides have, by definition, the same numerator. This implies that the maximum values for all three stresses are present in this element, as expected.

The absolute pointwise measure identifies the two elements in this crude finite element model with singularities as containing high discretization errors. Thus, we can conclude that the absolute pointwise refinement guide is doing the job for which it was designed, namely, identifying elements purely on the basis of discretization errors. As a consequence, we must consider it a more balanced measure than the relative measure.

The results for the initial mesh show that the two pointwise refinement guides are both doing the specific jobs for which they were designed. Both measures have a higher resolution than the strain energy measure. The relative measure identifies regions with both high stress levels and high discretization error. The absolute measure measures the level of discretization error without regard for the level of stress. These differences do not mean that one refinement guide is better than the other. It simply demonstrates that pointwise refinement guides have the flexibility to identify a wide variety of modeling deficiencies in finite element models.

Meshes 2 and 3

The characteristics of the refinement guides are now examined as the initial mesh is refined twice with a uniform refinement. The elemental values of the refinement guides for the two uniform refinements are shown in Figs. 4 and 5.

The pattern of behavior of the two relative refinement guides is the same in the refined models as in the original mesh. However, the differences in the magnitudes of the measures contain a danger. As before, the percentage values produced by the two relative refinement guides differ markedly. In Fig. 4, the pointwise measure reports a maximum value of over 31%. Even an inexperienced analyst would not be likely to interpret the finite element results as being acceptable after inspecting the pointwise error measures. The strain energy measure reports a maximum value of only 1.16%. This result makes the finite element result look deceptively good and an inexperienced user might find it acceptable. Thus, we can conclude that the higher resolution of the pointwise measure is a valuable analysis tool.

The relative pointwise refinement guide is performing the job for which it was designed. It identifies the critical point as being in the most need of refinement and it provides a percentage measure that cannot be construed as being acceptable by even inexperienced users. This does not imply that we necessarily want to use the refinement guide as a termination criteria, but it does not preclude the use of this refinement guide as a termination criteria. This is discussed in the next lesson.

The behavior of the two pointwise refinement guides identified in the analysis of the initial mesh is repeated in the results produced by the two refinement guides for the two uniform mesh refinements. The relative pointwise measure identifies the elements containing the singular points at the fixed end as the elements in the most need of

31.28	24.07	18.48	17.51	15.67	14.13	12.54	10.95	9.45	7.65
1.16	0.36	0.26	0.22	0.17	0.14	.11	0.09	0.08	0.03
31.28	28.91	26.37	28.33	29.82	32.37	36.16	42.60	56.08	100.00

31.28	24.07	18.48	17.51	15.67	14.13	12.54	10.95	9.45	7.65
1.16	0.36	0.26	0.22	0.17	0.14	.11	0.09	0.08	0.03
31.28	28.91	26.37	28.33	29.82	32.37	36.16	42.60	56.08	100.00

Figure 4. Three types of refinement guides for the second mesh.

18.70	10.69	9.77	10.02	9.44	9.09	8.71	8.34	7.97	7.60	7.24	6.87	6.50	6.13	5.76	5.39	5.02	4.57	4.42	4.28
.14	.03	.03	.03	.03	.03	.01	.01	.01	.01	.01	.01	.01	.01	.01	.01	.01	.00	.00	.00
18.72	14.70	14.30	15.37	16.44	15.85	16.28	16.79	17.36	18.06	18.91	19.34	21.22	22.56	25.06	28.12	32.67	39.45	53.99	86.30
12.96	13.20	9.45	8.06	7.77	7.38	7.01	6.64	6.27	5.90	5.53	5.16	4.79	4.42	4.06	3.68	3.23	3.14	2.51	2.90
.06	.04	.03	.03	.03	.01	.01	.01	.01	.01	.01	.01	.01	.00	.00	.00	.00	.00	.00	.00
29.63	29.24	23.06	21.36	21.70	21.89	22.18	22.50	22.87	23.30	23.81	24.42	25.16	26.09	27.27	28.53	31.11	36.33	36.69	64.62
12.96	13.20	9.45	8.06	7.77	7.38	7.01	6.64	6.27	5.90	5.53	5.16	4.79	4.42	4.06	3.68	3.23	3.14	2.51	2.90
.06	.04	.03	.03	.03	.01	.01	.01	.01	.01	.01	.01	.01	.00	.00	.00	.00	.00	.00	.00
29.63	29.24	23.06	21.36	21.70	21.89	22.18	22.50	22.87	23.30	23.81	24.42	25.16	26.09	27.27	28.53	31.11	36.33	36.69	64.62
18.70	10.69	9.77	10.02	9.44	9.09	8.71	8.34	7.97	7.60	7.24	6.87	6.50	6.13	5.76	5.39	5.02	4.57	4.42	4.28
.14	.03	.03	.03	.03	.03	.01	.01	.01	.01	.01	.01	.01	.01	.01	.01	.01	.00	.00	.00
18.72	14.70	14.30	15.37	16.44	15.85	16.28	16.79	17.36	18.06	18.91	19.34	21.22	22.56	25.06	28.12	32.67	39.45	53.99	86.30

Figure 5. Three types of refinement guides for the third mesh.

refinement. The absolute pointwise measure identifies the singularities at the corners where the shear stress is applied as the elements with the highest discretization error. Note that the absolute measure does not highlight the elements containing the singular points at the fixed end for special attention in the second uniform refinement shown in Fig. 5. The value of the refinement guides produced by the absolute measure in the boundary elements on the interior boundary are higher than those for the elements containing the singularities.

Improved Resolution

The range of values for the three refinement guides for the three meshes used to represent the example problem are shown in Fig. 6. This figure shows the higher resolution of the pointwise refinement guides as compared to the strain energy measure. The ranges for the three refinement guides as percentages are presented for the three different meshes.

For the initial mesh, the relative pointwise refinement guide ranges from 13.6% for the smallest to 54.8% for the largest. The range for the strain energy measure is from 0.4 to 11.4%. The range of the relative pointwise measure is approximately 3.75 times that of the strain-energy-based measure. This shows that the relative pointwise refinement guide has a higher resolution than the strain energy measure.

For the absolute pointwise refinement guide the range is 6.7 times that of the strain energy measure and 1.8 times that of the relative pointwise measure. This larger resolution for the absolute measure was designed into the measure in the following way. As we know, the denominator in the absolute measure depends on the stress levels in the individual elements and not on the largest stresses in the overall problem as is the case for the relative pointwise measure. This means that the pointwise discretization measure given by Eq. 18.2 is typically divided by a smaller number than is the case in the relative measure. As a result, the absolute measure will have a larger span of values, or a higher resolution, for a given problem than the relative measure.

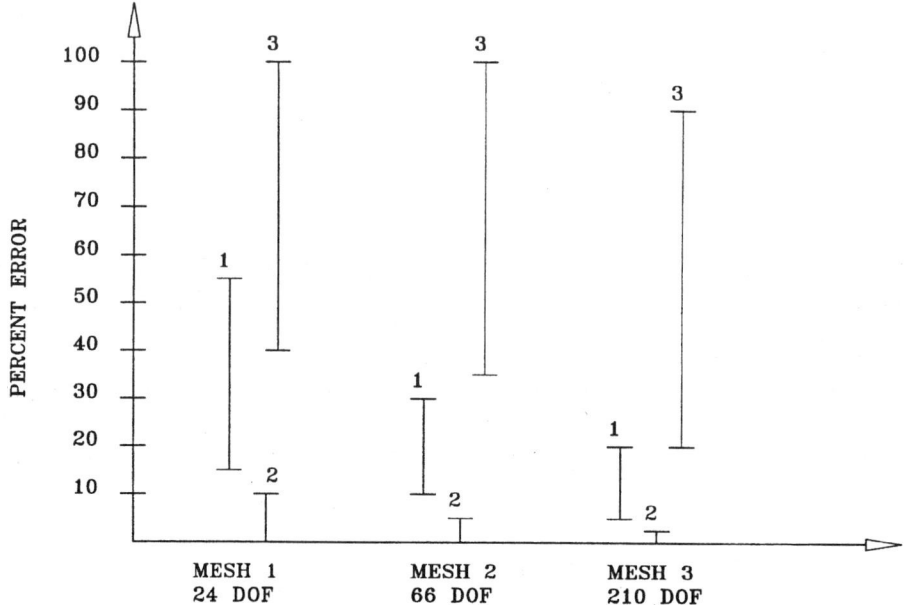

Figure 6. Resolution of the three refinement guides vs mesh refinement: line1, relative point wise; line 2, strain energy; line 3, absolute point wise.

When the initial mesh is refined, the resolution of the two pointwise measures is even higher with respect to the strain energy measures than for the initial mesh. The relative and the absolute pointwise guides are 21 times and 103 times larger than the strain energy measures, respectively, for the second mesh. Similarly, the multipliers for the third mesh are 86 and 483, respectively.

The relative pointwise refinement guide is designed to identify regions of high stress that are in need of refinement. As we have seen in this example, this goal has been accomplished. The relative refinement guide successfully identifies the elements in the problem that contain the points that are critical to the design of the beam as being in need of the most refinement. In this case, the critical elements contains singularities and the highest stresses in the problem.

The absolute pointwise refinement guide is designed to identify the element with the highest level of discretization error regardless of the level of stress in the element. This example problem shows that this refinement guide has successfully achieved this aim. The absolute pointwise measure has identified the free end of the beam as having the largest amount of discretization error. This is understandable because the applied load is a discontinuity in the stress distribution in the element at the free end. The absolute pointwise refinement guide identifies the inability of the four-node element to model the complex stress distribution associated with this discontinuity and the reentrant corners even though the stresses are low.

The resolution of the strain energy measure is very low as a result of the averaging process inherent in the formulation, as has been discussed. However, the strain energy measure identifies the elements containing the singular points at the fixed end as the critical regions. Even though the strain-energy-based refinement guides are not meant to serve as a termination criterion, one can imagine an inexperienced user observing the strain energy measure decreasing from a maximum of 11.4% in the initial mesh to a maximum of 0.14% in the third mesh and falsely concluding that the stresses at every point in this analysis are adequate for design purposes.

In contrast to this, the relative pointwise refinement guide indicates an error of 18.7% in the third mesh. Even an inexperienced user is not likely to find this level of error at a critical point acceptable. This contrast in the resolutions between the pointwise and the strain energy measure indicates the usefulness of the pointwise approach to error estimation in structural analysis problems. Thus, we can conclude that the goal of increasing the resolution of the refinement guides through the use of pointwise measures has been successfully achieved.

Example 2   Square Panel with a Circular Hole

The second example problem is a square shear panel with a circular hole, as shown in Fig. 7. This problem is chosen for analysis because it contains a stress concentration in a known location that is not a singular point. Furthermore, we have analyzed this problem in detail in the previous two lessons. It is refined beyond convergence in Lesson 16.

In the finite element analysis, this structure is represented with a quarter-section, as shown in Fig. 8 to reduce the number of unknowns in the set of equations that must be solved. It is possible to use the reduced problem because of symmetry conditions. In the quarter section model, the top surface and the right-hand side are supported with roller boundaries. The uniformly distributed tension load is applied to the left-hand edge. The problem is modeled with an original mesh of 8 elements and two refinements containing 32 and 128 elements, respectively. The three models contain 30, 90, and 306 degrees of freedom. The model is not refined to the point of convergence because convergence is achieved in Lesson 16. However, it is refined here to the point where the characteristics of the refinement guides in the region of a stress concentration can be ascertained.

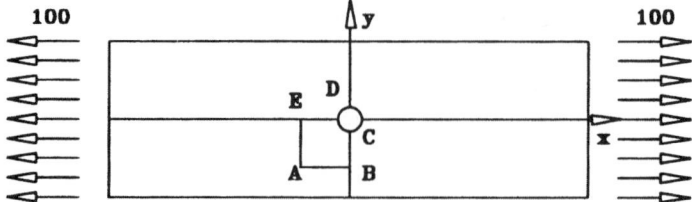

Figure 7. Problem definition for the square shear panel with a cirular hole.

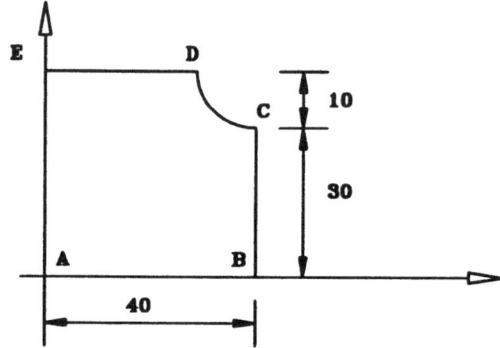

Figure 8. Symmetric portion of the problem analyzed.

Mesh 1

The elemental values of the three refinement guides for the initial mesh are shown in Fig. 9. The refinement guides are shown in the same order as in the previous example, namely, top, relative pointwise; middle, strain energy; and bottom, absolute pointwise.

All three of the refinement guides identify the same element as being in the most need of refinement. We know that this element is in the region of the stress concentration from previous analyses. It is at the location of the maximum discretization error found in

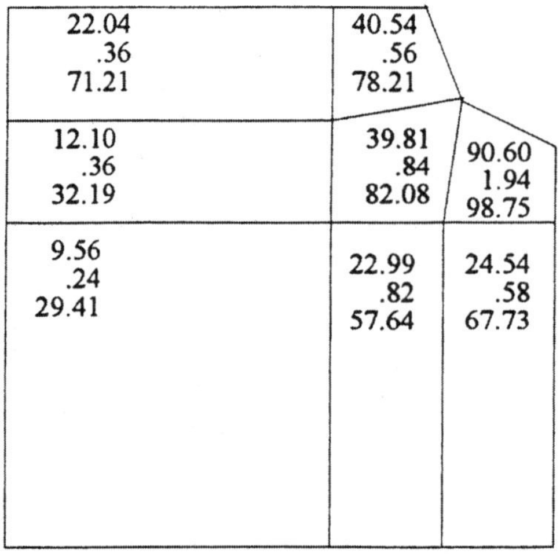

22.04 .36 71.21		40.54 .56 78.21	
12.10 .36 32.19		39.81 .84 82.08	90.60 1.94 98.75
9.56 .24 29.41		22.99 .82 57.64	24.54 .58 67.73

Figure 9. Three types of refinement guides for the initial mesh.

Lesson 16 using the strain energy measure for the fully converged model. We see that the other two elements on the boundary of the internal hole possess high levels of discretization error. The relative pointwise and the strain energy refinement guides drop off rapidly away from the region of the stress concentration. As we can see, the absolute pointwise measure has a more uniform distribution than the other two refinement guides. This is as expected because the absolute measure is independent of the stress levels in the individual elements. The resolution of the pointwise measures is higher than that for the strain energy refinement guides. This feature is discussed in more detail later in this lesson.

The results produced by the absolute pointwise measure for this model differ in one significant way from the previous example. The value of the absolute measure in the convex corner where the load is applied is significantly lower than its value in the region of the stress concentration. This is the case in this problem because the load is applied normal to the surface so it is not supported directly by shear stresses, as was the case in the previous example. As a result, the element in question does not contain a large discretization error. Note that if this problem did not contain an internal hole, the result produced by this coarse mesh would be close to the exact result.

The distribution of values produced by the relative pointwise and the strain energy refinement guides can be interpreted to mean that these measures identify the region critical to the structure. This contrasts to the result produced by the absolute pointwise measure, which indicates that a relatively even distribution of discretization errors exists in the model. Thus, we can conclude that the two pointwise refinement guides identify the types of deficiencies for which they are each designed.

Mesh 2

The results of evaluating the second mesh with the three elemental refinement guides are shown in Fig. 10. Each element in the initial mesh has been subdivided into 4 elements. The new mesh contains 32 elements and 90 degrees of freedom. Adaptive refinement was not used because the objective of this lesson is to study the behavior of pointwise refinement guides, not to exploit their capabilities.

The elements identified by all three refinement guides as being in the most need of refinement are still on the boundary of the internal hole. They are on a different section of the boundary than in the first refinement. As expected, the location of the element exhibiting the maximum value for the refinement guide is different for the relative and the absolute pointwise measures. The magnitude of the maximum value has increased for all three measures.

Again, the relative pointwise measure and the strain energy measure identify the same element as requiring the most refinement. This element is one of the elements formed from refining the element in the initial mesh with the maximum refinement guide. It is not the element at the stress concentration. However, in Lesson 16 this same element was reported as having the maximum relative strain energy error and we found the reason to be due to high discretization errors in the area of the applied shear stress. For the absolute measure, the reported value has its second highest value for this element. The highest measure for the absolute measure is reported for the element away from the stress concentration that is the mirror image of this point.

In this refinement, we noted that the value of the maximum refinement guides has increased for all three measures. This probably occurs because as the mesh is refined, the strain gradients are increasing. This would increase the discretization error in the element responsible for representing the high strain gradient. It is not likely that a mesh as coarse as that shown in Fig. 9 would be used in a practical problem. As a result, it is likely that the increase in the refinement guide seen from mesh 1 to mesh 2 would not be seen in

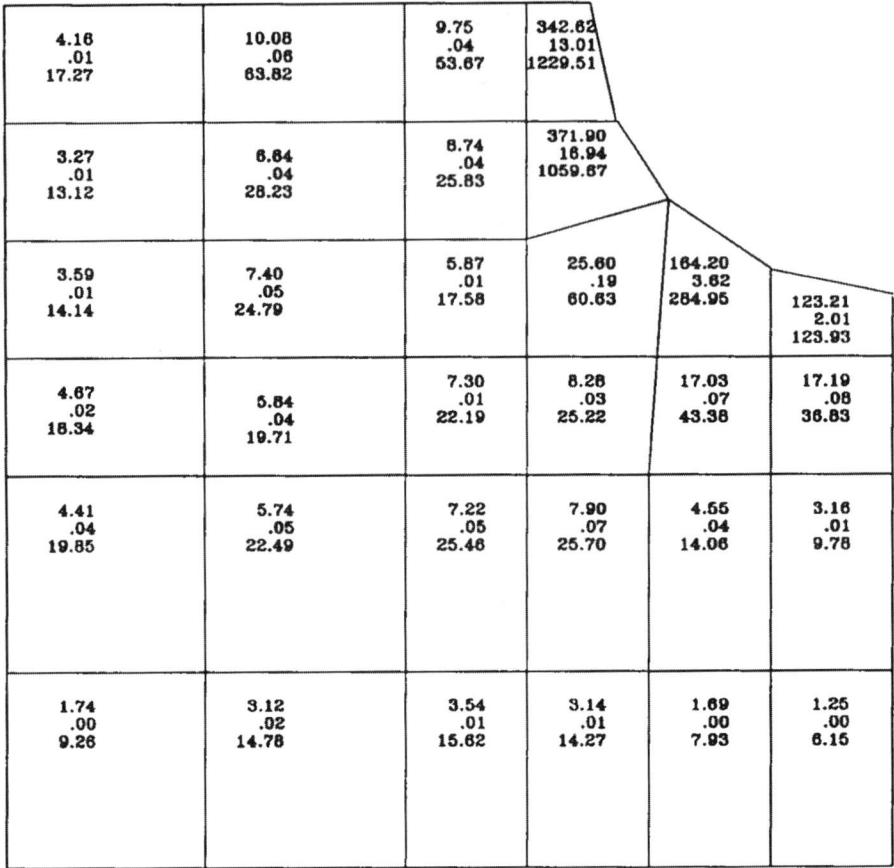

Figure 10. Three refinement guides for mesh 2.

practice. This is mentioned in passing to emphasize that the refinement guides are just that. They are refinement guides and not termination criteria. The size of the pointwise refinement guides ensures that even an inexperienced user would not assume that the result was acceptable with "error estimates" of approximately 100%. As we can see, the resolution of the pointwise measures is significantly higher than that for the strain energy measure. The resolutions of the elements are considered in more detail after the second mesh refinement is discussed.

Mesh 3

The elemental refinement guides for the second mesh refinement are shown in Fig. 11. Again, each element has been subdivided into 4 elements. The new mesh contains 128 elements and 309 degrees of freedom.

The maximum values for the three measures are all substantially reduced in this model. As a result, we can conclude that the refinement guides are starting to converge. We can say this with a certain amount of confidence for the relative pointwise measure and the strain energy refinement guide because the locations of the elements with the maximum refinement guides match those for the highly refined meshes of Lesson 16.

We now discuss the differences in resolution for the relative pointwise and the strain energy refinement guides. We will not discuss the absolute pointwise measure for two reasons. First, the level of resolution for the absolute measure is so high with respect to the strain energy measure that the scale of the figure comparing the relative measures

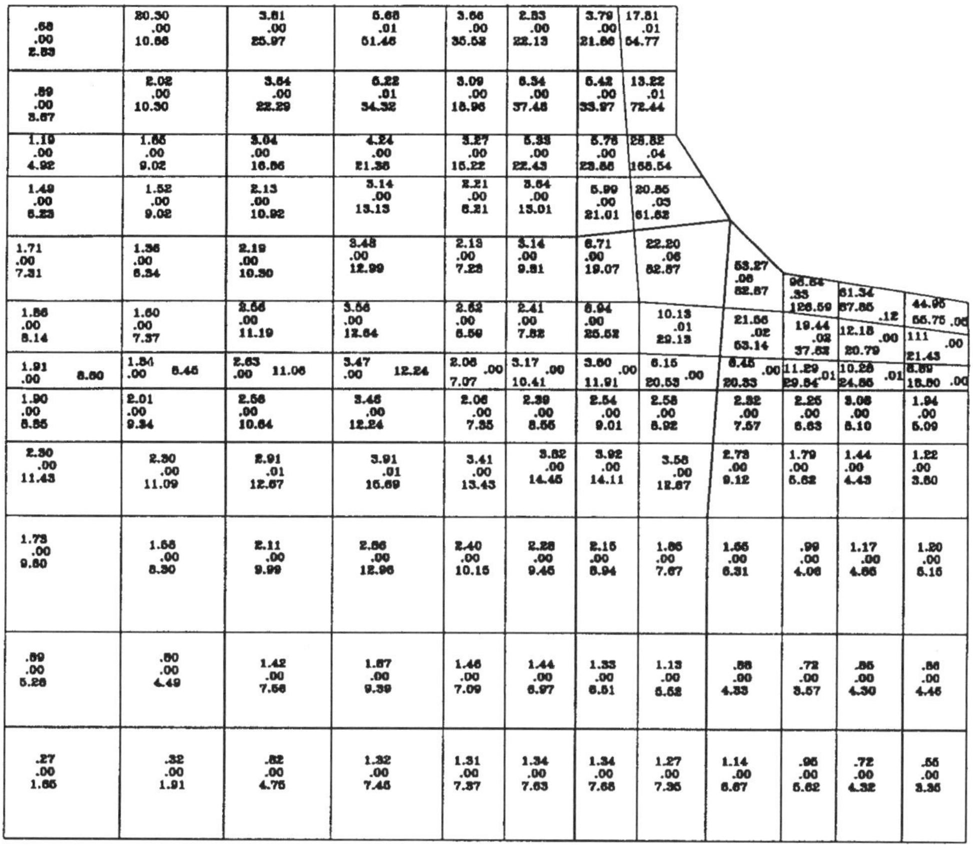

.68 .00 2.83	20.30 .00 10.68	3.81 .00 25.97	6.68 .01 51.46	3.66 .00 36.52	2.83 .00 22.13	3.79 .00 21.86	17.81 .01 54.77				
.89 .00 3.87	2.02 .00 10.30	3.54 .00 22.29	6.22 .01 34.32	3.09 .00 18.96	6.34 .00 37.48	6.42 .00 33.97	13.22 .01 72.44				
1.19 .00 4.92	1.65 .00 9.02	3.04 .00 16.56	4.24 .01 21.36	3.27 .00 15.22	5.33 .00 22.43	5.75 .00 23.86	28.82 .04 168.54				
1.49 .00 6.23	1.52 .00 9.02	2.13 .00 10.92	3.14 .00 13.13	2.21 .00 8.21	3.64 .00 13.01	5.99 .00 21.01	20.85 .03 61.62				
1.71 .00 7.31	1.36 .00 8.34	2.19 .00 10.30	3.48 .00 12.99	2.13 .00 7.23	3.14 .00 9.91	6.71 .00 19.07	22.20 .06 82.87	53.27 .06 82.87	95.84 .35 126.59	61.34 87.85	44.95 .12 55.75 .05
1.86 .00 8.14	1.60 .00 7.37	2.56 .00 11.19	3.56 .00 12.64	2.52 .00 8.59	2.41 .00 7.82	5.94 .00 25.52	10.13 .01 29.13	21.56 .02 53.14	19.44 .02 37.52	12.18 .00 20.79	111 .00 21.43
1.91 .00	1.54 .00 8.45	2.63 .00 11.06	3.47 .00 12.24	2.06 .00 7.07	3.17 .00 10.41	3.60 .00 11.91	6.15 .00 20.53	6.45 .00 20.33	11.29 .01 29.84	10.25 .01 24.85	6.99 .01 18.50 .00
1.90 .00 8.85	2.01 .00 9.34	2.56 .00 10.64	3.46 .00 12.24	2.06 .00 7.35	2.39 .00 8.55	2.54 .00 9.01	2.58 .00 8.92	2.32 .00 7.57	2.25 .00 6.63	3.06 .00 8.10	1.94 .00 6.09
2.30 .00 11.43	2.30 .00 11.09	2.91 .01 12.67	3.91 .01 16.69	3.41 .00 13.43	3.52 .00 14.45	3.92 .00 14.11	3.58 .00 12.67	2.73 .00 9.12	1.79 .00 5.62	1.44 .00 4.43	1.22 .00 3.80
1.73 .00 9.80	1.58 .00 8.30	2.11 .00 9.99	2.86 .00 12.96	2.40 .00 10.15	2.28 .00 9.45	2.15 .00 8.94	1.86 .00 7.67	1.66 .00 6.31	.99 .00 4.06	1.17 .00 4.66	1.20 .00 6.16
.69 .00 5.28	.80 .00 4.49	1.42 .00 7.56	1.87 .00 9.39	1.46 .00 7.09	1.44 .00 6.97	1.33 .00 6.51	1.13 .00 5.62	.88 .00 4.33	.72 .00 3.57	.85 .00 4.30	.86 .00 4.46
.27 .00 1.85	.32 .00 1.91	.82 .00 4.75	1.32 .00 7.46	1.31 .00 7.37	1.34 .00 7.63	1.34 .00 7.68	1.27 .00 7.36	1.14 .00 6.67	.95 .00 5.62	.72 .00 4.32	.55 .00 3.35

Figure 11. Three refinement guides for mesh 3.

would render the figure useless. Second, the absolute measure was included here more to demonstrate the flexibility of the pointwise approach than to forward the measure as a serious candidate for practical applications. This flexibility was demonstrated in the previous problem and its high value in this example substantiates the first demonstration.

The relative resolution of the two refinement guides being discussed is shown in Fig. 12 for the three meshes. For the initial mesh, the resolution of the pointwise measure is 47 times that of the strain energy measure. For the two subsequent refinements, the resolution of the pointwise measure is 22 and 292 times the resolution of the pointwise measures, respectively. These results show that the pointwise approach to error estimation has accomplished its goal of increasing the resolution of the refinement guides.

Closure

The primary objective of this lesson was to develop pointwise refinement guides and to demonstrate that they are improvements on the strain-energy-based measures discussed in Lessons 16 and 17. This objective was successfully accomplished by forming pointwise refinement guides that (1) have higher resolution than the strain energy measures; (2)

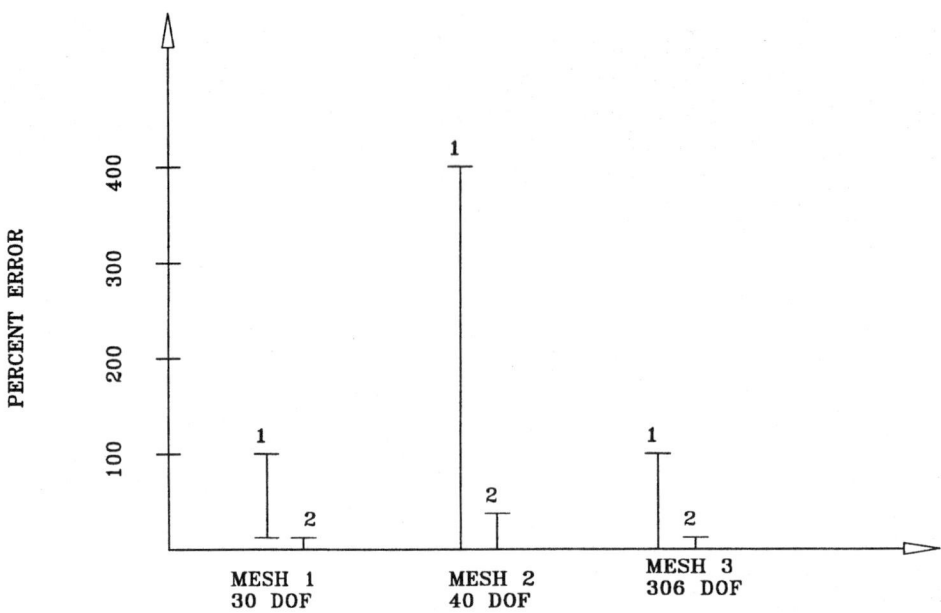

Figure 12. Resolution of the two refinement guides vs mesh refinement: line 1, relative point wise; line 2, strain energy.

have the flexibility to identify different combinations of deficiencies and characteristics, as desired by the analyst; and (3) require less computational effort to form than the strain energy measures. The third characteristic was not demonstrated directly. However, it is obviously true because the pointwise measures do not require the integration step needed to form the strain energy measure.

The flexibility of the pointwise measures is demonstrated by developing two refinement guides that identify modeling deficiencies with different characteristics. One of the refinement guides provides an absolute measure of discretization error in individual elements and the other provides a relative measure of this modeling deficiency. The absolute refinement guide quantifies the need for refinement purely in terms of the magnitude of the discretization error in the individual elements. The relative measure scales the discretization error measure in terms of the largest stresses in the overall problem. This focuses the relative refinement guide on elements with high discretization errors and high stress.

Elements exhibiting large values of this refinement guide exist in regions that contain points that are critical to the design of structures. St. Venant's principle makes it safe to use this type of refinement guide in structural applications. For example, if the region where the load is applied is not critical to the design, the finite element model need not always be highly refined in that region because the static equivalent of the load can be successfully transmitted to the region of critical interest with a finite element representation that is not fully converged.

Two example problems are used to demonstrate the effectiveness of the absolute and the relative pointwise refinement guides. A cantilever beam problem is chosen to demonstrate the ability of the pointwise measures to identify the type of condition for which they were designed. The absolute measure identifies lightly stressed elements with singularities and a stress discontinuity due to the applied load as containing the highest levels of discretization error. The relative measure identifies elements with singular points and the highest level of stresses as the critical regions in the finite element model. That is to say, the relative refinement guide identified the elements that must be refined if the model is to better represent the points that are critical to the design of the structure.

The second example, a square shear panel with a circular hole, demonstrates the characteristics of the two pointwise refinement guides in the presence of a simple stress concentration. That is to say, we can observe how the two refinement guides behave when the stress concentration is not as severe as a singularity. In this example, both the absolute and the relative measures identified elements on the circular hole as having the highest value of refinement guide.

As the mesh is refined, the value of the relative refinement guide for the element that contains the stress concentration is large, but it is not the largest. An element near the critical point is identified as having the greatest need for refinement. This element is in the same region as that identified by the strain energy measure in Lesson 16 for the fully converged model. When that case was investigated in detail, it was found that the element with the highest refinement guide contained a severe discontinuity in the shear stress. Thus, we can conclude that the relative refinement guide functions as designed. The element identified as most needing refinement is an element with both high stresses and high discretization error.

The absolute measure identifies an element on the interior hole that is away from the stress concentration as containing the highest discretization error. It does not identify the element containing the singular point on the surface where the load is applied for this distinction. In fact, this element has a low value for the discretization error measure. The reason for this is due to the orientation of the load. If the problem being solved did not contain the interior hole, it would be a simple uniform tension case. This finite element representation is capable of representing that constant stress problem exactly. As a result of St. Venant's principle, we can expect the loaded element on the boundary to contain a stress distribution that is nearly constant. This element is perfectly capable of representing this simple stress distribution, so we can conclude that the absolute refinement guide is doing its job.

In both example problems, the resolutions of the pointwise refinement guides are significantly higher than those of the strain energy refinement guides. The resolution of the relative pointwise measures are typically 4 to 10 times those of the relative strain energy refinement guides. The resolution of the absolute measures is even higher. The availability of higher resolution enables us to accomplish proportional mesh refinement in a simple manner. Thus, we can conclude that the pointwise refinement guides accomplished the goal of having a significantly higher refinement than the strain energy measures. It is clear from the results presented here that pointwise refinement guides provide the best way for evaluating discretization errors in finite element results.

Theoretical Basis for the Pointwise Refinement Guides

The pointwise refinement guides have been developed heuristically. They were developed to eliminate deficiencies contained in the strain energy refinement guides. Up to now, the capabilities of the pointwise refinement guides have been supported only by anecdotal evidence. The pointwise measures have not been given any theoretical support. This support is provided in Lesson 19, where the augmented finite element result is interpreted as an approximate finite difference result. This interpretation enables us to put the pointwise stresses and the pointwise refinement guides on a solid theoretical basis.

References

1. Hamernik, Jubal, ''A Unified Approach to Error Analysis in the Finite Element Method,'' Ph.D. Dissertation, University of Colorado, 1993.

2. Aarnes, Knut Axel, ''Extension and Applications for a Pointwise Error Estimator Based on the Finite Difference Method,'' Master's Thesis, University of Colorado, 1993.

Lesson 18   Problems

The objective of these problems is to provide a step-by-step approach for identifying the flexibility and capabilities of pointwise error measures or refinement guides.

1. Define two refinement measures in terms of capabilities and equations that are based on individual elements (ones from the text or your own). Define why these guides are more useful for identifying one condition than for another. That is to say, for example, explain why one guide is for use with ceramics and the other is to be designed to clearly identify singular points.

2. Define two refinement guides that are computed for nodal points instead of for individual elements.

3. Discuss which refinement guide from Problem 2 provides the greatest resolution as a refinement guide, i.e., which one differentiates the most between areas of high error and low error.

4. Discuss the role of the numerator in the refinement guide.

5. Discuss the role of the denominator in the refinement guide.

6. Discuss how one might form pointwise quantities for use in refinement guides at Gaussian points and what if any advantages or disadvantages this would have.

Lesson 19

A Theoretical Foundation for Pointwise Evaluation Measures

Purpose of Lesson   To provide a theoretical foundation for two types of pointwise evaluation measures: (1) those based on stress differences between the smoothed and the finite element stresses (see Lesson 18) and (2) those based on residuals contained in the augmented solution.

In this lesson, the displacements of the augmented finite element solution formed in Lesson 17 are interpreted as an approximate finite difference result. Since the finite element and the actual finite difference solutions must converge to the same result as the discrete model is refined, the amount by which the augmented finite element solution fails to represent an actual finite difference result, as measured by the residuals, identifies deficiencies in the discrete model. As a consequence, we can use the residuals contained in the augmented solution to identify the level of discretization error in the finite element model and the amount by which the augmented solution fails to satisfy the exact result. That is to say, the residuals, which must equal zero in an actual finite difference result, can be used as refinement guides and to form termination criteria.

Furthermore, the interpretation of the augmented solution as an approximate finite difference solution provides a theoretical basis for explaining the effectiveness of the refinement guides based on the pointwise differences between the smoothed and the finite element stresses demonstrated in Lesson 18. This interpretation enables us to compare the stresses on a pointwise basis at the nodes because of the Taylor series nature of the finite difference formulation. Then, as a consequence of the common Taylor series nature of the two solutions, we can show that the requirement for the finite element stresses to be continuous (zero interelement jumps) and equal to the smoothed stresses is a more stringent constraint on the accuracy of the approximate solution than is the requirement that the residuals in the augmented solution be zero.

We also use the fact that both the finite element and finite difference methods can be derived from a common Taylor series basis to identify the conditions for which the smoothed stresses are of an order higher than the finite element stresses. This means that the common practice of using the smoothed stresses as the output of finite element analyses has a solid theoretical basis. Furthermore, since the smoothed stresses are of the same order as the displacements in the finite element solution (more accurate than the finite element stresses), we can justify the use of the convergence of the smoothed stresses as termination criteria.

■   ■   ■

The primary objective of this lesson is to provide a theoretical foundation for two types of pointwise evaluation measures: (1) those based on residuals and (2) those based on stress quantities. The theoretical basis for the pointwise evaluation measures is provided by assuming that the augmented finite element result is an approximate finite difference solution. The augmented result is not an actual finite difference solution because the finite difference representation of the governing differential equations are not satisfied at the nodal points of the augmented result unless the solution is exact.

The relationships between the finite element (FE) result, the augmented result (the approximate finite difference), and an actual finite difference (FD) result are shown in Fig. 1. In these representations, the same mesh is used on the domain of the problem and the mesh need *not* be regular. The developments in Part IV of the book presented procedures for forming irregular or distorted nine-node finite difference templates.

Residual-Based Evaluation Measures

The amounts by which the approximate finite difference solution fail to satisfy the components of the governing differential equations identically at the nodes are called the residuals. Since the finite element and the finite difference solutions must be identical when the solution is exact, we can attribute the amount by which the augmented solution differs from an actual finite difference solution as being due to deficiencies in the discrete model. In other words, the residuals contained in the augmented finite element solution quantify the differences between the two approximate solutions and, consequently, can serve as both refinement guides and termination criteria.

The use of the residuals contained in the augmented finite element solution is a new approach for evaluating the accuracy of finite element results. The ability of the residuals contained in the augmented solution to identify the level of discretization error is demonstrated in this lesson. The success of this demonstration serves to validate the

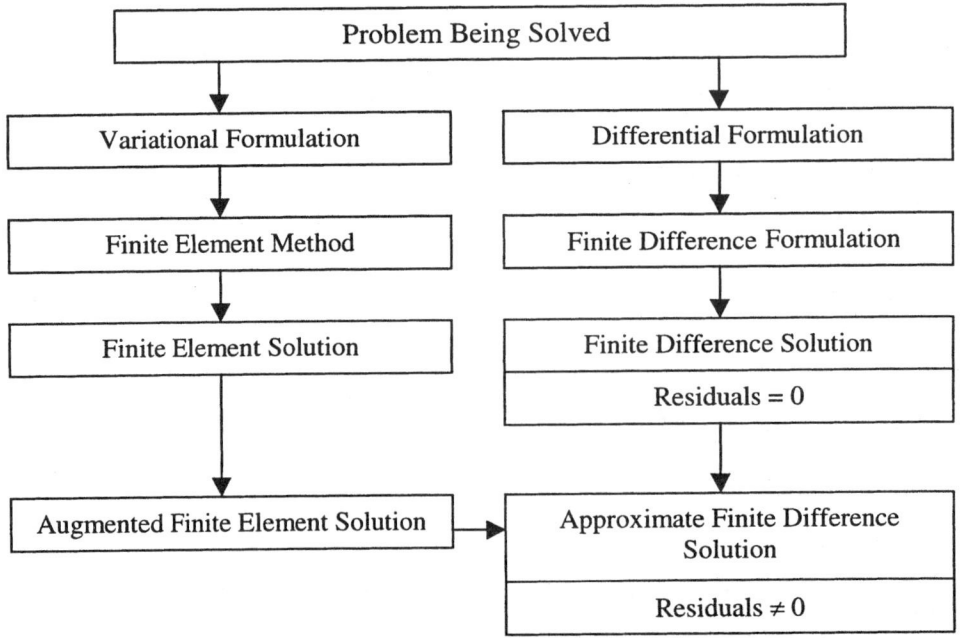

Figure 1. Relationship between the FE solution, the augmented FE solution (i.e., the approximate FD solution), and the actual FD solution.

interpretation of the augmented finite element result as an approximate finite difference result that converges as the mesh is refined.

Stress-Based Evaluation Measures

The interpretation of the augmented finite element result as an approximate finite difference result enables us to put evaluation measures based on stresses on a solid theoretical foundation. This interpretation enables us to compare the stresses on a pointwise basis because of the Taylor series nature of the finite difference formulation. In addition, we explain the effectiveness of the refinement guides demonstrated in the previous lesson by showing that the requirement for the interelement jumps to be zero is a more stringent constraint on convergence than is the requirement that the residuals in the augmented solution be zero.

Stress-based evaluation measures are of importance because of their practical utility. Having evaluation measures expressed in stress units makes them easy to relate to the quantities of most interest to analysts: the stresses at critical points. The smoothed stresses used in the formulation of the evaluation measures are not alien to current practice. The stresses contained in finite element models are often reported in the form of smoothed quantities. Furthermore, as we prove in this lesson, the smoothed stresses are of an order higher than the finite element stresses in many practical applications. This means that the smoothed stresses are more accurate than the finite element stresses, so the convergence of the smoothed stresses at critical points can legitimately be used as termination criteria. The use of the convergence of the stress representations as termination criteria also matches commonly used practice.

Further Background Information

In addition to providing a theoretical basis for the pointwise evaluation measures, the interpretation of the augmented finite element result as an approximate finite difference solution enables residual-based evaluation procedures developed by other researchers to be related to the two pointwise evaluation approaches developed here. In existing approaches using residuals, the interelement jumps are often taken as residual quantities without proof and the interelement jumps are sometimes combined with the residuals in the individual finite elements to evaluate the finite element result. We see that the new pointwise evaluation procedures developed here encompass other residual approaches and have advantages over these other residual approaches.

Contents of This Lesson

As a consequence of interpreting the augmented finite element result as an approximate finite difference solution, several previously unrecognized relationships between finite element and finite difference solutions are identified here. The identification of these relationships enables us to develop other new capabilities of importance to computational mechanics. Although some of these capabilities are not directly related to the evaluation of finite element analyses, they are discussed in this lesson because of the insights and research opportunities they provide.

The following nine items identify capabilities and relationships that can be demonstrated as a consequence of interpreting the augmented finite element result as an approximate finite difference result. The capabilities and relationships identified in these nine items are discussed in turn:

1. A new approach for extracting residuals from finite element results
2. A new iterative approach for solving finite difference problems
3. A new approach for using residuals in the evaluation of finite element results

4. The validation of the pointwise refinement guides developed in Lesson 18

5. The identification of the conditions for the smoothed strains to be of an order higher than the strain components in the finite element result

6. The identification of the relationship between the residuals in the finite element solution and those in the augmented result

7. A development of the relationship between interelement jumps in strain and the residuals in the augmented solution

8. A summary and comparison of the capabilities available for evaluating finite element results

9. An outline of a universal evaluation postprocessor for finite element results

Item 1 validates the interpretation of the augmented finite element result as an approximate finite difference solution. A procedure for computing the residuals contained in the approximate finite difference result is developed. Then the behavior of the residuals under mesh refinement is demonstrated for a cantilever beam with two loading conditions. The residuals are seen to converge as the mesh is refined except in regions of singularities. Thus, we can conclude that the augmented finite element result behaves as we would expect an approximate finite difference result to behave, so it is, indeed, an approximate finite difference result. The interpretation of the augmented solution as an approximate finite difference result provides the basis for the remainder of the developments presented here. This is the case because the smoothed stresses extracted from the approximate finite difference solution can be treated as pointwise Taylor series quantities and the residuals can be used as evaluation measures.

Item 2 presents a new approach for solving finite difference problems using finite element solution techniques. A finite element model is adaptively refined until the residuals in the associated augmented solution are acceptably close to zero. Then the refinement is stopped and the augmented solution is taken as the finite difference solution. The primary objective in presenting this new solution technique is to further clarify the relationships between (1) a finite element result, (2) an augmented finite element result, and (3) an actual finite difference result that is shown in Fig. 1. This figure contains a flow chart for the new approach to solving finite differential problems. If the residuals of the approximate finite difference problems do not equal zero, the finite element model is refined and the process is repeated until the residuals are reduced to a satisfactory level. The flow chart, which is presented in list form, for the new solution technique clearly shows how the residuals in the augmented solution are used to identify deficiencies in the discrete model. As a result of this solution technique, we can see that we do not have to claim that either the finite element or the finite difference solution is closer to the actual solution than the other to use the residuals as refinement guides. That is to say, the identification of the discretization errors can occur before we can say anything useful about the accuracy of the stresses. Operationally, this means that we have a wider latitude for forming refinement guides than we do for termination criteria. We use these ideas in later developments.

Item 3 outlines a procedure for using the residuals contained in the augmented finite element solution to guide the adaptive refinement of finite element results. The residuals are interpreted as fictitious applied loads that cause the existing solution to deviate from the actual result. This idea is used to further establish the pointwise nature of the quantities contained in the augmented finite element result.

Item 4 takes a step back from the *assumption* that the smoothed stresses are a better representation of the actual stresses than the stresses in the finite element result. This means that the differences between the finite element and the smoothed stresses are interpreted as deficiencies in the discrete model. At this point, the differences are not interpreted as estimates of the errors in the finite element stresses. The interpretation of the differences as indicating deficiencies in the discrete model is used to provide a theoretical basis for using the pointwise differences between the smoothed stresses and the finite element stresses as refinement guides. This is done in preparation for showing that the smoothed stresses are, indeed, more accurate than the finite element stresses. In this work, these differences and the interelement jumps in the finite element stresses are often used synonymously.

Item 5 uses the ideas of item 1 and item 4 to show the conditions that must hold for the smoothed stresses to be of the same order of accuracy as the displacements. As a consequence of this relationship, the smoothed stresses are of an order higher than the finite element stresses. This result is implicitly interpreted to mean that the smoothed stresses are more accurate than the finite element stresses after the mesh is refined.

Item 6 introduces the residual approach developed by Kelly et al. for evaluating and correcting finite element results. This background is used in item 7 to further relate the discontinuous stresses in the finite element results to the smoothed stresses in the augmented result.

Item 7 uses the conditions presented in Kelly's approach to show that the need for the interelement jumps in the stresses to be zero is a more stringent requirement for convergence than the requirement that the residuals in the augmented solution be zero. This demonstration provides a solid foundation for the idea that the smoothed stresses are more accurate than the discontinuous stresses contained in the finite element result.

Item 8 compares the merits of the refinement guides and termination criteria that are developed and discussed in developments presented here.

Item 9 uses the results of the previous items to outline the characteristics of a universal postprocessor for evaluating finite element results and for extracting smoothed stress and strain results. This has significant implications concerning the future development of finite element codes. It also provides a basis for certain lines of future research.

Although the developments presented here improve the smoothing approach to error analysis, the organization and contents of this lesson differ significantly from the previous lessons. The previous lessons contained specific and relatively closed-ended topics. The contents of this lesson are diverse and truly open ended. For example, the integration of two competing error analysis procedures, the smoothing and the residual approaches, presented in this lesson opens many interesting research questions. It might be found that further improvements to the smoothing approach can be made as a result of this integration. For example, improvements might be provided by incorporating residual quantities into the metrics used to terminate the analyses. Some of these open research questions are discussed in the text that follows.

The nine items just outlined are presented in turn.

Item 1 — A Method for Extracting Residuals from Augmented Finite Element Results

The developments contained in this lesson are based on interpreting the augmented finite element result formed in Lesson 17 as *an approximate finite difference result*. The augmented finite element result is not an actual finite difference result because, unless the solution is exact, its displacements will not satisfy the finite difference approximations of the governing differential equations. The differences between the approximate and the actual finite difference results are measured by the nodal values of the amount by which the displacements of the augmented solution fail to satisfy the finite difference approximations of the equilibrium equations. These differences are known as the residuals.

As we will see, the augmented result approaches a finite difference result as the finite element solution converges toward the exact answer. That is to say, the residuals decrease as the finite element model is improved.

Computation of Residuals

We now develop a procedure for computing the residuals contained in an augmented finite element result. The procedure is then demonstrated with the example of a cantilever beam loaded first with an end shear and then with an axial load. The finite element model of the beam is refined to show the behavior of the residuals as the mesh is improved.

The residuals are computed by substituting the displacements of the augmented solution into the finite difference approximations of the governing differential equations. The governing differential equations for the plane stress problem from Lesson 3, Eq. 3.26, can be rearranged and presented as

---

Governing Differential Equations for Plane Stress

$$\left(\frac{E}{1-\nu^2}\right)\left[u_{,xx} + \nu v_{,xy} + \left(\frac{1-\nu}{2}\right)(u_{,yy} + v_{,xy})\right] + p_x = 0$$

$$\left(\frac{E}{1-\nu^2}\right)\left[v_{,yy} + \nu u_{,xy} + \left(\frac{1-\nu}{2}\right)(v_{,xx} + u_{,xy})\right] + p_y = 0$$

(19.1)

We formalize the definition of the residuals by modifying Eq. 19.1 as follows:

Governing Differential Equations in Residual Form

$$R_x = \left(\frac{E}{1-\nu^2}\right)\left[u_{,xx} + \nu v_{,xy} + \left(\frac{1-\nu}{2}\right)(u_{,yy} + v_{,xy})\right] + p_x$$

$$R_y = \left(\frac{E}{1-\nu^2}\right)\left[v_{,yy} + \nu u_{,xy} + \left(\frac{1-\nu}{2}\right)(v_{,xx} + u_{,xy})\right] + p_y$$

(19.2)

When we compare Eq. 19.2 to Eq. 19.1, we can see that the only difference between the two equations is that the zeroes in Eq. 19.1 have been replaced by the symbols R_x and R_y in Eq. 19.2. These variables, known as residuals, are introduced to provide a measure of the differences between the augmented finite element result and an actual finite difference result.

Before discussing this equation, we put it in terms of strain gradient quantities so it is consistent with previous developments. When this change of variables is introduced, we can write the residuals in terms of strain gradient quantities as

Governing Differential Equations in Residual Form

$$R_x = \left(\frac{E}{1 - \nu^2}\right)\left[\varepsilon_{x,x} + \nu\varepsilon_{y,x} + \left(\frac{1 - \nu}{2}\right)\gamma_{xy,y}\right] + p_x$$

$$R_y = \left(\frac{E}{1 - \nu^2}\right)\left[\varepsilon_{y,y} + \nu\varepsilon_{x,y} + \left(\frac{1 - \nu}{2}\right)\gamma_{xy,x}\right] + p_y$$

$$(19.3)$$

The residuals defined by Eqs. 19.2 and 19.3 have the same units as the distributed loads, p_x and p_y. This recognition enables us to interpret the residuals as fictitious distributed loads. In fact, the residuals can be further interpreted as the fictitious loads that must be applied to the problem to satisfy equilibrium. If we move the residuals to the right-hand side of Eq. 19.2 or Eq. 19.3, the equations again equal zero, i.e., equilibrium is satisfied. Thus, we can conclude that the size of the residual is related to the amount by which the approximate solution differs from the exact result.

Another interpretation of the residuals is the following. We can say that the augmented solution solves a modified version of the original representation exactly. The modified representation consists of the original model plus a distributed load that is equal to the amount of the residuals contained in the augmented solution. As the residuals get smaller, the augmented solution approaches the exact solution of the original problem. We discuss the use of residuals in the evaluation of finite element results in a later section.

We can also express the residuals in terms of strain units. When Eq. 19.3 is rearranged, the residuals are expressed in terms of the rates of change of the strain components as

Governing Differential Equations in Residual Form

$$r_x = \varepsilon_{x,x} + \nu\varepsilon_{y,x} + \left(\frac{1 - \nu}{2}\right)\gamma_{xy,y} + \left(\frac{1 - \nu^2}{E}\right)p_x$$

$$r_y = \varepsilon_{y,y} + \nu\varepsilon_{x,y} + \left(\frac{1 - \nu}{2}\right)\gamma_{xy,x} + \left(\frac{1 - \nu^2}{E}\right)p_y$$

$$(19.4)$$

This form of the governing differential equations enables us to relate the residuals to the strain states existing in the problem being analyzed. In other words, we can attempt to assess the level of error in the approximate solution by comparing the residuals to the magnitude of the strains existing in the finite element solution. This use of the residuals is discussed in a later section.

To exhibit the nature of the residuals when they are expressed in terms of nodal displacements, the derivatives in Eq. 19.4 are replaced with the finite difference approximation for an evenly spaced mesh, as shown in Fig. 2. When these substitutions are made and the equations are rearranged, the result is

Residuals for a Nine-Node Finite Difference Model

$$r_x = 1.4u_5 + 4.0u_6 + 1.4u_7 + 4.0u_8 - 10.8u_9$$
$$+ 0.65v_1 - 0.65v_2 + 0.65v_3 - 0.65v_4$$
$$+ 3.64(h^2/E)p_x$$

$$r_y = 4.0v_5 + 1.4v_6 + 4.0v_7 + 1.4v_8 - 10.8v_9$$
$$+ 0.65u_1 - 0.65u_2 + 0.65u_3 - 0.65u_4$$
$$+ 3.64(h^2/E)p_y$$

(19.5)

To simplify these expressions we take ν as 0.3. Note that Eq. 19.5 contains the parameter h, which is the nodal spacing. This means that the equation is applicable to any size of mesh with even spacing. Although Eq. 19.5 applies only to evenly spaced meshes, this equation shows that we can compute the residuals with relatively simple computations.

The residual expressions for irregular meshes are similar to those given by Eq. 19.5. Each expression would contain extra displacement terms and the existing coefficients would be different. All we need to compute the residuals are the difference operators for the derivative terms contained in the differential equations. If we have formed the smoothed solution using the finite difference approach, these operators are readily available to us.

Example Problems

The procedure just developed is now be applied to the cantilever beam shown in Fig. 3 for two load cases: (1) the shear load shown and (2) the beam hanging vertically under its *own weight*. The initial mesh of five four-node elements is uniformly refined three times in both coordinate directions. The finite element models have 24, 66, 210, and 738 degrees of freedom, respectively. The first three meshes are shown as Figs. 3–5 in Lesson 18.

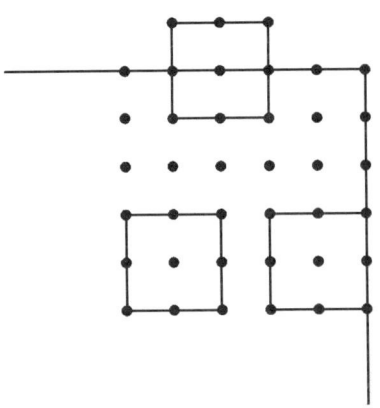

Figure 2. A nine-node template overlaid on a finite element mesh.

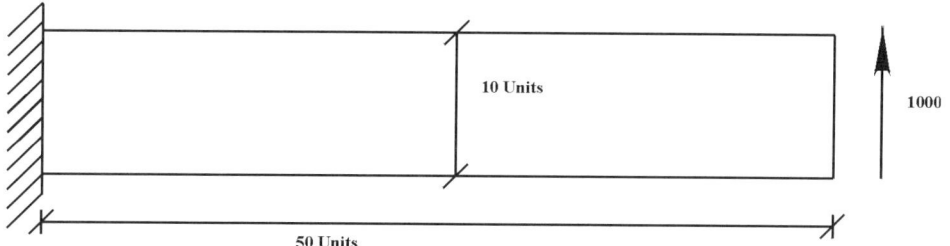

Figure 3. Cantilever beam problem with an end shear.

Shear Load.   The convergence characteristics of the residuals for the problem of the beam loaded with the end shear is discussed first. This problem was chosen as an example because it contains two well-defined singularities. Thus, this problem provides a challenge to the interpretation of the augmented finite element result as an approximate finite difference solution.

The residuals in the x direction for the top row of nodes in the finite element model produced by the augmented solutions for the four meshes are shown in Table 1. This reduced set of residuals was chosen for presentation because it contains the extremes of behavior for the residuals and it reduces the volume of data that must be presented. The residuals are computed using Eq. 19.3 so they can be interpreted as distributed loads needed to put the problem into equilibrium.

The sequence of residuals at positions 30.0 and 40.0 behave exactly as expected as the mesh is refined. They start large and decrease rapidly to small values. We quantify the meaning of "small" in a later discussion. The residuals at locations 10.0 and 20.0 behave almost as expected. After their initial increase in the second mesh, the residuals decrease to relatively small values. We will soon discuss the reason for this.

The values of the residuals are zero at location zero, the root of the beam. This occurs because we forced them to be so when we formed the augmented solution using the procedures developed in Lesson 17. These residuals are forced to zero in order to introduce point-wise equilibrium at points in the augmented solution where we know the displacements to be correct and we have no knowledge of the actual stresses. This procedure has the effect of increasing the error estimates in elements on fixed boundaries when stresses are used to evaluate finite element solutions.

However, if we are actually going to use the augmented solution as an approximate finite difference result or if we are going to use an error estimation scheme based on residuals, we would probably change our procedure for forming the augmented result on fixed boundaries. Instead of forcing the residuals to zero on fixed boundaries, we would probably force the boundary stresses to equal something like the average value of the finite element results. We would do this to enable the residuals to indicate whether equilibrium was actually being satisfied. In most cases, this would increase the magnitude of the residuals on the fixed boundary so deficiencies in the discrete model would be better indicated. Until this is done, we do not know for sure how well the residuals near the fixed boundary will identify modeling errors.

Because of the ambiguity existing on the fixed boundary, let us concentrate on the residuals present in the problem from the midpoint to the free end of the beam. The residual at location 25.0 in the final mesh is small enough so that we can consider it to be zero. As we proceed out the beam, the residuals in the final refinement are relatively small until location 38.75 is reached. Then they steadily increase. With the exception of the end node, which is a singularity, the residuals decrease at each node in this section of the beam as the mesh is refined.

Table 1. Residuals vs position for four meshes (a dash indicates that a node does not exist at this location in the given mesh).

Nodal locations	Residual values			
	Mesh 1	Mesh 2	Mesh 3	Mesh 4
0.00	0.0	0.0	0.0	0.0
1.25	—	—	—	− 2,747.73
2.50	—	—	− 2,479.75	973.53
3.75	—	—	—	− 90.68
5.00	—	− 1,927.53	628.34	16.98
6.25	—	—	—	− 21.28
7.50	—	—	− 107.70	− 5.79
8.75	—	—	—	− 3.63
10.00	21.10	607.42	28.30	− 1.067
11.25	—	—	—	− 0.38
12.50	—	—	− 8.41	− 0.075
13.75	—	—	—	− 0.0077
15.00	—	− 152.39	2.09	− 0.0079
16.25	—	—		0.0059
17.50	—	—	− 0.52	0.0032
18.75	—	—		0.0013
20.00	− 2.95	38.81	0.13	0.00051
21.25	—	—		0.00015
22.50	—	—	− 0.035	0.00013
23.75	—	—	—	− 0.0000055
25.00	—	− 9.41	0.008	− 0.00011
26.25	—	—	—	− 0.00023
27.50	—	—	0.00020	− 0.00042
28.75	—	—	—	− 0.00068
30.00	− 132.74	0.99	− 0.014	− 0.0010
31.25	—	—	—	− 0.0015
32.50	—	—	0.072	− 0.0021
33.75	—	—	—	− 0.0027
35.00	—	40.40	− 0.029	− 0.0015
36.25	—	—	—	0.0077
37.50	—	—	− 0.032	0.047
38.75	—	—	—	0.17
40.00	14,325.55	50.16	− 11.17	0.56
41.25	—	—	—	1.49
42.50	—	—	78.00	3.58
43.75	—	—	—	6.13
45.00	—	20,483.95	− 164.11	7.39
46.25	—	—	—	− 33.26
47.50	—	—	26,010.98	81.56
48.75	—	—	—	− 502.58
50.00	− 2,252.02	− 3,017.46	− 4,321.68	28,930.22

Let us examine the behavior of the residuals at the singular points as the mesh is refined. First, let us remember that the stresses at a singularity are "undefined." Quite often this means that the stresses are unbounded. As a result, high strain gradients can occur in the neighborhood of singularities. In our representation of the node at the singularity, we are forcing the stresses to take on finite values because of the boundary conditions. As the simple polynomials in our approximate representation are forced to satisfy these conflicting conditions, the amount by which equilibrium fails to be satisfied increases as the mesh is refined, so the residuals increase.

We anticipate that a similar process would occur at the fixed boundary if we enforced the stresses in the finite element model on the augmented solution. As a consequence, the residuals would identify modeling deficiencies on fixed boundaries.

This contention is backed up by the results near the singularities at the fixed boundary. As we can see in Table 1, the residuals in the nodes close to the singularity at the fixed end are also large. The residual at the corner is zero because we forced equilibrium to be satisfied at the node, i.e., we forced the residual to be zero. Thus, we can anticipate that the residuals at this singularity would also be large if we forced the finite element stresses to be satisfied at the singularity. That is to say, the residual approach to evaluating finite element results would identify the singularities on the fixed boundary.

In summary, we can conclude that the residuals are converging as the mesh is refined except at or near the singular points. The fact that the residuals are not converging in the region of the singular points is a reason in favor of using the residuals to evaluate the finite element result. The residuals are identifying locations where the discrete model cannot adequately represent the situation being modeled. In this case, the finite element model cannot represent a "perfectly square corner."

The evaluation procedure is identifying a design impossibility. The analyst is implicitly being told to replace the "square" corner with a curved boundary. Thus, the residuals contained in the augmented solution are performing as we would desire an evaluation measure to perform. Not only are they identifying where we should refine the mesh to better represent the stress distribution, they are informing us where we are attempting to represent a physical impossibility.

Gravity Load. Let us now study the convergence characteristics of the residuals for the problem of the bar loaded with a gravity load. This problem was chosen because it contains the same two well-defined singularities as the previous problem. However, the residuals behave differently in this case because of the difference in the stress distribution produced by the loading condition. The residuals for this problem are shown in Table 2.

The residuals for this problem exhibit the same pattern of convergence as they did in the shear loading case. The residuals are significantly smaller everywhere on the bar as compared to the previous example, but the pattern of behavior near the singularities is identical to that seen for the shear loading case. The residuals are small enough everywhere in the mesh so that any one of the meshes can be considered as an exact representation, but the residuals still show relatively significant increases at the singularities.

At this point, we can conclude that the residuals contained in the augmented finite element result behave as we would expect for an approximate finite difference result. The residuals converge toward zero in regions where the representations are accurate and they are large in regions where the representation is not accurate. In regions of large residuals, we can modify the representation to improve the model. At singularities, this may mean that we need to replace the sharp corners with a curved boundary. As we have seen in

Table 2. Residuals vs position for four meshes for bar hanging under its own weight (a dash indicates that a node does not exist at this location in the given mesh).

Nodal locations	Residual values			
	Mesh 1	Mesh 2	Mesh 3	Mesh 4
0.00	0.0	0.0	0.0	0.0
1.25	—	—	—	-2.112×10^{-3}
2.50	—	—	-7.185×10^{-4}	1.775×10^{-4}
3.75	—	—	—	2.098×10^{-4}
5.00	—	-2.244×10^{-4}	1.592×10^{-4}	1.042×10^{-4}
6.25	—	—	—	7.072×10^{-5}
7.50	—	—	9.755×10^{-5}	3.986×10^{-5}
8.75	—	—	—	2.199×10^{-5}
10.00	-9.826×10^{-6}	1.118×10^{-4}	2.798×10^{-5}	1.068×10^{-5}
11.25	—	—	—	4.450×10^{-6}
12.50	—	—	3.615×10^{-6}	1.278×10^{-6}
13.75	—	—	—	-1.074×10^{-7}
15.00	—	-1.802×10^{-5}	-1.749×10^{-6}	-5.681×10^{-7}
16.25	—	—	—	-6.034×10^{-7}
17.50	—	—	-1.385×10^{-6}	-4.828×10^{-7}
18.75	—	—	—	-3.343×10^{-7}
20.00	-3.642×10^{-5}	-1.906×10^{-6}	-5.330×10^{-7}	-2.088×10^{-7}
21.25	—	—	—	-1.190×10^{-7}
22.50	—	—	-1.120×10^{-7}	-6.147×10^{-8}
23.75	—	—	—	-2.785×10^{-8}
25.00	—	7.433×10^{-7}	-7.222×10^{-9}	-9.990×10^{-9}
26.25	—	—	—	-1.586×10^{-9}
27.50	—	—	1.770×10^{-8}	1.627×10^{-9}
28.75	—	—	—	2.252×10^{-9}
30.00	6.823×10^{-6}	-2.345×10^{-8}	4.823×10^{-9}	1.720×10^{-9}
31.25	—	—	—	7.327×10^{-10}
32.50	—	—	-1.204×10^{-8}	-4.166×10^{-10}
33.75	—	—	—	-1.594×10^{-9}
35.00	—	-6.661×10^{-7}	-3.027×10^{-8}	-2.619×10^{-9}
36.25	—	—	—	-3.028×10^{-9}
37.50	—	—	-1.253×10^{-8}	-1.792×10^{-9}
38.75	—	—	—	3.208×10^{-9}
40.00	6.507×10^{-6}	2.449×10^{-6}	1.678×10^{-7}	1.530×10^{-8}
41.25	—	—	—	4.121×10^{-8}
42.50	—	—	9.294×10^{-7}	8.429×10^{-8}
43.75	—	—	—	1.773×10^{-7}
45.00	—	8.601×10^{-6}	1.770×10^{-6}	2.724×10^{-7}
46.25	—	—	—	8.150×10^{-7}
47.50	—	—	6.941×10^{-6}	1.435×10^{-6}
48.75	—	—	—	6.797×10^{-6}
50.00	-9.235×10^{-6}	-3.384×10^{-7}	3.966×10^{-6}	1.263×10^{-5}

previous lessons, we have the capability to represent curves of any configuration with meshes with different spacing and irregular nodal arrangements.

Thus, we can conclude that the augmented solution can be considered as an approximate finite difference result. Furthermore, we can conclude that we can use the residuals contained in the approximate finite difference solutions as refinement guides and termination criteria. In the remainder of this lesson, we exploit the use of the interpretation of the augmented finite element result as an approximate finite difference result.

Item 2 — A New Finite Difference Solution Technique

In this section, we outline an iterative approach for solving finite difference problems with adaptively refined finite element models. This avenue is not explored in detail here. This discussion is designed to give further insights into the nature of the relationships between the three types of solution being considered in this lesson. The primary objective of this discussion is to provide the basis for bridging the gap between the smoothing and the residual approaches for evaluating finite element results. This iterative approach to solving finite difference problems consists of the following steps, which form a flow chart for the method:

1. Develop a finite difference mesh including fictitious nodes.
2. Perform a *finite element analysis* using the nodes of the finite difference mesh that are on the domain of the problem, i.e., the fictitious nodes are not included.
3. Form an augmented finite element result by introducing the boundary conditions using the fictitious nodes.
4. Compute the residuals at the nodes on the domain of the problem.
5. Determine if the analysis can be concluded by evaluating the size of the residuals, and if the analysis is not terminated, continue to step 6.
6. Refine the model in regions of high residuals to improve the representation of the problem. Singularities are removed as necessary in this step by altering the geometry of the problem.
7. Return to step 2.

The ability to directly relate the finite difference and the finite element methods through this iterative solution technique provides us with two insights that are useful in evaluating finite element results: (1) we see that the approximate finite difference solutions, i.e., the augmented finite element results, are bounded and (2) we argue that the residuals can be interpreted directly as discretization error measures. The second argument is a consequence of the argument proving the boundedness. The second argument bolsters the results seen in the examples of item 1 that the residuals exhibit the behavior desired for a refinement guide.

In item 7, a similar argument is made that the differences between the stresses in the augmented finite element solution (the smoothed solution) and the finite element result are directly related to deficiencies in the discrete model. Then, the similarity between the meaning of the results of the two approaches is used to claim that the two approaches for evaluating discrete models are equivalent.

This means that the relative merit of the evaluation techniques using stresses and those using residuals depends on the computational efficiency and the effectiveness of the two methods in identifying deficiencies in the model. It does not depend on some theoretical difference in the two approaches. We will see that both approaches try to minimize the strain energy in the problem being solved.

Although we are not going to pursue this iterative approach for solving finite difference problems here, there is one computational advantage to this approach that is worth noting, namely, the size of the associated finite element problem is smaller than the finite difference problem being solved. It does not contain the fictitious nodes in the primary computation. Also, there *may* be another computational advantage to this iterative procedure. The condition numbers of the associated finite element model *may* be smaller than that for the finite difference model.

We talk about these theoretical observations and computational considerations in turn.

A Strain Energy Bound for Finite Difference Results

As far as we know, no bounds on finite difference solutions exist that are equivalent to the Rayleigh-Ritz criteria for finite element solutions. If such bounds do exist, this presentation provides another approach for forming them. The iterative solution technique just outlined enables us to apply the Rayleigh-Ritz criteria to the approximate finite difference result. The argument is as follows.

The Rayleigh-Ritz criterion bounds the strain energy content contained in the underlying finite element result if the model is constructed with elements that satisfy interelement compatibility. As a result, the approximate finite difference solution is bounded at each iterative step. Since the augmented finite element solution converges toward a finite difference result in the limit, the finite difference result is bounded. In other words, the fully converged finite difference result contains the same amount of strain energy as the exact solution.

That is it — this is a statement of the obvious. The final conclusion is not of major importance in itself. This observation does not directly affect finite difference solution techniques. However, the idea behind this argument is important to the understanding and development of techniques for evaluating finite element solutions. It enables us to relate the residuals and, later, the differences between the finite element and the augmented finite element results directly to deficiencies in the discrete model.

The idea behind the formulation of this bound is the recognition that the finite element and the finite difference methods are not just competing solution techniques trying to solve the same problem. They are intimately related through the principle of minimum potential energy. The finite difference method attempts to minimize the potential energy of the problem being solved by finding an approximate solution to the differential form of the problem. The finite element method attempts to minimize the strain energy by finding an approximate solution that directly minimizes the strain energy. Thus, we can see the residuals as measuring the differences between an approximate variational solution and an approximate differential solution that are based on the same mesh.

Interpretation of the Residuals as Direct Measures of Model Deficiencies

The importance of recognizing that the residuals quantify the differences between an approximate variational solution and an approximate differential solution based on the same discretization is the following. Since both solution techniques have the same objective and both methods will exactly represent the problem being solved for an adequately refined mesh, the differences between the two approximate solutions must be due to deficiencies in the discrete model.

This point of view has the following two advantages. First, it relates the residuals in the augmented finite element solution directly to deficiencies in the discrete representation of the problem. Second, this point of view separates the evaluation of the discrete model from the evaluation of the accuracy of the stresses and strains. That is to say, we do not have to go through the intermediate step of deciding whether the augmented result is a "better solution" than the finite element result to evaluate the finite element model.

The identification of deficiencies in the discrete model precedes the question of which approximate solution produces better stress and strain results, i.e., which approximate solution is closer to the exact result. The residuals evaluate the model rather than the results produced by the model. This direct relationship between the residuals and deficiencies in the discrete model enables us to use the residuals to form practical pointwise refinement guides and pointwise termination criteria.

Reduced Problem Size

In the standard form of the finite difference method, the size of the problem is directly related to the number of nodes in the problem, i.e., the number of nodes on the domain and the number of fictitious nodes. In the iterative approach described here, the problem is broken into two parts. The displacements of the nodes on the domain are computed with a finite element analysis. The displacements of the fictitious nodes are computed with an auxiliary computation. Since both sets of equations need not be solved simultaneously, the problem size is reduced by the use of the iterative approach.

The iterative nature of the new solution technique is not a drawback to its efficiency. The deficiencies in the discrete model identified in the iterative approach also exist in the standard finite difference approach. This means that the representation in the standard approach must be adaptively refined to achieve a solution of acceptable accuracy. Since the mesh for the standard form of the problem must also be improved, the iterative nature of the new approach does not make it noncompetitive. The evaluation of the results of the standard form of the finite difference method will be discussed later in this lesson and demonstrated in Lesson 20.

Possibility of Improved Computational Characteristics

It is *possible* that the computational characteristics for the iterative approach are better than those for the standard finite difference formulation. We have already seen that the standard finite difference approach results in a larger sized coefficient matrix than does the augmented finite element approach. Furthermore, the condition number for the underlying finite element model *may* be lower than that for the standard finite difference representation. This may be the case since the order of the derivatives contained in the finite element model are lower than those contained in the finite difference model.

If such a proof shows that the computational characteristics of the finite difference approach are better, we have not lost. Since we have seen that practically any problem that can be represented with a finite element model can be represented with a finite difference model, we can abandon the finite element method and use the finite difference approach.

Although the new approach for solving finite difference problems developed here was not demonstrated, we have seen that it has the following advantages over the standard form of the finite difference method:

1. It reduces the size of the problem that must be solved. One large problem is solved as two smaller problems.
2. It produces refinement guides and termination criteria as part of the solution process.

3. The approach may have better computational characteristics than the standard finite difference approach.

This hybrid approach to solving finite difference problems seems an avenue worth exploring.

An important feature of this development is that we did not have to make reference to the accuracy of one solution type with respect to another. In fact, there has been no reference to the accuracy of the stresses or strains contained in either solution. The only topic that has entered is the quantification of deficiencies in the discrete model. This does not mean that we cannot reach meaningful conclusions concerning accuracy of the solution. The point is that we do not have to consider these topics at this stage. The relative accuracy of the stresses and strains in the two solutions can be considered when we want to examine the possibility of directly improving the stress and strain results for the problem being solved without further iterations.

The primary contribution of this section has been the clarification of the relationships that exist between the finite difference, the finite element and the augmented finite element results. We explore and exploit these relationships further in the sections that follow.

Item 3 — A New Residual-Based Finite Element Evaluation Procedure

In the previous section, we developed a new iterative approach for solving finite difference problems. With minor modifications, the approach for the iterative finite difference solution technique can be modified so it describes an adaptive refinement procedure for the finite element method using residual-based refinement guides and termination criteria. The changes in the flow chart from that for the finite difference solver are written in italics and underlined. When the flow chart presented in list form under item 2 is modified, it becomes:

1. Develop a *finite element mesh*.

2. Perform a finite element analysis.

3. *Develop a set of fictitious nodes* and form an augmented finite element result by introducing the boundary conditions.

4. Compute the residuals at the nodes on the domain of the problem.

5. Determine if the analysis can be concluded by evaluating the size of the residuals and if the analysis is not terminated continue to step 6.

6. Refine the model in regions of high residuals to improve the representation of the problem. Singularities are removed as necessary by altering the geometry of the problem in this step.

7. Return to step 2.

For this approach to be practical, the magnitude of the residuals must be related to the quantities of interest to the analyst. Such a relationship can be identified for this evaluation process by remembering that the residuals produced by Eq. 19.3 can be interpreted as fictitious loads that must be applied to the finite difference model to produce equilibrium. The residuals represent the magnitude of the distributed loads that must be applied at the nodes.

As an aside, the residuals can also be interpreted in a slightly different way. The solution that we do have can be interpreted as the exact solution to the problem we are solving with additional loads applied to it that are equal to the residuals. In linear problems, we can see that the errors in the displacement and stresses would be proportional to the "fictitious load."

Thus, we can deduce whether the fictitious loads are large or small by comparing them to the loads actually applied to the problem. When the relative size of the residuals are determined, we can identify regions that must be refined or determine if the adaptive refinement process can be terminated.[1]

We now demonstrate the process of evaluating the relative size of the residuals with the example of the cantilever beam problem with the two loading conditions presented earlier. The residuals for the case with the end shear load are evaluated first. The residuals for this case are given in Table 1.

As discussed earlier, the quantities contained in Table 1 can be interpreted as the values of a fictitious distributed load that would have to be applied at the nodal points of the discrete model to produce equilibrium. We can put the magnitudes of the residuals in perspective by comparing them to the applied load. The load at the end is equal to 1000 force units. If this load were distributed over the beam, the distributed load would be equal to 2 force/area units (1000 force units/500 area units).

For example, if the residual at the center node on the top surface in the fourth mesh were applied uniformly over the whole beam, the total load would equal 0.0055% (0.00011 × 100/2.0) of the total end shear load applied to the problem. This number is the residual (0.00011 force/area units) divided by the size of the applied load if it is uniformly distributed (2.0 force/area units) multiplied by 100%. We can conclude that the error in this region is minimal. If the distributed loads at the singularities on the free end of the beam were applied to the beam (28,930 force/area units), they would be essentially infinite loads. Thus, we can conclude that the errors in these regions are unacceptable.

As a further example, let us study the residuals for the case where the beam is hanging vertically and is loaded with a uniformly distributed load to simulate its own weight. The residuals for this problem are given in Table 2.

In this problem with its simpler stress distribution, the residuals are found to be small with respect to the applied load at every node after the first refinement except at the nodes next to the fixed boundary. The finite element results for this problem had essentially converged except at the fixed boundary after the first mesh refinement.

Thus, we can conclude that the residuals contained in the augmented finite element result can serve as refinement guides and termination criteria. We discuss the relative merit of the residual and the smoothing approaches to evaluating finite element results in a later section. It is important to note for later developments that the residuals are pointwise quantities. This is the case because of the Taylor series nature of the finite difference method.

Item 4 — A Theoretical Basis for Stress-Based Pointwise Refinement Guides

We validated the interpretation of the augmented finite element result as an approximate finite difference result in the previous three sections. This was accomplished by examining the behavior of the residuals contained in the augmented solution as the mesh is refined. We saw that the residuals quantified the deficiencies in the discrete model on a

---

[1] At this point there might be a temptation to compute a correction factor by applying the fictitious loads to the discrete model used to find the residuals. We must remember that if the current model could exactly represent the applied load, then there would be no errors. In other words, the question arises as to whether the residual loads are "orthogonal" to the model we are currently using. This question is discussed further in a later section. This approach might serve for an error estimator but it would be expensive compared to procedures currently available.

pointwise basis. This behavior was explained by arguing that the residuals in the augmented solution measure the differences between approximations of the variational and differential forms of the problem being solved. That is to say, the residuals measure how far the augmented finite element is from an actual finite difference result. Since both solutions must converge to the exact result, the pointwise differences indicate deficiencies in the discrete model.

In this section, we claim that this explanation of the behavior of the residuals in the augmented finite element result and the Taylor series nature of the two formulations provides a theoretical basis for pointwise refinement guides demonstrated in Lesson 18. That is to say, the pointwise refinement guides based on the differences in the stress components between the two stress representations at the nodes are effective for two reasons: (1) they are measuring the differences between stress representations contained in approximations to the variational and differential forms of the problems being solved and (2) the two approximate solution techniques are based on Taylor series expansions around the nodes.

The explanation just given for the success of the pointwise refinement guides based on the stress quantities is an obvious explanation at this point in our development. In fact, it may seem that we are almost back to where we started with the original Zeinkiewicz and Zhu approach. That is to say, Zeinkiewicz and Zhu revolutionized the evaluation of finite element results by implicitly considering the smoothed solution as an improved solution. By comparing the finite element stress results to the smoothed stresses, they were able to develop simple refinement guides that were successful in all of the practical applications to which they were applied.

We have rejected the *assumption* that the smoothed solution is an improved solution.[2] The rejection of this assumption may seem to be a step backward. However, this is not the case because we have improved our position in two ways by interpreting the smoothed solution as an approximate finite difference result. First, we do not have to *assume* that the smoothed solution is a better solution than the finite element result to provide a theoretical basis for the refinement guides. That is to say, we can make the observation that the development of discretization error measures precedes the question of the accuracy of the stresses and strains. Second, we can use the Taylor series nature of the approximate finite difference solution to justify the pointwise nature of the refinement guides.

Finally, as a result of the Taylor series nature of the smoothed solution, we are able to demonstrate the conditions that guarantee that the smoothed stresses are of an order higher than the finite element representations. This enables us to form pointwise termination criteria that measure the convergence of quantities of direct interest to analysts, namely, the stresses at critical points.

The demonstration that the stresses in the smoothed solution are of an order higher than the finite element results is presented in the next section.

Item 5 — The Smoothed Solution as a Higher Order Stress Result

Up to this point in the development of evaluation measures, we have not assumed that the stresses in the augmented finite element result are better than the stresses in the finite element result. In this section, we identify specific conditions that guarantee that the

---

[2] In the next section we are going to identify the conditions that define when the augmented solution is of an order higher than the finite element result. We are going to take this as a proof that the smoothed stresses are a better solution than the finite element stresses.

stresses in the smoothed solution are closer to the exact result than are the finite element representations. This is done with the following argument.

When the residuals are small, we can consider the smoothed solution as a good approximation to a finite difference solution. If we can then show that the approximate finite difference result is of an order higher than the finite element result, the stress representation in the smoothed solution must be closer to the exact result than the finite element result. That is to say, the error terms in the smoothed solution are of an order higher than the error terms in the finite element representation so they converge to zero faster. We use the Taylor series nature of the finite element and the finite difference methods to show when the smoothed result is of an order higher than the finite element result.

The higher-order nature of the augmented result is developed for the case where nine-node finite difference templates are superimposed on a mesh composed of four-node elements, as shown in Fig. 2. The four-node element is based on displacement polynomials of the following form:

Displacement Approximations for Four-Node Elements

$$u = a_1 + a_2 x + a_3 y + a_4 xy$$
$$v = b_1 + b_2 x + b_3 y + b_4 xy \tag{19.6}$$

The strain expressions for the four-node element produced by these displacement representations are the following:

Uncorrected Strain Approximations for Four-Node Elements

$$\varepsilon_x = \frac{\partial u}{\partial x} = a_2 + a_4 y$$

$$\varepsilon_y = \frac{\partial v}{\partial y} = b_3 + b_4 x \tag{19.7}$$

$$\gamma_{xy} = \frac{\partial u}{\partial y} + \frac{\partial v}{\partial x} = (a_3 + b_2) + a_4 x + b_4 y$$

As we can see, the normal strain in the x direction contains only one linear term. The same is true for the normal strain in the y direction. Thus, the normal strains are complete only through the constant terms. As shown in Eq. 19.7, the shear strain appears to contain both linear terms. However, these two terms are the source of parasitic shear and should be eliminated from the shear expression because they cause the model to be overly stiff.[3] If these terms are eliminated, the shear is left with only a constant term. Whether or not the parasitic shear terms are eliminated, the overall strain representation in a four-node element is complete only through the constant terms because of the missing terms in the normal strain components.

For the case of a 3×3 finite difference template, the displacement polynomials are given as

---

[3] The concept and the effects of parasitic shear are discussed in detail in Part II.

Displacement Approximation for Nine-Node Templates

$$u = a_1 + a_2x + a_3y + a_4x^2 + a_5xy + a_6y^2 + a_7x^2y + a_8xy^2 + a_9x^2y^2$$
$$v = b_1 + b_2x + b_3y + b_4x^2 + b_5xy + b_6y^2 + b_7x^2y + b_8xy^2 + b_9x^2y^2$$
(19.8)

When the strain representations are formed from Eq. 19.8, the result is

Uncorrected Strain Approximations for Nine-Node Templates

$$\varepsilon_x = \frac{\partial u}{\partial x} = a_2 + 2a_4x + a_5y + 2a_7xy + a_8y^2 + 2a_9xy^2$$

$$\varepsilon_y = \frac{\partial v}{\partial y} = b_3 + b_5x + 2b_6y + b_7x^2 + 2b_8xy + 2b_9x^2y$$
(19.9)

$$\gamma_{xy} = \frac{\partial u}{\partial y} + \frac{\partial v}{\partial x} = (a_3 + b_2) + (a_5 + 2b_4)x + (2a_6 + b_5)y + a_7x^2$$

$$+ (2a_8 + 2b_7)xy + b_8y^2 + 2a_9x^2y + 2b_9xy^2$$

The shear strain expression contained in Eq. 19.9 is not corrected for parasitic shear. Specifically, it contains the two parasitic shear terms, x^2y and xy^2. Even when these terms are removed from the shear strain expression, the strain representations are complete linear expressions. Thus, the strain models for the augmented finite element result are one order higher than the strain models for the four-node finite element. This also means that the strain models for the augmented solution are of the same order as the displacement models for the four-node finite element.

In this case, we can conclude that the smoothed stresses are more accurate than the finite element representation. This means that we could not use the differences between the smoothed stresses and the finite element stresses as estimates of the errors in the finite element stresses. At this point, the availability of this error estimate is not as important as it might seem. Since we have seen that the smoothed stresses are of an order higher than the finite element results, we want to use them as the output from the finite element analysis. Thus, we would like to have a termination criterion based on the accuracy of the smoothed results. We discuss this in item 8.

Termination criteria based on the pointwise differences in the stresses have been successfully used as termination criteria in Ref. 5. In this work, several different measures were used to demonstrate the flexibility of the pointwise approach.

In the next section, we see another interpretation of residuals that enables us to look at the results just developed from another point of view that enables us to develop other procedures for evaluating finite element results.

Item 6 — Another Residual Evaluation Technique

There is an approach for evaluating and improving finite element results developed by Kelly et al. that extracts residual quantities directly from *the finite element result* (see Refs. 1–4). In this procedure, two types of residual quantities are identified: (1) the amount by which the displacement approximation polynomials for the individual elements fail to satisfy the equilibrium equations and (2) the jumps in the interelement stresses. The objective of this section is to relate the residual quantities contained in the finite element results as identified by Kelly to the residuals contained in the augmented finite element result (an approximate finite difference result). This objective includes the validation of the idea that the jumps in the interelement stresses are indeed residual quantities.

As we will see, the relationships between these two evaluation procedures can be clarified by relating the residuals contained in the two approximate solutions. The identification of these relationships further extends our understanding of the smoothing approach for evaluating finite element results. This, in turn, will enable us to develop new evaluation procedures.

The first type of residual quantity identified by Kelly is similar to that computed for the augmented finite element result in item 1. In the procedure presented in item 1, the displacement polynomials for the finite difference template superimposed on the finite element mesh are substituted into Eq. 19.3 to find the residuals. In Kelly's approach, the displacement polynomials for the individual finite elements are substituted into Eq. 19.3. Both of these residual quantities identify the amount by which the respective approximation polynomials fail to satisfy the equilibrium equations of the problem being represented. Although the two residual quantities appear to be nearly identical, they have major differences.

The residuals computed in item 1 using the superimposed finite difference templates are pointwise quantities associated with an implicitly seamless domain. That is to say, the finite difference method does not subdivide the overall problem into separate regions as does the finite element method. The finite difference method approximates the field variables at individual points on the domain of the whole problem by approximating the governing differential equations at the nodal points.

In contrast, the first type of residual quantity computed by Kelly covers the domain of the problem, one element at a time. Note that for a three-node triangle, the residuals will be identically zero for a problem with no distributed loads because of the low order of the polynomial representation associated with this element. This piecewise approach leaves the residuals on the boundary to be accounted for in another way. Kelly accounts for the residuals on the boundaries between the elements by interpreting the interelement jumps in the stresses as residual quantities. The primary conclusions of this discussion are that the interelement jumps can legitimately be interpreted as residual quantities and that the residuals for the augmented finite element result encompasses both types of residuals identified by Kelly et al.

We demonstrate these conclusions with an example problem. We evaluate and then correct the finite element results for the example problem that follows with the two residual approaches. We find that both of the approaches produce identical results. This demonstrates that the interpretation by Kelly of the interelement jumps as residual quantities is correct. This validation is significant for two reasons: (1) the validation of the interpretation of the interelement jumps as residual quantities enables us to better understand the smoothing approach for evaluating finite element results and (2) we can show that the stresses in the smoothed solution definitively represent the actual stress better than the finite element result. We discuss these assertions after we demonstrate the two residual approaches for evaluating the same finite element result.

Example Problem — Application of Kelly's Residual Approach

The residual approach developed by Kelly et al. for evaluating and correcting finite element results can be outlined as follows (see Refs. 1–4). The finite element model is interpreted as the *exact* solution to a modified version of the problem being analyzed. The modified problem consists of the original problem with additional distributed fictitious loads and with additional fictitious concentrated loads.[4] The fictitious distributed load is due to the residuals on the individual elements and is equal to the amount by which the finite element interpolation polynomials fails to satisfy the governing differential equations. The fictitious concentrated loads are due to the jumps in the stresses between elements.

This idea can be used to evaluate and to correct finite element results. The finite element results for the example of a one-dimensional bar hanging under its own weight, as shown in Fig. 4a, is evaluated and corrected with this approach. This problem is taken from the Appendix of Ref. 3.

The governing differential equation and the boundary conditions for the problem are given as

---

Governing Differential Equation for a Bar

$$EA\frac{d^2u}{dx^2} - b = 0; \qquad 0 < x < 1$$

$$u(0) = 0 \tag{19.10}$$

$$\frac{du}{dx}(1) = 0$$

where b is equal to the value of the distributed load, which in this case is the density of the bar.

In this problem, the length is equal to unity as is the cross-sectional area. Note that this equation is the one-dimensional analog of Eq. 19.1. This equation can be written in residual form as

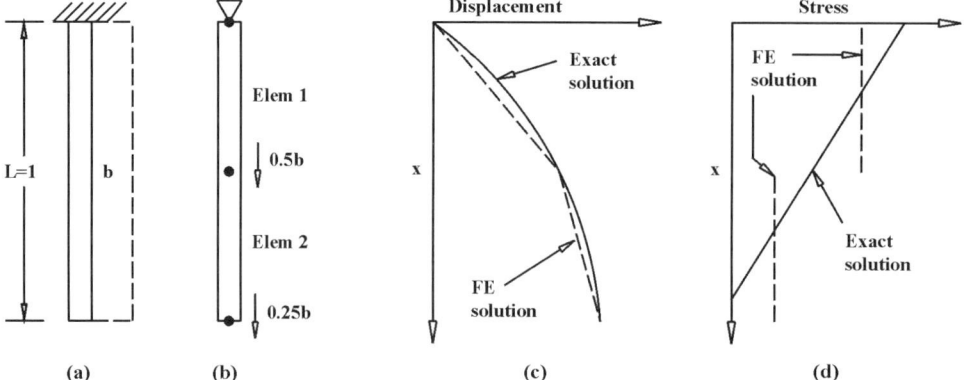

Figure 4. Problem definition, finite element model, and finite element results; (a) problem definition, (b) finite element model, (c) finite element and exact displacement vs position, and (d) finite element and exact stresses vs position.

---

[4] The idea used earlier of interpreting the residuals computed with Eq. 19.3 as fictitious distributed loads that could be applied to the actual problem being analyzed to produce equilibrium was taken from Refs. 1–4.

Equilibrium Equation for a Bar in Residual Form

$$R_x = EA\frac{d^2u}{dx^2} - b; \qquad 0 < x < 1 \tag{19.11}$$

The exact solution for the displacements of this problem is given as

Exact Displacement Solution

$$u(x) = \frac{b}{E}\left(x + \frac{1}{2}x^2\right) \tag{19.12}$$

The stress distribution for the exact solution is given as:

Exact Stress Solution

$$\sigma_x = E\varepsilon_x(x) = E\frac{du}{dx}(x) = b(1 - x) \tag{19.13}$$

The finite element model of the problem is shown in Fig. 4b. It consists of two two-node bar elements. The displacement representation in the finite element model of the bar is a linear function and the stress and strain representations are constants. The finite element displacements are compared to the actual displacements in Fig. 4c. As we can see, the finite element displacements are equal to the exact displacements at the nodes. In fact, for this loading condition, the finite element displacements at the nodes are equal to the exact result regardless of the location of the internal node.

The stresses produced by the finite element model are compared to the exact answer in Fig. 4d. As we can see, the finite element result contains two discontinuities. One exists between the two elements and one exists on the free boundary, i.e., the finite element result does not match the known boundary condition of zero stress that exists on a free boundary.

The residuals and pointwise errors for the problem are now found using the method of Refs. 1–4. The first step is to compute the residuals on the individual elements. This is accomplished by substituting the appropriate derivatives of the finite element displacement polynomial representations into the residual form of the governing differential equation given by Eq. 19.11.

When we take the required second derivative of the linear displacement polynomial used to form a two node bar element, the result is zero. When we substitute this zero value into Eq. 19.11, we find the residual on the domain of the element to be equal to the distributed load, which in this case is a constant equal to the gravity load. This component of the residual loads is shown applied to the individual elements in Fig. 5a.

The second step in the process is to determine the residuals due to the discontinuities in the stresses at the boundaries and due to the jumps in the interelement stresses. The discontinuities in the natural boundary conditions are the amounts by which the finite element result fails to satisfy the natural boundary conditions. The discontinuities at the fixed boundaries are the differences between the finite element results and the actual reactions. They are found in the process of putting the elements on fixed boundaries into equilibrium.

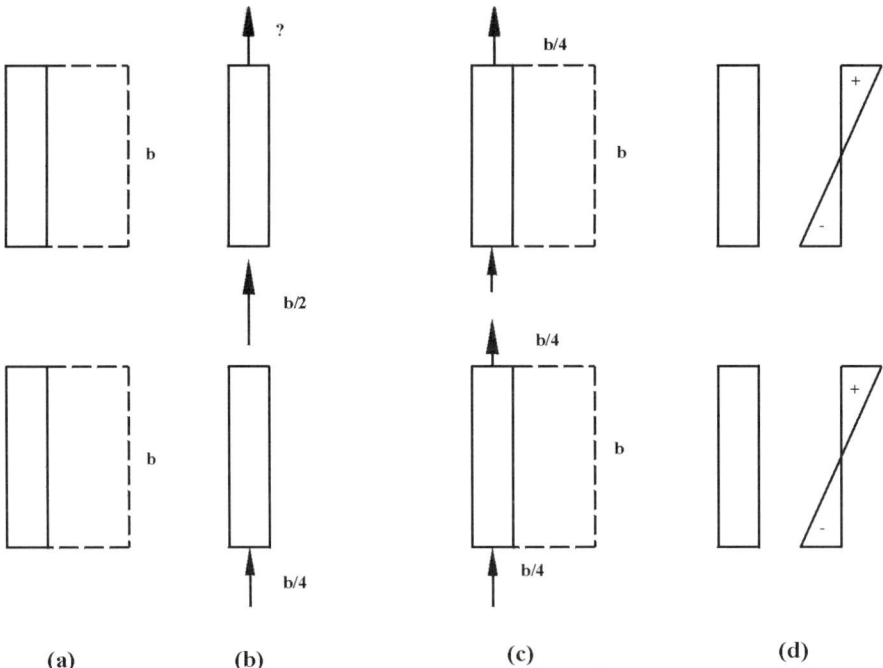

Figure 5. Residuals from the finite element model: (a) residuals on the domain of the individual elements, (b) undifferentiated inter-element residuals and errors in boundary conditions, (c) self-equilibrating residuals, and (d) stress errors in the finite element model.

The determination of these residual quantities is demonstrated in the example. The boundary condition residuals and the residuals due to the interelement jump in the stress between the elements are shown in Fig. 5b. The actual reaction at the fixed end is shown as a question mark because it is not known at this point in the analysis. The interelement jump is shown in undistributed form in Fig. 5b. That is to say, we do not yet know how to divide the residuals between the elements because the elements have not yet been put into equilibrium.

The initial computational effort in the approach being discussed is expended in finding the residuals shown in Fig. 5b. The next computational effort is expended in finding the distribution of the interelement loads that puts the individual elements into equilibrium. This is a simple task for this one-dimensional example but it requires techniques that are not standard to the finite element method for two- and three-dimensional problems. These techniques are discussed in Refs. 1–4.

For element 2 (the bottom element in the figure), a portion of the interelement residual equal to $b/4$ is required to put the element into equilibrium. The remaining portion of the interelement residual is applied to element 1. The boundary load required to put element 1 in equilibrium is equal to $b/4$. The individual elements are shown with the residual forces that put them in equilibrium in Fig. 5c.

Now that we have computed and distributed the interelement residuals so the individual elements have been put into equilibrium, we can compute the errors in the finite element solution. These errors could be estimated by subdividing the individual elements into several smaller finite elements and applying the residual loads to these subdivided elements. That is to say, we could treat each element as an individual finite element problem. However, that is not how the errors are computed in the example shown here.

We determine the errors in the original finite element solution by finding the exact solution to the governing differential equations given by Eq. 19.10, one element at a time (see also the Appendix of Ref. 3). The stress distributions for the exact solution due to the loading conditions shown in Fig. 5c are shown in Fig. 5d. These are the errors contained in the individual elements. When we combine these errors with the finite element results shown in Fig. 4d, the exact stress results for the original problem as shown in Fig. 4d are produced.

Before discussing the results just found for this example problem, we evaluate and correct the finite element result for this problem using the augmented finite element result.

Example Problem — Augmented Finite Element Approach

We now evaluate the finite element result for the bar problem using the augmented finite element approach. The first step in forming the augmented result is to superimpose three-node central difference templates on the nodes of the finite element model. This is shown in Fig. 6 where the templates are superimposed on the individual nodes. When a finite difference template is overlaid on the node at the fixed boundary, a fictitious node is introduced into the problem. The displacement of this node is found by forcing the finite difference approximation of the governing differential equation to be satisfied at this node. As mentioned earlier, the rationale behind this constraint is that the error in the displacements on the fixed boundary is zero so the governing differential equations must be satisfied.

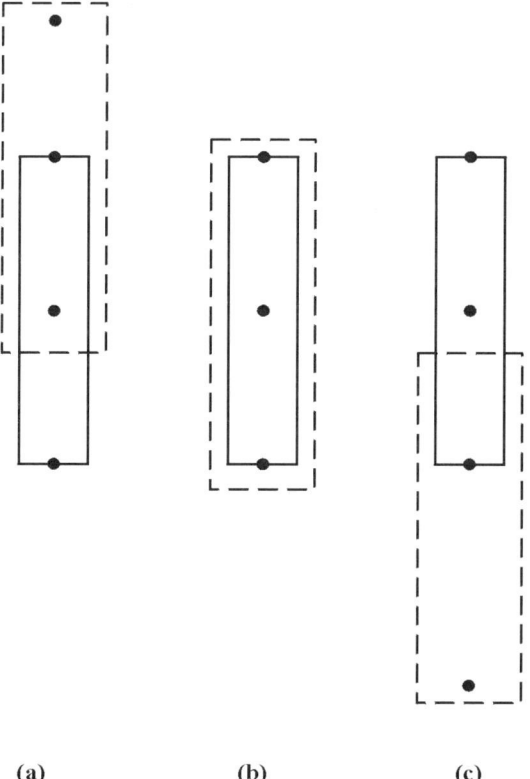

(a)   (b)   (c)

Figure 6. Three-node central difference templates superimposed on the finite element model: (a) at node 1, (b) at node 2, and (c) at node 3.

When we superimpose a finite difference template on the node at the free end, a second fictitious node is introduced into the problem. We find the displacement of this node by forcing the stress boundary condition at the free end to be satisfied. The augmented finite element solution consists of the displacements found in the finite element analysis and the displacements of the fictitious nodes. As mentioned on other occasions, we can consider the augmented finite element result to be an approximate finite difference solution.

To introduce the boundary stresses at the free end and to force the residuals at the fixed end to be zero, the finite difference approximations of the first and second derivatives must be available. For a one-dimensional problem with even spacing, the finite difference operators that approximate these derivatives are

---

Finite Difference Derivative Approximation

$$\left(\frac{du}{dx}\right)_0 = \frac{1}{2h}(u_3 - u_1)$$

$$\left(\frac{d^2u}{dx^2}\right)_0 = \frac{1}{h^2}(u_1 - 2u_2 + u_3)$$

(19.14)

where the subscript 0 indicates that these are Taylor series values evaluated at the local origin.

This is significant because it means we can consider these quantities on a pointwise basis. The nodal designations indicate the nodal locations in a standard finite difference templates. For example, the second derivative at the fixed node is found by identifying node 1 as the fictitious node, node 2 as the fixed node, and node 3 as the interior node of the finite element model.

These derivative approximations are used in the process of introducing the boundary conditions into the augmented solution and solving for the displacements of the fictitious nodes. With the availability of the displacements of the fictitious nodes, the residuals at the nodes of the finite element model can be computed. This is accomplished by substituting the finite difference approximation for the second derivative given by Eq. 19.14 into the equilibrium equation given by Eq. 19.11. The residuals at the nodes are the amounts by which the differential equation is not satisfied by the augmented result.

When we compute the residuals for this problem, we find them to be zero at all three nodes. This computation is shown for the case of the interior node of the example problem. The displacements for the three nodes are $u_1 = 0$, $u_2 = 3b/8$, and $u_3 = b/2$. When we substitute these quantities into the second derivative expression given in Eq. 19.14 for $h = 1/2$, the result is $-b$. When we substitute this result into the expression for the residual given in Eq. 19.11, we find the residual to be zero.

This means that the augmented finite element solution is an actual finite difference result and that the displacements and the stress results for the augmented result are the exact results at the nodes. This also means that the nodal displacements of the finite element result are exact even if the stress results in the finite element model do not match the exact result. This shows that the stresses in the augmented result are one order higher than those for the finite element representation.

Interpretation of Results

As stated earlier, the primary purpose of this section is to validate the interpretation of the interelement jumps as residual quantities and to relate the residuals in Kelly's approach to those found in the augmented finite element result. We have seen that Kelly's approach produces exact results. This means that the quantities extracted from the finite element model measure the difference between the finite element stress results and the exact stresses. Specifically, the use of the residuals on the individual elements and the interelement jumps as loads to put the discrete system into pointwise equilibrium is correct. This means that we can call the interelement jumps residuals because they are analogous in function to the usual definition of the residuals. Note that the interelement jumps have the same units as the residuals, i.e., a delta stress is divided by a delta distance, which in this case is zero.

Now that we have seen that the discontinuities in the interelement stresses can be legitimately interpreted as residual quantities, we have another interpretation for the smoothing approach. The refinement guides and the termination criteria based on the differences between the smoothed solution and the finite element result are directly measuring one component of the force system that separates the finite element representation from an exact solution of the problem being studied. That is to say, the smoothing approach to evaluating the finite element result is not measuring some "soft" quantity defined as "a lack of continuity," it is measuring a "hard" quantity, an equilibrating force.

Besides providing us with another interpretation of the smoothing approach to evaluating finite element results, the example problem clarified the capabilities of the two residual approaches for evaluating and correcting finite element results. The residual approaches can be used in the following three ways: 1) to directly form refinement guides and termination criteria, 2) to compute estimates for errors in the stresses and 3) to compute corrections for the finite element stress representations. We will now discuss these three uses in turn.

Residual-Based Evaluation Quantities.   The direct connection between the residual quantities extracted using Kelly's approach and the deficiencies in the model are apparent if Kelly's approach is described in slightly different terms. In this alternate interpretation, the finite element solution is taken as the *exact* solution of a modified version of the problem being analyzed. The modified problem consists of the original problem to which an additional set of fictitious loads has been applied. The added loads put the modified problem into pointwise equilibrium. The fictitious additional loads are precisely equal to the residuals on the individual elements and the interelement jumps in the stresses. That is to say, the residual components quantify the amount by which the finite element stresses fail to satisfy the problem being solved. As the mesh is refined and the results are improved, the residuals approach zero. Thus, we can use the magnitude of the residuals as both refinement guides and termination criteria. As we see in the next section, the requirement that the interelement jumps be zero is sufficient to guarantee convergence, but they are overly stringent, i.e., they are not necessary conditions.

A similar interpretation applies to the residuals extracted from the augmented finite element result. The residuals measure how far the stresses in the augmented result differ from a finite difference result. As the mesh is refined, the residuals approach zero. As the augmented result approaches an actual finite difference result, it also approaches an exact solution. This last condition holds because the result extracted from the discrete model is exact if the finite element and the finite difference results are identical.

This subsection contains an important idea. The two types of residuals measure different things. The residuals extracted using Kelly's approach quantify how far the finite element stresses are from the exact result. The residuals extracted from the augmented result quantify how far the smoothed stresses differ from a finite difference result and, hence, from an exact result. These differences are used in the next section (item 7) to show that the smoothed stresses are closer to the exact result than the finite element stresses.

The extraction of the residuals from the augmented result has two advantages over Kelly's approach. The smoothing approach requires the calculation of only one type of residual, i.e., the residuals are calculated at the nodes with Eq. 19.3. In addition to performing a similar calculation on an element-by-element basis, the Kelly approach requires that the interelement jumps in the stresses be computed. In some applications of the Kelly approach, only the interelement jumps are used in the evaluation process. This reduces the Kelly approach to a Zienkiewicz and Zhu (Z/Z) type of approach. That is to say, both methods quantify the interelement jumps. The Kelly approach does this directly and the Z/Z approach does this indirectly using a comparison with a smoothed solution.

As we will see, the requirement that the interelement jumps be zero is a more stringent measure of convergence than is the requirement that the residuals extracted from the augmented result be zero. This means that quantities based on the interelement jumps are good refinement guides but that they may be too conservative to serve alone as termination criteria.

Computation of Stress Error Estimates.   Estimates of the errors in the stress results for the finite element model can be computed in terms of the residuals found by Kelly's approach. The two types of residuals are applied as loads to the original finite element model. This approach has a certain merit for linear problems because the solution of the finite element model loaded with the residual loads is not computationally expensive because the problem has been solved for the original loading condition. However, this approach is being discussed not so much for its merit but because of the limitations it shows for this approach.

The application of the residual loads to the original finite element model can *only* produce estimates of the errors in the stresses. This procedure *cannot* produce the actual errors in the solution. The reason for this is that the distributed residual load has the same structure as the original load. As a result, the finite element model cannot represent the errors any better than it represented the original loads. However, the estimations are directly related to the quantities of primary interest to the analyst, namely, the stresses. This gives us another way to interpret the residual loads.

The example problem just solved can be used to clarify this concept. The load applied to the original problem is a constant distributed load as shown in Fig. 4a. The finite element model did not give the exact stress result for this loading or the residuals would have been zero. As can be seen in Figs. 5a or 5c, the fictitious loadings due to the distributed residuals are also constant distributed loadings. Thus, the existing finite element model cannot give the exact result for the loading due to the residuals any more than it did for the original linear loading condition. In Refs. 1–4, this idea is embodied by the statement that the finite element model is orthogonal to the residual loads.

Computation of Corrections for the Finite Element Stresses.   Corrections to the finite element stresses can be computed by applying the residual loads found by Kelly's approach to improved representations of the individual elements. We can find an *estimated* correction by applying the residual loading to a finite element model of the *individual elements*. We can find an exact correction by solving the governing differential equation, as was done in the example problem. This might be simpler than trying to solve the governing differential equations for the original problem because the boundary conditions are simpler.

This procedure is of interest because we can view it as a totally new solution technique. The individual elements could be refined one at a time until the error in the element is zero. This means that to obtain an exact solution we would have to solve as many smaller problems as there are elements in the original problem. More significantly, this approach provides a means for refining only the individual elements located in regions considered critical to the analysis. We can view this focused solution technique as a localized application of St. Venant's principle. This idea is not pursued here, but it seems worth exploring. The crux of the problem is to divide the interelement jumps into a self-equilibrating load system.

Summary

In this section, we have presented the Kelly approach for evaluating and correcting finite element results using residual quantities extracted from the finite element results. On the one hand, this approach was presented for completeness. The reader is introduced to an alternate residual method for evaluating finite element results. On the other hand, the Kelly approach was presented as a way to identify the relationship between the residuals contained in the finite element result and the augmented finite element result. The identification of the relationship between these residual quantities is used in the next section to show that the stress results contained in the smoothed solution are indeed more representative of the actual results than the finite element stresses.

Item 7 — Necessary and Sufficient Conditions for Convergence

In the previous section, we identified the sufficient conditions for the finite difference stresses to be of an order higher than the finite difference stresses. Specifically, we saw that when a higher order finite difference template is superimposed on a finite element model consisting of lower order elements, we find the smoothed stresses to have the same order as the displacements in the finite element results. This means that the error terms in the smoothed solution approach zero faster than the error terms in the finite element representation. We can interpret this to mean that the smoothed stresses are more accurate than the finite element stresses. This result coincides with the underlying assumption of the Zeinkiewicz and Zhu approach for evaluating finite element results.

We would like to be able to prove that the smoothed stresses are always at least as good as the finite element stresses for the general case. That is to say, we would like to be able to prove this same result when the finite difference template is of the same order as the finite element stiffness representation. For example, when a nine-node template is superimposed on a mesh consisting of six-, eight-, or nine-node finite elements. That the smoothed stresses are better in this case seems to have ''intuitive'' support. The displacement quantities in the nine-node finite difference template surround the node at which the stresses are being computed. This enables us to compute ''two-sided'' approximations of the derivatives.

At this time, it is not clear how to prove this result for the general case using nine-node templates or if, indeed, a general proof is possible. We could always superimpose a 5×5 finite difference template on these meshes and proceed as before to get a proof. However, this approach does not seem necessary in the actual computation of the evaluation measures. This could, of course, be tested empirically.

In this section, we demonstrate a somewhat weaker result. We show for three specific examples that the smoothed result is the better result as the finite element result approaches a finite difference result.[5] This result implies that the smoothed stresses are always at least as good as the finite element stresses.

This development proceeds as follows. Kelly's conditions for identifying an exact solution[6] is imposed on a patch of finite elements that make up a higher order central difference template. That is to say, we require the finite element model to exhibit interelement stress (or strain) compatibility and the residuals on the individual elements to be satisfied.[7] We then show that these requirements impose conditions that are more stringent than those required for the residuals in the augmented solution to be zero. We demonstrate this for the following cases: (1) one-dimensional bar elements, (2) four-node planar elements, and (3) three-node planar elements. We then briefly discuss the relationship between the interelement jumps and the residuals in a six-node model. The four-node elements are discussed before the three-node elements because there are fewer elements covered by the superimposed nine-node template.

The developments presented in this section are important for three reasons: (1) they show that the smoothed solution is, indeed, a better solution; (2) they identify the relationships between the interelement jumps, the residuals on individual elements, and the residuals in the augmented finite element result; and (3) they provide us with insights into the capabilities of the different approaches for evaluating finite element results. We compare the characteristics of the different quantities used in error analyses procedures in the next section.

Analysis of the One-Dimensional Model

Let us consider the patch of two two-node bar elements shown in Fig. 7. In this figure, a three-node central difference template is superimposed on the patch of two elements.

The expressions for the strains in the two bar elements are given in terms of the nodal numbering shown in Fig. 7 in the following equations:

Strain Expressions in Bar Elements

$$[\varepsilon_x]_1 = \frac{1}{h}(u_2 - u_1)$$

$$[\varepsilon_x]_2 = \frac{1}{h}(u_3 - u_2)$$

(19.15)

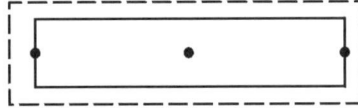

SEE ALSO FIG. 6

Figure 7. A patch of two two-node bar elements with a superimposed three-node central difference template.

---

[5] These demonstrations apply to problems with simple or no distributed loads.
[6] Kelly's criteria for an exact result states that a result is exact if the interelement jumps and the residuals on the individual finite elements are zero.
[7] We actually impose interelement strain compatibility instead of stress compatibility. The two conditions are essentially the same and the proof is more compact using strains because the finite difference operators are expressed in terms of strain quantities.

In this case, the stresses and strains are directly related, so we consider the strains directly because the finite difference templates directly produce strain quantities. An expression for the interelement jump in the strain between the two finite elements can be formed by subtracting the strain expression for bar 1 from that for bar 2. When this is done, the result is

---

Expression for Interelement Jumps in Strain

$$\Delta \varepsilon_x = [\varepsilon_x]_2 - [\varepsilon_x]_1$$

$$= \frac{1}{h}(u_3 - u_2) - \frac{1}{h}(u_2 - u_1) \qquad (19.16)$$

$$= \frac{1}{h}(u_1 - 2u_2 + u_3)$$

Let us now compare this expression for strain compatibility to the difference operator contained in the finite difference model of the residual form of the governing differential for a bar. The difference operator is given in Eq. 19.14. As we can see, the numerator of the difference operator is identical to the expression for the jump in the interelement strain. This means that the three-node difference operator is zero if interelement strain compatibility is satisfied.

Let us now evaluate the meaning of this relationship between the interelement jump and the central difference operator on the accuracy of the finite element result by studying the governing differential equation for a bar, as given in Eq. 19.11. We see that the condition of a zero jump in the interelement strain is identical to the satisfaction of equilibrium *if the problem does not contain a distributed load*. If the problem contains a distributed load, this result shows that a bar represented with constant strain elements cannot represent the exact solution *with elements of finite length*. We could argue that in the limit as the elements approach zero length that the results for the constant strain model converge.

It is useful if we interpret this relationship in a different way. We can say that Kelly's criteria for the finite element stresses to be an exact solution is more stringent than the requirement for the smoothed stresses to be an exact result. Specifically, this means that the residuals in the augmented solution can be zero even if the interelement jumps in the finite element stresses are not zero. This is an indirect way of saying that the stresses in the augmented stresses can represent the exact result when the finite element stresses do not represent the exact result. In other words, the smoothed stress result is a better solution than the finite element stresses when the displacement results have converged. We saw such a case in the examples presented in the previous section.

Note that the question of interelement compatibility for the augmented result does not arise because it is a finite difference result. This means that the result is "guaranteed" only at the nodes. Pointwise equilibrium is not addressed at any other point in a finite difference representation. Furthermore, we do not have to worry about the residuals on the individual elements when there is no distributed load because the difference operator in Eq. 19.11 is identically zero for a two-node bar.

Analysis of the Two-Dimensional Four-Node Finite Element Model

Let us consider the patch of four four-node elements shown in Fig. 8. In this figure, a nine-node central difference template is superimposed on the square patch of finite elements as shown.

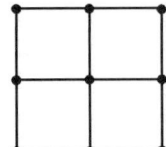

Figure 8. A patch of four-node elements with a superimposed nine-node central difference template.

As can be seen in Fig. 8, four four-node elements intersect at the central node of the finite difference template. This means that the jumps in the interelement stresses and strains at this point must all be zero. In this analysis, we evaluate the interelement jumps in terms of strains instead of in terms of stresses. This is the course followed because the finite difference operators are given directly in terms of strain quantities. The expressions for the three strain components in the four elements are given in terms of the nodal numbering shown in Fig. 8 with the following three equations as

Strain Expression in the Four Four-Node Finite Elements

$$(\varepsilon_x)_2 = \frac{1}{h}(u_9 - u_8) \qquad (\varepsilon_x)_1 = \frac{1}{h}(u_6 - u_9)$$

$$(\varepsilon_x)_3 = \frac{1}{h}(u_9 - u_8) \qquad (\varepsilon_x)_4 = \frac{1}{h}(u_6 - u_9)$$

(19.17)

Strain Expression in the Four Four-Node Finite Elements

$$(\varepsilon_y)_2 = \frac{1}{h}(v_7 - v_9) \qquad (\varepsilon_y)_1 = \frac{1}{h}(v_7 - v_9)$$

$$(\varepsilon_y)_3 = \frac{1}{h}(v_9 - v_5) \qquad (\varepsilon_y)_4 = \frac{1}{h}(v_9 - v_5)$$

(19.18)

Strain Expression in the Four Four-Node Finite Elements

$$(\gamma_{xy})_2 = \frac{1}{h}(-v_8 - u_9 + v_9 + u_7) \qquad (\gamma_{xy})_1 = \frac{1}{h}(-u_9 - v_9 + v_6 + u_7)$$

$$(\gamma_{xy})_3 = \frac{1}{h}(-u_5 + u_9 + v_9 - v_8) \qquad (\gamma_{xy})_4 = \frac{1}{h}(-u_5 + v_6 + u_9 - v_9)$$

(19.19)

The local origin for each of the four-node elements and for the nine-node central difference template are at node 9, the central node of the finite difference template. This origin is chosen so the expressions for interelement jumps at the central node can be easily formed. The conditions necessary for the interelement jumps in the strains to be zero at the intersection of the four four-node finite elements are given as

Expressions for Interelement Jumps in Strain

$$(\varepsilon_x)_1 - (\varepsilon_x)_2 = \frac{1}{h}(u_6 + u_8 - 2u_9)$$

$$(\varepsilon_y)_1 - (\varepsilon_y)_3 = \frac{1}{h}(v_5 + v_7 - 2v_9)$$

$$(\gamma_{xy})_1 - (\gamma_{xy})_2 = \frac{1}{h}(v_6 + v_8 - 2v_9)$$

$$(\gamma_{xy})_2 - (\gamma_{xy})_3 = \frac{1}{h}(u_7 + u_5 - 2u_9)$$

(19.20)

The first condition comes from Eq. 19.17. This expression quantifies the differences between the two right-hand elements and the two left-hand elements. The second condition comes from Eq. 19.18. This expression quantifies the difference in the normal stresses between the two upper elements and the two lower elements. The third and fourth conditions in Eq. 19.20 come from Eq. 19.19. In the case of the shear strain, there are six possibilities for interelement jumps in the strains. Element 1 must have the same shear strain as elements 2, 3, and 4 etc.

The residual form of the governing differential equations for the plane stress problem are given in Eq. 19.3. These equations contain six strain gradient terms that can be represented by the difference operators for a nine-node template. For an evenly spaced mesh, these operators are given as

Central Difference Operators for an Evenly Spaced Mesh

$$\varepsilon_{x,x} = \frac{1}{h^2}(u_6 + u_8 - 2u_9)$$

$$\varepsilon_{y,y} = \frac{1}{h^2}(v_5 + v_7 - 2v_9)$$

$$\varepsilon_{y,x} = \frac{1}{4h^2}(v_1 - v_2 + v_3 - v_4)$$

$$\varepsilon_{x,y} = \frac{1}{4h^2}(u_1 - u_2 + u_3 - u_4)$$

$$\gamma_{xy,x} = \frac{1}{4h^2}(u_1 - u_2 + u_3 - u_4) + \frac{1}{h^2}(v_6 + v_8 - 2v_9)$$

$$\gamma_{xy,y} = \frac{1}{4h^2}(v_1 - v_2 + v_3 - v_4) + \frac{1}{h^2}(u_5 + u_7 - 2u_9)$$

(19.21)

When interelement strain compatibility is satisfied, the conditions in Eq. 19.20 are all equal to zero. When these terms are zero, we see that four of the eight terms in Eq. 19.21 are zero. Specifically, the first two difference operators are zero and the final term in the two shear strain quantities is zero.

When the remaining four terms are inspected, we see that only two independent terms need to be considered. We can show that these two terms are zero when the residuals on the individual elements are zero. We now proceed to demonstrate this.

Only two of the six central difference operators contained in Eq. 19.3 or Eq. 19.21 exist for a four-node finite element. The two flexure terms, $\varepsilon_{x,y}$ and $\varepsilon_{y,x}$, have nonzero values. The other difference operators do not exist because of the low order of the polynomial representations contained in the four-node element. When these two nonzero operators are evaluated for the four four-node elements in terms of the nodal numbering shown in Fig. 8, the results are given by the following two sets of equations as

---

One Flexure Term for a Four-Node Element

$$(\varepsilon_{x,y})_1 = \frac{1}{h^2}(u_3 - u_7 - u_6 + u_9)$$

$$(\varepsilon_{x,y})_2 = \frac{1}{h^2}(u_7 - u_4 - u_9 + u_8)$$

$$(\varepsilon_{x,y})_3 = \frac{1}{h^2}(u_1 - u_5 - u_8 + u_9)$$

$$(\varepsilon_{x,y})_4 = \frac{1}{h^2}(u_6 - u_2 - u_9 + u_5)$$

(19.22)

One Flexure Term for a Four-Node Element

$$(\varepsilon_{y,x})_1 = \frac{1}{h^2}(v_3 - v_7 - v_6 + v_9)$$

$$(\varepsilon_{y,x})_2 = \frac{1}{h^2}(v_7 - v_4 - v_9 + v_8)$$

$$(\varepsilon_{y,x})_3 = \frac{1}{h^2}(v_1 - v_5 - v_8 + v_9)$$

$$(\varepsilon_{y,x})_4 = \frac{1}{h^2}(v_6 - v_2 - v_9 + v_5)$$

(19.23)

When we average the operators in Eqs. 19.22 and 19.23, the result is

---

Average Central Difference Operators

$$(\varepsilon_{x,y})_{ave} = \frac{1}{4h^2}(u_3 - u_4 - u_2 + u_1)$$

$$(\varepsilon_{y,x})_{ave} = \frac{1}{4h^2}(v_3 - v_4 - v_2 + v_1)$$

(19.24)

When we compare these expressions to the two terms that are still unaccounted for in Eq. 19.21, we see that they match. Our task is to show that these two terms must be zero.

One of Kelly's criteria for an exact finite element stress result is that the residuals on the individual finite elements be zero. For the case of a problem with no distributed load, this means that each of the terms in Eqs. 19.22 and 19.23 must be zero. Since the components of Eq. 19.24 are the average of these four quantities, the components of Eq. 19.24 must be zero.

This means that the remaining terms in Eq. 19.21 are zero. Thus, for the case where the distributed load is zero, the conditions specified by Kelly mean that the residuals for the augmented finite element result are zero.

Let us now consider the meaning of this result. If the interelement jumps in the stresses are zero, the difference operators contained in the governing differential equations for the problem are zero. If the problem does not have a distributed load, the condition of zero-interelement jumps in stresses means that the equilibrium equations are identically zero. However, equilibrium in the two-dimensional problem can be satisfied without each of the differential operators being individually zero. Only the sum of the terms must be zero.

Now let us consider the case of a problem with a distributed load. In this case, the four-node finite element representation cannot satisfy the nonhomogeneous elemental equilibrium equations because the elements do not represent the complete set of linear strain terms. Thus, the finite element model cannot represent the exact solution *with elements of finite length*. The four-node model can represent the exact result in the limit as the element dimensions approach zero.

We can interpret this result in exactly the same way we interpreted the result for the constant strain bar elements. That is, we can interpret this relationship in a different way. We can say that Kelly's criteria for the finite element stresses to be an exact solution is more stringent than the requirement for the smoothed stresses to be an exact result. Specifically, this means that the residuals in the augmented result can be zero even if the interelement jumps in the finite element stresses are not zero. This is an indirect way of saying that the stresses in the augmented stresses can represent the exact result when the finite element stresses do not represent an exact result. In other words, the smoothed stress result is a better solution than the finite element stresses when the displacement results have converged.

This result is not unexpected because the finite difference model is of an order higher than the finite element model. Now, let us develop a similar argument for the case of a finite element model based on three-node elements. This case is interesting because the equilibrium equations on the individual elements are satisfied by definition for the case of a problem with no distributed load because the approximate displacement polynomials are linear, i.e., their second derivatives are zero. This means that all of the residuals in the three-node model for such a problem are embodied in the inter-element jumps.

Analysis of the Two-Dimensional Three-Node Finite Element Model

Let us consider the patch of eight three-node finite elements shown in Fig. 9. In this figure, a nine-node central difference template is superimposed on the square finite element model shown. The local origin for each of the three-node elements is at the central node of the finite difference template shown in Fig. 9.

The expression for the three strain components in the eight elements in terms of the nodal displacements of the superimposed nodal displacements of the superimposed nine-node finite difference template are given in the following three sets of equations as

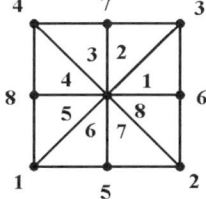

Figure 9. A patch of eight three-node elements with a superimposed nine-node central difference templates (see also Fig. 2).

Strain Expressions in the Eight Three-Node Finite Elements

$$(\varepsilon_x)_1 = u_6 - u_9 \qquad (\varepsilon_x)_5 = u_9 - u_8$$
$$(\varepsilon_x)_2 = u_3 - u_7 \qquad (\varepsilon_x)_6 = u_5 - u_1$$
$$(\varepsilon_x)_3 = u_7 - u_4 \qquad (\varepsilon_x)_7 = u_2 - u_5 \qquad (19.25)$$
$$(\varepsilon_x)_4 = u_9 - u_8 \qquad (\varepsilon_x)_8 = u_6 - u_9$$

Strain Expressions in the Eight Three-Node Finite Elements

$$(\varepsilon_y)_1 = v_3 - v_6 \qquad (\varepsilon_y)_5 = v_8 - v_1$$
$$(\varepsilon_y)_2 = v_7 - v_9 \qquad (\varepsilon_y)_6 = v_9 - v_5$$
$$(\varepsilon_y)_3 = v_7 - v_9 \qquad (\varepsilon_y)_7 = v_9 - v_5 \qquad (19.26)$$
$$(\varepsilon_y)_4 = v_4 - v_8 \qquad (\varepsilon_y)_8 = v_6 - v_2$$

Strain Expressions in the Eight Three-Node Finite Elements

$$(\gamma_{xy})_1 = v_6 - v_9 + u_3 - u_6 \qquad (\gamma_{xy})_5 = v_9 - v_8 + u_8 - u_1$$
$$(\gamma_{xy})_2 = v_3 - v_7 + u_7 - u_9 \qquad (\gamma_{xy})_6 = v_5 - v_1 + u_9 - u_5$$
$$(\gamma_{xy})_3 = v_7 - v_4 + u_7 - u_9 \qquad (\gamma_{xy})_7 = v_2 - v_5 + u_9 - u_5 \qquad (19.27)$$
$$(\gamma_{xy})_4 = v_9 - v_8 + u_4 - u_8 \qquad (\gamma_{xy})_8 = v_6 - v_9 + u_6 - u_2$$

The goal of this development is the same as that for the four-node finite element case. We want to show that the difference operators contained in the residual form of the governing differential equations for the nine-node template are zero if the interelement jumps in the strains between the elements at the nodes are zero. For this model, there are eight elements intersecting at the central node. If we enforce interelement strain compatibility in each of the three strains, there are seven factorial relations possible that must be zero for each strain component. Out of this number of relationships there are several paths for showing that the six difference operators given in Eq. 19.21 are zero if interelement strain compatibility is satisfied.[8] We now present one path that provides the desired results.

The first term in Eq. 19.21 is the rate of change of ε_x in the x direction. We can show that this term must be zero by forming an interelement compatibility relationship for ε_x between two adjacent elements along the x axis. That is to say, we are comparing the difference between ε_x in the direction of a change in x. Specifically, we show this relationship between the interelement compatibility and the difference operator by enforcing compatibility for ε_x between elements 1 and 4. When we force the interelement jump in ε_x to be zero for these elements, we get the following:

[8] There are fewer constraint equations for the four-node case. This may indicate the difference in the amount of residual value that is contained in the interelement jumps.

Compatibility Relationship

$$(\varepsilon_x)_1 - (\varepsilon_x)_4 = 0$$
$$(u_6 - u_9) - (u_9 - u_8) = 0 \qquad (19.28)$$
$$(u_6 + u_8 - 2u_9) = 0$$

The second term in Eq. 19.21 is the rate of change of ε_y in the y direction. When we form a compatibility relationship between two elements oriented along the y axis, we see that the corresponding difference operator is identically zero. This relationship is formed as follows:

Compatibility Relationship

$$(\varepsilon_y)_2 - (\varepsilon_y)_6 = 0$$
$$(v_7 - v_9) - (v_9 - v_5) = 0 \qquad (19.29)$$
$$(v_5 + v_7 - 2v_9) = 0$$

The third term in Eq. 19.21 can be satisfied by forming an expression for the change in ε_y in the x direction. This is accomplished by summing the strain expressions for elements 1 and 8 and subtracting this total from the sum of the strain expressions for elements 4 and 5. When this is done, the result is

Forming the Third Term in Eq. 19.21

$$[(\varepsilon_y)_1 + (\varepsilon_y)_8] - [(\varepsilon_y)_4 + (\varepsilon_y)_5] = 0$$
$$[(v_3 - v_6) + (v_6 - v_2)] - [(v_4 - v_8) + (v_8 - v_1)] = 0$$
$$[v_3 - v_2] - [v_4 - v_1] = 0 \qquad (19.30)$$
$$(v_1 - v_2 + v_3 - v_4) = 0$$

The fourth term in Eq. 19.21 can be satisfied by forming an expression for the change in ε_x in the y direction. This is accomplished by summing the strain expressions for elements 2 and 3 and subtracting this total from the sum of the strain expressions for elements 6 and 7. When this is done, the result is

Forming the Fourth Term in Eq. 19.21

$$[(\varepsilon_x)_2 + (\varepsilon_x)_3] - [(\varepsilon_x)_6 + (\varepsilon_x)_7] = 0$$
$$[(u_3 - u_7) + (u_7 - u_4)] - [(u_5 - u_1) + (u_2 - u_5)] = 0$$
$$(u_3 - u_4) - (u_2 - u_1) = 0 \qquad (19.31)$$
$$(u_1 - u_2 + u_3 - u_4) = 0$$

The fifth term in Eq. 19.21 can be satisfied by forming an expression for the change in γ_{xy} in the x direction. This is accomplished by summing the strain expressions for elements 4 and 5 and subtracting this total from the sum of the strain expressions for elements 1 and 8. When this is done, the result is

Forming the Fifth Term in Eq. 19.21

$$[(\gamma_{xy})_1 + (\gamma_{xy})_8] - [(\gamma_{xy})_4 + (\gamma_{xy})_5] = 0$$
$$[(v_6 - v_9 + u_3 - u_6) + (v_6 - v_9 + u_6 - u_2)] - [(v_9 - v_8 + u_4 - u_8) + (v_9 - v_8 + u_8 - u_1)] = 0$$
$$(2v_6 - 2v_9 + u_3 - u_2) - (2v_9 - 2v_8 + u_4 - u_1) = 0$$
$$\frac{1}{2}(v_6 + v_8 - 2v_9) + (u_1 - u_2 + u_3 - u_4) = 0 \tag{19.32}$$

We know that the group of four terms is zero because of the result presented in Eq. 19.31. Therefore, the desired result is zero.

The sixth term in Eq. 19.21 can be satisfied by forming an expression for the change in the shear strain in the y direction. This is accomplished by summing the strain expressions for elements 6 and 7 and subtracting this total from the sum of the strain expressions for elements 2 and 3. When this is done, the result is

Forming the Sixth Term in Eq. 19.21

$$[(\gamma_{xy})_2 + (\gamma_{xy})_3] - [(\gamma_{xy})_6 + (\gamma_{xy})_7] = 0$$
$$[(v_3 - v_7 + u_7 - u_9) + (v_7 - v_4 + u_7 - u_9)] - [(v_5 - v_1 + u_9 - u_5) + (v_2 - v_5 + u_9 - u_5)] = 0$$
$$(v_3 - v_4 + 2u_7 - 2u_9) - (v_2 - v_1 + 2u_9 - 2u_5) = 0$$
$$(v_1 - v_2 + v_3 - v_4) + \frac{1}{2}(u_5 + u_7 - 2u_9) = 0 \tag{19.33}$$

We know that the group of four terms is zero because of the result presented in Eq. 19.30. Therefore, the desired difference operator is zero when these interelement compatibility relationships are satisfied.

We can interpret this result in exactly the same way we interpreted the result for the constant strain bar elements. We can say that Kelly's criteria for the finite element stresses to be an exact solution are more stringent than the requirement for the smoothed stresses to be an exact result. Specifically, this means that the residuals in the augmented result can be zero even if the interelement jumps in the finite element stresses are not zero. This is an indirect way of saying that the stresses in the augmented stresses can represent the exact result when the finite element stresses do not represent an exact result. In other words, the smoothed stress result is a better solution than the finite element stresses when the displacement results have converged.

When we compare this result to the earlier result for the four-node element, we can make one more observation. The interelement jumps for the three-node model contain a larger portion of the residual content than do the interelement jumps for a four-node model. This means that the stress-based refinement guides for a model based on three-node elements may have a resolution higher than those for a model based on four-node elements. For this reason and others, the three-node model may be a better basis element than the four-node element. The "other" reasons consist of the following. The three-node model does not contain parasitic shear. Meshes consisting of three node elements are easier to refine than meshes consisting of four-node elements. Finally, we do not need to directly use the stress representations contained in the three-node finite elements. We can use the smoothed stresses to represent the stresses in the finite element model.

Final Observations

As mentioned earlier, it is not clear how to proceed with a similar argument for six-, eight-, or nine-node elements in terms of nine-node finite difference templates. This is the case since the finite elements are of the same order as finite difference template. Also as mentioned earlier, we could superimpose 5×5 finite difference templates on the finite element mesh for these higher order finite elements and proceed as before. However, this is not done because the point has been made. There is a direct connection between the residuals and the interelement jumps. In fact, we have shown that the interelement jumps are a more stringent requirement on convergence than the residuals, so they should be more sensitive to discretization errors than the residuals. The final decision about whether to use interelement jumps or residuals can be made from empirical grounds because we have seen the theoretical relationship between the two types of quantities. We discuss the characteristics of the various evaluation measures at length in the next section.

Item 8 — Assessment of Deficiency Measures

The objective of this section is to clearly identify the capabilities we have available to us for evaluating finite element results as a consequence of the developments presented in this part of the book. We first discuss termination criteria. Then we discuss refinement guides. In the following section, a universal postprocessor for finite element software packages is proposed that is based on these capabilities.

Termination Criteria

We discuss four types of termination criteria: (1) strain energy measures, (2) estimated errors in finite element stresses at nodal points, (3) convergence of smoothed stresses, and (4) residual measures.

Strain Energy Measures.   In Lesson 16, we saw that estimates of the errors in strain energy quantities did not provide good termination criteria. The estimated errors in the global strain energy were not correlated to the errors in the *finite element stresses* at critical points. In Lesson 17, we saw that the same lack of correlation existed between the estimated errors in the strain energy for the individual elements and the *finite element stresses*. We reached this conclusion concerning the use of strain energy measures by observing that the stresses at critical points converged while the strain energy measures continued to get smaller as the mesh was refined.

   We discussed the reason for this lack of correlation in Lesson 18. We noted that the integration required to form strain energy quantities diluted the connection between the errors in the finite element stresses at critical points and the differences in the strain energy quantities. The problem was exacerbated when the global strain energy measures were formed by summing the strain energy quantities to form the global measures. With this understanding of the reason for the lack of correlation between the strain energy measures and the level of error in the finite element stresses, there is no reason to attempt to use global or local strain energy measures as termination criteria.

Estimated Errors in Nodal Stresses.   As just mentioned in the discussion of strain energy measures as termination criteria, the estimated errors in the elemental strain energy are not well correlated to the errors in the *finite element stresses* because the strain energy measures are aggregated quantities. We can overcome this deficiency by forming termination criteria based on the pointwise differences in the two stress representations at the nodes. In fact, termination criteria of this type have been successfully demonstrated (see Ref. 5).

The reason for the success of these pointwise measures is easily explained. The initial assumption of Zienkiewicz and Zhu is correct. The smoothed stresses are indeed better solutions than the finite element stresses. This was shown in item 5 when we demonstrated that the smoothed stresses are of an order higher than the finite element stresses. Thus, the differences between the two stress representations estimates the errors in the finite element stresses. However, the demonstration of the different orders between the stress representation raises the question of whether we actually want to use these measures as termination criteria.

Since the smoothed stresses are of an order higher than the finite element stresses, the smoothed stresses are typically used as the output of finite element analyses. However, the differences between the two stress representations are estimates of the errors in the *finite element stress representations*. So the question arises as to whether we can estimate the errors in the smoothed stresses.

We could make an estimate of the errors in the smoothed stresses if we could make some kind of argument about the discontinuous finite element stresses at a node bounding the actual solution. However, such a proof seems unlikely for two reasons: (1) the discontinuous stresses do not always bound the smoothed stresses when irregular central difference templates are required (see Ref. 5) and (2) the stresses and strains are not invariant unless the finite elements are based on complete polynomials, e.g., three-node and six-node elements (see Lessons 8 and 10).

For these reasons, it is doubtful that a direct relationship between the interelement jumps and the errors in the smoothed solutions can be put on a solid theoretical foundation. However, this does not preclude the development of a practical heuristic termination criteria based on interelement jumps. The jumps are indicative of how far the finite element stress results are from convergence, so they must be indicative of the approximate error in the smoothed solution.

Convergence of Smoothed Stresses.   Just because it does not seem possible for us to form a proof that the discontinuous finite element stresses bracket the actual result does not mean that we cannot form a termination criteria based on the smoothed stresses. The original observation that the strain energy measures are not good termination criteria provides an obvious candidate for a termination criteria. That is to say, we disqualified strain energy measures as termination criteria because we have seen cases where the smoothed stresses have converged while the strain energy measures continue to decrease, albeit slowly. This suggests the use of the convergence of the smoothed stresses at critical points as termination criteria.

The use of a sequence of smoothed stresses at critical points as termination criteria is supported by the behavior just discussed. In addition, we saw in the previous section that the smoothed stresses are of the same order as the displacement results. Therefore, the convergence of the smoothed stresses is directly connected to the convergence of the displacements, the most accurate quantities in a finite element result.[9] Furthermore, the direct use of the smoothed stresses as termination criteria keeps the quantities of interest before the analyst.

---

[9] As an aside, the effect of mesh refinement on the level of error in finite element results has not been analyzed using strain gradient notation, i.e., the rate of change of error due to the reduction in mesh size. This notation provides a new tool for such an analysis, as demonstrated in the previous section. There are no guarantees that the approach will be fruitful. However, every use of this physically based notation to date has provided valuable insights into the question addressed.

An obvious problem with using a sequence of stresses as a termination criteria is the case where refinement has not occurred in regions of high error for some reason. We can overcome this problem by checking the refinement guides in regions of critical importance before terminating an analysis under the authority of the convergence of the smoothed stresses. For example, a termination criteria based on the convergence of smoothed stresses can be coupled with a refinement guide based on interelement jumps to form a composite termination criteria.

Residual Quantities.   The residuals contained in the augmented finite element result are obvious candidates for termination criteria. They are direct measures of the differences between the finite element result and the actual result. Specifically, zero residuals indicate zero errors. This means that the accuracy of the stresses can be easily scaled to the level of residuals.

Since the magnitude of the residuals can be interpreted as fictitious distributed loads, these discretization error measures can be related to the magnitude of the actual loads applied to the problem. For example, if the total magnitude of the fictitious distributed load is small with respect to the applied load, we can terminate the analysis. Furthermore, in linear problems, we can apply the fictitious applied load due to the residuals to the finite element model to get an estimate of the errors in the stresses and/or strains at critical points.

The residual measures *seem* to be nearly the ideal termination criteria. However, the residual measures have four impediments to immediate and universal applications: (1) the inclusion of the approach might be difficult for some problems, (2) the inclusion of the approach might be difficult in some codes, (3) the other measures provide satisfactory termination criteria, and (4) the other measures do not require the direct inclusion of the differential form of the governing differential equations into the finite element code.

This discussion has presented three strong candidates for practical use as termination criteria: (1) pointwise estimates of errors in finite element stresses at the nodes, (2) the convergence of the smoothed stresses at the nodes, and (3) the magnitude of the residuals at the nodes. The relative merits of these termination criteria are considered further after we discuss the refinement guides.

Refinement Guides

We have discussed three kinds of refinement guides up to now, namely, (1) strain energy measures, (2) differences between the finite element stresses at the nodes, and (3) residuals in augmented finite element solutions. Any of these refinement guides are capable of adequately guiding the adaptive refinement process. This reduces the task before us to comparing the characteristics of the various refinement guides so we can choose the measure that best applies to a particular problem.

The refinement guides can be broken into two categories. Those based on differences between the stress components in a smoothed solution and the finite element result (i.e., those based on inter-element jumps) and those based on residual quantities. We discuss the refinement guides contained in the two categories in turn.

Refinement Guides based on Smoothed Solutions.   The original refinement guides developed by Zienkiewicz and Zhu compute the strain energy contained in the differences between a smoothed stress result and the finite element stresses. As shown in Lesson 16, this refinement guide satisfactorily guides the adaptive refinement of finite element results to convergence. However, we saw in Lesson 18 that pointwise refinement guides based on stress differences have a higher resolution than the refinement guides based on strain energy measures.

The strain energy measures have the advantage of being invariant with respect to the orientation of the coordinate system regardless of the element used in the analysis. The same is not true for the pointwise differences in the stress components. However, the individual pointwise stress components are invariant for elements that are based on complete polynomials. Even if the elements are not based on complete polynomials, it is not clear whether the invariance in the stresses has any bearing on the effectiveness of the refinement guides. This is said because the changes in the refinement guides for one stress component will probably be compensated for by changes in the other components. Thus, we can conclude that the pointwise refinement guides developed in Lesson 18 are effective even if they are not invariant.

In the previous section, we saw that the requirement for the interelement jumps to be zero was a more stringent requirement on the convergence of the finite element result than the requirement for the convergence of the smoothed stresses or the residuals. That is to say, the smoothed stresses and/or the residuals could have converged before the differences between the smoothed solution and the finite element stresses indicated convergence, i.e., they are equal or the interelement jumps are zero. This occurs because the refinement guides based on comparing the finite element stresses to a set of smoothed stresses are striving to improve a result that is less accurate than the smoothed stresses. This means that the refinement guides based on the differences between the two types of stress representations should be more conservative than the refinement guides based on residuals. This suggests that the refinement guides based on the differences in the finite element stresses and the smoothed solution would be the best to use in practice. Only a careful study can validate this speculation.

Although this discussion of refinement guides has focussed on the use of augmented finite element solutions, results presented here (see Lesson 16) ensure that the Zeinkiewicz and Zhu procedures (old or new) can adequately provide refinement guides. This implies that the pointwise differences in the two stress representations at the nodes in the Zienkiewicz and Zhu approaches also successfully identify discretization errors. This means that pointwise stress differences formed with nodal averaging or through the use of smoothing patches of Gaussian points lead to converged results. The refinement guides based on the differences between the averaged nodal values and the finite element stresses might be a little slow to identify elements with large errors on boundaries for early meshes. However, after a few refinements, the refinement guides formed using a smoothed solution based on nodal averaging should approach those produced by the augmented finite element result.

Refinement Guides Based on Residuals. As we saw in Lesson 18, we can also use the residuals contained in the augmented finite element result as refinement guides. This is the case because these residual quantities are directly related to the error in the finite element result. Furthermore, we should be able to easily scale the level of refinement to these residuals because of this direct relationship to the level of error. Such a scaling has yet to be developed. As just mentioned, the relative merit of the pointwise refinement guides based on residuals and those based on the differences in stress representations has not been determined.

The residual quantities in the individual finite elements (as defined for use in Kelly's approach) are not a good candidate for refinement measures. This is the case because of the difference in the orders of the polynomials on which the individual elements are based. For example, in a three-node model with no distributed load, the residuals are identically zero. This is the case because the polynomials in this element cannot represent the derivatives contained in the residual form of the governing differential equations. In the case of four-node elements, only some of the terms in the residual form of the governing differential equations are present. The six-node element is the lowest order element that can fully represent the residuals.

Recommendations for the Use of Evaluation Measures

The capabilities of the various refinement guides and termination criteria have been outlined. We now make recommendations for using these capabilities. Two situations must be considered to cover the situations that can be expected. In the first, we assume that finite difference smoothing will not be used. In the second, we assume that finite difference smoothing is to be used.

Without Finite Difference Smoothing. The interpretation of the augmented finite element result as an approximate finite difference result provides the basis for developing the theoretical foundation for the developments presented here. Furthermore, finite difference smoothing provides the basis for the most effective evaluation measures. However, situations may exist where the use of finite difference is inconvenient. For example, the use of finite difference smoothing might not be desired because an interim evaluation package is to be quickly implemented. Effective evaluation procedures can be developed without using finite difference smoothing. In this case, the evaluation measures on the boundaries are not as effective as when finite difference smoothing is used and residual measures are not available.[10]

If finite difference smoothing is not used, effective refinement guides and termination criteria can be formed using either of the two forms of smoothing introduced by Zeinkiewicz and Zhu, namely, (1) nodal averaging of the stress components or (2) interpolation and extrapolation of the stresses at the Gaussian points as used in the modified Zienkiewicz and Zhu approach. It is recommended that pointwise stress differences at the nodes (or the Gaussian points) be used as the refinement guides. We recommend that the convergence of the stresses at critical points be used as termination criteria.

If these procedures are in place, the analyst must ensure that adequate refinement takes place on the boundaries during early refinements. If not, it is possible that a region with high initial errors might possibly not be refined at all. As the results get ''reasonably'' close to the exact results, these pointwise evaluation measures approach the values produced by finite difference smoothing. Note that strain energy measures are not recommended.

With Finite Difference Smoothing. Wherever possible, we recommend that finite difference smoothing be used as the basis for the evaluation of finite element results. The use of the approximate finite difference solution means that we have the full range of refinement guides and termination criteria available to us. That is to say, we can use refinement guides based on stress differences and/or on residuals. Similarly, we can form termination criteria based on the convergence of the stresses and/or on the convergence of the residuals.

In these two sentences the ''and/or'' option is not written to show only the range of options that are available to us. It is meant to also recommend an approach with built-in redundancies. We can use either stress-based evaluation measures or residual-based evaluation measures. Or we can use both types of measures simultaneously. The use of

---

[10] Residual measures could be used. However, if they are used, finite difference smoothing is a relatively simple addition.

both measures might be advised if the evaluation procedures are being used in a "fully automated" adaptive refinement scheme where there is little human intervention. The residual measures could be used to check the stress-based measures or vice versa.

Finally, the use of finite difference smoothing makes the creation of a universal evaluation postprocessor possible for finite element analyses. That is to say, a postprocessor for forming refinement guides and for computing termination criteria that can be appended to most existing and future finite element codes can be written if finite difference smoothing is used. Such a postprocessor can be written because the stresses in the underlying finite element model are never actually needed in the procedures developed here. This is obvious if we use the residual-based evaluation measures and the smoothed stresses for output. It is not so obvious if we use the stress-based evaluation measures.

If we are using stress-based refinement guides, we need some form of discontinuous stress representation against which to compare the smoothed stresses. We need not retrieve these quantities from the finite element model directly. We can compute a set of discontinuous stresses by subdividing the nine-node finite difference template into "fictitious" finite elements with either three or four nodes (see Figs. 8 and 9). The three-node approach is demonstrated in the next lesson. We can then extract a set of discontinuous stresses from these pseudo-elements for use in the refinement guides. This does not affect the output of the program because we are using the smoothed stresses for this function.

This approach has merit for several reasons. The resultant postprocessor can be used to evaluate the results for any finite element representation of a given class, i.e., plane stress, plane strain, plates, and shells. That is why it is called a universal evaluation postprocessor. It is independent of the underlying finite element model. Such a postprocessor and its ramifications are discussed in the next sub-section of this lesson.

Item 9 — A Universal Postprocessor for Finite Element Programs

In the previous section, the idea of a universal postprocessor for evaluating finite element results was introduced. The idea is quite simple. An augmented finite element result is formed from the displacements of a finite element analysis. If any discontinuous stress representations are needed to form evaluation measures, they are extracted from subdivisions of the nine-node finite difference templates superimposed on the finite element mesh. This means that the only information needed from the underlying finite element model for the evaluation of the finite element results are the nodal locations, the nodal displacements, the material properties, and the boundary conditions.

This approach has several advantages. It separates the evaluation process from the specific finite element model used in the analysis. This enables us to apply the evaluation postprocessor to a wide variety of finite element codes. All that is needed is a transition program that supplies the information concerning the finite element model given in the previous paragraph. This structure enables us to easily modify the evaluation measures. In fact, one can envision a set of preprogrammed refinement guides and termination criteria. Then the analyst simply chooses the evaluation measures desired for a particular problem from a menu. Finally, the use of a universal evaluation postprocessor means that the job can be specialized. One can envision specially prepared software for evaluating finite element results. This means that every analyst does not have to stay current with the latest developments in evaluation procedures. This is done by the specialist who produce the evaluation postprocessors.

Note that the procedure of subdividing the nine-node finite difference templates into three- or four-node ''pseudo-elements'' dodges the question of whether the smoothed stresses are of an order higher than the finite element stresses if the underlying finite element model uses six-, eight-, or nine-node elements. The smoothed stresses are of an order higher than the stresses extracted from the three- or four-node subdivisions. If higher order elements actually make up the model, the results are more accurate than we would get if we actually extracted the finite element stresses. Thus, the evaluation measures computed by this approach are more conservative than we would have if we used the actual finite element stresses. Note that this process does not affect the output if we are using the smoothed stresses to find the stresses in the problem being analyzed.

We can give the process of using stresses extracted from subdivisions of the finite difference templates another interpretation. We can view the derivatives extracted from the subdivisions as one-sided derivatives. Thus, instead of thinking that we are comparing the discontinuous finite element stresses to the smoothed solution, we can consider that we are comparing the one-sided derivatives and the two-sided derivatives in the pointwise Taylor series representation at the nodes. Thus, we can interpret the procedure as evaluating *some kind* of continuity relationship in the Taylor series representation. This point of view is not addressed in detail here.

The creation of a universal evaluation postprocessor can also bring significant changes to the analysis process. As most experienced analysts know, the displacements produced by analyses using different elements are very similar. In the stress extraction process, the results produced by the different elements differ the most. The use of the smoothed stresses as the output of the finite element analysis reduces the number of elements that need to be made available to an analyst. Instead of having pet elements for different classes of problems, the same elements can be used for any problem. In this way, the computational effectiveness of the element can be the primary criterion used to choose an element. That is to say, we want an element that is easy to refine and that produces the ''most accuracy per node.''

From the results contained in Part III of this book, elements based on complete polynomials seem the most likely candidates for the underlying finite element models if we use the smoothed stresses and a universal evaluation postprocessor. That is to say, three- and six-node elements seem to be the most likely candidates for use as basic elements in finite element models. The added stiffness existing in four-, eight-, and nine-node elements resulting from the inability to represent the Poisson effects and the presence of parasitic shear terms seem to disqualify them from consideration.

Evaluation of Finite Difference Results

We can invert the ideas just presented to produce procedures for evaluating finite difference results. The various nine-node central difference templates that make up the finite difference models are subdivided into three- or four-node ''pseudo-elements.'' Then refinement guides based on the interelement jumps in the stresses are formed. This can be interpreted as overlaying a finite element model on the finite difference model.

Although this book is primarily aimed at the analysis of the finite element method, this discussion of evaluation procedures for finite difference results is presented for two reasons. On one hand, the unification of the finite element and the finite difference methods is then complete. We can formulate the two methods from the same basis. After analyzing a problem with one of the methods, we can use the other method in the evaluation of the results. That is to say, we can evaluate the results of one method in terms of the other method. And finally, it shows that we can evaluate the results of the two methods in similar ways.

As an aside, it should be obvious that we cannot evaluate the finite difference results in terms of residuals. That is to say, since the finite difference method forces the residuals to be zero, there is no way to use these quantities at the nodes to evaluate the results. The idea of tracking the strain energy content in a sequence of finite difference results has not been pursued in this work. It does not seem overly promising because it is a global quantity, but it cannot simply be dismissed as impractical.

In the next lesson, we evaluate the results of *finite difference* analyses with the ideas just presented. We use the results of the refinement guides to locally refine the finite element mesh. This is done for three reasons. One, it completes the unification of the finite element and the finite difference methods, i.e., we can use one method to analyze the other. Two, it shows that the two methods have the same capabilities. That is to say, we cannot just dismiss the finite difference method because we are not familiar with it. We must choose the method that best suits the problem being solved. Finally, the local refinement of a finite difference mesh lets us emphasize one more time that the nine-node finite difference templates do not have to be uniform. The need for this emphasis has been seen in the reviews of the papers from which this book is written. Many reviewers were so locked into the idea that a finite difference template is a uniform creature that saying and showing that it is not in figures was often missed and the procedure was criticized for not being capable of being applied to non-uniform meshes.

Closure

This lesson has completed the integration of the finite element and the finite difference methods. In Part III of the book, we saw that the two methods could be formulated from the same basis using strain gradient notation. In Lesson 18, we used techniques from the finite difference method to introduce additional information concerning the problem into finite element displacement results. In this lesson, we have shown that this augmented finite element result can be interpreted as an approximate finite difference result.

The primary significance of this interpretation is that it justifies the use of pointwise quantities from the finite element result. That is to say, if the augmented finite element result is an approximate finite difference result, it has a definite Taylor series basis. Thus, pointwise quantities have meaning. The exploitation of the pointwise nature of the augmented result has resulted in the development of pointwise evaluation metrics and the extraction of higher order pointwise stress components. This has resulted in the identification of new evaluation metrics, the extension of old evaluation metrics to pointwise measures and validation that the smoothed stresses are, indeed, of an order higher than the finite element stress results.

The solid theoretical basis provided by the pointwise interpretation of the augmented finite element result suggests the development of a universal evaluation postprocessor for finite element results. This has the advantage of making the evaluation process easily accessible to a wide range of finite element codes. In addition, the postprocessor can include a wide variety of evaluation procedures that can be tailored to the problem being analyzed. Furthermore, the availability of a universal postprocessor enables us to incorporate new knowledge concerning evaluation procedures into existing finite element codes.

Finally, the idea of using a universal postprocessor for evaluating and extracting the accuracy of the stress results enables us to analyze the finite element method from a fresh point of view. We can ignore the confusion created by looking at the stress results produced by the finite element models (we are using the smoothed stresses) and concentrate on the primary issues, namely, (1) the accuracy of the displacements, (2) the computational robustness of the finite element model, and (3) the computational efficiency of the formulation and solution procedures. As discussed in item 9, if the efficiency of the finite element representation is used as a point of comparison, this strongly suggests the use of elements based on complete polynomials, i.e., three- and six-node elements. Furthermore, this criterion also suggests a close look at replacing the isoparametric approach with the strain gradient approach.

References

1. Kelly, D. W., "The Self-Equilibration of Residuals and Complementary *A Posteriori* Error Estimates in the Finite Element Method," *International Journal for Numerical Method in Engineering*, Vol. 20, 1984, pp. 1491–1506.

2. Kelly, D. W., and J. D. Isles, "Procedures for Residual Equilibration and Local Error Estimation in the Finite Element Method," *Communications in Applied Numerical Methods*, Vol. 5, 1989, pp. 497–505.

3. Kelly, D. W., and J. D. Isles, "A Procedure for *A Posteriori* Error Analysis for the Finite Element Method which Contains a Bounding Measure," *Computers and Structures*, Vol. 31, No. 1, 1989, pp. 63–71.

4. Yang, J. D., D. W. Kelly, and J. D. Isles, "*A Posteriori* Point-wise Upper Bound Error Estimates in the Finite Element Method," *International Journal for Numerical Method in Engineering*, Vol. 36, 1993, pp. 1279–1298.

5. Aarnes, K. A., "Extension and Application for a Pointwise Error Estimator Based on the Finite Difference Method," Master's Thesis, University of Colorado, 1993.

Lesson **20**

Application of a Universal Evaluation Post-Processor

Purpose of Lesson   To demonstrate the application of a universal post-processor for quantifying the local discretization errors in both finite element and finite difference solutions.

A procedure for forming refinement guides for finite element and finite difference results is demonstrated that requires a minimum of information concerning the underlying discrete model. The procedure requires only the nodal displacements, the nodal locations, the material properties, and the boundary conditions in order to form refinement guides and termination criteria. Knowledge of the number of nodes in the elements used to form the underlying finite element model is not needed. The discontinuous stresses needed to form the stress-based refinement guides are extracted from the displacements of the finite element model by superimposing three-node finite elements on the mesh. Other elements could be used if so desired. As a result of this independence from the underlying model, a post-processor that can be applied to a wide variety of finite element and finite difference codes can be developed.

The effectiveness of the universal post-processor is demonstrated by forming refinement guides for finite element models constructed using four-, six-, and nine-node elements. The resulting refinement guides are compared to those produced by the standard approach. Since the refinement guides do not depend on the source of the displacement results, the idea of a universal post-processor is extended by forming refinement guides for *finite difference results* with an analogous process. The discontinuous stresses needed to form the refinement guides are extracted from the finite difference displacements by superimposing three-node finite elements on the finite difference mesh. The finite difference result serves as the smoothed solution. Termination criteria are available for both the finite element models and the finite difference models in terms of the convergence of the smoothed stresses.

The universal evaluation postprocessor is applied to finite difference results for several reasons: (1) to demonstrate the universality of the approach, (2) to demonstrate that adaptive refinement is available to finite difference analyses, and (3) to reiterate the idea that finite difference models can represent any geometry that can be represented by finite element models. This demonstration shows that the error analysis procedures developed in Part V can be used in practical, production applications.

■   ■   ■

This lesson demonstrates the feasibility of developing a post-processor for evaluating finite element results that can be appended to a wide variety of finite element codes. This universality of application occurs because this postprocessor requires a minimum of input data from the underlying finite element program. Only the displacements from the analysis, the nodal locations, the material properties, and the boundary conditions are needed to compute refinement guides and termination criteria.

If the finite element results are to be evaluated using residuals, the evaluation measures can be computed directly from the data just outlined. However, if either point-wise or strain energy based evaluation measures are to be used, a set of discontinuous stresses and a set of smoothed stresses must be extracted from the displacements of the underlying finite element model. Note that both of these stress representations can be computed without knowledge of the details of the finite elements used to construct the underlying model, e.g., the number of nodes and the associated strain representations. Thus, error estimators can be computed that are largely independent of the underlying model.

As outlined in the previous lesson, we can form such an independent set of discontinuous stresses by overlaying a finite element model consisting of three-node elements on the mesh of the augmented finite element solution and extracting the stresses from the individual elements. The obvious approach for overlying the three-node elements is to subdivide each of the nine-node finite difference templates used to form the augmented finite element result. Note that any mesh can be subdivided into three-node sectors.

The refinement guides used to illustrate the effectiveness of the universal postprocessor are strain energy measures. The preferred pointwise measures are not used because the work reported here was performed in 1989–1990 (see Ref. 1). This means that these results were found before the ideas of finite difference smoothing and pointwise refinement guides had been perfected. That is to say, the results presented in this lesson are for low-resolution error measures. This work was not updated for two reasons not based on resources: (1) the original work validates the concept and (2) the use of elemental strain-energy-based refinement guides is the hardest test for the postprocessor. Thus, the validation presented here means that the universal postprocessor performs well with either elemental or pointwise error measures.

The universal postprocessor is first applied to a finite element representation based on different elements, i.e., four-, six-, and nine-node elements. Then, it is applied to finite difference results. The finite difference results are analyzed for two reasons. First, the successful application of the postprocessor to finite difference results definitively demonstrates that the approach does not depend on the underlying finite element model because there is no finite element model. It is applied to the results of a finite difference analysis. Second, the application to the finite difference results shows that the two solution techniques have another similarity. The discretization errors in either approximation can be quantified by evaluating the results in terms of quantities derived from the complementary solution technique. This is equivalent to comparing an approximate variational result and an approximate differential result in each case. Thus, the evaluation procedure can be compactly described as a process where an approximate variational result is compared to an approximate differential result.

As a result of this demonstration, we can claim that we have totally integrated the finite element and the finite difference methods. In Part IV, we saw that both methods could be formed in the same way using [Φ] matrices. In the previous lesson, we saw that the finite difference problems could be solved iteratively using the finite element method and vice versa. This lesson shows that we can evaluate the discretization errors in either type of model by comparing it to quantities derived from its complementary solution technique.

Because of their common and/or complimentary characteristics, the finite element and the finite difference methods are virtually interchangeable. This means that the finite difference method can be used to solve problems that are traditionally considered the domain of the finite element method. The inverse is true for problems that have been routinely solved using the finite difference method. Such reconsiderations may improve the solution of selected problems. At the very least, the integration of the two solution techniques has produced new evaluation procedures for both methods.

Application to Finite Element Models

This section presents the results of applying the universal postprocessor to a finite element problem modeled with four-, six-, and nine-node elements. The refinement guides found using the universal approach are compared to those produced by applying the standard approach to the problems. That is to say, we compare refinement guides found using the discontinuous stresses extracted from the overlaid three-node finite elements to the refinement guides found using the underlying finite element model actually used in the solution as the source of the discontinuous stresses. Nodal averaging is used to form the smoothed solution in both cases.

The example problem solved is a uniform cantilever beam bending under its own weight as shown in Fig. 1. We solve the problem using an original mesh with 66 degrees of freedom and two subsequent uniform refinements of 210 and 738 degrees of freedom, respectively. These uniform refinements subdivide each original element into four elements in each refinement step. These meshes are shown in Figs. 2, 4, and 6 for the four-, six-, and nine-node models.

Four-Node Analysis

The refinement guides for the standard approach of evaluating the finite element results for the three meshes formed using four-node elements are shown in Fig. 2. The level of discretization error corresponds to the degree of shading. The darker the region, the higher the refinement guide. The refinement guides in these figures are normalized to the maximum error in each individual mesh. As a result, the magnitude of the refinement guide represented by each shade of gray is not comparable between these meshes.

The refinement guides produced by the superimposed three-node elements are shown in Fig. 3. When we compare Figs. 2 and 3, we can see that both techniques identify the same region as being in the most need of refinement. Thus, we can conclude that the use of the three-node elements produces a distribution of refinement guides that is at least as good as the standard approach.

We can make a comparison of the magnitudes of the refinement guides by computing the estimates of the error in the global strain energy for the two approaches. The total estimated error in the strain energy and the maximum displacements for each mesh are given in Table 1. As expected from Rayleigh's principle, the maximum displacements increase with mesh refinement, i.e., the model with the highest strain energy content is the best model.

As we can see, the estimate of the error in the global strain energy given by the approach using the superimposed triangles is much higher than that provided by the

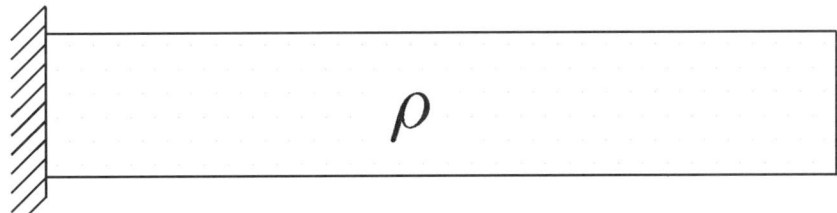

Figure 1. Cantilever beam bent under its own weight.

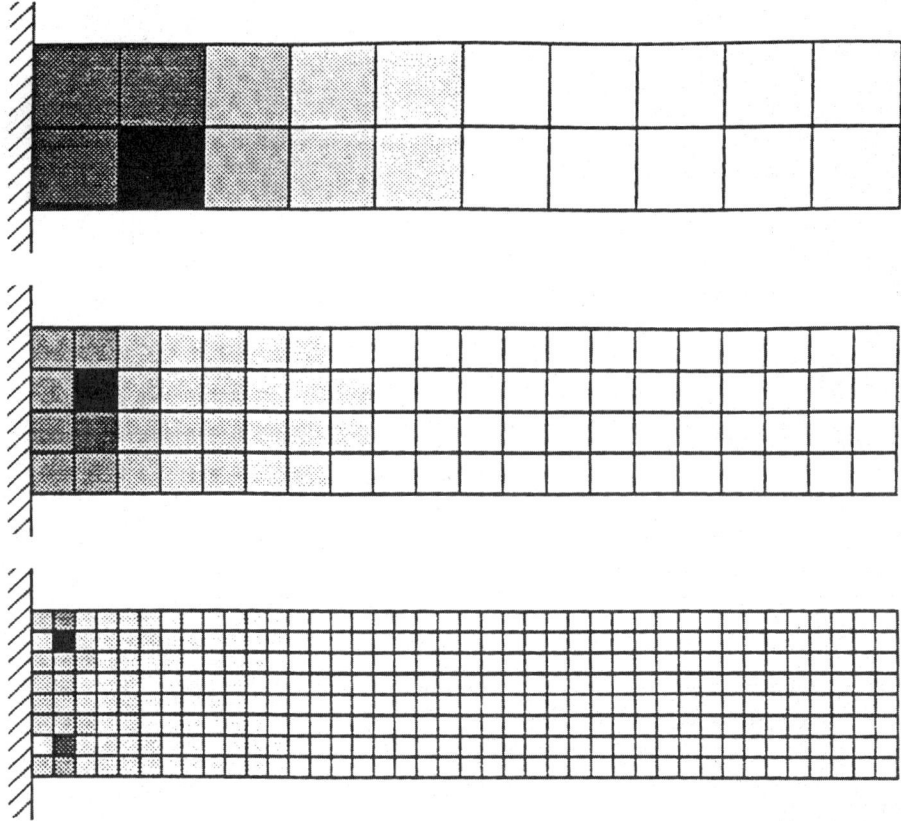

Figure 2. Standard evaluation of cantilever beam bent under its own weight, four-node elements.

standard approach. Both evaluation techniques exhibit significant decreases in the global error estimate as the mesh is refined. This decrease corresponds with the increase in the maximum displacements.

The fact that the universal approach provides a more conservative result should not be surprising. From an intuitive point of view there are simply more interelement boundaries to produce interelement jumps and the stress representations do not represent the residuals on the individual elements very well.

Note that the use of the global error estimate is not being recommended for use as a termination criterion because of its use here. The global error is being used here because when this work was performed, we did not have a solid reason for using other termination criteria. However, the global measure is a valid measure here because all of the problems

Table 1. Analysis results for cantilever beam problem, four-node finite elements.

Degrees of freedom	Maximum displacement	Universal error estimate	Standard error estimate
66	28.740	64.77	29.46
210	31.336	46.89	16.86
738	32.107	27.86	8.96

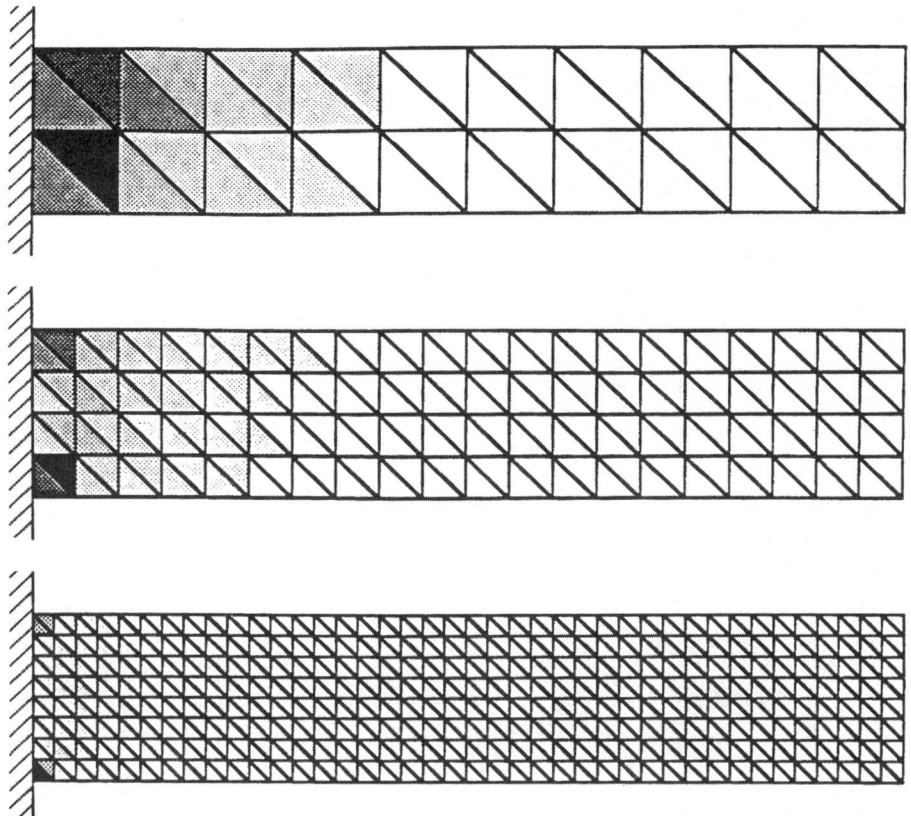

Figure 3. Universal evaluation of cantilever beam bent under its own weight, three-node elements superimposed on the four-node model.

have the same "volume" and we are not focusing on the accuracy of the stresses at critical points.

Six-Node Analysis

The refinement guides for the standard approach of evaluating the finite element results for the three meshes found using six-node elements are shown in Fig. 4. The refinement guides produced using the superimposed three-node elements are shown in Fig. 5.

We can make a comparison of the magnitudes of the refinement guides by computing the global error measures for the two approaches. The total estimated error in the strain energy and the maximum displacements for each mesh are given in Table 2.

The results for the universal postprocessor in this case are almost identical to those produced for the examples modeled with the four-node elements. The error estimates for

Table 2. Analysis results for cantilever beam problem, six-node finite elements.

Degrees of freedom	Maximum displacement	Universal error estimate	Standard error estimate
66	31.826	64.21	9.64
210	32.274	46.53	4.43
738	32.360	27.61	2.30

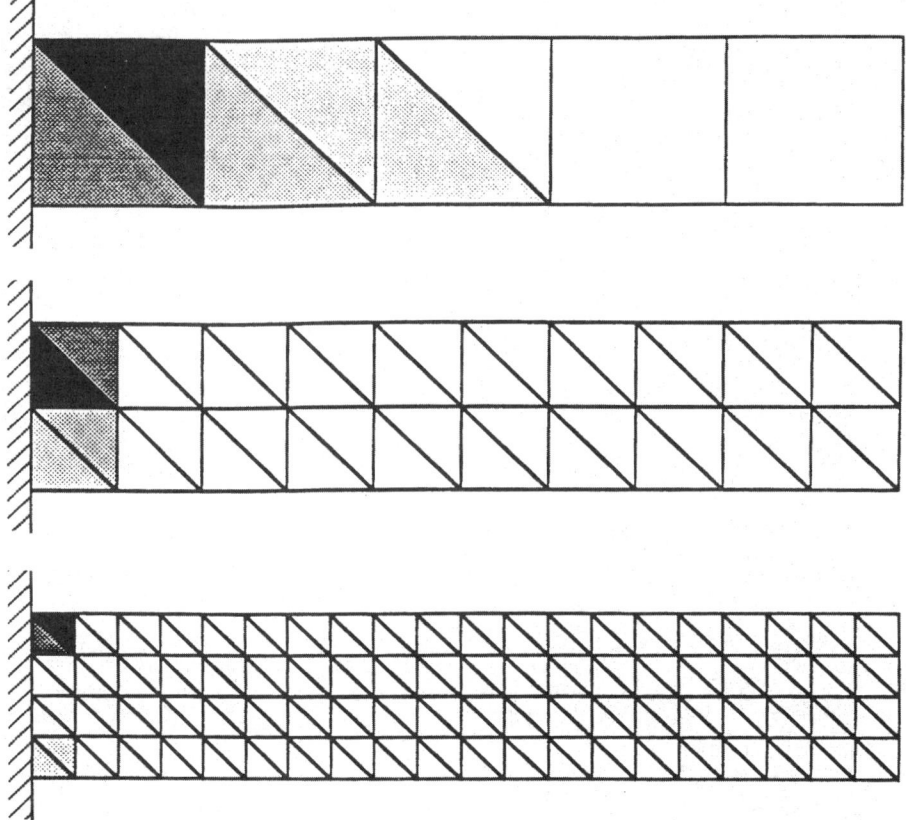

Figure 4. Standard evaluation of cantilever beam bent under its own weight, six-node elements.

the global strain energy produced by the standard approach are smaller than those in the previous example except for the first refinement. These similarities and differences can be explained as follows.

The error estimates for the universal approach are approximately the same for the two cases because the displacements in the two models must be approximately the same. Since the total lengths of the interelement boundaries are identical for the two cases, the error estimates are similar. The estimated errors for the standard approach for the six-node models are smaller than those for the four-node models because the length of the interelement boundaries is shorter. The anomaly in the error estimate for the initial mesh probably occurs because of effects on the boundary discussed in Lesson 18. We can determine this only by performing similar analyses using finite difference smoothing.

Nine-Node Analysis

The refinement guides for the standard approach of evaluating the finite element results for the three meshes are shown in Fig. 6. The refinement guides produced by the superimposed three-node elements are shown in Fig. 7.

We can make a comparison of the magnitudes of the refinement guides by computing the global error measures for the two approaches. The total estimated error in the strain energy and the maximum displacements for each mesh are given in Table 3.

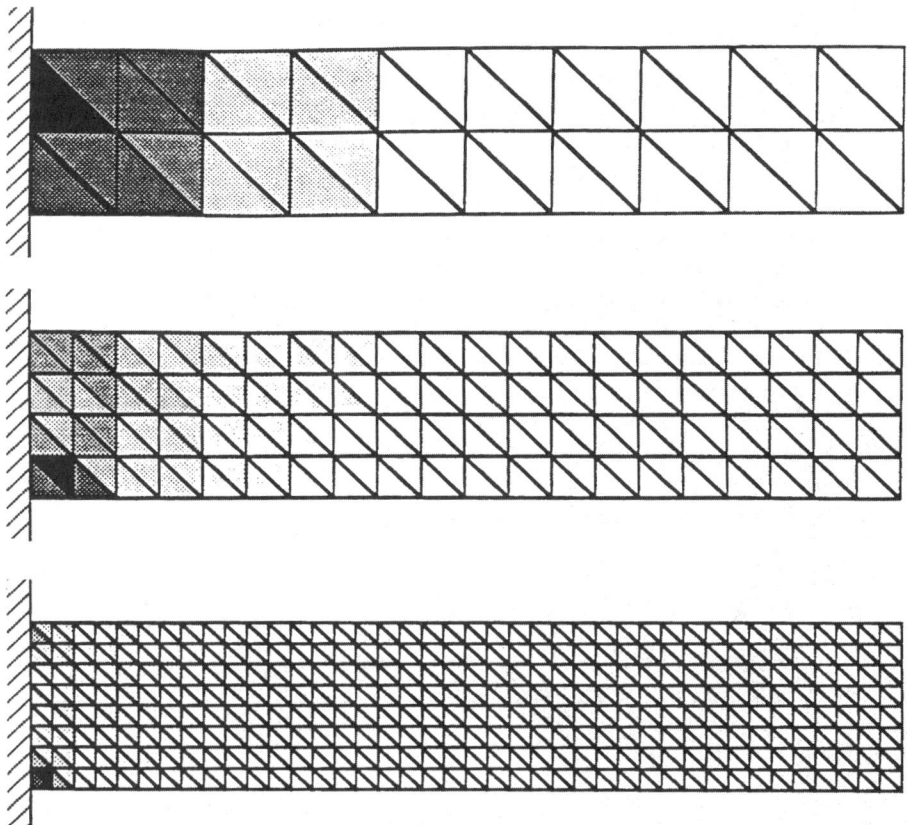

Figure 5. Universal evaluation of cantilever beam bent under its own weight, three-node element superimposed on the six-node model.

These results continue the trends observed when we compared the six-node examples to the four-node examples.

Discussion

The results produced by the universal approach behave as expected. The magnitudes of the estimated errors in the global strain energies are similar regardless of the underlying model. This means that we have a technique for evaluating the discretization errors that is correlated with the discretization errors and not with the stress extraction process. Thus, we can conclude that the implementation of a universal evaluation postprocessor has been successfully demonstrated. The exact form that the universal postprocessor should take in

Table 3. Analysis results for cantilever beam problem, nine-node finite elements.

Degrees of freedom	Maximum displacement	Universal error estimate	Standard error estimate
66	32.171	64.67	2.39
210	32.352	47.01	1.68
738	32.383	28.06	1.02

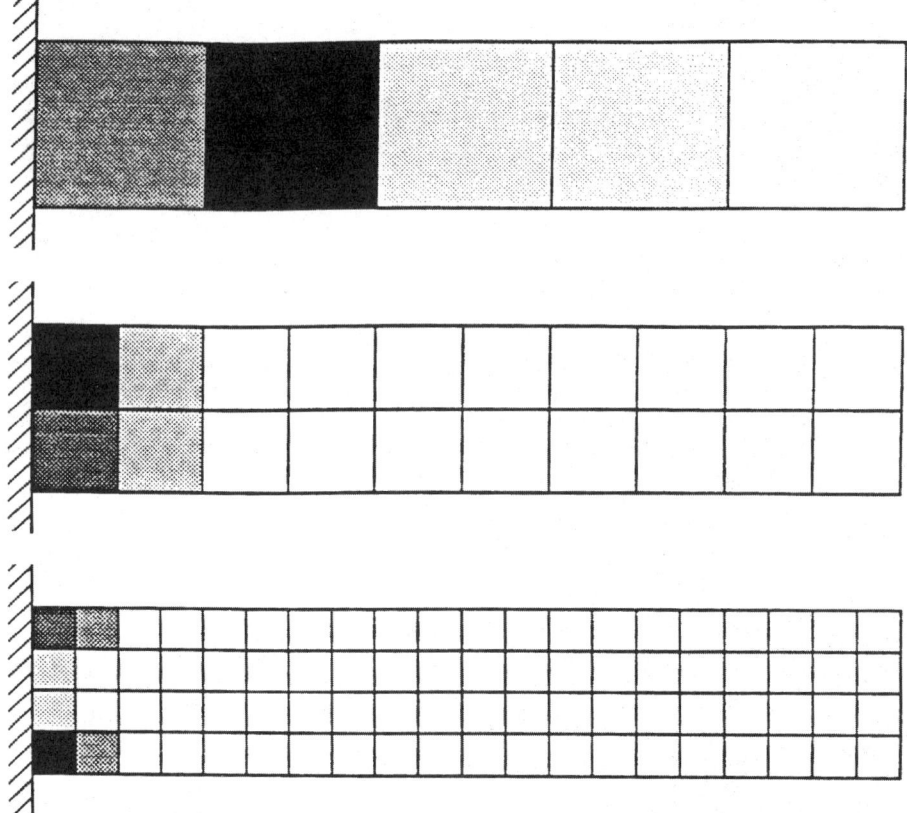

Figure 6. Standard evaluation of cantilever beam bent under its own weight, nine-node elements.

practice must be determined with detailed studies comparing pointwise refinement guides and finite difference smoothing for application to finite element results.

Application to Finite Difference Models

This section presents the results of applying the universal postprocessor to a problem solved using the finite difference method. The problem is an L-shaped panel with body forces directed along a line rotated 45 degrees, as shown in Fig. 8. The solution is approximated using three finite difference models containing 42, 130, and 450 degrees of freedom, respectively. The models consist of an initial mesh and two uniformly refined models. We compare the results of solving and evaluating the finite difference result to the results for a similar analysis of the problem using nine-node finite element representations.

The refinement guides produced by applying the universal postprocessor to the finite difference models are shown in Fig. 9. The results of a similar analysis of the nine-node finite element models are shown in Fig. 10. As we can see, the refinement guides for the two solution techniques identify the same areas as needing refinement. The relative magnitudes of the evaluations are shown along with the maximum displacements produced by the analyses in Table 4.

As we can see, the two results are similar. The maximum displacements produced by the finite element analysis are larger than those produced by the finite difference analysis.

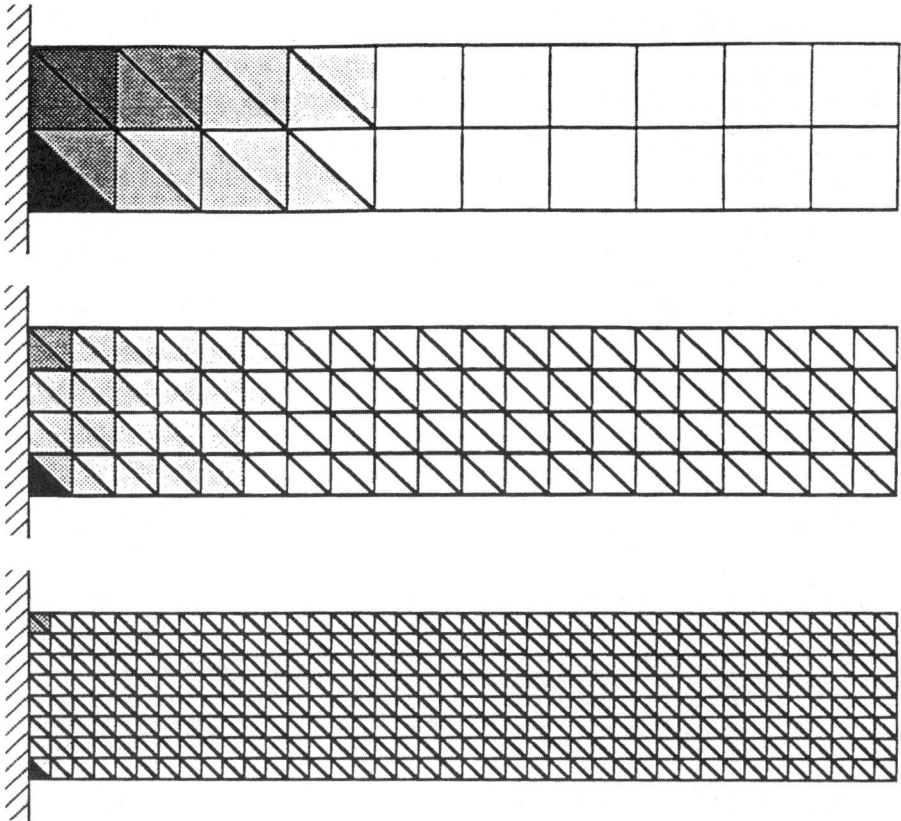

Figure 7. Universal evaluation of cantilever beam bent under its own weight, three-node elements superimposed on the nine-node model.

However, the global estimate of the strain energy error is correspondingly smaller. Since the objective of this presentation is not to compare the relative merits of the two solution techniques but to assess the capabilities of the universal evaluation postprocessor, we can conclude from the evaluation results that it is acceptable to apply the universal postprocessor to both types of analyses. Since the idea behind the universal postprocessor is to apply it to finite element models regardless of the underlying model, we can conclude that this demonstration with displacements produced by a finite difference result shows the robustness of the approach.

Since we can conclude that the procedure for forming refinement guides applies to finite difference results, we have a tool for guiding the adaptive refinement of finite

Table 4. Analysis results for the L-shaped panel, finite difference vs finite element.

	Finite element results		Finite difference results	
Degrees of freedom	Maximum displacement	Universal error estimate	Maximum displacement	Triangle error estimate
42	0.0105	68.39	0.0129	61.43
130	0.0150	46.84	0.0140	47.67
450	0.0156	29.69	0.0145	34.29

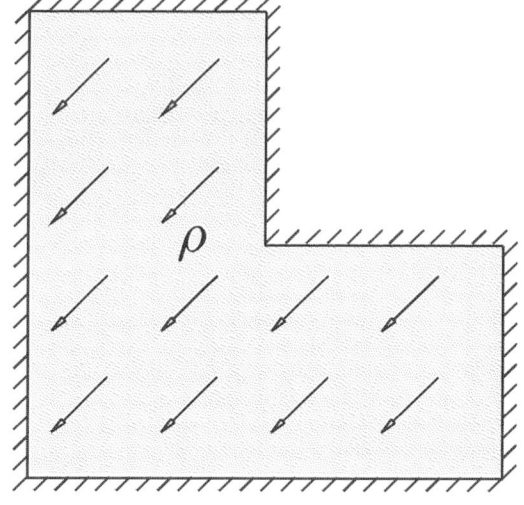

Figure 8. An L-shaped panel with a body force.

difference models. For adaptive refinement to be feasible for finite difference analyses, we must have the capability of locally refining finite difference meshes. This capability is demonstrated in the next subsection for the L-shaped panel.

Application to Locally Refined Finite Difference Models

In this section, the local refinement of finite difference models is demonstrated with the L-shaped panel analyzed in the previous section. The corners in the initial mesh and in the first uniform refinement are locally refined. We do this so that the two locally refined meshes are bracketed by uniform meshes with fewer and larger numbers of degrees of freedom. The refinement guides produced by applying the universal postprocessor to these two meshes are shown in Fig. 11. The magnitudes of the estimates of the error in the global strain energy and the maximum displacements for the five models of the L-shaped problem are presented in Table 5.

As we can see, the maximum displacements and the estimated errors for the locally refined meshes are bracketed by the results produced by the uniform meshes. This means

Table 5. Analysis results for the locally refined finite difference model of the L-shaped panel.

Degrees of freedom	Maximum displacement	Universal error estimate
42[a]	0.0129	61.43
94	0.0139	55.43
130[a]	0.0141	47.67
224	0.0143	42.67
450[a]	0.0145	34.29

[a] Uniformly refined meshes.

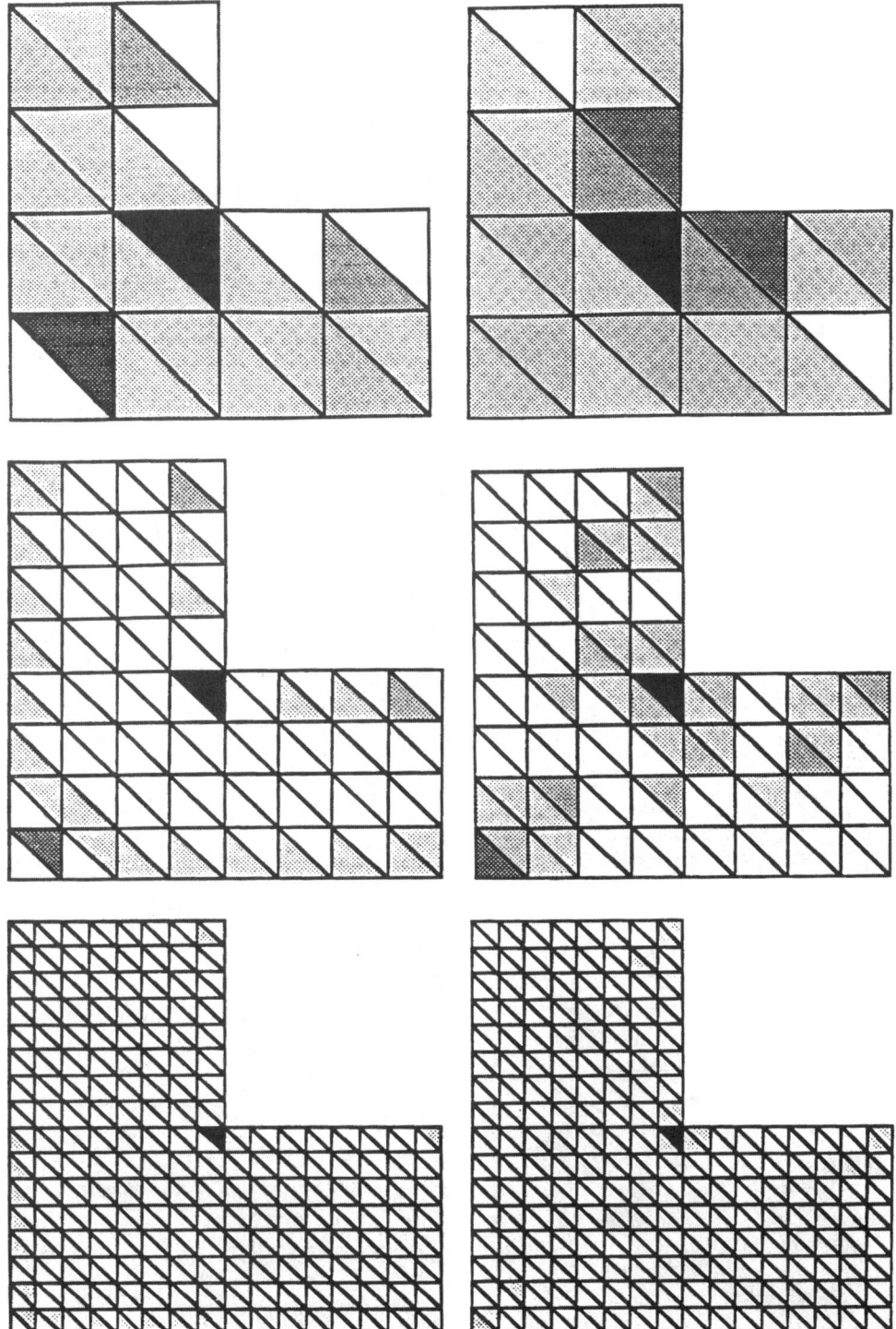

Figure 9. Universal analysis of finite difference results of an L-shaped panel.

Figure 10. Universal analysis of finite element analysis of an L-shaped panel, nine-node elements.

Figure 11. Universal analysis for two locally refined finite difference models of the L-shaped panel.

that the locally refined meshes with their distorted nine-node finite difference templates produce results that are consistent with the uniform meshes (see Note 1). Thus, we can conclude that finite difference models can be successfully refined locally and that refinement guides can be computed for them with a postprocessor based on the idea presented here. Or for a wider conclusion, we can use these techniques to develop adaptive refinement procedures for the finite difference method.

Closure

This lesson has presented three significant results: (1) a universal postprocessor for analyzing finite element and finite difference results has been successfully demonstrated, (2) the final piece in the unification of the finite element and the finite difference methods has been presented (see Note 2), and (3) a new set of tools for adaptively refining finite difference problems has been presented (see Note 3). Each of these three results identifies opportunities for significant research and practical developments.

Research Opportunities

The demonstration of the feasibility of a universal evaluation postprocessor raises the question of what an optimum version of such a postprocessor should look like. Some of the questions that must be answered were raised in Lesson 19 when we compared the evaluation capabilities of various pointwise error measures. The form that is envisioned is a postprocessor that is connected to a very flexible "transition" program. The transition program will have the capability to extract the input needed for the postprocessor from a variety of finite element codes.

The point of view that the demonstration of the universal postprocessor is the final step in the unification of the finite element and the finite difference methods raises some intriguing questions. Since the two methods are nearly interchangeable, we have the opportunity to ask which solution technique should be the primary approach for a particular class of problems. This question can now be asked without preconceived notions about limitations in geometry or boundary conditions, or changes in material type.

The developments presented in this book have provided the means for asking and answering the appropriate questions to resolve the question concerning solution type in a definitive manner. The choice of which method to use need no longer be made by superficial concerns about the difficulties in incorporating general boundary conditions in finite difference models or the inability to locally refine finite difference models. Instead, the decision of which method to use can be answered in terms of solution efficiency, solution robustness, solution accuracy, and the quantities sought in the analysis.

In the previous lesson, there were speculations about certain advantages of the finite element solution having to do with smaller problem size, i.e., there are no fictitious nodes in the primary representation of the problem, and the lower order of the derivatives involved in the derivatives contained in the problem formulation. However, two areas in which the finite difference method may have significant advantages are in shell problems and in problems with nonlinear material properties. These topics are not discussed here, but these are clearly open research questions.

The ability to use these techniques to adaptively refine finite difference models has significance on two levels. First, it shows that the finite difference method has the same capabilities as the finite element method in every way (see Note 4). Second, it opens the door to a new approach to finite difference analysis. Whether this avenue should be pursued is, again, an open research question.

Notes

1. Finite element templates with unevenly spaced nodes are developed in Lesson 13. We can use these "distorted" finite difference templates to form locally refined finite difference meshes or to match complex boundary geometries. We can also accomplish the transitions from coarse to fine meshes using finite difference templates with a number of nodes other than nine. We can also use these "augmented templates" to represent point loads in finite difference models. This is demonstrated in Lesson 15.

2. We have seen that we can evaluate finite element results by superimposing finite difference templates on finite element models. Now we have seen that we can evaluate finite difference models by superimposing finite elements on finite difference meshes. In both cases, we are comparing an approximate solution to the differential form of the problem to an approximate solution of the variational form of the problem. This means that we can solve any of the problems analyzed by either method. Furthermore, these capabilities make it possible to develop models that solve one part of the problem using one method and another part of the problem with the other method, i.e., we can form hybrid models.

3. The new tools are the following: (1) new approaches for quantifying the discretization errors in finite element models and (2) new approaches for locally refining finite difference models.

4. The idea that the finite difference method is entirely interchangeable with the finite element method can be challenged on two points concerning modeling capabilities. The finite element method is more facile than the finite difference method in representing (1) point loads and (2) "sharp" corners. In this context, a sharp corner means where two surfaces meet with a discontinuity in the slope. Examples are corners.

In truth, we can use both of these cases to vindicate the finite difference method. Neither of these situations can actually exist in nature. So the finite difference method can be said to have a "nature" closer to that of the problems being represented. However, the point load and the sharp corner are convenient idealizations of physical situations. If the regions containing these convenient representations are not critical to the problem, then the simplifications embedded in point loads and "sharp corners" are acceptable.

However, we have seen that the finite difference method can represent both of these situations in a number of ways. It is simply a matter of chosing a representation that accomplishes that which is desired for a particular problem. Submodels that represent these two cases can be envisioned that remove any inconvenience that these artificial constructs may introduce into the modeling process.

References and Other Reading

1. Stevenson, Ian, "A Generalized Adaptive Refinement Procedure for Finite Element and Finite Difference Analyses," Master's Thesis, University of Colorado, 1990.

2. Dow, John O., and Stevenson, Ian, "An Adaptive Refinement Procedure for the Finite Difference Method," *The International Journal of Numerical Methods in Partial Differential Equations*, Vol. 8, 1992, pp. 537–550.

Index